Springer Series in Computational Mathematics

46

For further volumes:
www.springer.com/series/797

Boško S. Jovanović · Endre Süli

Analysis of Finite Difference Schemes

For Linear Partial Differential Equations with Generalized Solutions

Springer

Boško S. Jovanović
Faculty of Mathematics
University of Belgrade
Belgrade, Serbia

Endre Süli
Mathematical Institute
University of Oxford
Oxford, UK

ISSN 0179-3632 Springer Series in Computational Mathematics
ISBN 978-1-4471-7259-8 ISBN 978-1-4471-5460-0 (eBook)
DOI 10.1007/978-1-4471-5460-0
Springer London Heidelberg New York Dordrecht

Mathematics Subject Classification: 65M06, 65M08, 65M12, 65M15, 65N06, 65N08, 65N12, 65N15

Printed on acid-free paper

Springer is part of Springer Science+Business Media (www.springer.com)

Preface

Boundary-value problems and initial-boundary-value problems for partial differential equations of continuum mechanics and mathematical physics that arise in applications in the physical sciences and engineering frequently contain 'nonsmooth' or 'singular' data, such as jumps in the coefficients in the equation, caused by discontinuities in material properties, or concentrated loads that are modelled as point sources, or indeed discontinuities in the solution at interfaces in transmission problems. There is a wealth of such practical examples. The present book, which arose from series of lectures given by the authors over a number of years at the University of Belgrade and the University of Oxford, respectively, is devoted to the construction and the mathematical analysis of numerical methods for the approximate solution of such problems. More specifically, we focus on the numerical solution of linear partial differential equations by variously generalized finite difference schemes in instances when the coefficients, source terms or initial or boundary data belong to spaces of weakly differentiable functions, e.g. Sobolev, Besov or Bessel-potential spaces of nonnegative order, or certain spaces of distributions, such as negative-order Sobolev, Besov or Bessel-potential spaces.

The fundamental mathematical result that underpins the convergence analysis of discretization methods for linear partial differential equations, and finite difference methods in particular, is the Lax equivalence theorem (cf. [156], Sect. 3.5), which, loosely speaking, states that a sequence of numerical solutions, generated on a family of meshes by means of a *consistent* finite difference approximation of a well-posed initial/boundary-value problem for a linear partial differential equation, *converges* to the analytical solution of the problem if, and only if, the finite difference method is *stable*.

Consistency of a finite difference scheme amounts to the requirement that the *truncation error*, defined by inserting the unknown analytical solution to the partial differential equation into the finite difference approximation of the equation, when measured in a suitable mesh-dependent norm, converges to zero, possibly at a certain rate, which is typically a positive power of the maximum mesh-size h, in the limit of h converging to zero.

The conventional mathematical tool for investigating the consistency of a finite difference approximation to a partial differential equation is multivariate Taylor series expansion. The truncation error is expanded to terms of order as high as is necessary so as to extract the highest possible power of h admitted by the finite difference scheme; the power of h in question is referred to as the *order of accuracy* or *order of consistency* of the finite difference method. The underlying assumption in such, frequently tedious, but completely elementary calculations based on Taylor series expansions is that the solution to the partial differential equation is sufficiently smooth, to the extent that it admits a Taylor series expansion up to derivatives whose order is as high as is needed in order to extract the highest possible power of h from the truncation error.

When confronted with partial differential equations whose solutions are known not to be differentiable or even continuous, and Taylor series expansion of the analytical solution, and thereby of the truncation error of the finite difference scheme, fails to make sense due to lack of regularity in the classical sense, a natural question is whether there are alternative mathematical tools one can resort to in a systematic fashion. A second, closely related and even more basic question is, of course, how, in the first place, should one construct finite difference approximations to partial differential equations whose coefficients, source terms or initial or boundary data are so 'rough' that sampling them at the points of the computational mesh is, quite evidently, a meaningless endeavour.

It is the mathematical analysis of these two questions that the present monograph is devoted to. The second question posed above, concerning the construction of finite difference schemes for partial differential equations with nonsmooth data, is addressed by mollifying the data through convolution (possibly in the sense of distributions) with suitable functions with compact support, which are typically (multivariate) B-splines whose support is commensurate with the mesh-size h. As for the first question, regarding the analysis of consistency in the absence of meaningful Taylor series expansions, we resort to a technique that is familiar in the realm of finite element methods but is seemingly alien to the world of finite difference schemes: interpreting the truncation error as a linear functional on a suitable function space (typically a certain Sobolev space of nonnegative order), scaling to a canonical 'element', which is chosen to be a scaled-up version of the support of the B-spline used in the definition of the mollification, followed by an application of a result known as the Bramble–Hilbert lemma and, finally, rescaling. The Bramble–Hilbert lemma plays the role of Taylor series expansion with remainder of the truncation error up to the highest possible derivative, with the lower-order terms in the Taylor polynomial cancelling: it simply states that a bounded linear functional on a Sobolev space with the property that the linear functional vanishes on polynomials of degree one less than the (positive) differentiability index of the Sobolev space, can be bounded by the highest-order Sobolev seminorm of the space. The subsequent rescaling from the canonical element then relies on the fact that the highest-order Sobolev seminorm is a homogeneous function of a certain degree in the mesh-size h (the homogeneity index of the Sobolev seminorm being dependent on the differentiability and integrability indices of the Sobolev seminorm and the number of dimensions).

Our objective throughout the book is to systematically develop this methodology based on the combination of mollification of the nonsmooth data on the one hand and the application of variants of the Bramble–Hilbert lemma in conjunction with scaling arguments on the other, for a range of linear elliptic, parabolic and hyperbolic partial differential equations.

Chapter 1 provides a brief survey of some basic results from linear functional analysis, the theory of distributions and function spaces, Fourier multipliers and mollifiers in function spaces, and function space interpolation. Chapter 2 is concerned with the construction and the convergence analysis of finite difference schemes for elliptic boundary-value problems. One of the key contributions of the chapter is the derivation of optimal-order bounds on the error between the analytical solution and its finite difference approximation for elliptic equations with variable coefficients under minimal regularity hypotheses on the coefficients and the solution, the minimal regularity hypotheses on the coefficients being expressed in terms of spaces of multipliers in Sobolev spaces. In Chaps. 3 and 4 of the book we then pursue an analogous programme for some model linear parabolic and hyperbolic equations.

We shall consider finite difference methods on both uniform and nonuniform computational meshes. In order to avoid cluttering the presentation with the inclusion of technical details that are secondary to the central theme of the book, we shall confine ourselves throughout to boundary-value problems and initial-boundary-value problems on axiparallel domains. Curved boundaries give rise to additional complexities, which we do not address. Having said this, the starting point of a convergence analysis for any finite difference method is a stability result, which is typically a discrete counterpart of a stability or regularity result for the differential problem under consideration. For elliptic equations in arbitrary domains discrete versions of interior regularity results in L_2 and, more generally, L_p type norms were developed by Thomée and Westergren [179] and Shreve [166], respectively. Discrete versions of interior Schauder estimates were proved by Thomée [175]. For Lipschitz domains, discrete versions of elliptic regularity results, up to the boundary, were established by Hackbusch in [66] and [67]. For parabolic problems discrete interior regularity results in arbitrary spatial domains were proved by Brandt [22] and Bondesson [18, 19]. These, and related results, can be seen as a starting point for the development of a theoretical framework in arbitrary domains, analogous to the one considered here on axiparallel domains.

There are of course several excellent books concerned with the mathematical theory of finite difference schemes for partial differential equations. A classical source in the field is the influential monograph by R.D. Richtmyer and K.W. Morton: *Difference Methods for Initial-Value Problems* [156]; some other significant books include the following: A.A. Samarskiĭ: *The Theory of Difference Schemes* [159], J. Strikwerda: *Finite Difference Schemes and Partial Differential Equations* [170], B. Gustafsson, H.-O. Kreiss and J. Oliger: *Time Dependent Problems and Difference*

Methods [64], the short monograph by P. Brenner, V. Thomée and L.B. Wahlbin entitled *Besov Spaces and Applications to Difference Methods for Initial Value Problems* [24], the monograph by A.A. Samarskiĭ, R.D. Lazarov and L. Makarov: *Finite Difference Schemes for Differential Equations with Weak Solutions* (in Russian) [160], and Chap. 4 and Sects. 9.2, 10.2.2 and 11.3 of the book by W. Hackbusch entitled *Elliptic Differential Equations: Theory and Numerical Treatment* [68]. Instead of replicating the material contained in those and other books on the analysis of finite difference schemes for partial differential equations, our aim here has been to focus on ideas that have not been covered elsewhere in the literature previously, at least not in the form of a book. While we have made every effort to ensure that the text is reasonably accessible and self-contained, a disclaimer is in order: it is fair to say that this monograph has been written with a mathematical audience in mind. Some of the material we have included here has been successfully used in third- and fourth-year mathematics undergraduate courses on the numerical analysis of partial differential equations (e.g. Chap. 1, Sects. 1.1–1.4; Chap. 2, Sects. 2.1–2.4; Chap. 3, Sects. 3.1, 3.2; Chap. 4, Sects. 4.1, 4.2); however, the vast majority of the theoretical questions we discuss are firmly beyond the scope of the undergraduate numerical analysis syllabus, and will be of primary interest to graduate students, researchers and specialists working in the field of numerical analysis of partial differential equations. Readers will certainly find it helpful to possess prior knowledge of elements of linear functional analysis, the theory of linear partial differential equations, and basic concepts from the theory of distributions and function spaces. Although we chose to focus on linear problems throughout, it is nevertheless hoped that the methodology that is systematically developed here in the case of linear partial differential equations has some bearing on the mathematical analysis of finite difference approximations of nonlinear partial differential equations with nonsmooth solutions, particularly those that arise from continuum mechanics and the sciences in general. The recent upsurge of interest in numerical algorithms for atomistic models of crystalline materials, such as quasi-continuum methods, whose analysis relies on techniques from the theory of finite difference methods [14, 15, 29, 132, 133, 149, 150, 194], has provided added impetus to this book: we hope that some of the technical tools developed here will also prove useful to researchers working in that field.

Acknowledgements

We are grateful to Joerg Sixt at Springer London for his constructive and helpful suggestions and to Edita Baronaitė at VTeX for technical assistance.

The second author wishes to thank his colleagues in the Mathematical Institute at the University of Oxford, particularly Mike Giles, Nick Trefethen and Andy Wathen, for numerous stimulating discussions, and to Iain Smears for helpful comments on the draft manuscript.

This preface would be incomplete without our words of deep and heartfelt gratitude to our families. Their patience, support and encouragement to complete this book are gratefully acknowledged.

Belgrade, Serbia
Oxford, UK
18 October 2013

Boško S. Jovanović
Endre Süli

Contents

Chapter 1
Distributions and Function Spaces

Numerous linear partial differential equations that arise in mathematical models of physical phenomena possess discontinuous coefficients or nonsmooth forcing terms. Such lack of smoothness of the data and the resulting loss of regularity of the solution give rise to conceptual difficulties that are hard to resolve within the classical theory of partial differential equations. The theory of distributions has been developed with the aim to overcome these limitations by weakening the notion of differentiability, and to provide a general tool for the study of linear partial differential equations with nonsmooth solutions. In this chapter we give a brief overview of this theory and present a collection of results concerning function spaces.

In Sect. 1.1 we review some elementary ideas from linear functional analysis. Section 1.2 states the definitions of basic function spaces, such as those of continuously differentiable and Lebesgue-integrable functions. Section 1.3 concentrates on simple tools from the theory of distributions. In Sects. 1.4 and 1.5 we define Sobolev spaces and review their crucial properties. Section 1.6 is devoted to Besov spaces, while Sects. 1.7 and 1.8 discuss interpolation properties of Sobolev spaces and point multipliers (or, simply, multipliers) in Sobolev spaces, respectively. We conclude, in Sect. 1.9, by considering Fourier multipliers and their application to the construction of smoothing operators (mollifiers) in Bessel-potential spaces, Sobolev spaces and Besov spaces. For a detailed account of the theory of distributions we refer to Gel'fand and Shilov [52], Hörmander [72], Rudin [158], Schwartz [163], and Vladimirov [184]; for details of the theory of function spaces the reader may wish to consult Adams [1], Adams and Fournier [2], Kufner, John and Fučik [116], Maz'ya [136], and Triebel [181, 183], for example.

1.1 Elements of Functional Analysis

Much of numerical analysis is concerned with the approximate solution of equations. Regardless of the type of equation under consideration, the construction of a numerical method for its approximate solution is frequently preceded by a mathematical analysis of the problem, with the aim to ascertain useful information about

B.S. Jovanović, E. Süli, *Analysis of Finite Difference Schemes*,
Springer Series in Computational Mathematics 46,
DOI 10.1007/978-1-4471-5460-0_1, © Springer-Verlag London 2014

the existence and uniqueness of the solution, and its sensitivity to perturbations of the data. Functional analysis provides a general framework for studying such questions in an abstract setting; the purpose of this section is to introduce the basic concepts of this theory.

1.1.1 A Survey of Abstract Spaces

There is a hierarchy of abstract spaces, ranging from the most general to those with the most structure. Here we shall concentrate on linear spaces that are particularly relevant in applications: normed linear spaces, Banach spaces and Hilbert spaces.

1.1.1.1 Normed Linear Spaces

Suppose that $\mathcal{U}$ is a linear space over $\mathbb{R}$, the field of real numbers (or the field $\mathbb{C}$ of complex numbers), and let $\mathbb{R}_+$ denote the set of nonnegative real numbers. A function $\|\cdot\| : \mathcal{U} \to \mathbb{R}_+$, whose value at u is denoted by $\|u\|$, is called a *norm* on $\mathcal{U}$ provided that it satisfies the following axioms:

❶ $\|u\| = 0$ if, and only if, $u = 0$;
❷ $\|\lambda u\| = |\lambda|\,\|u\|$ for all $\lambda \in \mathbb{R}$ (or $\lambda \in \mathbb{C}$), and all u in $\mathcal{U}$; (homogeneity);
❸ $\|u + v\| \le \|u\| + \|v\|$ for all u and v in $\mathcal{U}$; (the triangle inequality).

If $\|\cdot\|$ satisfies the last two axioms only then it is called a *seminorm*. A linear space $\mathcal{U}$ equipped with a norm is called a *normed linear space* (over the field $\mathbb{R}$ or $\mathbb{C}$, as the case may be). Let $\mathcal{U}$ be a normed linear space, let u_0 belong to $\mathcal{U}$ and suppose that r is a positive real number. The set

$$B(u_0, r) := \{u \in \mathcal{U} : \|u - u_0\| < r\}$$

is called an *open ball* with centre u_0 and radius r. Let $\varepsilon > 0$; for the sake of brevity we shall write B_ε instead of $B(0, \varepsilon)$. We define

$$A^\varepsilon := A + B_\varepsilon = \{u + v : u \in A, v \in B_\varepsilon\},$$

the *ε-neighbourhood* of the set A. A subset M of a normed linear space $\mathcal{U}$ is said to be *open* in $\mathcal{U}$ if, for every u_0 in M, there exists a real number $r = r(u_0) > 0$ such that $B(u_0, r) \subset M$, i.e. $B(u_0, r)$ is contained in M. A *neighbourhood* of a point u in $\mathcal{U}$ is any open set in $\mathcal{U}$ that contains u. A subset M of a normed linear space $\mathcal{U}$ is said to be *closed* in $\mathcal{U}$ if $M^c := \mathcal{U} \setminus M$, the *complement* of M in $\mathcal{U}$, is open in $\mathcal{U}$. The *closure* of a set M, denoted by $\overline{M}$, in a normed linear space $\mathcal{U}$ is defined as the intersection of all closed sets in $\mathcal{U}$ containing M. Suppose that M is a subset of a normed linear space $\mathcal{U}$; we say that M is *dense* in $\mathcal{U}$ provided that $\overline{M} = \mathcal{U}$. A normed linear space is called *separable* if it contains a countable, dense subset. A subset M of a normed linear space is said to be *bounded* if there exists a positive real number r such that $B(0,r)$, the open ball of radius r centred at the zero of the linear space, contains the set M. A subset M of a normed linear space is said to be *convex* if, whenever u and v belong to M also $\theta u + (1-\theta)v$ belongs to M for all $\theta \in [0,1]$. For example, thanks to the triangle inequality, any open ball in a normed linear space is a convex set.

Example 1.1 The n-dimensional Euclidean space $\mathbb{R}^n$ of all ordered n-tuples of real numbers is a normed linear space with the norm $\|\cdot\|$ defined by

$$\|x\| := \left(\sum_{i=1}^{n} |x_i|^2\right)^{1/2}, \quad x = (x_1, \ldots, x_n) \in \mathbb{R}^n.$$

Example 1.2 The linear space $C([0,1])$ of (real- or complex-valued) functions u defined and continuous on the closed interval $[0,1]$ of the real line is a normed linear space with the norm

$$\|u\|_{C([0,1])} := \max_{x\in[0,1]} \big|u(x)\big|, \quad u \in C\big([0,1]\big).$$

1.1.1.2 Inner Product Spaces

Let $\mathcal{U}$ be a linear space over the field of real (or complex) numbers. A real- (or complex-) valued function $(\cdot,\cdot)$ defined on the Cartesian product $\mathcal{U} \times \mathcal{U}$ is called an *inner product* on $\mathcal{U}$ provided that it satisfies the following axioms:

❶ $(u,u) > 0$ for every u in $\mathcal{U} \setminus \{0\}$;
❷ $(\lambda u, v) = \lambda(u,v)$ for all λ in $\mathbb{R}$ (or $\mathbb{C}$), and all u and v in $\mathcal{U}$;
❸ $(u+v,z) = (u,z) + (v,z)$ for all u, v and z in $\mathcal{U}$;
❹ $\overline{(u,v)} = (v,u)$ for all u and v in $\mathcal{U}$.

The overline in the last axiom signifies complex conjugation. The linear space $\mathcal{U}$ with inner product $(\cdot,\cdot)$ is called an *inner product space*. If $(u,v) = 0$ for u and v in $\mathcal{U}$, we say that u and v are *orthogonal*. For u in $\mathcal{U}$, we define

$$\|u\| := (u,u)^{1/2}.$$

It is left to the reader to show that, with such a definition of $\|\cdot\|$, one has

$$\big|(u,v)\big| \leq \|u\|\,\|v\| \quad \forall u,v \in \mathcal{U} \quad \text{(the Cauchy–Schwarz inequality)},$$

and the triangle inequality holds; i.e.

$$\|u+v\| \leq \|u\| + \|v\| \quad \forall u,v \in \mathcal{U}.$$

Consequently $\|\cdot\|$ is a norm on $\mathcal{U}$, induced by the inner product $(\cdot,\cdot)$, and $\mathcal{U}$ is a normed linear space. It is easy to show that if $\mathcal{U}$ is an inner product space with the induced norm $\|\cdot\|$ then the following *parallelogram identity* holds:

$$\|u+v\|^2 + \|u-v\|^2 = 2\|u\|^2 + 2\|v\|^2 \quad \forall u,v \in \mathcal{U}. \tag{1.1}$$

Example 1.3 The n-dimensional Euclidean space $\mathbb{R}^n$ is an inner product space with

$$(x, y) := \sum_{i=1}^{n} x_i y_i,$$

where $x = (x_1, \dots, x_n)$ and $y = (y_1, \dots, y_n)$ are any elements in $\mathbb{R}^n$.

Example 1.4 The linear space $C([0, 1])$ of continuous real-valued functions defined on the closed real interval $[0, 1]$ is an inner product space with

$$(u, v) := \int_0^1 u(x)v(x)\,\mathrm{d}x, \quad u, v \in C\big([0, 1]\big).$$

For complex-valued functions u and v, defined and continuous on $[0, 1]$, the definition of the inner product above is modified to

$$(u, v) := \int_0^1 u(x)\overline{v(x)}\,\mathrm{d}x, \quad u, v \in C\big([0, 1]\big),$$

where, as in the fourth axiom of inner product, the overline denotes complex conjugation.

1.1.1.3 Convergence and Cauchy Sequences

Let $\mathcal{U}$ be a normed linear space with norm $\|\cdot\|$, and suppose that $\{u_n\}_{n=1}^{\infty}$ is a sequence in $\mathcal{U}$. We say that $\{u_n\}_{n=1}^{\infty}$ *converges* to u in $\mathcal{U}$ (and write $\lim_{n\to\infty} u_n = u$ in $\mathcal{U}$, or simply $u_n \to u$ in $\mathcal{U}$), if

$$\lim_{n\to\infty} \|u_n - u\| = 0.$$

In this case, the sequence $\{u_n\}_{n=1}^{\infty}$ is said to be *convergent in* $\mathcal{U}$, and $u \in \mathcal{U}$ is called the *limit* of $\{u_n\}_{n=1}^{\infty}$ in $\mathcal{U}$. A sequence $\{u_n\}_{n=1}^{\infty}$ in $\mathcal{U}$ is called a *Cauchy sequence* if

$$\lim_{n,m\to\infty} \|u_n - u_m\| = 0.$$

Obviously every convergent sequence in $\mathcal{U}$ is a Cauchy sequence in $\mathcal{U}$; however, as is indicated by the next example, the converse is not true in general.

Example 1.5 Let $C([0, 1])$ be the linear space of all real-valued functions that are defined and continuous on the interval $[0, 1]$, equipped with the norm

$$\|u\|_1 := \int_0^1 \big|u(x)\big|\,\mathrm{d}x.$$

The sequence $\{u_n\}_{n=1}^{\infty}$ defined by

$$u_n(x) := \begin{cases} (2x)^n & \text{if } 0 \le x < 1/2, \\ 1 & \text{if } 1/2 \le x \le 1, \end{cases}$$

is a Cauchy sequence in $C([0, 1])$, but it does not converge to any element of $C([0, 1])$.

1.1.1.4 Completeness, Banach Space, Hilbert Space

A normed linear space $\mathcal{U}$ is said to be *complete* if every Cauchy sequence in $\mathcal{U}$ is convergent in $\mathcal{U}$. A complete normed linear space is called a *Banach space*. Let $\mathcal{U}$ be an inner product space, with inner product $(\cdot, \cdot)$. If $\mathcal{U}$ is complete with respect to the norm $\|u\| := (u, u)^{1/2}$ induced by this inner product, then $\mathcal{U}$ is called a *Hilbert space*. A Hilbert space over the field $\mathbb{R}$ (respectively, $\mathbb{C}$) is called a real (respectively, complex) Hilbert space.

Example 1.6 The set $\mathbb{R}$ of real numbers, equipped with the norm $|\cdot|$ (absolute value), is a Banach space. The set $\mathbb{R}^n$, with the inner product defined in Example 1.3, is a real Hilbert space; the norm induced by this inner product is the Euclidean norm, appearing in Example 1.1.

Example 1.7 Let $\mathcal{U}$ denote the linear space $C([0, 1])$, equipped with the inner product defined in Example 1.4; then, $\mathcal{U}$ is not a Hilbert space. This is easily seen by noting that the sequence $\{u_n\}_{n=1}^{\infty}$ from Example 1.5 is a Cauchy sequence in $\mathcal{U}$; if, however, it converged in the norm induced by the inner product from Example 1.4, then by the Cauchy–Schwarz inequality it would also converge in the norm $\|\cdot\|_1$ from Example 1.5, resulting in a contradiction.

1.1.1.5 Compactness

A set U in a normed linear space $\mathcal{U}$ is said to be *sequentially compact* if it is sequentially relatively compact (i.e. every sequence in U contains a subsequence that is convergent in $\mathcal{U}$) and closed. Henceforth we shall omit the attribute "sequential", and will simply write *compact* and *relatively compact* instead of *sequentially compact* and *sequentially relatively compact*, respectively.

Example 1.8 Let $\mathcal{U}$ denote the set of real numbers with norm $|\cdot|$ (absolute value). Then, the open interval $U = (0, 1)$ is a relatively compact set in $\mathcal{U}$, and its closure, $\overline{U} = [0, 1]$, is compact in $\mathcal{U}$.

It is left as an exercise to show that every relatively compact set in a normed linear space is bounded, and that a closed subset of a compact set in a normed linear space is compact.

$U \Subset V$ will signify that the closure of U is compact and contained in V. A normed linear space $\mathcal{U}$ is said to be *locally compact* if 0 has a neighbourhood whose closure is a compact set in $\mathcal{U}$. The following theorem will be useful in the subsequent discussion (cf. Rudin [158], Theorem 1.22 on p. 17).

Theorem 1.1 *A normed linear space $\mathcal{U}$ is finite-dimensional if, and only if, it is locally compact.*

The next section is devoted to linear operators in normed linear spaces.

1.1.2 Linear Operators in Normed Linear Spaces

Let $\mathcal{U}$ and $\mathcal{V}$ be two normed linear spaces with norms $\|\cdot\|_{\mathcal{U}}$ and $\|\cdot\|_{\mathcal{V}}$, respectively, and let U be a set in $\mathcal{U}$. Suppose further that a rule A is given, which to every element u in U assigns a uniquely determined element in $\mathcal{V}$; we denote this element by Au and say that the rule defines an *operator* A on U. The set U is called the *domain* of the operator A and it is denoted by $D(A)$. The set

$$R(A) := \{v \in \mathcal{V} : v = Au, u \in D(A)\}$$

is called the *range* of the operator A. The *inverse image* of a set $V \subset \mathcal{V}$ is the set

$$A^{-1}(V) := \{u : u \in D(A), Au \in V\}.$$

An operator from $\mathcal{U}$ into $\mathcal{V}$ is called an *injection* (or a *one-to-one mapping*) if for each $v \in R(A)$ there exists a unique $u \in D(A)$ such that $Au = v$. An operator A from $\mathcal{U}$ into $\mathcal{V}$ is called a *surjection* (or a mapping *onto* $\mathcal{V}$) if $R(A) = \mathcal{V}$. An operator A from $\mathcal{U}$ into $\mathcal{V}$ is called a *bijection* if it is both an injection and a surjection.

If A is an injection from $\mathcal{U}$ into $\mathcal{V}$ then every $v \in R(A)$ is assigned a uniquely determined element u in $D(A)$ by the rule $Au = v$. This is written as $u = A^{-1}v$ and A^{-1} is called the *inverse operator* of A. Clearly we have that $D(A^{-1}) = R(A)$ and $R(A^{-1}) = D(A)$. If, in addition, A is a bijection from $\mathcal{U}$ onto $\mathcal{V}$, then $D(A^{-1}) = \mathcal{V}$.

Let $\mathcal{U}$ and $\mathcal{V}$ be two normed linear spaces and let A be an operator from $\mathcal{U}$ into $\mathcal{V}$; A is said to be a *continuous operator* if whenever $u_n \to u$ in $\mathcal{U}$ then also $Au_n \to Au$ in $\mathcal{V}$, for every sequence $\{u_n\}_{n=1}^{\infty}$ such that $u_n \in D(A)$ and $u \in D(A)$.

A set U in a normed linear space $\mathcal{U}$ is said to be a *linear subset* of $\mathcal{U}$ if $\alpha u + \beta v \in U$ for every u and v in U and every α and β in $\mathbb{R}$ (or $\mathbb{C}$). An operator A from a normed linear space $\mathcal{U}$ into a normed linear space $\mathcal{V}$ whose domain $D(A)$ is a linear subset of $\mathcal{U}$ is called a *linear operator* from $\mathcal{U}$ into $\mathcal{V}$ provided that

$$A(\alpha u + \beta v) = \alpha Au + \beta Av \quad \forall u, v \in D(A),\ \forall \alpha, \beta \in \mathbb{R}\ (\text{or } \mathbb{C}).$$

A linear operator A from a normed linear space $\mathcal{U}$ into a normed linear space $\mathcal{V}$ is said to be *bounded* if there exists a positive real number K such that

$$\|Au\|_{\mathcal{V}} \le K\|u\|_{\mathcal{U}} \quad \forall u \in D(A).$$

The norm $\|\cdot\| = \|\cdot\|_{\mathcal{U}\to\mathcal{V}}$ of a bounded linear operator $A : \mathcal{U} \to \mathcal{V}$ is defined by

$$\|A\| := \sup_{0 \neq u \in D(A)} \frac{\|Au\|_{\mathcal{V}}}{\|u\|_{\mathcal{U}}}. \tag{1.2}$$

Note that $\|Au\|_{\mathcal{V}} \leq \|A\| \|u\|_{\mathcal{U}}$ for all $u \in D(A)$. It is easily seen that a linear operator is bounded if, and only if, it is continuous. We shall therefore use the terms *bounded linear operator* and *continuous linear operator* interchangeably. The set of all bounded linear operators $A : \mathcal{U} \to \mathcal{V}$ will be denoted by $\mathcal{L}(\mathcal{U}, \mathcal{V})$.

Let A be a linear operator from a normed linear space $\mathcal{U}$ into a normed linear space $\mathcal{V}$ with $D(A) = \mathcal{U}$; we shall say that A is a *compact operator* if every bounded set in $\mathcal{U}$ is mapped by A into a relatively compact set in $\mathcal{V}$.

1.1.2.1 Embedding Operators

Let $\mathcal{U}$ and $\mathcal{V}$ be two normed linear spaces and let $\mathcal{U} \subset \mathcal{V}$. We define the *identity* operator I from $\mathcal{U}$ into $\mathcal{V}$, with $D(I) = R(I) = \mathcal{U}$, as the operator that assigns every element u in $\mathcal{U}$ to itself, i.e. $Iu = u$, regarded as an element of $\mathcal{V}$. Clearly the identity operator is linear. If it is, in addition, a continuous operator, then we call it an *embedding* from $\mathcal{U}$ into $\mathcal{V}$. If an embedding from $\mathcal{U}$ into $\mathcal{V}$ exists, we shall say that $\mathcal{U}$ is embedded in $\mathcal{V}$ and will write this as

$$\mathcal{U} \hookrightarrow \mathcal{V}.$$

The continuity of the embedding operator from $\mathcal{U}$ into $\mathcal{V}$ implies the existence of a positive constant K such that

$$\|u\|_{\mathcal{V}} \leq K \|u\|_{\mathcal{U}} \quad \forall u \in \mathcal{U}.$$

If $\mathcal{U}$ is embedded in $\mathcal{V}$ and the embedding operator is a compact linear operator, we shall say that $\mathcal{U}$ is *compactly embedded* in $\mathcal{V}$ and will write this as $\mathcal{U} \hookrightarrow\hookrightarrow \mathcal{V}$.

1.1.2.2 Continuous Linear Functionals

Let $\mathcal{U}$ be a normed linear space and suppose that $\mathcal{V} = \mathbb{C}$ or $\mathcal{V} = \mathbb{R}$; then, any operator $A : \mathcal{U} \to \mathcal{V}$ is called a *functional*. Let us denote by $\mathcal{U}'$ the set of all bounded (or, equivalently, continuous) linear functionals defined on a normed linear space $\mathcal{U}$. Clearly $\mathcal{U}'$ is a linear space provided that we define addition of linear functionals and multiplication of a linear functional by a scalar (in $\mathbb{R}$ or $\mathbb{C}$) in the usual way; that is,

$$(f+g)(u) := f(u) + g(u), \quad f, g \in \mathcal{U}', u \in \mathcal{U},$$
$$(\lambda f)(u) := \lambda f(u), \quad f \in \mathcal{U}', \lambda \in \mathbb{R} \text{ (or } \mathbb{C}\text{)}, u \in \mathcal{U}.$$

In addition, $\mathcal{U}'$ can be equipped with a norm $\|\cdot\|_{\mathcal{U}'}$ defined by

$$\|f\|_{\mathcal{U}'} := \sup_{0 \neq u \in \mathcal{U}} \frac{|f(u)|}{\|u\|_{\mathcal{U}}}.$$

The resulting normed linear space $\mathcal{U}'$ is called the *dual space* of $\mathcal{U}$, and $\|\cdot\|_{\mathcal{U}'}$ is usually referred to as the *dual norm*. It is a simple matter to show that the dual, $\mathcal{U}'$, of a normed linear space $\mathcal{U}$ is a Banach space with the dual norm.

Example 1.9 Let $\mathcal{U}$ denote the linear space of all continuous real-valued functions on the closed interval $[-1, 1]$ of the real line equipped with the norm

$$\|u\|_{C([-1,1])} := \max_{x \in [-1,1]} \big|u(x)\big|.$$

The functional $\delta : \mathcal{U} \to \mathbb{R}$, defined by

$$\delta(u) := u(0),$$

is contained in $\mathcal{U}'$. Indeed, δ is a linear functional on $\mathcal{U}$, and

$$\big|\delta(u)\big| = \big|u(0)\big| \leq \|u\|_{C([-1,1])} \quad \forall u \in C\big([-1, 1]\big).$$

Thus δ is a bounded linear functional on $\mathcal{U} = C([0, 1])$. Clearly $\|\delta\|_{\mathcal{U}'} = 1$.

$\mathcal{U}' = (C([0, 1]))'$ can be shown to coincide with the linear space $rca([0, 1])$ of all regular countably additive scalar-valued set functions defined on the σ-algebra of all Borel subsets of $[0, 1]$ (see, Theorem 3 on p. 265 of [36]).

We shall need the following result regarding the extension of a linear functional defined on a linear subspace M of a linear space $\mathcal{U}$ to the entire space.

Theorem 1.2 (Hahn–Banach Theorem) *Let $\mathcal{U}$ be a real (or complex) linear space, let M be a linear subspace of $\mathcal{U}$, and $p(\cdot)$ a seminorm on $\mathcal{U}$. Let l_M be a linear functional from M to $\mathbb{R}$ (or $\mathbb{C}$) such that $|l_M(v)| \leq p(v)$ for all $v \in \mathcal{U}$. Then, there exists a linear functional l on $\mathcal{U}$ such that $l(v) = l_M(v)$ for all v in M and $|l(v)| \leq p(v)$ for all v in $\mathcal{U}$.*

For a proof of Theorem 1.2 we refer to Theorem 3.3 on p. 57 of Rudin [158].

Corollary 1.3 *Suppose that $\mathcal{U}$ is a normed linear space and $u \in \mathcal{U}$; then, there exists a $y_* \in \mathcal{U}'$ such that $y_*(u) = \|u\|_{\mathcal{U}}$ and $\|y_*\|_{\mathcal{U}'} = 1$.*

Proof If $u = 0$, then we take $y_* := 0$. If $u \neq 0$, then we apply Theorem 1.2 with M chosen as the one-dimensional space spanned by u, $p(\cdot) := \|\cdot\|_{\mathcal{U}}$ on $\mathcal{U}$, $l_M(\alpha u) := \alpha\|u\|_{\mathcal{U}}$ on M, $\alpha \in \mathbb{R}$ (or $\mathbb{C}$), and we take $y_* := l$. □

Let $\mathcal{U}$ be a Banach space; then, each u in $\mathcal{U}$ defines a linear functional on $\mathcal{U}'$ by the correspondence $l_u(y) = y(u)$, $y \in \mathcal{U}'$. Clearly, l_u is a bounded linear functional on $\mathcal{U}'$; thus $l_u \in \mathcal{U}''$, where $\mathcal{U}''$ denotes the dual space of $\mathcal{U}'$. Indeed, $\|l_u\|_{\mathcal{U}''} \leq \|u\|_{\mathcal{U}}$. By Corollary 1.3 there exists a y_* in $\mathcal{U}'$ with $y_*(u) = \|u\|_{\mathcal{U}}$ and $\|y_*\|_{\mathcal{U}'} = 1$. This implies that $\|l_u\|_{\mathcal{U}''} = \|u\|_{\mathcal{U}}$; hence we have an *isometric isomorphism* $u \mapsto l_u$, defined by $l_u(y) = y(u)$, from $\mathcal{U}$ onto a closed linear subspace of $\mathcal{U}''$. A Banach space $\mathcal{U}$ is called *reflexive* if the mapping $u \mapsto l_u$ from $\mathcal{U}$ into $\mathcal{U}''$ is a surjection.

Let $\mathcal{U}$ be a normed linear space and $\{u_n\}_{n=1}^{\infty}$ a sequence in $\mathcal{U}$. We say that u_n *converges weakly* to $u \in \mathcal{U}$ if

$$\lim_{n\to\infty} f(u_n) = f(u)$$

for all f in $\mathcal{U}'$. In this case, u is called the *weak limit* of the sequence $\{u_n\}_{n=1}^{\infty}$. It is easy to see that if a sequence $\{u_n\}_{n=1}^{\infty}$ converges to u in $\mathcal{U}$ (in the norm of $\mathcal{U}$), then it also converges weakly to u in $\mathcal{U}$.

1.1.3 Sublinear Functionals

This section is devoted to an abstract result, due to Dražić [32], that is a useful tool in the error analysis of finite difference methods in various function spaces under minimum smoothness requirements on the data; the result is stated in Theorem 1.9. We begin by introducing the necessary concepts and by proving some preliminary results.

Definition 1.4 Let $\mathcal{U}$ be a linear space. A mapping $S : \mathcal{U} \to \mathbb{R}_+$ such that

$$S(\alpha u + \beta v) \leq |\alpha| S(u) + |\beta| S(v),$$

for all α, β in $\mathbb{R}$ (or in $\mathbb{C}$) and all u and v in $\mathcal{U}$, is called a *sublinear functional.* A sublinear functional $S : \mathcal{U} \to \mathbb{R}_+$ is said to be *bounded* if there exists a positive constant C such that

$$S(u) \leq C\|u\|_{\mathcal{U}} \quad \forall u \in \mathcal{U}.$$

For a bounded sublinear functional we define

$$\|S\| := \sup_{0\neq u\in\mathcal{U}} \frac{S(u)}{\|u\|_{\mathcal{U}}}.$$

We note that any norm or seminorm on a linear space $\mathcal{U}$ is a sublinear functional on $\mathcal{U}$ in the sense of Definition 1.4.

Lemma 1.5 *Let $\mathcal{U}$ be a linear space and suppose that $S : \mathcal{U} \to \mathbb{R}_+$ is a sublinear functional; then,* $S(0) = 0$.

Proof This is easily seen by noting that, for any u in $\mathcal{U}$,

$$0 \leq S(0) = S(0 \cdot u) \leq |0| \cdot S(u) = 0.$$

Hence $S(0) = 0$. □

Lemma 1.6 *Let $\mathcal{U}$ be a linear space and suppose that $S : \mathcal{U} \to \mathbb{R}_+$ is a sublinear functional. Then, the kernel of S, defined by*

$$\mathrm{Ker}(S) := \{u \in \mathcal{U} : S(u) = 0\},$$

is a linear subset of $\mathcal{U}$.

Proof According to the previous lemma the kernel of S is nonempty as $0 \in \mathrm{Ker}(S)$. Suppose that $u, v \in \mathrm{Ker}(S)$; then, for any $\alpha, \beta \in \mathbb{R}$,

$$0 \leq S(\alpha u + \beta v) \leq |\alpha| S(u) + |\beta| S(v) = 0.$$

Hence $\alpha u + \beta v \in \mathrm{Ker}(S)$. □

Theorem 1.7 *Let $\mathcal{U}_1$ be a Banach space that is compactly embedded in a normed linear space $\mathcal{U}_0$, and let $S_i : \mathcal{U}_i \to \mathbb{R}_+$, $i = 0, 1$, be two bounded sublinear functionals such that*

$$\|u\|_{\mathcal{U}_1} \leq S_0(u) + S_1(u)$$

for all $u \in \mathcal{U}_1$. Then,

$$\mathcal{P} := \mathrm{Ker}(S_1)$$

is a finite-dimensional closed linear subspace of $\mathcal{U}_1$.

Proof Lemma 1.6 implies that $\mathcal{P}$ is a linear space. The fact that $\mathcal{P}$ is closed follows from the boundedness of S_1; indeed, suppose that $u_m \in \mathcal{P}$, $m = 1, 2, \ldots$, and let $\lim_{m\to\infty} u_m = u$ in $\mathcal{U}_1$. Then,

$$\begin{aligned} 0 \leq S_1(u) &= S_1(u - u_m + u_m) \leq S_1(u - u_m) + S_1(u_m) \\ &= S_1(u - u_m) \\ &\leq \|S_1\| \, \|u - u_m\|_{\mathcal{U}_1}. \end{aligned}$$

Since the expression on the right-hand side converges to zero as $m \to \infty$, it follows that $S_1(u) = 0$, and therefore $u \in \mathcal{P}$. This implies that $\mathcal{P}$ is a closed linear subspace of $\mathcal{U}_1$. It remains to show that $\mathcal{P}$ is finite-dimensional; we shall do so by proving that the linear space $\mathcal{P}$ is locally compact in $\mathcal{U}_1$. Consider

$$\hat{\mathcal{P}} := \{u \in \mathcal{P} : \|u\|_{\mathcal{U}_1} \leq 1\}.$$

The set $\hat{\mathcal{P}}$ is bounded and closed in $\mathcal{U}_1$; thus, $\hat{\mathcal{P}}$ is a compact subset of $\mathcal{U}_0$. Consequently, we can extract a sequence $\{u_m\}_{m=1}^{\infty} \subset \hat{\mathcal{P}}$ that converges in $\mathcal{U}_0$; let u denote its limit. Because $\hat{\mathcal{P}}$ is closed, it follows that u belongs to $\hat{\mathcal{P}}$; in addition, $\{u_m\}_{m=1}^{\infty}$

is a Cauchy sequence in $\mathcal{U}_0$. Now,

$$\begin{aligned}\|u_m - u_n\|_{\mathcal{U}_1} &\le S_0(u_m - u_n) + S_1(u_m - u_n)\\ &= S_0(u_m - u_n)\\ &\le \|S_0\|\,\|u_m - u_n\|_{\mathcal{U}_0}.\end{aligned}$$

Thus, $\{u_m\}_{m=1}^{\infty}$ is a Cauchy sequence in $\mathcal{U}_1$ also. Since $\mathcal{U}_1$ is a Banach space, it follows that $\{u_m\}_{m=1}^{\infty}$ converges in the norm of $\mathcal{U}_1$, and since $\mathcal{U}_1 \subset \mathcal{U}_0$, by the uniqueness of the limit $\lim_{m\to\infty} u_m = u$ in $\mathcal{U}_1$. As $u \in \hat{\mathcal{P}}$, we have thus shown that $\hat{\mathcal{P}}$ is compact in $\mathcal{U}_1$; therefore the closed linear space $\mathcal{P}$ is locally compact in $\mathcal{U}_1$. By Theorem 1.1 it then follows that $\mathcal{P}$ is finite-dimensional. That completes the proof of the theorem. □

Theorem 1.8 *Under the same hypotheses as in Theorem* 1.7, *there exists a positive constant C such that, for all $u \in \mathcal{U}_1$,*

$$\inf_{p\in\mathcal{P}} \|u - p\|_{\mathcal{U}_1} \le C S_1(u),$$

where $\mathcal{P} := \mathrm{Ker}(S_1)$.

Proof Let $N := \dim \mathcal{P}$ and let f_i, $1 \le i \le N$, be a basis in the dual space of $\mathcal{P}$. Thus, for any $p \in \mathcal{P}$, we have that

$$f_i(p) = 0 \quad \forall i \in \{1, \dots, N\} \quad \Leftrightarrow \quad p = 0.$$

By the Hahn–Banach theorem, each f_i can be extended from $\mathcal{P}$ to a bounded linear functional, still denoted by f_i, on the whole of $\mathcal{U}_1$, $i = 1, \dots, N$. Let us suppose for a moment that we have proved the following statement: *there exists a constant $C > 0$ such that, for all $u \in \mathcal{U}_1$,*

$$\|u\|_{\mathcal{U}_1} \le C\left(S_1(u) + \sum_{i=1}^{N} |f_i(u)|\right). \tag{1.3}$$

The desired result then easily follows. Indeed, let $u \in \mathcal{U}_1$ and choose $q \in \mathcal{P}$ such that

$$f_i(u - q) = 0 \quad \forall i \in \{1, \dots, N\}.$$

We note in passing that there is a unique such $q \in \mathcal{P}$, which can be found by seeking $q = \alpha_1 p_1 + \cdots + \alpha_N p_N$, where $\{p_1, \dots, p_N\}$ is a basis of the linear space $\mathcal{P}$, and solving the system of linear equations $\alpha_1 f_i(p_1) + \cdots + \alpha_N f_i(p_N) = f_i(u)$, $i = 1, \dots, N$, for the scalars α_j, $j = 1, \dots, N$.

Now, by (1.3),

$$\inf_{p\in\mathcal{P}} \|u - p\|_{\mathcal{U}_1} \le \|u - q\|_{\mathcal{U}_1}$$

$$\le C\left(S_1(u-q)+\sum_{i=1}^{N}\left|f_i(u-q)\right|\right)$$
$$= CS_1(u-q) \le CS_1(u),$$

which then completes the proof.

It remains to prove (1.3). Suppose that (1.3) is false; then, there exists a sequence $\{u_m\}_{m=1}^{\infty}$ in $\mathcal{U}_1$ such that

$$\|u_m\|_{\mathcal{U}_1} = 1, \quad m = 1, 2, \ldots, \tag{1.4}$$

and

$$\lim_{m\to\infty}\left(S_1(u_m)+\sum_{i=1}^{N}\left|f_i(u_m)\right|\right) = 0. \tag{1.5}$$

From (1.4), thanks to the assumed compact embedding of $\mathcal{U}_1$ in $\mathcal{U}_0$, there exists a subsequence $\{u_{m_k}\}_{k=1}^{\infty}$ of $\{u_m\}_{m=1}^{\infty}$, which converges in $\mathcal{U}_0$; let us denote the corresponding limit by u. Hence $\{u_{m_k}\}_{k=1}^{\infty}$ is a Cauchy sequence in $\mathcal{U}_0$. On the other hand, by the assumptions of the theorem,

$$\begin{aligned}\|u_{m_k}-u_{m_l}\|_{\mathcal{U}_1} &\le S_0(u_{m_k}-u_{m_l}) + S_1(u_{m_k}-u_{m_l})\\ &\le \|S_0\|\,\|u_{m_k}-u_{m_l}\|_{\mathcal{U}_0} + S_1(u_{m_k}) + S_1(u_{m_l}).\end{aligned}$$

Thus, thanks to (1.5), $\{u_{m_k}\}_{k=1}^{\infty}$ is a Cauchy sequence in $\mathcal{U}_1$. Since, by assumption, $\mathcal{U}_1$ is a Banach space, $\{u_{m_k}\}_{k=1}^{\infty}$ is convergent in $\mathcal{U}_1$, and by uniqueness of the limit, $\lim_{k\to\infty} u_{m_k} = u$ in $\mathcal{U}_1$. Therefore, by passing to the limit over this subsequence in (1.4) and (1.5), we have that

$$\|u\|_{\mathcal{U}_1} = 1 \quad\text{and}\quad S_1(u)+\sum_{i=1}^{N}\left|f_i(u)\right| = 0.$$

Consequently,

$$\|u\|_{\mathcal{U}_1} = 1,$$
$$u \in \mathcal{P} \quad\text{and}\quad f_i(u) = 0 \quad \forall i \in \{1,\ldots,N\},$$

and therefore both $\|u\|_{\mathcal{U}_1} = 1$ and $u = 0$, which is a contradiction. □

Theorem 1.9 (Dražić [32]) *Under the assumptions of Theorem* 1.7, *and assuming in addition that* $S : \mathcal{U}_1 \to \mathbb{R}_+$ *is a bounded sublinear functional such that*

$$\mathrm{Ker}(S_1) \subset \mathrm{Ker}(S),$$

there exists a constant $C_1 > 0$ *such that*

$$S(u) \le C_1 S_1(u) \quad \forall u \in \mathcal{U}_1.$$

Proof We begin by noting that, for any $u \in \mathcal{U}_1$ and any $p \in \mathcal{P} := \mathrm{Ker}(S_1)$,

$$\begin{aligned} S(u) &= S(u - p + p) \leq S(u - p) + S(p) \\ &= S(u - p) \leq \|S\| \|u - p\|_{\mathcal{U}_1}. \end{aligned}$$

Thus, by Theorem 1.8,

$$S(u) \leq \|S\| \inf_{p \in \mathcal{P}} \|u - p\|_{\mathcal{U}_1} \leq C_1 S_1(u),$$

with $C_1 = C\|S\|$, where C is as in Theorem 1.8. □

In what follows we shall consider special cases of this abstract result in various function spaces, such as the Bramble–Hilbert lemma in integer-order and fractional-order Sobolev spaces, and use these to derive sharp bounds on the error between analytical solutions to partial differential equations and their numerical approximations.

1.1.4 Linear Functionals on Hilbert Spaces

This section is devoted to a fundamental result in Hilbert space theory, the Riesz representation theorem; its proof requires some preliminary results, and establishing these is our first task. We begin with the following simple lemma.

Lemma 1.10 *Let $\mathcal{U}$ be a (real or complex) Hilbert space, equipped with the norm $\|\cdot\|$, and let M be a closed convex subset in $\mathcal{U}$. For u in $\mathcal{U}$, we define the distance from u to M by*

$$\mathfrak{d}(u, M) := \inf_{v \in M} \|u - v\|.$$

Then, there exists a unique element v_* in M such that $\|u - v_*\| = \mathfrak{d}(u, M)$.

Proof If $u \in M$ then the proof is trivial: we simply take $v_* = u$. Let us therefore assume that u does not belong to M. According to the definition of $\mathfrak{d}(u, M)$, there exists a sequence $\{v_m\}_{m=1}^{\infty}$ in M such that $\lim_{m\to\infty} \|u - v_m\| = \mathfrak{d}(u, M)$. By recalling the parallelogram identity (1.1), we have that

$$\|2u - v_n - v_m\|^2 + \|v_n - v_m\|^2 = 2\|u - v_n\|^2 + 2\|u - v_m\|^2.$$

Thanks to the convexity of the set M, $(v_n + v_m)/2$ belongs to M; thus, the first term on the left-hand side is $\geq 4[\mathfrak{d}(u, M)]^2$. Consequently,

$$\|v_n - v_m\|^2 \leq 2\|u - v_n\|^2 + 2\|u - v_m\|^2 - 4\big[\mathfrak{d}(u, M)\big]^2.$$

Since the right-hand side converges to 0 as $n, m \to \infty$, we deduce that $\{v_m\}_{m=1}^{\infty}$ is a Cauchy sequence in $\mathcal{U}$; let v_* denote its limit in $\mathcal{U}$. As M is closed, it follows that

v_* belongs to M. Moreover, because $\|\cdot\|:\mathcal{U}\to\mathbb{R}$ is a continuous function (thanks to the triangle inequality), we deduce from the definition of the sequence $\{v_m\}_{m=1}^{\infty}$ that $\|u-v_*\|=\eth(u,M)$.

It remains to show the uniqueness of such a v_*. Let us suppose that v_*' is another element in M with the same property: $\|u-v_*'\|=\eth(u,M)$. Then, by the parallelogram identity,

$$\left\|u-\frac{1}{2}(v_*+v_*')\right\|^2=\left[\eth(u,M)\right]^2-\frac{1}{4}\left\|v_*-v_*'\right\|^2.$$

Since $(v_*+v_*')/2$ belongs to M, the left-hand side of this inequality is $\geq[\eth(u,M)]^2$. Therefore, by the first axiom of norm, $v_*'=v_*$. □

Let $\mathcal{S}$ be a closed linear subspace of a Hilbert space $\mathcal{U}$; we define the orthogonal complement $\mathcal{S}^\perp$ of $\mathcal{S}$ by

$$\mathcal{S}^\perp:=\left\{u\in\mathcal{U}:(u,v)=0\ \ \forall v\in\mathcal{S}\right\}.$$

With this definition and using the previous lemma we can prove the following result.

Theorem 1.11 *Suppose that $\mathcal{S}$ is a closed linear subspace of a (real or complex) Hilbert space $\mathcal{U}$. Then, $\mathcal{U}=\mathcal{S}\oplus\mathcal{S}^\perp$; i.e. every element u in $\mathcal{U}$ can be written uniquely as $u=f+g$, where $f\in\mathcal{S}$ and $g\in\mathcal{S}^\perp$.*

Proof Since $\mathcal{S}$ is a closed convex set, according to Lemma 1.10 f can be defined as the unique element in $\mathcal{S}$ that minimizes $\eth(u,\mathcal{S})$ for a given u in $\mathcal{U}$. We define g in $\mathcal{U}$ by $g:=u-f$. The rest of the proof is devoted to showing that $g\in\mathcal{S}^\perp$, and that f and g are the unique such elements.

Let v be any element of $\mathcal{S}$, and consider the convex linear combination $\theta v+(1-\theta)f$ of v and f in $\mathcal{S}$, with $\theta\in(0,1)$. Then,

$$\begin{aligned}\left[\eth(u,\mathcal{S})\right]^2&\leq\left\|u-\left(\theta v+(1-\theta)f\right)\right\|^2=\left\|u-f-\theta(v-f)\right\|^2\\&=\|u-f\|^2-2\theta\Re(u-f,v-f)+\theta^2\|v-f\|^2\\&=\left[\eth(u,\mathcal{S})\right]^2-2\theta\Re(u-f,v-f)+\theta^2\|v-f\|^2;\end{aligned}$$

here $\Re z$ denotes the real part of the complex number z. Hence

$$\Re(u-f,v-f)\leq\frac{1}{2}\theta\|v-f\|^2.$$

Letting $\theta\to 0$ and recalling the definition of g, it follows that

$$\Re(g,v-f)\leq 0\quad\forall v\in\mathcal{S}.\tag{1.6}$$

Let z be any element in $\mathcal{S}$ and let α be a complex number such that $\overline{\alpha}(g, z) = |(g, z)|$. By taking $v = f + \alpha z$, we deduce from (1.6) that

$$\big|(g, z)\big| = \Re\big|(g, z)\big| = \Re\big(\overline{\alpha}(g, z)\big) = \Re(g, \alpha z) = \Re(g, v - f) \leq 0 \quad \forall z \in \mathcal{S}.$$

Consequently $(g, z) = 0$ for all z in $\mathcal{S}$, which implies that g belongs to $\mathcal{S}^{\perp}$.

The uniqueness of the representation $u = f + g$ is easy to establish: let us suppose that there exist f_1 in $\mathcal{S}$ and g_1 in $\mathcal{S}^{\perp}$ such that also $u = f_1 + g_1$. Now since $f - f_1 = g_1 - g$, with $f - f_1 \in \mathcal{S}$ and $g_1 - g \in \mathcal{S}^{\perp}$, we have that

$$\|f - f_1\|^2 = (f - f_1, f - f_1) = (f - f_1, g_1 - g) = 0,$$

and therefore $f = f_1$; hence also $g = g_1$, which proves uniqueness. □

We are now ready to state the main result of this section.

Theorem 1.12 (The Riesz Representation Theorem) *Let f be a bounded linear functional on a (real or complex) Hilbert space $\mathcal{U}$; then, there exists a unique element u in $\mathcal{U}$, called the* Riesz representer *of f, such that*

$$f(v) = (v, u) \quad \forall v \in \mathcal{U}.$$

Proof The uniqueness of the Riesz representer u is obvious, provided that it exists. Indeed, let us suppose that u' is another element in $\mathcal{U}$ such that $f(v) = (v, u')$ for all $v \in \mathcal{U}$. Subtracting this equality from $f(v) = (v, u)$, we deduce that $(v, u - u') = 0$ for all v in $\mathcal{U}$; therefore $u = u'$.

It remains to establish the existence of the Riesz representer. In the trivial case when $f(v) = 0$ for all v in $\mathcal{U}$, we take $u = 0$, so let us suppose that we are dealing with the nontrivial case when the kernel M of the linear functional f is not the whole of $\mathcal{U}$; then, M is a proper closed linear subspace of the Hilbert space $\mathcal{U}$ and, by Theorem 1.11, $M^{\perp}$ is nontrivial. In fact, $M^{\perp}$ is a one-dimensional linear subspace of $\mathcal{U}$. Indeed, if v_1 and v_2 are any two (nonzero) elements of $M^{\perp}$, we shall prove that they are linearly dependent. For this purpose we consider $v = v_1 - \alpha v_2$, where $\alpha = f(v_1)/f(v_2)$. Then, v belongs to $M^{\perp}$ and $f(v) = 0$, and hence also $v \in M$. The only element that belongs to both M and $M^{\perp}$ is $v = 0$; hence we deduce that $v_1 = \alpha v_2$. Thus any two elements in $M^{\perp}$ differ only by a scalar factor.

Let u_0 be an arbitrary nonzero element in $M^{\perp}$ and let v be an element of $\mathcal{U}$. Then, by Theorem 1.11, there exists a v_M in M and a complex number β such that

$$v = v_M + \beta u_0.$$

Clearly $\beta = f(v)/f(u_0)$, whereby

$$v = v_M + \frac{f(v)}{f(u_0)} u_0.$$

We define $u = \alpha u_0$, where

$$\alpha = \frac{\overline{f(u_0)}}{\|u_0\|^2}.$$

Combining these we obtain the desired result:

$$(v, u) = (v_M, u) + \frac{f(v)}{f(u_0)}(u_0, u) = \frac{f(v)}{f(u_0)}\overline{\alpha}\|u_0\|^2 = f(v),$$

and that completes the proof. □

Suppose that $\mathcal{U}$ is a *real* Hilbert space. Given $u \in \mathcal{U}$, consider the linear functional $f_u \in \mathcal{U}'$ defined by $f_u(v) = (v, u)$, $v \in \mathcal{U}$. According to the Riesz representation theorem, the mapping $u \mapsto f_u$ that takes $\mathcal{U}$ into its dual space $\mathcal{U}'$ is linear, bijective, and it is an isometry (that is, $\|u\|_{\mathcal{U}} = \|f_u\|_{\mathcal{U}'}$). Thus any Hilbert space is reflexive.

1.1.4.1 Adjoint of a Linear Operator on a Hilbert Space

Suppose that A is a bounded linear operator from a Hilbert space $\mathcal{U}$, with inner product $(\cdot, \cdot)$ and induced norm $\|\cdot\|$, into itself and let $D(A) = \mathcal{U}$. For a fixed element v in $\mathcal{U}$, consider the linear functional f_v defined on $\mathcal{U}$ by

$$f_v(u) := (Au, v), \quad u \in \mathcal{U}.$$

By the Riesz representation theorem, there exists a unique w in $\mathcal{U}$ such that

$$f_v(u) = (u, w) = (Au, v).$$

The mapping $v \mapsto w$ is linear and bounded on $\mathcal{U}$. It therefore defines a bounded linear operator from $\mathcal{U}$ into itself, denoted by A^* and called the *adjoint* of the linear operator A; hence, $w = A^*v$. With this definition, $D(A^*) = \mathcal{U}$ and

$$(Au, v) = \big(u, A^*v\big) \quad \forall u, v \in \mathcal{U}.$$

If $A^*u = Au$ for all $u \in \mathcal{U}$, then A is called a *selfadjoint* (bounded) linear operator on the Hilbert space $\mathcal{U}$.

In many cases of interest a linear operator A is only partially defined on a Hilbert space $\mathcal{U}$ in the sense that its domain $D(A)$ is a strict subset of $\mathcal{U}$, although $D(A)$ is dense in $\mathcal{U}$, i.e. the closure of $D(A)$ in the norm of $\mathcal{U}$ coincides with $\mathcal{U}$. We then say that A is *densely defined* on $\mathcal{U}$. If a linear operator A on a Hilbert space $\mathcal{U}$ with domain $D(A) \subset \mathcal{U}$ is not bounded on $\mathcal{U}$, then we say that A is an *unbounded linear operator* on $\mathcal{U}$.

Let A be an unbounded linear operator on a Hilbert space $\mathcal{U}$ whose domain $D(A)$ is dense in $\mathcal{U}$. An element $v \in \mathcal{U}$ is said to belong to the domain $D(A^*)$ of the *adjoint operator* A^* if there exists a $w \in \mathcal{U}$ such that

$$(Au, v) = (u, w) \quad \forall u \in D(A).$$

In this case the adjoint operator A^* maps the element v into w, i.e. $A^*v = w$. The Riesz representation theorem implies that the domain $D(A^*)$ of the adjoint operator A^* is equal to the set of all $v \in \mathcal{U}$ such that $|(Au, v)| \leq C_v \|u\|$ for all $u \in D(A)$, where C_v is a positive constant, which may depend on v, but not on u. The adjoint operator is defined (uniquely) only if the original operator is densely defined, and is then a linear operator on $D(A^*)$.

A linear operator A on a Hilbert space $\mathcal{U}$ with inner product $(\cdot, \cdot)$ is called *symmetric* if

$$(Au, v) = (u, Av) \quad \forall u, v \in D(A).$$

If A is a densely defined symmetric linear operator on a Hilbert space $\mathcal{U}$, then $D(A) \subset D(A^*)$. If $D(A) = D(A^*)$ and $Au = A^*u$ for all $u \in D(A)$, then we say that the (densely defined) linear operator A is *selfadjoint*. A (densely defined) selfadjoint linear operator is clearly symmetric; the converse of this statement is however not true in general: a symmetric densely defined linear operator on a Hilbert space need not be selfadjoint.

In contrast, a symmetric everywhere defined linear operator on a Hilbert space is selfadjoint. Also, according to the Hellinger–Toeplitz theorem (cf. [154], Sect. III.5, p. 84), a symmetric everywhere defined linear operator on a Hilbert space is bounded. In most situations of relevance in the theory of differential equations A is a densely defined symmetric or selfadjoint linear operator on a Hilbert space $\mathcal{U}$, but A is unbounded, and its domain is therefore a strict subset of $\mathcal{U}$.

1.1.4.2 Bilinear Functionals on Real Hilbert Spaces

Let $\mathcal{U}$ be a real Hilbert space with norm $\|\cdot\|$, and let $a(\cdot, \cdot)$ be a real-valued functional defined on the Cartesian product $\mathcal{U} \times \mathcal{U}$ such that $a(\cdot, \cdot)$ is:

❶ *bilinear*, i.e. $a(w, v)$ is linear in w for v fixed, and linear in v for w fixed;
❷ *bounded*, i.e. there exists a positive real number c_1 such that

$$\left|a(w, v)\right| \leq c_1 \|w\| \|v\| \quad \forall w, v \in \mathcal{U};$$

❸ *$\mathcal{U}$-coercive*, i.e. there exists a positive real number c_0 such that

$$a(v, v) \geq c_0 \|v\|^2 \quad \forall v \in \mathcal{U}.$$

A bilinear functional is also called a *bilinear form*. Variational formulations of boundary-value problems for differential equations often have the following form: given a bounded linear functional f on a real Hilbert space $\mathcal{U}$ and a $\mathcal{U}$-coercive bounded bilinear functional $a(\cdot, \cdot)$ on $\mathcal{U} \times \mathcal{U}$, find u in $\mathcal{U}$ such that

$$a(u, v) = f(v) \quad \forall v \in \mathcal{U}.$$

The next theorem provides a useful device for verifying the existence and uniqueness of a solution to a problem of this kind.

Theorem 1.13 (Lax–Milgram Theorem) *Let f be a real-valued bounded linear functional on a real Hilbert space $\mathcal{U}$ with norm $\|\cdot\|$ and let $a(\cdot,\cdot)$ be a real-valued $\mathcal{U}$-coercive bounded bilinear functional on $\mathcal{U}\times\mathcal{U}$. Then, there exists a unique element $u\in\mathcal{U}$ such that*

$$a(u,v)=f(v) \quad \forall v\in\mathcal{U}. \tag{1.7}$$

In addition,

$$\|u\|\le\frac{1}{c_0}\|f\|_{\mathcal{U}'}.$$

Proof By the Riesz representation theorem, there exists a unique element b in $\mathcal{U}$ such that

$$f(v)=(v,b) \quad \forall v\in\mathcal{U},$$

and, for any z in $\mathcal{U}$, there exists a unique element Az in $\mathcal{U}$ such that

$$a(z,v)=(v,Az) \quad \forall v\in\mathcal{U}.$$

Thus (1.7) can be rewritten in the equivalent form

$$Au=b.$$

Clearly the mapping $A: z\in\mathcal{U}\mapsto Az\in\mathcal{U}$ is a linear operator on $\mathcal{U}$; furthermore, $\|Az\|\le c_1\|z\|$ and $\|Az\|\ge c_0\|z\|$ for all z in $\mathcal{U}$. Thus A is an *injective* bounded linear operator on $\mathcal{U}$. Next, we show that $R(A)$, the range of A, is closed in $\mathcal{U}$. Suppose that $\{Au_n\}_{n=1}^{\infty}$ is a sequence in $R(A)$ that converges in $\mathcal{U}$. Then, $\{Au_n\}_{n=1}^{\infty}$ is a Cauchy sequence in $\mathcal{U}$ and

$$\|Au_n-Au_m\|=\big\|A(u_n-u_m)\big\|\ge c_0\|u_n-u_m\|.$$

Thus $\{u_n\}_{n=1}^{\infty}$ is a Cauchy sequence in $\mathcal{U}$. As $\mathcal{U}$ is a Hilbert space, $\{u_n\}_{n=1}^{\infty}$ converges in $\mathcal{U}$. Letting $u\in\mathcal{U}$ be the limit of this sequence and noting that

$$\|Au_n-Au\|=\big\|A(u_n-u)\big\|\le c_1\|u_n-u\|,$$

it follows that $\{Au_n\}_{n=1}^{\infty}$ converges to Au in $\mathcal{U}$, which implies that the range of A is closed. Finally we will show that $R(A)=\mathcal{U}$, which will imply that A is also *surjective*. Suppose that this is not the case; then, by Theorem 1.11, there exists a $z_0\ne 0$ in the orthogonal complement $R(A)^{\perp}$ of the closed linear space $R(A)$. For such a z_0,

$$0=(z_0,Az)=a(z,z_0)$$

for all z in $\mathcal{U}$. In particular, $a(z_0,z_0)=0$, which is a contradiction, since $a(\cdot,\cdot)$ is $\mathcal{U}$-coercive and $z_0\ne 0$.

Thus we have shown that A is *bijective* and $\|Az\| \geq c_0\|z\|$ for all z in $\mathcal{U}$. Therefore A is invertible, and A^{-1} is a bounded linear operator with $\|A^{-1}\| \leq 1/c_0$. Consequently $u = A^{-1}b$ is the unique solution of (1.7), and

$$\|u\| \leq \frac{1}{c_0}\|b\| = \frac{1}{c_0}\|f\|_{\mathcal{U}'};$$

that completes the proof. □

In Chap. 2 we shall use the Lax–Milgram theorem to show that, under suitable assumptions, boundary-value problems for elliptic partial differential equations have a unique solution in appropriate function spaces, which will be introduced in Sect. 1.2. Before doing so, we shall discuss the abstract idea of Banach space interpolation.

1.1.5 Interpolation of Banach Spaces

By Banach space interpolation we refer to a process, which for two given Banach spaces constructs a family of 'intermediate' spaces. In this section we shall be concerned with one particular method of Banach space interpolation, called the K-method.

Let A_1 and A_2 be two Banach spaces, linearly and continuously embedded in a topological linear space $\mathcal{A}$ (i.e. a linear space with a topology that makes the operations of addition in the linear space and multiplication by a scalar continuous,—a relevant special case of a topological linear space being a normed linear space). Two such spaces are called an *interpolation pair* $\{A_1, A_2\}$. Consider also the space $A_1 \cap A_2$, equipped with the norm

$$\|a\|_{A_1\cap A_2} := \max\big\{\|a\|_{A_1}, \|a\|_{A_2}\big\},$$

and the space

$$A_1 + A_2 := \{a \in \mathcal{A} : a = a_1 + a_2, a_j \in A_j, j = 1, 2\},$$

with the norm

$$\|a\|_{A_1+A_2} := \inf_{a=a_1+a_2, a_j\in A_j} \big\{\|a_1\|_{A_1} + \|a_2\|_{A_2}\big\}.$$

Obviously, $A_1 \cap A_2 \subset A_j \subset A_1 + A_2$, $j = 1, 2$.

In order to proceed, we require the following basic definition from category theory; see, for example, Definition 1.1 on p. 9 of Jacobsen [77].

Definition 1.14 A *category* $\mathcal{C}$ consists of the following three ingredients:

❶ a class ob($\mathcal{C}$) of *objects* (usually denoted by A, B, C, etc.);

❷ for each ordered pair (A, B) of objects $A, B \in \mathcal{C}$ a set $\hom(A, B)$ whose elements are called *morphisms* with domain A and range (codomain) B. For $f \in \hom(A, B)$, we shall write $f : A \to B$, and will say that f is a morphism from A to B;

❸ for every three objects A, B and C contained in $\mathcal{C}$, a binary operation $\hom(A, B) \times \hom(B, C) \to \hom(A, C)$ called *composition of morphisms*, the composition of $f : A \to B$ and $g : B \to C$ being denoted by $g \circ f$, such that the following axioms hold:

① if $(A, B) \neq (C, D)$, then $\hom(A, B)$ and $\hom(C, D)$ are disjoint;

② (associativity): if $f : A \to B$, $g : B \to C$ and $h : C \to D$, then

$$h \circ (g \circ f) = (h \circ g) \circ f;$$

③ (identity): for every object A, there exists a morphism $1_A \in \hom(A, A)$ such that $f \circ 1_A = f$ for every $f \in \hom(A, B)$ and $1_A \circ g = g$ for every $g \in \hom(B, A)$. (The morphism 1_A is unique.)

Let us consider the category $\mathcal{C}_1$, where the objects $A, B, C, \ldots$ are Banach spaces and the morphisms are bounded linear operators $L \in \mathcal{L}(A, B)$. Let, also, $\mathcal{C}_2$ be a category where the objects are interpolation pairs $\{A_1, A_2\}, \{B_1, B_2\}, \ldots$ while the morphisms L belong to the set $\mathcal{L}(\{A_1, A_2\}, \{B_1, B_2\})$ of bounded linear operators from $A_1 + A_2$ into $B_1 + B_2$, whose restrictions to A_j belong to the set $\mathcal{L}(A_j, B_j)$, $j = 1, 2$.

By an *interpolation functor* from $\mathcal{C}_2$ to $\mathcal{C}_1$ we mean a rule, which to every interpolation pair $\{A_1, A_2\}$ from $\mathcal{C}_2$ assigns an object $\mathbb{F}(\{A_1, A_2\})$ from $\mathcal{C}_1$, with $A_1 \cap A_2 \subset \mathbb{F}(\{A_1, A_2\}) \subset A_1 + A_2$, and to every morphism $L \in \mathcal{L}(\{A_1, A_2\}, \{B_1, B_2\})$ from $\mathcal{C}_2$ it assigns a morphism $\mathbb{F}(L)$ from $\mathcal{C}_1$, which is the restriction of the operator L to $\mathbb{F}(\{A_1, A_2\})$.

The corresponding Banach space $A = \mathbb{F}(\{A_1, A_2\})$ is called an *interpolation space*. We note in particular that $A_1 \cap A_2$ and $A_1 + A_2$ are interpolation spaces.

Suppose that there exist real numbers $C \geq 1$ and $\theta \in (0, 1)$ such that

$$\|L\|_{\mathbb{F}(\{A_1,A_2\}) \to \mathbb{F}(\{B_1,B_2\})} \leq C \|L\|^{1-\theta}_{A_1 \to B_1} \|L\|^{\theta}_{A_2 \to B_2}$$

is satisfied for all $L \in \mathcal{L}(\{A_1, A_2\}, \{B_1, B_2\})$; then, the interpolation functor $\mathbb{F}$ is said to be of *type* θ; in particular if $C = 1$, then we say that the interpolation functor $\mathbb{F}$ is *exact, of type* θ. Here $\|L\|_{A_j \to B_j}$ denotes the usual operator norm of $L : A_j \to B_j$, with an analogous definition of $\|L\|_{\mathbb{F}(\{A_1,A_2\}) \to \mathbb{F}(\{B_1,B_2\})}$ (cf. (1.2)).

One of the most frequently used interpolation methods is the so-called K-method (cf. Bergh and Löfström [9], Sect. 3.1; or Triebel [182], Sect. 1.3). Let $\{A_1, A_2\}$ be an interpolation pair, and define the function

$$K(t, a, A_1, A_2) := \inf_{a \in A_1 + A_2, a = a_1 + a_2, a_j \in A_j} \left\{ \|a_1\|_{A_1} + t \|a_2\|_{A_2} \right\}.$$

Clearly, for any fixed $t \in (0, \infty)$, $a \mapsto K(t, a, A_1, A_2)$ is a norm in $A_1 + A_2$, equivalent to the norm $a \mapsto \|a\|_{A_1 + A_2}$. For $0 < \theta < 1$ and $1 \leq q \leq \infty$ we define the space

$(A_1, A_2)_{\theta,q}$ as the set of all elements $a \in A_1 + A_2$ for which $\|a\|_{(A_1,A_2)_{\theta,q}}$ is finite, where

$$\|a\|_{(A_1,A_2)_{\theta,q}} := \left\{ \int_0^\infty \left[t^{-\theta} K(t,a,A_1,A_2)\right]^q \frac{dt}{t} \right\}^{1/q} \quad \text{if } 1 \le q < \infty,$$

$$\|a\|_{(A_1,A_2)_{\theta,\infty}} := \sup_{0<t<\infty} t^{-\theta} K(t,a,A_1,A_2) \quad \text{if } q = \infty.$$

The normed linear space $(A_1, A_2)_{\theta,q}$ thus defined is an interpolation space. The following relations hold:

$$(A_1, A_2)_{\theta,q} = (A_2, A_1)_{1-\theta,q},$$

$$(A, A)_{\theta,q} = A,$$

$$(A_1, A_2)_{\theta,1} \subset (A_1, A_2)_{\theta,q} \subset (A_1, A_2)_{\theta,\tilde{q}} \subset (A_1, A_2)_{\theta,\infty},$$
$$1 \le q \le \tilde{q} \le \infty,$$

$$(A_1, A_2)_{\theta,q} \subset (A_1, A_2)_{\tilde{\theta},\tilde{q}},$$
$$\text{if } A_1 \subset A_2,\ 0 < \theta < \tilde{\theta} < 1,\ 1 \le q \le \tilde{q} \le \infty,$$

$$\exists C_{\theta,q} > 0\ \forall a \in A_1 \cap A_2 \quad \|a\|_{(A_1,A_2)_{\theta,q}} \le C_{\theta,q} \|a\|_{A_1}^{1-\theta} \|a\|_{A_2}^{\theta}.$$

The corresponding interpolation functor

$$\mathbb{F}(\{A_1, A_2\}) = (A_1, A_2)_{\theta,q}$$

is exact, of type θ, i.e. for any $L \in \mathcal{L}(\{A_1, A_2\}, \{B_1, B_2\})$,

$$\|L\|_{(A_1,A_2)_{\theta,q} \to (B_1,B_2)_{\theta,q}} \le \|L\|_{A_1 \to B_1}^{1-\theta} \|L\|_{A_2 \to B_2}^{\theta}. \tag{1.8}$$

We refer to Theorem 3.4.1 on p. 46 of Bergh and Löfström [9] and Theorem 1.3.3 on p. 25 of the monograph of Triebel [182] for proofs of these statements.

1.2 Basic Function Spaces

In this section, we recall the definitions of some standard function spaces, including those of continuously differentiable and Lebesgue-integrable functions.

1.2.1 Spaces of Continuous Functions

Let $\mathbb{N}$ denote the set of nonnegative integers. An n-tuple $\alpha = (\alpha_1, \dots, \alpha_n)$ in $\mathbb{N}^n$ is called a *multi-index*. The nonnegative integer $|\alpha| := |\alpha_1| + \cdots + |\alpha_n|$ is called the *length* of α. We shall write $0 := (0, \dots, 0)$, and let

$$\alpha! := \alpha_1! \cdots \alpha_n!,$$

$$\partial^\alpha := \partial_1^{\alpha_1} \cdots \partial_n^{\alpha_n}, \qquad \partial_j := \frac{\partial}{\partial x_j}, \quad j = 1, \dots, n.$$

For $x \in \mathbb{R}^n$ and $\alpha \in \mathbb{N}^n$, we define

$$x^\alpha := x_1^{\alpha_1} \cdots x_n^{\alpha_n},$$

with the convention that $0^0 := 1$. For $x, y \in \mathbb{R}^n$, we shall write

$$x \pm y := (x_1 \pm y_1, \dots, x_n \pm y_n).$$

Let $\mathbb{Z}$ denote the set of all integers. The set $\mathbb{Z}^n$ can be partially ordered by lexicographical ordering; that is, for α and β in $\mathbb{Z}^n$,

$$\alpha \le \beta \quad \Leftrightarrow \quad \alpha_j \le \beta_j, \quad j = 1, \dots, n.$$

For $\alpha, \beta \in \mathbb{N}^n$, such that $0 \le \beta \le \alpha$, we define

$$\binom{\alpha}{\beta} := \binom{\alpha_1}{\beta_1} \cdots \binom{\alpha_n}{\beta_n} = \frac{\alpha!}{\beta!(\alpha - \beta)!}.$$

Leibniz's formula in multi-index notation exemplifies the usefulness of this compact symbolism: assuming that u and v are two (sufficiently smooth) functions and α is a multi-index, then

$$\partial^\alpha (uv) = \sum_{0 \le \beta \le \alpha} \binom{\alpha}{\beta} \partial^{\alpha - \beta} u \, \partial^\beta v. \tag{1.9}$$

The proof, by induction, is easy and is left to the reader.

Suppose that Ω is an open subset of $\mathbb{R}^n$. For $k \in \mathbb{N}$, we denote by $C^k(\Omega)$ the set of all continuous (real- or complex-valued) functions u, defined on Ω, such that $\partial^\alpha u$ is continuous on Ω for every multi-index α, $|\alpha| \le k$. Further, we define

$$C^\infty(\Omega) := \bigcap_{k \ge 0} C^k(\Omega).$$

$C^0(\Omega)$ is abbreviated to $C(\Omega)$. $BC(\Omega)$ denotes the set of all bounded continuous functions defined on Ω, with the norm $\|u\|_{BC(\Omega)} := \sup_{x \in \Omega} |u(x)|$.

For $k \in \mathbb{N}$, we denote by $C^k(\overline{\Omega})$ the set of all $u \in C^k(\Omega)$ such that $\partial^\alpha u$ can be continuously extended from Ω onto $\overline{\Omega}$ (the closure of Ω), for every multi-index α, $|\alpha| \le k$. Further, we define

$$C^\infty(\overline{\Omega}) := \bigcap_{k \ge 0} C^k(\overline{\Omega}).$$

$C^0(\overline{\Omega})$ is abbreviated to $C(\overline{\Omega})$.

Assuming that Ω is a bounded open set in $\mathbb{R}^n$ and $k \in \mathbb{N}$, the linear space $C^k(\overline{\Omega})$ is a Banach space equipped with the norm

$$\|u\|_{C^k(\overline{\Omega})} := \max_{|\alpha| \leq k} \sup_{x \in \Omega} \left| \partial^\alpha u(x) \right|.$$

For $k \in \mathbb{N}$ and $0 < \lambda \leq 1$, we denote by $C^{k,\lambda}(\overline{\Omega})$ the set of all $u \in C^k(\overline{\Omega})$ such that the quantity

$$|u|_{C^{k,\lambda}(\overline{\Omega})} := \max_{|\alpha|=k} \sup_{x \neq y,\ x,y \in \Omega} \frac{|\partial^\alpha u(x) - \partial^\alpha u(y)|}{|x-y|^\lambda}$$

is finite. $C^{k,\lambda}(\overline{\Omega})$ is a Banach space with the norm

$$\|u\|_{C^{k,\lambda}(\overline{\Omega})} := \|u\|_{C^k(\overline{\Omega})} + |u|_{C^{k,\lambda}(\overline{\Omega})}.$$

When u belongs to $C^{0,\lambda}(\overline{\Omega})$, $0 < \lambda < 1$, we say that u is *Hölder-continuous* on $\overline{\Omega}$ with exponent λ; if $\lambda = 1$, the function u is said to be *Lipschitz-continuous* on $\overline{\Omega}$.

Example 1.10 Let $\Omega := B_1$, the unit ball in $\mathbb{R}^n$ centred at the origin, and let

$$u(x) := |x|^\lambda, \quad x \in B_1,$$

where $|\cdot| = \|\cdot\|$ is the Euclidean norm from Example 1.1. For $0 < \lambda < 1$, the function u is Hölder-continuous on $\overline{\Omega}$ with exponent λ; when $\lambda = 1$, u is Lipschitz-continuous on $\overline{\Omega}$.

The *support*, $\operatorname{supp} u$, of a continuous function u, defined on an open set Ω contained in $\mathbb{R}^n$, is the closure in Ω of the set $\{x \in \Omega : u(x) \neq 0\}$; in other words, $\operatorname{supp} u$ is the smallest closed subset of Ω such that $u = 0$ on $\Omega \setminus \operatorname{supp} u$. If $\operatorname{supp} u \Subset \Omega$, we say that u has *compact support* in Ω.

For $k \in \mathbb{N} \cup \{\infty\}$, $C_0^k(\Omega)$ denotes the set of all $u \in C^k(\Omega)$ with compact support in Ω. In the theory of distributions the elements of $C_0^\infty(\Omega)$ are called *test functions*. Our next example demonstrates the existence of test functions.

Example 1.11 Consider the real-valued function ω defined on $\mathbb{R}^n$ by

$$\omega(x) = \begin{cases} C\exp((|x|^2 - 1)^{-1}) & \text{if } |x| < 1, \\ 0 & \text{otherwise}, \end{cases}$$

where C is a constant chosen so that $\int_{\mathbb{R}^n} \omega(x)\,\mathrm{d}x = 1$. For $\varepsilon > 0$ we define $\omega_\varepsilon(x) = \varepsilon^{-n}\omega(x/\varepsilon)$. Then, ω_ε belongs to $C_0^\infty(\mathbb{R}^n)$, $\operatorname{supp} \omega_\varepsilon = B_\varepsilon := B(0,\varepsilon)$, and $\int_{\mathbb{R}^n} \omega_\varepsilon(x)\,\mathrm{d}x = 1$.

The next lemma encapsulates the properties of a special test function, which will be required in our subsequent arguments.

Lemma 1.15 *For an open set* $A \subset \mathbb{R}^n$ *and* $\varepsilon > 0$, *there exists a function* $\varphi_\varepsilon \in C_0^\infty(\mathbb{R}^n)$, *such that*

$$\varphi_\varepsilon(x) = 1, \quad x \in A^\varepsilon; \qquad \varphi_\varepsilon(x) = 0, \quad x \notin A^{3\varepsilon};$$

$$0 \le \varphi_\varepsilon(x) \le 1, \quad \left|\partial^\alpha \varphi_\varepsilon(x)\right| \le C_\varepsilon \varepsilon^{-|\alpha|} \quad \forall x \in \mathbb{R}^n,$$

where C_ε *is a positive constant, and* A^ε *and* $A^{3\varepsilon}$ *denote, respectively, the* ε*- and* 3ε*-neighbourhood of the set* A *(cf. Sect.* 1.1.1.1).

Proof Let ω_ε be the function defined in Example 1.11. The function

$$\varphi_\varepsilon(x) := \int_{A^{2\varepsilon}} \omega_\varepsilon(x-y)\,\mathrm{d}y$$

then possesses the required properties. □

1.2.2 Spaces of Integrable Functions

For a real number $p \ge 1$ and an open set $\Omega \subset \mathbb{R}^n$, let $L_p(\Omega)$ denote the set of all (real- or complex-valued) Lebesgue-measurable functions u defined on Ω such that $|u|^p$ is integrable on Ω with respect to the Lebesgue measure $\mathrm{d}x = \mathrm{d}x_1 \cdots \mathrm{d}x_n$ (see, for example, Bartle [8]); we assume here that any two functions that are equal almost everywhere (i.e. equal, except perhaps on a set of zero Lebesgue measure) are identified. With this convention, $L_p(\Omega)$ is a Banach space with the norm

$$\|u\|_{L_p(\Omega)} := \left(\int_\Omega \left|u(x)\right|^p \mathrm{d}x\right)^{1/p}.$$

In particular when $p = 2$, $L_2(\Omega)$ is a Hilbert space with the inner product

$$(u, v) := \int_\Omega u(x)\overline{v(x)}\,\mathrm{d}x.$$

$L_\infty(\Omega)$ denotes the set of all Lebesgue-measurable functions u defined on Ω such that $|u|$ has finite essential supremum; the *essential supremum* of $|u|$ is defined as the infimum of the set of all positive real numbers M such that $|u| \le M$ almost everywhere on Ω. Again, any two functions that are equal almost everywhere on Ω are identified. $L_\infty(\Omega)$ is a Banach space with the norm

$$\|u\|_{L_\infty(\Omega)} := \text{ess.sup}_{x\in\Omega}\left|u(x)\right|.$$

Hölder's inequality. Let $u \in L_p(\Omega)$ and $v \in L_q(\Omega)$, where $1/p + 1/q = 1$, $1 \le p, q \le \infty$. Then, $uv \in L_1(\Omega)$ and

$$\left|\int_\Omega u(x)v(x)\,\mathrm{d}x\right| \le \|u\|_{L_p(\Omega)}\|v\|_{L_q(\Omega)}.$$

For $p = q = 2$, this yields as a special case the *Cauchy–Schwarz inequality*:

$$\left|(u, v)\right| \le \|u\|_{L_2(\Omega)} \|v\|_{L_2(\Omega)}.$$

When $u \in L_1(\mathcal{O})$ for every set $\mathcal{O} \Subset \Omega$, u is said to be *locally integrable* on Ω. The set of all locally integrable functions defined on Ω is denoted by $L_{1,loc}(\Omega)$. Clearly, $C(\Omega)$ is contained in $L_{1,loc}(\Omega)$ but not in $L_1(\Omega)$.

Example 1.12 The function $u(x) = \exp|x|$ is continuous and locally integrable on $\mathbb{R}^n$, but it does not belong to $L_p(\mathbb{R}^n)$ for any p, $1 \le p \le \infty$.

Lemma 1.16 (du Bois-Reymond's Lemma) *Let u and v be locally integrable functions on Ω and suppose that*

$$\int_\Omega u(x)\varphi(x)\,\mathrm{d}x = \int_\Omega v(x)\varphi(x)\,\mathrm{d}x \quad \forall \varphi \in C_0^\infty(\Omega);$$

then, $u = v$ almost everywhere on Ω.

Proof Let us define $w := u - v$. Then,

$$\int_\Omega w(x)\varphi(x)\,\mathrm{d}x = 0 \quad \forall \varphi \in C_0^\infty(\Omega). \tag{1.10}$$

Further, as $w \in L_{1,loc}(\Omega)$, according to a strengthened version of Lebesgue's differentiation theorem (cf. Theorem 7.7 on p. 138 of Rudin [157]),

$$\lim_{\varepsilon \to 0} \varepsilon^{-n} \int_{|x-y|<\varepsilon} \left|w(x) - w(y)\right| \mathrm{d}y = 0, \tag{1.11}$$

for almost every x. By recalling the definition of the function ω from Example 1.11 and assuming that ε is sufficiently small, (1.10) implies that

$$w(x) = \varepsilon^{-n} \int_{|x-y|<\varepsilon} \big(w(x) - w(y)\big)\omega\left(\frac{x-y}{\varepsilon}\right) \mathrm{d}y, \quad x \in \Omega.$$

By noting that $\max_{|x-y|\le\varepsilon} |\omega((x-y)/\varepsilon)| = \max_{|z|\le 1} \omega(z) = C/\mathrm{e}$, with C as in Example 1.11, and then letting $\varepsilon \to 0$ and applying (1.11), we deduce that $w(x) = 0$ for almost every $x \in \Omega$. □

Du Bois-Reymond's lemma is frequently referred to as the fundamental lemma of the Calculus of Variations.

The *support*, $\operatorname{supp} u$, of a measurable function u defined on Ω is the smallest closed subset of Ω such that $u = 0$ almost everywhere in $\Omega \setminus \operatorname{supp} u$. This definition is a consistent extension of our earlier definition of the support of a continuous function in Sect. 1.2.1.

1.3 Distributions

This section introduces various classes of distributions on an open set $\Omega \subseteq \mathbb{R}^n$ and surveys their main properties.

1.3.1 Test Functions and Distributions

To give an informal definition, a distribution is a continuous linear functional on the space $C_0^\infty(\Omega)$ of infinitely differentiable functions with compact support. In order to state the precise definition of a distribution, we have to qualify the word *continuous*. This is achieved by introducing a topology on $C_0^\infty(\Omega)$, or simply by defining the concept of *convergence* in $C_0^\infty(\Omega)$.

Definition 1.17 A sequence $\{\varphi_m\}_{m=1}^\infty \subset C_0^\infty(\Omega)$ is said to converge to φ in $C_0^\infty(\Omega)$ if there exists a set $\mathcal{O} \Subset \Omega$ such that $\operatorname{supp} \varphi_m \subset \mathcal{O}$ for every m, and $\partial^\alpha \varphi_m$ converges to $\partial^\alpha \varphi$, uniformly on Ω, as $m \to \infty$, for every multi-index $\alpha \in \mathbb{N}^n$.

When equipped with this definition of convergence the linear space $C_0^\infty(\Omega)$ is denoted by $\mathcal{D}(\Omega)$; thus we write $\varphi_m \to \varphi$ in $\mathcal{D}(\Omega)$ as $m \to \infty$.

Now suppose that u is a linear functional on $\mathcal{D}(\Omega)$, i.e. to every φ in $\mathcal{D}(\Omega)$, u assigns a (complex) number denoted by $\langle u, \varphi\rangle$ (instead of $u(\varphi)$), and

$$\langle u, \lambda\varphi + \mu\psi\rangle = \lambda\langle u, \varphi\rangle + \mu\langle u, \psi\rangle, \quad \lambda, \mu \in \mathbb{C}, \ \varphi, \psi \in \mathcal{D}(\Omega).$$

We shall say that u is a continuous linear functional on $\mathcal{D}(\Omega)$ if $\langle u, \varphi_m\rangle \to \langle u, \varphi\rangle$ as $m \to \infty$, whenever $\varphi_m \to \varphi$ in $\mathcal{D}(\Omega)$ as $m \to \infty$.

Definition 1.18 A continuous linear functional on $\mathcal{D}(\Omega)$ is called a *distribution* on Ω. The set of all distributions on Ω is denoted by $\mathcal{D}'(\Omega)$.

The next theorem provides a useful characterization of distributions.

Theorem 1.19 *Suppose that u is a linear functional on* $\mathcal{D}(\Omega)$; *then, the following statements are equivalent*:

(a) $u \in \mathcal{D}'(\Omega)$;
(b) *for every open set* $\mathcal{O} \Subset \Omega$ *there exists a real number* $K = K(\mathcal{O})$ *and a nonnegative integer* $m = m(\mathcal{O})$ *such that*

$$\left|\langle u, \varphi\rangle\right| \le K\|\varphi\|_{C^m(\overline{\mathcal{O}})} \quad \forall \varphi \in \mathcal{D}(\mathcal{O}). \tag{1.12}$$

Proof It is clear that (b) implies (a). The converse implication is established by *reductio ad absurdum*. Let us therefore assume that $u \in \mathcal{D}'(\Omega)$ and that (b) is false. Then, there exists a set $\mathcal{O} \Subset \Omega$ and a sequence $\{\varphi_m\}_{m=1}^\infty \subset \mathcal{D}(\mathcal{O})$ such that

$$\left|\langle u, \varphi_m\rangle\right| \ge m\|\varphi_m\|_{C^m(\overline{\mathcal{O}})}, \quad m = 1, 2, \ldots.$$

Let $\psi_m := \varphi_m/(m\|\varphi_m\|_{C^m(\overline{\mathcal{O}})})$ and note that $\sup_{x\in\mathcal{O}}|\partial^\alpha\psi_m(x)| \le 1/m$ for all $\alpha \in \mathbb{N}^n$ such that $|\alpha| \le m$. Thus, $\{\psi_m\}_{m=1}^\infty$ converges to zero in $\mathcal{D}(\Omega)$ as $m \to \infty$, whereas $|\langle u, \psi_m\rangle| \ge 1$ for all $m \ge 1$, and therefore u is not a continuous linear functional on $\mathcal{D}(\Omega)$. This contradicts (a). □

Suppose that u is a distribution on Ω. If the integer m appearing in (1.12) is independent of the choice of $\mathcal{O}$, we say that u is of *finite order*; the smallest such integer m is called the *order* of the distribution u. If such an integer does not exist, we say that u is of *infinite order*.

Example 1.13 The linear functional δ, defined by

$$\langle\delta,\varphi\rangle := \varphi(0), \quad \varphi \in \mathcal{D}(\mathbb{R}^n),$$

is a distribution on $\mathbb{R}^n$ of order 0; δ is called the *Dirac distribution* concentrated at 0.

Example 1.14 The linear functional u, defined by

$$\langle u,\varphi\rangle := \sum_{\alpha\in\mathbb{Z}^n}\varphi(\alpha), \quad \varphi \in \mathcal{D}(\mathbb{R}^n),$$

is a distribution of infinite order.

Definition 1.20 Two distributions $u, v \in \mathcal{D}'(\Omega)$ are said to be equal on $\mathcal{O} \subset \Omega$ if $\langle u,\varphi\rangle = \langle v,\varphi\rangle$ for every $\varphi \in \mathcal{D}(\mathcal{O})$. In particular, a distribution $u \in \mathcal{D}'(\Omega)$ is said to be equal to 0 on $\mathcal{O} \subset \Omega$ if $\langle u,\varphi\rangle = 0$ for every $\varphi \in \mathcal{D}(\mathcal{O})$.

We can now define the support of a distribution $u \in \mathcal{D}'(\Omega)$. Let Ω_u denote the union of all open sets $\mathcal{O} \subset \Omega$ such that u is equal to 0 on $\mathcal{O}$. Then, Ω_u is the largest open subset of Ω on which u is equal to 0. The complement of Ω_u with respect to Ω is called the *support* of u and is denoted by $\operatorname{supp} u$. By definition, the support of a distribution is a closed set, relative to Ω. If $\operatorname{supp} u \Subset \Omega$, we say that u has compact support in Ω. For example, the Dirac distribution concentrated at 0 has compact support $\operatorname{supp}\delta = \{0\}$ in $\mathbb{R}^n$, whereas the distribution considered in Example 1.14 has $\mathbb{Z}^n$ as its support, which, being an unbounded set in $\mathbb{R}^n$, is not compact in $\mathbb{R}^n$.

Next we show that a distribution whose support is a compact subset of an open set $\Omega \subset \mathbb{R}^n$ can be extended from $C_0^\infty(\Omega)$ to a continuous linear functional on $C^\infty(\Omega) \supset C_0^\infty(\Omega)$. For this purpose the linear space $C^\infty(\Omega)$ is equipped with a definition of *convergence*.

Definition 1.21 A sequence $\{\varphi_m\}_{m=1}^\infty \subset C^\infty(\Omega)$ is said to converge to φ in $C^\infty(\Omega)$ if, for every $\mathcal{O} \Subset \Omega$ and every multi-index α, $\partial^\alpha\varphi_m$ converges to $\partial^\alpha\varphi$, uniformly on $\mathcal{O}$, as $m \to \infty$.

The linear space $C^\infty(\Omega)$ equipped with convergence in this sense will be denoted by $\mathcal{E}(\Omega)$. Clearly $\mathcal{E}(\Omega) \supset \mathcal{D}(\Omega)$, as a topological inclusion, meaning that a sequence in $\mathcal{D}(\Omega)$ that converges to an element of $\mathcal{D}(\Omega)$ converges to the same element, when considered as a sequence in $\mathcal{E}(\Omega)$.

We denote by $\mathcal{E}'(\Omega)$ the linear space of continuous linear functionals on $\mathcal{E}(\Omega)$. According to the next theorem each element of $\mathcal{E}'(\Omega)$ can be identified with a distribution with compact support.

Theorem 1.22 *A distribution $u \in \mathcal{D}'(\Omega)$ has compact support in Ω if, and only if, it admits an extension from $\mathcal{D}(\Omega)$ to a continuous linear functional on $\mathcal{E}(\Omega)$.*

Proof Suppose that $u \in \mathcal{D}'(\Omega)$ and $K = \operatorname{supp} u \Subset \Omega$. Further, let $\eta \in \mathcal{D}(\Omega)$, with $\eta(x) = 1$ in a neighbourhood of K; the existence of such a function η is guaranteed by Lemma 1.15. We define $\tilde{u}$ by

$$\langle \tilde{u}, \varphi \rangle = \langle u, \eta\varphi \rangle, \quad \varphi \in \mathcal{E}(\Omega).$$

This definition is correct in the sense that it is independent of the choice of η in $\mathcal{E}(\Omega)$. Clearly $\tilde{u}$ is a continuous linear functional on $\mathcal{E}(\Omega)$, and $\langle \tilde{u}, \varphi \rangle = \langle u, \varphi \rangle$ for all $\varphi \in \mathcal{D}(\Omega)$. Thus $\tilde{u}$ is a *continuous extension* of u to $\mathcal{E}(\Omega)$.

We note in passing that $\tilde{u}$ is the unique continuous extension of u from $\mathcal{D}(\Omega)$ to $\mathcal{E}(\Omega)$. Suppose that $\tilde{\tilde{u}} \in \mathcal{E}'(\Omega)$ is another continuous extension of u from $\mathcal{D}(\Omega)$ to $\mathcal{E}(\Omega)$. We consider a sequence of open sets $\Omega_1 \Subset \Omega_2 \Subset \cdots$ such that $\Omega = \bigcup_{m=1}^\infty \Omega_m$, and a sequence of test functions $\{\eta_m\}_{m=1}^\infty \subset \mathcal{D}(\Omega)$ such that $\eta_m = 1$ on Ω_m, and $\eta_m \to 1$ in $\mathcal{E}(\Omega)$ as $m \to \infty$. Then, for every $\varphi \in \mathcal{E}(\Omega)$, $\eta_m\varphi \to \varphi$ in $\mathcal{E}(\Omega)$ as $m \to \infty$. Consequently,

$$\begin{aligned}\langle \tilde{u}, \varphi \rangle &= \left\langle \tilde{u}, \lim_{m\to\infty} \eta_m\varphi \right\rangle = \lim_{m\to\infty} \langle \tilde{u}, \eta_m\varphi \rangle = \lim_{m\to\infty} \langle u, \eta_m\varphi \rangle \\ &= \lim_{m\to\infty} \langle \tilde{\tilde{u}}, \eta_m\varphi \rangle = \left\langle \tilde{\tilde{u}}, \lim_{m\to\infty} \eta_m\varphi \right\rangle = \langle \tilde{\tilde{u}}, \varphi \rangle, \quad \varphi \in \mathcal{E}(\Omega),\end{aligned}$$

and therefore $\tilde{u} = \tilde{\tilde{u}}$.

Conversely, suppose that $u \in \mathcal{D}'(\Omega)$ admits an extension to a continuous linear functional $\tilde{u}$ in $\mathcal{E}'(\Omega)$ and assume that u does not have compact support in Ω. Then, we can construct a sequence of sets, $\Omega_1 \Subset \Omega_2 \Subset \cdots$, $\Omega = \bigcup_{m=1}^\infty \Omega_m$, and a sequence of test functions $\{\varphi_m\}_{m=1}^\infty \subset \mathcal{D}(\Omega)$ such that $\operatorname{supp} \varphi_m \subset \Omega \setminus \overline{\Omega_m}$, and $\langle u, \varphi_m \rangle = 1$. Because for any $\mathcal{O} \Subset \Omega$ one can choose m_0 so large that $\mathcal{O} \cap \operatorname{supp} \varphi_m$ is an empty set for $m \geq m_0$, it follows that $\varphi_m \to 0$ in $\mathcal{E}(\Omega)$ as $m \to \infty$; hence $\langle \tilde{u}, \varphi_m \rangle \to 0$, $m \to \infty$. On the other hand,

$$\langle \tilde{u}, \varphi_m \rangle = \langle u, \varphi_m \rangle = 1, \quad m = 1, 2, \ldots,$$

which is a contradiction. Therefore u must have compact support. □

So far, we have treated functions and distributions as disparate mathematical objects. We shall now show that every locally integrable function can be identified with a distribution in a unique fashion. For $f \in L_{1,loc}(\Omega)$, consider the linear functional u_f on $\mathcal{D}(\Omega)$ defined by

$$\langle u_f, \varphi \rangle := \int_\Omega f(x)\varphi(x)\,\mathrm{d}x, \quad \varphi \in \mathcal{D}(\Omega). \tag{1.13}$$

By applying Theorem 1.19 it is easy to show that u_f is a distribution on Ω of order 0.

A distribution $u_f \in \mathcal{D}'(\Omega)$ associated with a locally integrable function $f \in L_{1,loc}(\Omega)$ through (1.13) is called a *regular distribution*. By Lemma 1.16, (1.13) establishes a one-to-one correspondence $f \mapsto u_f$ between locally integrable functions and regular distributions. In particular, the support of a locally integrable function coincides with the support of the associated regular distribution. In the following, for the sake of notational simplicity, a locally integrable function will be identified with the associated regular distribution and the same symbol will be used to signify both. If a distribution is not regular, it is called *singular*.

Example 1.15 The Heaviside function H, defined by

$$H(x) := \begin{cases} 1 & \text{if } x > 0, \\ 0 & \text{otherwise,} \end{cases}$$

is associated with a regular distribution (also denoted by H) through

$$\langle H, \varphi \rangle := \int_0^\infty \varphi(x)\,\mathrm{d}x, \quad \varphi \in \mathcal{D}(\mathbb{R}).$$

We have seen above that every regular distribution is of order 0. The next example shows that the converse statement is false; therefore, regular distributions constitute a proper subset of the set of distributions of order 0.

Example 1.16 The Dirac distribution δ is a singular distribution of order 0. Indeed, Lemma 1.16 implies that there is no function $f \in L_{1,loc}(\mathbb{R}^n)$ such that

$$\varphi(0) = \int_{\mathbb{R}^n} f(x)\varphi(x)\,\mathrm{d}x, \quad \varphi \in \mathcal{D}(\mathbb{R}^n).$$

For $u, v \in \mathcal{D}'(\Omega)$ and $\lambda, \mu \in \mathbb{C}$, we define the linear combination $\lambda u + \mu v$ by

$$\langle \lambda u + \mu v, \varphi \rangle := \lambda \langle u, \varphi \rangle + \mu \langle v, \varphi \rangle, \quad \varphi \in \mathcal{D}(\Omega).$$

By recalling Definition 1.18, it is easily seen that $\lambda u + \mu v$ belongs to $\mathcal{D}'(\Omega)$. Thus we have equipped $\mathcal{D}'(\Omega)$ with the structure of a linear space. Next we define convergence in $\mathcal{D}'(\Omega)$.

Definition 1.23 A sequence of distributions $\{u_m\}_{m=1}^\infty \subset \mathcal{D}'(\Omega)$ is said to converge to a distribution u in $\mathcal{D}'(\Omega)$ if $\langle u_m, \varphi \rangle \to \langle u, \varphi \rangle$, $m \to \infty$, for every $\varphi \in \mathcal{D}(\Omega)$.

In order to illustrate the difference between convergence in the sense of distributions and pointwise convergence, we consider two examples.

Example 1.17 (Convergence in $\mathcal{D}'$ by oscillation) Consider the sequence $\{u_m\}_{m=1}^\infty$ in $\mathcal{D}'(\mathbb{R})$, where $u_m(x) = \sin m\pi x$, $m = 1, 2, \ldots$. Then, for each $\varphi \in \mathcal{D}(\mathbb{R})$,

$$\begin{aligned}\langle u_m, \varphi\rangle &= \int_{-\infty}^{\infty} \varphi(x) \sin m\pi x \,\mathrm{d}x \\ &= \frac{1}{m\pi} \int_{-\infty}^{\infty} \varphi'(x) \cos m\pi x \,\mathrm{d}x \to 0, \quad m \to \infty.\end{aligned}$$

Hence $\{u_m\}_{m=1}^\infty$ converges to 0 in $\mathcal{D}'(\mathbb{R})$. Note, however, that the sequence of real numbers $\{u_m(x)\}_{m=1}^\infty$ does not converge, unless x is an integer.

Example 1.18 (Convergence in $\mathcal{D}'$ by concentration) Suppose that u is a continuous function with compact support, defined on $\mathbb{R}$, such that $\operatorname{supp} u = [0, 1]$ and $\int_{\mathbb{R}} u(x)\,\mathrm{d}x = 1$. Consider the sequence $\{u_m\}_{m=1}^\infty$, with $u_m(x) := mu(mx)$, $m = 1, 2, \ldots$. Then, for any $\varphi \in \mathcal{D}(\mathbb{R})$,

$$\begin{aligned}\big|\langle u_m, \varphi\rangle - \varphi(0)\big| &= \left|\int_0^1 u(x)\big[\varphi(x/m) - \varphi(0)\big]\,\mathrm{d}x\right| \\ &\le \sup_{0\le x\le 1/m} \big|\varphi(x) - \varphi(0)\big| \int_0^1 \big|u(x)\big|\,\mathrm{d}x \to 0, \quad m \to \infty.\end{aligned}$$

Hence, as a sequence of distributions, $\{u_m\}_{m=1}^\infty$ converges to δ, the Dirac distribution concentrated at 0. In contrast with this behaviour, the sequence of functions $\{u_m\}_{m=1}^\infty$ converges pointwise to 0; that is, $u_m(x) \to 0$ for each fixed $x \in \mathbb{R}$ as $m \to \infty$.

1.3.2 Operations with Distributions

In this section we introduce further operations, including multiplication of a distribution by a smooth function, differentiation, translation, reflection, tensor product and convolution of distributions.

(A) Multiplication by a Smooth Function Suppose that $u \in L_{1,loc}(\Omega)$ and $a \in C^\infty(\Omega)$; then au is also locally integrable on Ω. By identifying a locally integrable function with the associated regular distribution we can write

$$\begin{aligned}\langle au, \varphi\rangle &= \int_\Omega a(x)u(x)\varphi(x)\,\mathrm{d}x \\ &= \int_\Omega u(x)a(x)\varphi(x)\,\mathrm{d}x = \langle u, a\varphi\rangle, \quad \varphi \in \mathcal{D}(\Omega).\end{aligned}$$

This identity motivates the following definition: for $u \in \mathcal{D}'(\Omega)$ and $a \in C^\infty(\Omega)$, we define the linear functional au on $\mathcal{D}(\Omega)$ by

$$\langle au, \varphi\rangle := \langle u, a\varphi\rangle, \quad \varphi \in \mathcal{D}(\Omega).$$

By recalling Definition 1.18 (or Theorem 1.19) and the Leibniz formula (1.9) it is easy to show that $au \in \mathcal{D}'(\Omega)$.

For this definition to be meaningful it is necessary that a is in $C^\infty(\Omega)$, and if this is not the case then the product au is not defined within the present theoretical framework. Concerning various extensions of the theory of distributions that overcome this limitation the reader may consult, for example, the monograph [145] or the survey paper [146].

Example 1.19 Let $a(x) = (1 + |x|^2)^s$, where s is a real number. Then, $a\delta = \delta$; indeed, $a \in C^\infty(\mathbb{R}^n)$ and

$$\langle a\delta, \varphi\rangle = \langle \delta, a\varphi\rangle = (a\varphi)(0) = \varphi(0) = \langle \delta, \varphi\rangle, \quad \varphi \in \mathcal{D}(\mathbb{R}^n).$$

(B) Differentiation Suppose that $u \in C^k(\Omega)$; then $\partial^\alpha u$ is a locally integrable function on $\Omega \subseteq \mathbb{R}^n$ for each $\alpha \in \mathbb{N}^n$ with $|\alpha| \le k$. By identifying a locally integrable function with the associated regular distribution we have that

$$\begin{aligned}\langle \partial^\alpha u, \varphi\rangle &= \int_\Omega (\partial^\alpha u)(x)\varphi(x)\,\mathrm{d}x \\ &= (-1)^{|\alpha|}\int_\Omega u(x)(\partial^\alpha \varphi)(x)\,\mathrm{d}x = (-1)^{|\alpha|}\langle u, \partial^\alpha\varphi\rangle, \quad \varphi \in \mathcal{D}(\Omega),\end{aligned}$$

where integration by parts has been performed to transfer the derivatives from u to φ.

Motivated by this identity, we *define* the (distributional) derivative $\partial^\alpha u$ of a distribution $u \in \mathcal{D}'(\Omega)$ by

$$\langle \partial^\alpha u, \varphi\rangle := (-1)^{|\alpha|}\langle u, \partial^\alpha \varphi\rangle, \quad \varphi \in \mathcal{D}(\Omega). \tag{1.14}$$

By recalling Definition 1.18 it is easy to show that $\partial^\alpha u \in \mathcal{D}'(\Omega)$. In addition, because our test functions are infinitely many times differentiable, it follows from (1.14) that a distribution admits derivatives of any order.

Example 1.20 Consider the Heaviside function H defined in Example 1.15. Since H is locally integrable on $\mathbb{R}$, it can be identified with a regular distribution, also denoted by H. The first (distributional) derivative of H, denoted by H', is δ, the Dirac distribution concentrated at 0. Indeed,

$$\langle H', \varphi\rangle = -\langle H, \varphi'\rangle = -\int_0^\infty \varphi'(x)\,\mathrm{d}x = \varphi(0) = \langle \delta, \varphi\rangle, \quad \varphi \in \mathcal{D}(\mathbb{R}),$$

where $\varphi'(x) = \frac{\mathrm{d}\varphi(x)}{\mathrm{d}x}$. If then follows by Definition 1.20 that $H' = \delta$.

(C) Translation Suppose that a is a fixed element of $\mathbb{R}^n$ and $u \in L_{1,loc}(\mathbb{R}^n)$. The translation $\tau_a u$ of u is defined by

$$(\tau_a u)(x) := u(x-a), \quad x \in \mathbb{R}^n.$$

Clearly, $\tau_a u$ is also locally integrable on $\mathbb{R}^n$. By identifying a locally integrable function with the associated regular distribution we obtain

$$\begin{aligned}\langle \tau_a u, \varphi\rangle &= \int_{\mathbb{R}^n} u(x-a)\varphi(x)\,\mathrm{d}x \\ &= \int_{\mathbb{R}^n} u(x)\varphi(x+a)\,\mathrm{d}x = \langle u, \tau_{-a}\varphi\rangle, \quad \varphi \in \mathcal{D}(\mathbb{R}^n).\end{aligned}$$

Motivated by this identity, we *define* the translation of a distribution $u \in \mathcal{D}'(\mathbb{R}^n)$ by

$$\langle \tau_a u, \varphi\rangle := \langle u, \tau_{-a}\varphi\rangle, \quad \varphi \in \mathcal{D}(\mathbb{R}^n).$$

Thanks to Definition 1.18, $\tau_a u \in \mathcal{D}'(\mathbb{R}^n)$.

Example 1.21 For $a \in \mathbb{R}^n$ consider the distribution δ_a defined by

$$\langle \delta_a, \varphi\rangle := \varphi(a), \quad \varphi \in \mathcal{D}(\mathbb{R}^n).$$

δ_a is called the Dirac distribution concentrated at a. By noting the definition of τ_a we can write $\delta_a = \tau_a\delta_0 = \tau_a\delta$.

(D) Reflection The reflection u_- of $u \in L_{1,loc}(\mathbb{R}^n)$ is defined by $u_-(x) = u(-x)$. Thus, by identifying a locally integrable function with the associated regular distribution, we obtain

$$\begin{aligned}\langle u_-, \varphi\rangle &= \int_{\mathbb{R}^n} u(-x)\varphi(x)\,\mathrm{d}x \\ &= \int_{\mathbb{R}^n} u(x)\varphi(-x)\,\mathrm{d}x = \langle u, \varphi_-\rangle, \quad \varphi \in \mathcal{D}(\mathbb{R}^n).\end{aligned}$$

This identity motivates the following definition of the reflection u_- of a distribution $u \in \mathcal{D}'(\mathbb{R}^n)$:

$$\langle u_-, \varphi\rangle := \langle u, \varphi_-\rangle, \quad \varphi \in \mathcal{D}(\mathbb{R}^n).$$

Example 1.22 The Dirac distribution concentrated at 0 is its own reflection; that is, $\delta_- = \delta$. More generally, $(\delta_a)_- = \delta_{-a}$ for any a in $\mathbb{R}^n$.

(E) Tensor Product Consider two functions, u and v, defined on Ω_1 and Ω_2, respectively, where Ω_1 and Ω_2 are open sets in $\mathbb{R}^{n_1}$ and $\mathbb{R}^{n_2}$, respectively, and

$u \in L_{1,loc}(\Omega_1)$ and $v \in L_{1,loc}(\Omega_2)$. The arguments of u and v will be denoted by x and y, respectively. The tensor product $u \times v$ of u and v is defined by

$$(u \times v)(x, y) := u(x)v(y)\big(= v(y)u(x)\big),$$

and is clearly locally integrable on $\Omega_1 \times \Omega_2$. By identifying a locally integrable function with the associated regular distribution we have that

$$\begin{aligned}
\langle u \times v, \varphi \rangle &= \int_{\Omega_1 \times \Omega_2} u(x)v(y)\varphi(x, y)\,\mathrm{d}x\,\mathrm{d}y \\
&= \int_{\Omega_1} u(x)\left(\int_{\Omega_2} v(y)\varphi(x, y)\,\mathrm{d}y\right)\mathrm{d}x \\
&= \int_{\Omega_2} v(y)\left(\int_{\Omega_1} u(x)\varphi(x, y)\,\mathrm{d}x\right)\mathrm{d}y \\
&= \int_{\Omega_1 \times \Omega_2} v(y)u(x)\varphi(x, y)\,\mathrm{d}x\,\mathrm{d}y = \langle v \times u, \varphi \rangle
\end{aligned}$$

for all φ in $\mathcal{D}(\Omega_1 \times \Omega_2)$, where Fubini's theorem has been used to change the order of integration (cf. Theorem 8.8 in Rudin [157]). Because the functions

$$x \mapsto \int_{\Omega_2} v(y)\varphi(x, y)\,\mathrm{d}y = \big\langle v, \varphi(x, \cdot)\big\rangle,$$

$$y \mapsto \int_{\Omega_1} u(x)\varphi(x, y)\,\mathrm{d}x = \big\langle u, \varphi(\cdot, y)\big\rangle$$

belong to $\mathcal{D}(\Omega_1)$ and $\mathcal{D}(\Omega_2)$, respectively, we can write

$$\langle u \times v, \varphi \rangle = \big\langle u, \langle v, \varphi \rangle\big\rangle, \quad \varphi \in \mathcal{D}(\Omega_1 \times \Omega_2), \tag{1.15}$$

$$\langle v \times u, \varphi \rangle = \big\langle v, \langle u, \varphi \rangle\big\rangle, \quad \varphi \in \mathcal{D}(\Omega_1 \times \Omega_2). \tag{1.16}$$

More generally, if $u \in \mathcal{D}'(\Omega_1)$ and $v \in \mathcal{D}'(\Omega_2)$, the functions $x \mapsto \langle v, \varphi(x, \cdot)\rangle$ and $y \mapsto \langle u, \varphi(\cdot, y)\rangle$ still belong to $\mathcal{D}(\Omega_1)$ and $\mathcal{D}(\Omega_2)$, respectively. In this case, we *define* $u \times v$ and $v \times u$ by (1.15) and (1.16), respectively. We note that $\times$ is a commutative operation, that is $u \times v = v \times u$.

The tensor product of m distributions $u_i \in \mathcal{D}'(\Omega_i)$, $i = 1, \ldots, m$, where Ω_i is an open subset of $\mathbb{R}^{n_i}$, $i = 1, \ldots, m$, is defined recursively, starting from the case $m = 2$ discussed above.

Example 1.23 Let $a = (a_1, \ldots, a_n)$, and consider the Dirac functions $\delta_a \in \mathcal{D}'(\mathbb{R}^n)$ and $\delta_{a_j} \in \mathcal{D}'(\mathbb{R})$, $j = 1, \ldots, n$, concentrated at the points a and a_j, $j = 1, \ldots, n$, respectively; then,

$$\delta_a = \delta_{a_1} \times \cdots \times \delta_{a_n}.$$

(F) Convolution The convolution $u * v$ of two functions u and v on $\mathbb{R}^n$ is defined by

$$(u * v)(x) := \int_{\mathbb{R}^n} u(y)v(x-y)\,dy, \quad x \in \mathbb{R}^n,$$

whenever the integral exists; below, we describe two instances when this is the case, and $u * v$ is locally integrable on $\mathbb{R}^n$.

(i) Suppose that $u \in L_{1,loc}(\mathbb{R}^n)$, $v \in L_{1,loc}(\mathbb{R}^n)$, $\operatorname{supp} u \subset A$, $\operatorname{supp} v \subset B$, where A and B are two subsets of $\mathbb{R}^n$ such that

$$T_M = \{(x, y) \in A \times B : |x + y| \le M\}$$

is a bounded set in $\mathbb{R}^{2n}$ for every $M > 0$; in particular, T_M is bounded in $\mathbb{R}^{2n}$ if either A or B is a bounded set in $\mathbb{R}^n$.

We shall prove that, under these hypotheses, $u * v$ belongs to $L_{1,loc}(\mathbb{R}^n)$. Indeed, for any $M > 0$, Fubini's theorem implies that

$$\begin{aligned}\int_{|x|\le M} |(u * v)(x)|\,dx &\le \int_{|x|\le M}\int_{\mathbb{R}^n} |u(y)||v(x-y)|\,dy\,dx \\ &= \int_{T_M} |v(\xi)||u(y)|\,d\xi\,dy < \infty.\end{aligned}$$

In particular, $u * v \in L_{1,loc}(\mathbb{R}^n)$ if either u or v has compact support in $\mathbb{R}^n$.

(ii) Suppose that $u \in L_p(\mathbb{R}^n)$ and $v \in L_q(\mathbb{R}^n)$, where $1/p + 1/q \ge 1$. Then, $u * v \in L_r(\mathbb{R}^n)$, where $1/r = 1/p + 1/q - 1$.

When $r = \infty$, the claim is a simple consequence of Hölder's inequality. If $1 \le r < \infty$, we choose $\alpha \in (0, 1]$, $\beta \in (0, 1]$, $s > 1$, $t > 1$ such that

$$1/r + 1/s + 1/t = 1, \qquad \alpha r = p = (1-\alpha)s, \qquad \beta r = q = (1-\beta)t.$$

Since $r \ge \max(p, q)$, such α, β, s and t always exist; we shall adopt the convention that $s = \infty$ if $\alpha = 1$ and $t = \infty$ if $\beta = 1$.

Hölder's inequality, Fubini's theorem and the translation-invariance of the Lebesgue measure yield the following sequence of estimates:

$$\begin{aligned}\|u * v\|^r_{L_r(\mathbb{R}^n)} &= \int_{\mathbb{R}^n}\left|\int_{\mathbb{R}^n} u(y)v(x-y)\,dy\right|^r dx \\ &\le \int_{\mathbb{R}^n}\left[\int_{\mathbb{R}^n} |u(y)|^\alpha |v(x-y)|^\beta |u(y)|^{1-\alpha} |v(x-y)|^{1-\beta}\,dy\right]^r dx \\ &\le \int_{\mathbb{R}^n}\int_{\mathbb{R}^n} |u(y)|^{\alpha r}|v(x-y)|^{\beta r}\,dy\,\|u\|^{r(1-\alpha)}_{L_p(\mathbb{R}^n)}\|v(x-\cdot)\|^{r(1-\beta)}_{L_q(\mathbb{R}^n)}\,dx \\ &= \|u\|^r_{L_p(\mathbb{R}^n)}\|v\|^r_{L_q(\mathbb{R}^n)}.\end{aligned}$$

This establishes *Young's inequality*:

$$\|u * v\|_{L_r(\mathbb{R}^n)} \le \|u\|_{L_p(\mathbb{R}^n)} \|v\|_{L_q(\mathbb{R}^n)},$$
$$u \in L_p(\mathbb{R}^n),\ v \in L_q(\mathbb{R}^n), \quad 1/r = 1/p + 1/q - 1. \tag{1.17}$$

In each of the two cases considered above $u * v$ is locally integrable on $\mathbb{R}^n$ and gives rise to a regular distribution, still denoted by $u * v$. Moreover,

$$\begin{aligned}
\langle u * v, \varphi\rangle &= \int_{\mathbb{R}^n} (u * v)(x)\varphi(x)\,\mathrm{d}x \\
&= \int_{\mathbb{R}^n} \varphi(x) \int_{\mathbb{R}^n} u(y)v(x-y)\,\mathrm{d}y\,\mathrm{d}x \\
&= \int_{\mathbb{R}^n} u(y) \int_{\mathbb{R}^n} v(x-y)\varphi(x)\,\mathrm{d}x\,\mathrm{d}y \\
&= \int_{\mathbb{R}^n} u(y) \int_{\mathbb{R}^n} v(\xi)\varphi(y+\xi)\,\mathrm{d}\xi\,\mathrm{d}y \\
&= \int_{\mathbb{R}^n}\int_{\mathbb{R}^n} u(x)v(y)\varphi_+(x,y)\,\mathrm{d}x\,\mathrm{d}y, \quad \varphi \in \mathcal{D}(\mathbb{R}^n),
\end{aligned}$$

where $\varphi_+ : (x, y) \in \mathbb{R}^n \times \mathbb{R}^n \mapsto \varphi(x + y)$, and Fubini's theorem has been used. Motivated by this identity, we *define* the convolution $u * v$ of $u \in \mathcal{D}'(\mathbb{R}^n)$ and $v \in \mathcal{D}'(\mathbb{R}^n)$ by

$$\langle u * v, \varphi\rangle := \langle u \times v, \varphi_+\rangle, \quad \varphi \in \mathcal{D}(\mathbb{R}^n),$$

whenever the right-hand side makes sense. We note that the hypothesis about the meaningfulness of the expression on the right-hand side is an essential part of the definition: for $\varphi \in \mathcal{D}(\mathbb{R}^n)$, φ_+ does not have compact support in $\mathbb{R}^{2n}$, and therefore the defining expression may not make sense for certain pairs of distributions u and v.

An important class of 'convolvable' distributions is singled out by the next theorem (see Sect. 4.3 of Chap. I in Vladimirov [185]); it is a natural generalization of case (i) considered above.

Theorem 1.24 *Suppose that $u \in \mathcal{D}'(A)$ and $v \in \mathcal{D}'(B)$, where A and B are two open sets in $\mathbb{R}^n$ such that*

$$T_M = \{(x, y) \in A \times B : |x + y| \le M\}$$

is a bounded set in $\mathbb{R}^{2n}$ for every $M > 0$. Then, $u * v$ exists as an element of $\mathcal{D}'(\mathbb{R}^n)$ and $\operatorname{supp}(u * v) \subset \overline{A + B}$; *furthermore,*

$$\langle u * v, \varphi\rangle = \langle u \times v, (\psi\eta)\varphi_+\rangle, \quad \varphi \in \mathcal{D}(\mathbb{R}^n),$$

where ψ and η are two functions in $C^\infty(\mathbb{R}^n)$ that are equal to 1 in A^ε and B^ε and equal to 0 in the complement of $A^{3\varepsilon}$ and $B^{3\varepsilon}$, respectively, for some $\varepsilon > 0$. In particular $u * v$ exists in $\mathcal{D}'(A + B)$ if either u or v has compact support.

We note that the existence of the functions ψ and η appearing in Theorem 1.24 is guaranteed by Lemma 1.15.

Example 1.24 Let $u \in \mathcal{D}'(\mathbb{R}^n)$. Since the Dirac distribution δ has compact support, $u * \delta$ exists in $\mathcal{D}'(\mathbb{R}^n)$. Moreover, $u * \delta = u$; indeed,

$$\langle u * \delta, \varphi\rangle = \langle u \times \delta, \varphi_+\rangle = \big\langle u, \langle \delta, \varphi_+\rangle\big\rangle = \langle u, \varphi\rangle, \quad \varphi \in \mathcal{D}\big(\mathbb{R}^n\big).$$

If the convolution $u * v$ exists then $v * u$ also exists, and because $\times$ is a commutative operation so is $*$; that is, $u * v = v * u$, whenever $u * v$ or $v * u$ exists. A particularly important property of convolution is that it commutes with differentiation. More precisely, if $u * v$ exists in $\mathcal{D}'(\mathbb{R}^n)$, then

$$\partial^\alpha u * v = \partial^\alpha (u * v) = u * \partial^\alpha v. \tag{1.18}$$

Example 1.25 For $h > 0$ let ψ_h denote the continuous piecewise linear function defined on $\mathbb{R}$ by

$$\psi_h(x) := \begin{cases} \frac{1}{h}(1 - |\frac{x}{h}|) & \text{if } |x| \leq h, \\ 0 & \text{otherwise,} \end{cases}$$

and let $u \in \mathcal{D}'(\mathbb{R}^n)$. Since ψ_h has compact support, $u * \psi_h$ is correctly defined in $\mathcal{D}'(\mathbb{R}^n)$ and (1.18) applies. In particular,

$$u'' * \psi_h = u * \psi_h'' = u * \frac{\delta_{-h} - 2\delta_0 + \delta_h}{h^2} = \left(\frac{\tau_h - 2 + \tau_{-h}}{h^2}\right) u,$$

where u'' and ψ_h'' denote the second distributional derivative of u and ψ_h, respectively. The expression on the far right is called the *second divided difference* of u.

The convolution of several distributions is defined analogously. For example, if u, v and w are three distributions on $\mathbb{R}^n$ and $\varphi_+ : (x, y, z) \in \mathbb{R}^n \times \mathbb{R}^n \times \mathbb{R}^n \mapsto \varphi(x + y + z)$, the convolution $u * v * w$ is defined by

$$\langle u * v * w, \varphi\rangle := \langle u \times v \times w, \varphi_+\rangle, \quad \varphi \in \mathcal{D}\big(\mathbb{R}^n\big),$$

whenever the right-hand side makes sense. The convolution $u * v * w$ is correctly defined if at least two of the three distributions u, v, w have compact support.

A further class of 'convolvable' distributions, whose properties mimic those described in case (ii) above, is discussed in the next section.

1.3.3 Tempered Distributions

In this section we consider a class of distributions with 'limited growth-rate', called tempered distributions. One of their key properties is that they have a well-defined

Fourier transform, which is of significance in the theory of partial differential equations. We begin by describing the associated test space of rapidly decreasing functions.

Let $\mathcal{S}(\mathbb{R}^n)$ denote the set of all functions $\varphi \in C^\infty(\mathbb{R}^n)$ such that, loosely speaking, $\partial^\alpha \varphi$ decays faster than any nonnegative power of $|x|^{-1}$ as $|x| \to \infty$, for every multi-index $\alpha = (\alpha_1, \ldots, \alpha_n) \in \mathbb{N}^n$. The elements of the set $\mathcal{S}(\mathbb{R}^n)$ are called *rapidly decreasing functions*. Thus φ is rapidly decreasing if

$$|\varphi|_{N,\alpha} := \sup_{x \in \mathbb{R}^n} |x|^N \left|\partial^\alpha \varphi(x)\right|$$

is finite for every $N \geq 0$ and every multi-index $\alpha \in \mathbb{N}^n$.

Example 1.26 The function $\varphi : x \in \mathbb{R}^n \mapsto x^\alpha \mathrm{e}^{-a|x|^2} \in \mathbb{R}$ is an element of $\mathcal{S}(\mathbb{R}^n)$ for any multi-index $\alpha \in \mathbb{N}^n$ and any real number $a > 0$.

The set $\mathcal{S}(\mathbb{R}^n)$ can be supplied with the structure of a linear space in the usual way. Next, we introduce the notion of convergence in $\mathcal{S}(\mathbb{R}^n)$.

Definition 1.25 A sequence $\{\varphi_m\}_{m=1}^\infty \subset \mathcal{S}(\mathbb{R}^n)$ is said to converge to φ in $\mathcal{S}(\mathbb{R}^n)$ if $|\varphi_m - \varphi|_{N,\alpha} \to 0$ as $m \to \infty$, for every $N \geq 0$ and every multi-index $\alpha \in \mathbb{N}^n$.

When equipped with convergence in this sense, the linear space $\mathcal{S}(\mathbb{R}^n)$ is called the *space of rapidly decreasing functions*, or *Schwartz class*. Clearly $\mathcal{D}(\mathbb{R}^n) \subset \mathcal{S}(\mathbb{R}^n)$; in fact $\mathcal{D}(\mathbb{R}^n)$ is dense in $\mathcal{S}(\mathbb{R}^n)$. This is easily seen by considering, for any $\varphi \in \mathcal{S}(\mathbb{R}^n)$, the sequence $\{\varphi_m\}_{m=1}^\infty \subset \mathcal{D}(\mathbb{R}^n)$ defined by

$$\varphi_m(x) := \omega\left(\frac{x}{m}\right)\varphi(x), \quad m = 1, 2, \ldots,$$

where $\omega \in \mathcal{D}(\mathbb{R}^n)$ with $\omega(0) = 1$, which converges to φ in $\mathcal{S}(\mathbb{R}^n)$ as $m \to \infty$.

Given a linear functional $u : \varphi \in \mathcal{S}(\mathbb{R}^n) \mapsto \langle u, \varphi \rangle \in \mathbb{C}$, we say that it is continuous on $\mathcal{S}(\mathbb{R}^n)$ if, whenever $\varphi_m \to \varphi$ in $\mathcal{S}(\mathbb{R}^n)$ as $m \to \infty$, it follows that $\langle u, \varphi_m \rangle \to \langle u, \varphi \rangle$ as $m \to \infty$.

Definition 1.26 A continuous linear functional on $\mathcal{S}(\mathbb{R}^n)$ is called a *tempered distribution*. The set of all tempered distributions is denoted by $\mathcal{S}'(\mathbb{R}^n)$.

Similarly to $\mathcal{S}(\mathbb{R}^n)$, the set $\mathcal{S}'(\mathbb{R}^n)$ can be equipped with the structure of a linear space in the usual way. Next we define convergence in $\mathcal{S}'(\mathbb{R}^n)$.

Definition 1.27 A sequence $\{u_m\}_{m=1}^\infty \subset \mathcal{S}'(\mathbb{R}^n)$ is said to converge to u in $\mathcal{S}'(\mathbb{R}^n)$ if $\langle u_m, \varphi \rangle \to \langle u, \varphi \rangle$ as $m \to \infty$ for every $\varphi \in \mathcal{S}(\mathbb{R}^n)$.

When equipped with convergence in this sense, the linear space $\mathcal{S}'(\mathbb{R}^n)$ is called the *space of tempered distributions*. It is clear from these definitions that if $u \in \mathcal{S}'(\mathbb{R}^n)$ then its restriction from $\mathcal{S}(\mathbb{R}^n)$ to $\mathcal{D}(\mathbb{R}^n)$ belongs to $\mathcal{D}'(\mathbb{R}^n)$.

Example 1.27 Suppose that f is a Lebesgue-measurable function on $\mathbb{R}^n$ such that

$$\int_{\mathbb{R}^n} \big(1+|x|\big)^{-m} \big|f(x)\big| \,\mathrm{d}x < \infty$$

for some $m \geq 0$; then, f defines a tempered distribution $u_f \in \mathcal{S}'(\mathbb{R}^n)$ via

$$\langle u_f, \varphi\rangle := \int_{\mathbb{R}^n} f(x)\varphi(x)\,\mathrm{d}x, \quad \varphi \in \mathcal{S}\big(\mathbb{R}^n\big).$$

In the following, any such function f will be identified with the induced tempered distribution u_f. Thus, in particular, by Hölder's inequality, $L_p(\mathbb{R}^n) \subset \mathcal{S}'(\mathbb{R}^n)$ for every p, $1 \leq p \leq \infty$.

Loosely speaking, Example 1.27 indicates that any function that has at most polynomial growth can be identified with a tempered distribution. There are, however, regular distributions that do not belong to $\mathcal{S}'(\mathbb{R}^n)$; a simple example is $f(x) = \exp|x|$, $x \in \mathbb{R}^n$.

Example 1.28 Suppose that μ is a positive Borel measure on $\mathbb{R}^n$ (cf. Ch. 2 of [157]), such that $\int_{\mathbb{R}^n}(1+|x|)^{-m}\,\mathrm{d}\mu(x) < \infty$ for some $m \geq 0$; then

$$\langle \mu, \varphi\rangle := \int_{\mathbb{R}^n} \varphi(x)\,\mathrm{d}\mu(x), \quad \varphi \in \mathcal{S}\big(\mathbb{R}^n\big),$$

defines a tempered distribution on $\mathbb{R}^n$.

Example 1.29 If $u \in \mathcal{E}'(\mathbb{R}^n)$, then its restriction from $\mathcal{E}(\mathbb{R}^n)$ to $\mathcal{S}(\mathbb{R}^n)$ is a tempered distribution. Thus $\mathcal{E}'(\mathbb{R}^n) \subset \mathcal{S}'(\mathbb{R}^n)$.

The basic operations on $\mathcal{D}'(\Omega)$ introduced in Sect. 1.3.2 can be carried across to the space $\mathcal{S}'(\mathbb{R}^n)$ by replacing $\mathcal{D}(\Omega)$ and $\mathcal{D}(\mathbb{R}^n)$ in the definitions of those operations with the Schwartz space $\mathcal{S}(\mathbb{R}^n)$.

(A) Multiplication by a Smooth Function Suppose that $a \in C^\infty_M(\mathbb{R}^n)$; that is, $a \in C^\infty(\mathbb{R}^n)$ and for every multi-index $\alpha \in \mathbb{N}^n$ there exist nonnegative real numbers K_α and m_α such that

$$\big|\partial^\alpha a(x)\big| \leq K_\alpha\big(1+|x|\big)^{m_\alpha}, \quad x \in \mathbb{R}^n.$$

Then, for $u \in \mathcal{S}'(\mathbb{R}^n)$, we define au in $\mathcal{S}'(\mathbb{R}^n)$ by

$$\langle au, \varphi\rangle := \langle u, a\varphi\rangle, \quad \varphi \in \mathcal{S}\big(\mathbb{R}^n\big).$$

As $a\varphi$ belongs to $\mathcal{S}(\mathbb{R}^n)$, this definition is correct.

(B) Differentiation For a multi-index $\alpha \in \mathbb{N}^n$, the derivative $\partial^\alpha u$ of $u \in \mathcal{S}'(\mathbb{R}^n)$ is defined by

$$\langle \partial^\alpha u, \varphi \rangle := (-1)^{|\alpha|} \langle u, \partial^\alpha \varphi \rangle, \quad \varphi \in \mathcal{S}(\mathbb{R}^n);$$

clearly $\partial^\alpha u$ is an element of $\mathcal{S}'(\mathbb{R}^n)$ for each $\alpha \in \mathbb{N}^n$.

(C) Translation and reflection The translation $\tau_a u$ and the reflection u_- of $u \in \mathcal{S}'(\mathbb{R}^n)$ are defined by

$$\langle \tau_a u, \varphi \rangle := \langle u, \tau_{-a} \varphi \rangle, \quad \varphi \in \mathcal{S}(\mathbb{R}^n),$$

and

$$\langle u_-, \varphi \rangle := \langle u, \varphi_- \rangle, \quad \varphi \in \mathcal{S}(\mathbb{R}^n),$$

respectively.

(D) Tensor Product The tensor product $u \times v$ of two tempered distributions $u \in \mathcal{S}'(\mathbb{R}^n)$ and $v \in \mathcal{S}'(\mathbb{R}^m)$ is defined by

$$\langle u \times v, \varphi \rangle := \langle u, \langle v, \varphi \rangle \rangle, \quad \varphi \in \mathcal{S}(\mathbb{R}^{n+m}).$$

Tensor product is a commutative operation: $u \times v = v \times u \in \mathcal{S}'(\mathbb{R}^{n+m})$.

(E) Convolution We see from Theorem 1.24 that if $u \in \mathcal{S}'(\mathbb{R}^n)$ and $v \in \mathcal{E}'(\mathbb{R}^n)$ then the convolution $u * v$ exists in $\mathcal{D}'(\mathbb{R}^n)$ and is given by

$$\langle u * v, \varphi \rangle := \langle u \times v, \eta \varphi_+ \rangle, \quad \varphi \in \mathcal{S}(\mathbb{R}^n), \tag{1.19}$$

where η is an arbitrary function in $\mathcal{D}(\mathbb{R}^n)$ such that $\eta(x) = 1$ on $(\operatorname{supp} v)^\varepsilon$ and $\eta(x) = 0$ in the complement of $(\operatorname{supp} v)^{3\varepsilon}$. In fact, since $u \times v$ belongs to $\mathcal{S}'(\mathbb{R}^n)$ and the mapping $\varphi \in \mathcal{S}(\mathbb{R}^n) \mapsto \eta \varphi_+ \in \mathcal{S}(\mathbb{R}^{2n})$ is linear and continuous, it follows that the right-hand side of (1.19) defines a continuous linear functional on $\mathcal{S}(\mathbb{R}^n)$, and therefore $u * v$ belongs to $\mathcal{S}'(\mathbb{R}^n)$.

We have seen in Sect. 1.3.2(F), case (ii), that the convolution of two locally integrable functions may exist when neither has compact support: all that is required is that they have appropriate rates of decay as $|x| \to \infty$. We now consider the generalization of this result to distributions.

If $u \in \mathcal{S}'(\mathbb{R}^n)$ and $v \in \mathcal{S}(\mathbb{R}^n)$, then the convolution $u * v$ exists in $C_M^\infty(\mathbb{R}^n)$ and is given by

$$\langle u * v, \varphi \rangle = \langle u \times v, \varphi_+ \rangle = \langle u, \langle v, \varphi_+ \rangle \rangle = \langle u, (v * \varphi)_- \rangle, \quad \varphi \in \mathcal{S}(\mathbb{R}^n).$$

Using this result it can be shown that $\mathcal{S}(\mathbb{R}^n)$ is dense in $\mathcal{S}'(\mathbb{R}^n)$: if $u \in \mathcal{S}'(\mathbb{R}^n)$, then, by recalling the definition of ω_ε from Example 1.11, the convolution $u_\varepsilon = \omega_\varepsilon * u$ belongs to $C_M^\infty(\mathbb{R}^n)$ and $u_\varepsilon \to u$ in $\mathcal{S}'(\mathbb{R}^n)$ as $\varepsilon \to 0$; in addition if $a \in C_M^\infty(\mathbb{R}^n)$ then the function a_ε defined by

$$a_\varepsilon(x) := a(x) \exp(-\varepsilon |x|^2), \quad x \in \mathbb{R}^n,$$

belongs to $\mathcal{S}(\mathbb{R}^n)$ and converges to a in $\mathcal{S}'(\mathbb{R}^n)$ as $\varepsilon \to 0$. Consequently the function

$$x \in \mathbb{R}^n \mapsto (\omega_\varepsilon * u)(x) \exp\left(-\varepsilon |x|^2\right)$$

belongs to $\mathcal{S}(\mathbb{R}^n)$ and converges to u in $\mathcal{S}'(\mathbb{R}^n)$ as $\varepsilon \to 0$. It is easy to show by a similar argument that $\mathcal{S}(\mathbb{R}^n)$ is dense in $L_p(\mathbb{R}^n)$ for $1 \le p < \infty$.

We note however that the closure of $\mathcal{S}(\mathbb{R}^n)$ in $L_\infty(\mathbb{R}^n)$ is a proper closed linear subspace of $L_\infty(\mathbb{R}^n)$ consisting of uniformly continuous functions on $\mathbb{R}^n$ that tend to zero at infinity. Therefore $\mathcal{S}(\mathbb{R}^n)$ is *not* dense in $L_\infty(\mathbb{R}^n)$.

1.3.4 Fourier Transform of a Tempered Distribution

We begin by considering the Fourier transform of a rapidly decreasing function. Suppose that $\varphi \in \mathcal{S}(\mathbb{R}^n)$; the *Fourier transform* $F\varphi$ of φ is defined by

$$(F\varphi)(\xi) := \int_{\mathbb{R}^n} \varphi(x) \mathrm{e}^{-\imath x \cdot \xi} \, \mathrm{d}x, \quad \xi \in \mathbb{R}^n,$$

where, for $x = (x_1, \ldots, x_n)$ in $\mathbb{R}^n$ and $\xi = (\xi_1, \ldots, \xi_n)$ in $\mathbb{R}^n$,

$$x \cdot \xi := x_1 \xi_1 + \cdots + x_n \xi_n.$$

It is clear from this definition that $F\varphi$ is a bounded continuous function on $\mathbb{R}^n$. Moreover, $F\varphi$ is infinitely many times continuously differentiable on $\mathbb{R}^n$ and

$$\partial^\alpha (F\varphi) = F\left[(-\imath x)^\alpha \varphi\right].$$

Furthermore, integration by parts yields the following identity:

$$F\left(\partial^\alpha \varphi\right) = (\imath \xi)^\alpha (F\varphi).$$

Consequently $F\varphi \in \mathcal{S}(\mathbb{R}^n)$ whenever $\varphi \in \mathcal{S}(\mathbb{R}^n)$. In fact, F maps $\mathcal{S}(\mathbb{R}^n)$ *onto* itself and any φ in $\mathcal{S}(\mathbb{R}^n)$ can be expressed in terms of its Fourier transform $F\varphi$ in $\mathcal{S}(\mathbb{R}^n)$ by means of the *Fourier inversion formula*

$$\varphi(x) = \frac{1}{(2\pi)^n} \int_{\mathbb{R}^n} F\varphi(\xi) \mathrm{e}^{\imath x \cdot \xi} \, \mathrm{d}\xi.$$

It follows from this formula that the inverse Fourier transform F^{-1} is defined on the whole of $\mathcal{S}(\mathbb{R}^n)$, $\varphi = F^{-1} F\varphi = F F^{-1} \varphi$, and F^{-1} is given by

$$\left(F^{-1}\varphi\right)(x) = \frac{1}{(2\pi)^n} \int_{\mathbb{R}^n} \varphi(\xi) \mathrm{e}^{\imath x \cdot \xi} \, \mathrm{d}\xi, \quad \varphi \in \mathcal{S}\left(\mathbb{R}^n\right).$$

The Fourier transform on $L_1(\mathbb{R}^n)$ is defined in the same way as on $\mathcal{S}(\mathbb{R}^n)$; given a function $u \in L_1(\mathbb{R}^n)$, we define

$$(Fu)(\xi) := \int_{\mathbb{R}^n} u(x) \mathrm{e}^{-\imath x \cdot \xi} \, \mathrm{d}x, \quad \xi \in \mathbb{R}^n.$$

According to the Riemann–Lebesgue lemma (cf. Theorem 1.2 in Chap. I of Stein and Weiss [168]) Fu is a bounded and continuous function on $\mathbb{R}^n$. By recalling Example 1.27 with $m > n$, we deduce that Fu defines a tempered distribution on $\mathbb{R}^n$, still denoted by Fu, via

$$\langle Fu, \varphi \rangle = \int_{\mathbb{R}^n} Fu(\xi)\varphi(\xi) \, \mathrm{d}\xi, \quad \varphi \in \mathcal{S}(\mathbb{R}^n).$$

By applying Fubini's theorem we deduce that

$$\begin{aligned} \int_{\mathbb{R}^n} Fu(\xi)\varphi(\xi) \, \mathrm{d}\xi &= \int_{\mathbb{R}^n} \left[\int_{\mathbb{R}^n} u(x) \mathrm{e}^{-\imath x \cdot \xi} \, \mathrm{d}x \right] \varphi(\xi) \, \mathrm{d}\xi \\ &= \int_{\mathbb{R}^n} u(x) \int_{\mathbb{R}^n} \varphi(\xi) \mathrm{e}^{-\imath x \cdot \xi} \, \mathrm{d}\xi \, \mathrm{d}x = \int_{\mathbb{R}^n} u(x) F\varphi(x) \, \mathrm{d}x \end{aligned}$$

for every $\varphi \in \mathcal{S}(\mathbb{R}^n)$. Thus

$$\langle Fu, \varphi \rangle = \langle u, F\varphi \rangle, \quad \varphi \in \mathcal{S}(\mathbb{R}^n). \tag{1.20}$$

Similarly,

$$\langle F^{-1}u, \varphi \rangle = \langle u, F^{-1}\varphi \rangle, \quad \varphi \in \mathcal{S}(\mathbb{R}^n). \tag{1.21}$$

These identities motivate the definitions of F and F^{-1} on $\mathcal{S}'(\mathbb{R}^n)$: for $u \in \mathcal{S}'(\mathbb{R}^n)$, we *define* Fu and $F^{-1}u$ by (1.20) and (1.21), respectively. Obviously, if $u \in \mathcal{S}'(\mathbb{R}^n)$ then Fu and $F^{-1}u$ are also tempered distributions, and $u = F^{-1}Fu = FF^{-1}u$.

The properties of the Fourier transform on $\mathcal{S}(\mathbb{R}^n)$ imply the following identities on $\mathcal{S}'(\mathbb{R}^n)$:

$$\begin{aligned} (\partial^\alpha F\varphi)(\xi) &= F[(-\imath x)^\alpha \varphi], \\ F(\partial^\alpha \varphi)(\xi) &= (\imath \xi)^\alpha F\varphi(\xi), \\ F\delta &= 1. \end{aligned}$$

If $u \in \mathcal{S}'(\mathbb{R}^n)$ and $v \in \mathcal{S}'(\mathbb{R}^m)$, then

$$F(u \times v) = Fu \times Fv;$$

here the letter F on the left-hand side signifies the Fourier transform on $\mathcal{S}'(\mathbb{R}^{n+m})$, whereas F in the first and second factor on the right-hand side denotes the Fourier transform on $\mathcal{S}'(\mathbb{R}^n)$ and $\mathcal{S}'(\mathbb{R}^m)$, respectively.

Since $\mathcal{E}'(\mathbb{R}^n)$ is contained in $\mathcal{S}'(\mathbb{R}^n)$, a distribution u with compact support has a well-defined Fourier transform Fu in $\mathcal{S}'(\mathbb{R}^n)$. However Fu can be shown to be more regular: when extended from $\mathbb{R}^n$ to $\mathbb{C}^n$, the Fourier transform of a distribution with compact support is holomorphic on the whole of $\mathbb{C}^n$; in other words, it is an *entire function*.

Theorem 1.28 *The Fourier transform of a distribution $u \in \mathcal{E}'(\mathbb{R}^n)$ is the function*

$$Fu : \xi \mapsto \langle u, \mathrm{e}_{-\xi} \rangle, \tag{1.22}$$

where $\mathrm{e}_\xi(x) = \exp(\imath x \cdot \xi)$. The right-hand side of (1.22) *is correctly defined for every complex vector $\xi \in \mathbb{C}^n$ and is an entire function of ξ, called the* Fourier–Laplace transform *of u.*

Proof The theorem is obviously true if u is a regular distribution. To prove it in general, let us recall the test function ω_ε from Example 1.11. As $u * \omega_\varepsilon \to u$ in $\mathcal{S}'(\mathbb{R}^n)$ when $\varepsilon \to 0$, it follows that $F(u * \omega_\varepsilon) \to F(u)$ in $\mathcal{S}'(\mathbb{R}^n)$ as $\varepsilon \to 0$. Now, the Fourier transform of $u * \omega_\varepsilon$ is the holomorphic function $\xi \in \mathbb{C}^n \mapsto \langle u * \omega_\varepsilon, \mathrm{e}_{-\xi} \rangle \in \mathbb{C}$, and

$$\langle u * \omega_\varepsilon, \mathrm{e}_{-\xi} \rangle = \big\langle u, (\omega_\varepsilon)_- * \mathrm{e}_{-\xi} \big\rangle = (F\omega)(\varepsilon\xi) \langle u, \mathrm{e}_{-\xi} \rangle.$$

Recall that if a sequence of holomorphic functions on an open set $\Omega \subset \mathbb{C}^n$ converges uniformly on compact subsets of Ω, then the limiting function is holomorphic on Ω (cf. Proposition 5 on p. 7 of Narasimhan [142]). Since $(F\omega)(\varepsilon\xi) \to (F\omega)(0) = 1$ as $\varepsilon \to 0$, uniformly on compact subsets of $\mathbb{C}^n$, and therefore the sequence $\{\xi \in \mathbb{C}^n \mapsto \langle u * \omega_\varepsilon, \mathrm{e}_{-\xi} \rangle \in \mathbb{C}\}_{\varepsilon > 0}$ converges uniformly on compact subsets of $\mathbb{C}^n$ as $\varepsilon \to 0$, it follows that $\langle u, \mathrm{e}_{-\xi} \rangle$ is an entire function of ξ (whose restriction to $\mathbb{R}^n$ is the Fourier transform of u). □

The next theorem will play an important role later in this chapter when we consider smoothing operators based on convolution with distributions possessing compact support (see Theorem 1.7.6 on p. 21 of Hörmander [72] and Theorem 7.19, part (c), on p. 179 of Rudin [158]).

Theorem 1.29 *Let $u \in \mathcal{S}'(\mathbb{R}^n)$ and $v \in \mathcal{E}'(\mathbb{R}^n)$; then, the convolution $u * v$ exists in $\mathcal{S}'(\mathbb{R}^n)$ and its Fourier transform satisfies the identity*

$$F(u * v) = Fu \cdot Fv.$$

This identity is also valid if $u \in \mathcal{S}'(\mathbb{R}^n)$ and $v \in \mathcal{S}(\mathbb{R}^n)$.

The next theorem encapsulates the key relationship between analyticity and growth of the Fourier–Laplace transform of a distribution with compact support. For a proof, we refer to Theorem 1.7.7 on p. 21 of Hörmander [72].

Theorem 1.30 (Paley–Wiener Theorem) *An entire function U is the Fourier–Laplace transform of a distribution with support in the closed ball $\overline{B}_\rho$ of radius ρ if, and only if, for some constants $C > 0$ and N, we have that*

$$\left|U(\zeta)\right| \le C\left(1+|\zeta|\right)^N e^{\rho|\Im\zeta|}, \quad \zeta \in \mathbb{C}^n;$$

here $\Im\zeta$ denotes the imaginary part of the complex vector ζ.

Finally we consider the Fourier transform on $L_2(\mathbb{R}^n)$. Example 1.27 implies that any function $u \in L_2(\mathbb{R}^n)$ can be identified with a tempered distribution, and as such u has a well-defined Fourier transform in $\mathcal{S}'(\mathbb{R}^n)$. In fact, according to *Plancherel's Theorem* (cf. Sect. 2 in Chap. I of [168]),

$$Fu(\xi) = \lim_{M\to\infty} \int_{|x|<M} u(x)e^{-\imath x\cdot\xi}\,dx, \quad \xi \in \mathbb{R}^n,$$

where the limit is to be understood with respect to the norm of the space $L_2(\mathbb{R}^n)$; moreover, F is a one-to-one mapping of $L_2(\mathbb{R}^n)$ onto itself, and the following *Parseval's identities* hold (cf. Sect. 2 in Chap. I of [168]):

$$(2\pi)^n(u,v) = (Fu, Fv), \quad u, v \in L_2\left(\mathbb{R}^n\right),$$

$$(2\pi)^n \|u\|^2_{L_2(\mathbb{R}^n)} = \|Fu\|^2_{L_2(\mathbb{R}^n)}, \quad u \in L_2\left(\mathbb{R}^n\right),$$

which will be used in subsequent sections. For $u, v \in \mathcal{S}(\mathbb{R}^n)$ the first identity follows from (1.20) by taking $\varphi = \overline{Fv}$ and noting that $F\varphi = F((2\pi)^n F^{-1}\overline{v}) = (2\pi)^n\overline{v}$. For $u, v \in L_2(\mathbb{R}^n)$ the first identity is then implied by the density of $\mathcal{S}(\mathbb{R}^n)$ in $L_2(\mathbb{R}^n)$ (cf. the penultimate paragraph of (E) in Sect. 1.3.3). The second identity follows from the first with $v = u$.

1.4 Sobolev Spaces

Now we introduce a class of function spaces, called Sobolev spaces (after the Russian mathematician S.L. Sobolev), which play an important role in modern differential equation theory; see, [1, 2, 116, 162, 181–183]. Suppose that Ω is an open set in $\mathbb{R}^n$. For a nonnegative integer k and $1 \le p \le \infty$, we define

$$W_p^k(\Omega) := \left\{u \in L_p(\Omega) : \partial^\alpha u \in L_p(\Omega), |\alpha| \le k\right\}.$$

We equip $W_p^k(\Omega)$ with the *Sobolev norm* defined by

$$\|u\|_{W_p^k(\Omega)} := \left(\sum_{|\alpha|\le k} \|\partial^\alpha u\|^p_{L_p(\Omega)}\right)^{1/p}$$

when $1 \le p < \infty$, and by

$$\|u\|_{W^k_\infty(\Omega)} := \max_{|\alpha|\le k} \left\|\partial^\alpha u\right\|_{L_\infty(\Omega)}$$

when $p = \infty$. The associated *Sobolev seminorm* is defined by

$$|u|_{W^k_p(\Omega)} := \left(\sum_{|\alpha|=k} \left\|\partial^\alpha u\right\|^p_{L_p(\Omega)} \right)^{1/p}$$

when $1 \le p < \infty$, and by

$$|u|_{W^k_\infty(\Omega)} := \max_{|\alpha|=k} \left\|\partial^\alpha u\right\|_{L_\infty(\Omega)}$$

when $p = \infty$. In these definitions the derivatives are to be understood in the sense of distributions, with the usual convention that locally integrable functions are identified with regular distributions; it is also understood that any two locally integrable functions that differ only on a set of zero measure are identified with each other, as in the case of the Lebesgue spaces $L_p(\Omega)$ discussed in Sect. 1.2.2.

The Sobolev space $W^k_p(\Omega)$ can be shown to be a Banach space with the norm $\|\cdot\|_{W^k_p(\Omega)}$, $1 \le p \le \infty$, $k \ge 0$. An important special case is when $p = 2$; the normed linear space $W^k_2(\Omega)$ is a Hilbert space with the inner product

$$(u, v)_{W^k_2(\Omega)} := \sum_{|\alpha|\le k} \left(\partial^\alpha u, \partial^\alpha v\right),$$

where $(\cdot,\cdot)$ is the inner product in $L_2(\Omega)$.

When boundary-value problems are considered for partial differential equations it is convenient to incorporate the boundary condition into the definition of the function space in which a solution is sought. We consider a class of Sobolev spaces that are particularly well suited to the study of partial differential equations with Dirichlet boundary conditions: we define $\mathring{W}^k_p(\Omega)$ as the closure of $C^\infty_0(\Omega)$ in the norm of $W^k_p(\Omega)$; $\mathring{W}^k_p(\Omega)$ is a Banach subspace of $W^k_p(\Omega)$. In particular, the function space $\mathring{W}^k_2(\Omega)$ is a Hilbert subspace of $W^k_2(\Omega)$. We shall see that when the boundary of Ω is sufficiently smooth, in a sense that will be made precise in the next two definitions, the elements of the function space $\mathring{W}^k_p(\Omega)$ satisfy appropriate boundary conditions.

Definition 1.31 Suppose that Ω is a bounded open set in $\mathbb{R}^n$. The boundary $\partial\Omega$ of Ω is said to be *Lipschitz-continuous* if, for every $x \in \partial\Omega$, there is an open set $\mathcal{O} \subset \mathbb{R}^n$ with $x \in \mathcal{O}$ and a local orthogonal co-ordinate system with co-ordinate $\zeta = (\zeta_1, \ldots, \zeta_n) =: (\zeta', \zeta_n)$ and $a \in \mathbb{R}^n_+$, such that

$$\mathcal{O} = \{\zeta : -a_j < \zeta_j < a_j,\ 1 \le j \le n\},$$

and there is a Lipschitz-continuous function φ defined on

$$\mathcal{O}' = \{\zeta' \in \mathbb{R}^{n-1} : -a_j < \zeta_j < a_j,\ 1 \le j \le n-1\},$$

with

$$|\varphi(\zeta')| \le a_n/2 \quad \text{for } \zeta' \in \mathcal{O}',$$

$$\Omega \cap \mathcal{O} = \{\zeta : \zeta_n < \varphi(\zeta'), \zeta' \in \mathcal{O}'\} \quad \text{and} \quad \partial\Omega \cap \mathcal{O} = \{\zeta : \zeta_n = \varphi(\zeta'), \zeta' \in \mathcal{O}'\}.$$

A bounded open set with a Lipschitz-continuous boundary is referred to as a *Lipschitz domain*.

An important property of a Lipschitz domain Ω is that (as a consequence of Rademacher's theorem; Theorem 3.1.6 on p. 216 of [46]) the unit outward normal to $\partial\Omega$ is defined almost everywhere with respect to the $(n-1)$-dimensional surface measure on $\partial\Omega$. A simple example of a Lipschitz domain is a bounded convex polyhedron in $\mathbb{R}^n$, $n \ge 2$.

Occasionally we shall have to work on domains with smooth boundaries. The next definition assigns a precise meaning to the word "smooth".

Definition 1.32 Suppose that Ω is a bounded open set in $\mathbb{R}^n$. We shall say that $\partial\Omega$ is of class C^m, $m \ge 1$, if, for every $x \in \partial\Omega$, there is an open set $\mathcal{O} \subset \mathbb{R}^n$ with $x \in \mathcal{O}$ and a local orthogonal co-ordinate system with co-ordinate $\zeta = (\zeta_1, \dots, \zeta_n) =: (\zeta', \zeta_n)$ and $a \in \mathbb{R}^n$, such that

$$\mathcal{O} = \{\zeta : -a_j < \zeta_j < a_j,\ 1 \le j \le n\},$$

and there is an m-times continuously differentiable function φ defined on

$$\mathcal{O}' = \{\zeta' \in \mathbb{R}^{n-1} : -a_j < \zeta_j < a_j,\ 1 \le j \le n-1\},$$

with

$$|\varphi(\zeta')| \le a_n/2 \quad \text{for } \zeta' \in \mathcal{O}',$$

$$\Omega \cap \mathcal{O} = \{\zeta : \zeta_n < \varphi(\zeta'), \zeta' \in \mathcal{O}'\} \quad \text{and} \quad \partial\Omega \cap \mathcal{O} = \{\zeta : \zeta_n = \varphi(\zeta'), \zeta' \in \mathcal{O}'\}.$$

A bounded open set with boundary of class C^m, $m \ge 1$, will be called a *domain of class C^m*.

We note in passing that every $(n-1)$-dimensional submanifold of $\mathbb{R}^n$ that is homeomorphic to the $(n-1)$-dimensional sphere decomposes $\mathbb{R}^n$ into two components and is their common boundary. This result is known as the Jordan–Brouwer theorem (cf. Corollary 6.4 on p. 66 of Massey [135]).

Theorem 1.33 *Let $\Omega \subset \mathbb{R}^n$ be a Lipschitz domain and $1 \le p < \infty$. Then,*

❶ *$C^\infty(\overline{\Omega})$ is dense in $W^k_p(\Omega)$ for $k \ge 0$; and*

❷ (Extension Theorem) *If* $k \geq 1$, *then there exists a continuous linear operator* E *from* $W_p^k(\Omega)$ *to* $W_p^k(\mathbb{R}^n)$, *called the* extension operator, *with the property that*

$$(Eu)|_\Omega = u, \quad u \in W_p^k(\Omega).$$

For the proofs of the two parts of this theorem we refer to Adams and Fournier [2] (Theorem 3.22 on p. 68) and Stein [167] (Theorem 5 on p. 181), respectively. We note that while $C^\infty(\overline{\Omega})$ is dense in $W_p^k(\Omega)$ for $k \geq 0$ and $1 \leq p < \infty$ on any Lipschitz domain Ω, the set $C_0^\infty(\Omega)$ is not dense in $W_p^k(\Omega)$ for any positive integer k (although it is dense in $L_p(\Omega) = W_p^0(\Omega)$, $1 \leq p < \infty$). Thus $\mathring{W}_p^k(\Omega)$ is a proper subspace of $W_p^k(\Omega)$.

In connection with the extension theorem we remark that since the *extension operator* E is continuous and linear it is also bounded in the sense that there exists a positive constant C such that

$$\|Eu\|_{W_p^k(\mathbb{R}^n)} \leq C\|u\|_{W_p^k(\Omega)}$$

for all $u \in W_p^k(\Omega)$ with $k \geq 1$ and $1 \leq p < \infty$.

Earlier in this chapter, in Sect. 1.1.2, we introduced the concept of *embedding* of a normed linear space $\mathcal{U}$ in another normed linear space $\mathcal{V}$, and we denoted this by $\mathcal{U} \hookrightarrow \mathcal{V}$. Here we present a brief overview of some important embedding results for Sobolev spaces; the collection of these is generally referred to as the *Sobolev embedding theorem* (see Theorem 5.4 on p. 97 of Adams [1] or Theorem 4.12 on p. 85 in Adams and Fournier [2]).

Theorem 1.34 (The Sobolev Embedding Theorem) *Suppose that* Ω *is a Lipschitz domain in* $\mathbb{R}^n$. *Let* $j \geq 0$ *and* $m \geq 1$ *be integers and let* k *be an integer,* $1 \leq k \leq n$. *Let* Ω^k *denote the intersection of* Ω *with a* k*-dimensional hyperplane; in particular,* $\Omega^k = \Omega$ *when* $k = n$. *For* $1 \leq p < \infty$ *the following embeddings hold:*

❶ *if* $mp < n$ *and either* $n - mp < k \leq n$ *or* $p = 1$ *and* $n - m \leq k \leq n$, *then, for* $p \leq q \leq kp/(n - mp)$, $W_p^{j+m}(\Omega) \hookrightarrow W_q^j(\Omega^k)$;
❷ *if* $mp = n$, *then, for* $1 \leq k \leq n$ *and* $p \leq q < \infty$, $W_p^{j+m}(\Omega) \hookrightarrow W_q^j(\Omega^k)$;
❸ *if* $mp > n > (m-1)p$, *then* $W_p^{j+m}(\Omega) \hookrightarrow C^{j,\lambda}(\overline{\Omega})$ *for* $0 < \lambda \leq m - (n/p)$. *The embedding also holds for* $n = (m-1)p$ *and* $0 < \lambda < 1$, *and for* $n = m - 1$, $p = 1$ *and* $0 < \lambda \leq 1$.

When $k < n$, the embeddings stated in ❶ and ❷ of Theorem 1.34 are to be understood as follows: by part ❶ of Theorem 1.33, $C^\infty(\overline{\Omega})$ is dense in $W_p^{j+m}(\Omega)$ for $j + m \geq 0$ and $1 \leq p < \infty$; hence, any u in $W_p^{j+m}(\Omega)$ is a limit (in $W_p^{j+m}(\Omega)$) of a sequence of functions $\{u_n\}_{n=1}^\infty$ contained in $C^\infty(\overline{\Omega})$; each u_n has a well-defined restriction to the hyperplane Ω^k. The embedding $W_p^{j+m}(\Omega) \hookrightarrow W_q^j(\Omega^k)$ means that the sequence of restrictions $\{u_n|_{\Omega^k}\}_{n=1}^\infty$ converges in $W_q^j(\Omega^k)$ to a function, which we shall call the *trace* of u on Ω^k. Thus, the notion of trace is the generalization of the concept of restriction of a continuous function defined on Ω to a lower-dimensional hypersurface Ω^k. The embedding stated in part ❸ of Theorem 1.34 has to be understood in the sense that any equivalence class of functions in $W_p^{j+m}(\Omega)$ contains an element with the required number of Hölder-continuous (or, if $\lambda = 1$, Lipschitz-continuous) derivatives.

In Sect. 1.1.2 we introduced the idea of *compact embedding* of a normed linear space $\mathcal{U}$ in another normed linear space $\mathcal{V}$, denoted by $\mathcal{U} \hookrightarrow\hookrightarrow \mathcal{V}$. The next theorem, generally known as the *Rellich–Kondrashov theorem*, is a collection of such compact embedding results for Sobolev spaces (cf. Theorem 6.2 on p. 144 of Adams [1] or Theorem 6.3 on p. 168 of Adams and Fournier [2]).

Theorem 1.35 (The Rellich–Kondrashov Theorem) *Suppose that Ω is a Lipschitz domain in $\mathbb{R}^n$, let $j \geq 0$ and $m \geq 1$ be integers, and let k be an integer, $1 \leq k \leq n$. For $1 \leq p < \infty$ the following compact embeddings hold:*

❶ *if $mp < n$ and $0 < n - mp < k \leq n$, then, for $1 \leq q < kp/(n - mp)$, $W_p^{j+m}(\Omega) \hookrightarrow\hookrightarrow W_q^j(\Omega^k)$;*

❷ *if $mp = n$, then, for $1 \leq k \leq n$ and $1 \leq q < \infty$, $W_p^{j+m}(\Omega) \hookrightarrow\hookrightarrow W_q^j(\Omega^k)$;*

❸ *if $mp > n$, then $W_p^{j+m}(\Omega) \hookrightarrow\hookrightarrow C^j(\overline{\Omega})$; and if $mp > n \geq (m-1)p$ and $0 < \lambda < m - (n/p)$, then $W_p^{j+m}(\Omega) \hookrightarrow\hookrightarrow C^{j,\lambda}(\overline{\Omega})$.*

In order to capture finer regularity properties of integrable functions, we consider fractional-order Sobolev spaces defined in the following way: given a positive real number s, $s \notin \mathbb{N}$, let us write $s = m + \sigma$, where $0 < \sigma < 1$ and $m = [s]$ is the integer part of s. The fractional-order Sobolev space $W_p^s(\Omega)$, $1 \leq p < \infty$, is the set of all $u \in L_p(\Omega)$ such that

$$|u|_{W_p^s(\Omega)} := \left\{ \sum_{|\alpha|=m} \int_\Omega \int_\Omega \frac{|\partial^\alpha u(x) - \partial^\alpha u(y)|^p}{|x-y|^{n+\sigma p}} \,\mathrm{d}x\,\mathrm{d}y \right\}^{1/p} < \infty,$$

with the usual modification when $p = \infty$. When equipped with the norm

$$\|u\|_{W_p^s(\Omega)} := \left(\|u\|_{L_p(\Omega)}^p + |u|_{W_p^s(\Omega)}^p \right)^{1/p}, \quad \text{if } 1 \leq p < \infty,$$

or the norm

$$\|u\|_{W_\infty^s(\Omega)} := \|u\|_{L_\infty(\Omega)} + |u|_{W_\infty^s(\Omega)}, \quad \text{if } p = \infty,$$

$W_p^s(\Omega)$ is a Banach space.

The Sobolev embedding theorem is still valid in the case of fractional-order Sobolev spaces, and a result completely analogous to that in Theorem 1.34 holds with the integers $m + j \geq 1$ and $j \geq 0$ replaced by real numbers $m + j > 0$ and $j \geq 0$ in parts ❶ and ❷; in part ❸ $m + j$ can be taken to be a positive real number and j a nonnegative integer. Also, Theorem 1.33 holds in fractional-order Sobolev spaces, with k denoting a nonnegative real number in part ❶, and a real number ≥ 1 in part ❷. The Rellich–Kondrashov theorem, Theorem 1.35, has the following counterpart when $p = q$ and $k = n$.

Theorem 1.36 *Suppose that Ω is a Lipschitz domain in $\mathbb{R}^n$, and let s and t be two real numbers, $0 \leq t < s$. Then, $W_p^s(\Omega) \hookrightarrow\hookrightarrow W_p^t(\Omega)$ for $1 \leq p < \infty$.*

It is a straightforward consequence of this theorem that the following norms are equivalent on $W_p^s(\Omega)$, with Ω a Lipschitz domain, $s > 0$ and $1 \leq p < \infty$:

$$\|u\|_{W_p^s(\Omega)} := \left(\|u\|_{L_p(\Omega)}^p + |u|_{W_p^s(\Omega)}^p\right)^{1/p},$$

$$\|u\|_{W_p^s(\Omega)}^* := \left(\sum_{|\alpha| \leq \lfloor s \rfloor} \|\partial^\alpha u\|_{L_p(\Omega)}^p + |u|_{W_p^s(\Omega)}^p\right)^{1/p},$$

where $\lfloor s \rfloor$ denotes the largest integer that is strictly smaller than s.

As in the case of integer-order Sobolev spaces, for noninteger $s > 0$ we define $\mathring{W}_p^s(\Omega)$ as the closure of $C_0^\infty(\Omega)$ in the norm of $W_p^s(\Omega)$; $\mathring{W}_p^s(\Omega)$ is a Banach subspace of $W_p^s(\Omega)$. By the next theorem, known as the *Friedrichs inequality*, when $\|\cdot\|_{W_p^s(\Omega)}^*$ is considered as a norm on $\mathring{W}_p^s(\Omega)$, the term

$$\sum_{|\alpha| \leq \lfloor s \rfloor} \|\partial^\alpha u\|_{L_p(\Omega)}^p$$

can be omitted from the norm (cf. Adams and Fournier, Theorem 6.30 on p. 183 and Corollary 6.31 on p. 184) in the case of integer $s > 0$ and Theorem 1.1 in [40] for $1/p < s < 1$, $1 < p < \infty$.

Theorem 1.37 (Friedrichs Inequality) *Let Ω be a Lipschitz domain in $\mathbb{R}^n$ with diameter d, and suppose that $s > 0$, $s - \lfloor s \rfloor > 1/p$ and $1 \leq p < \infty$. Then, there exists a constant $c_\star = c_\star(s, p, d)$ such that*

$$\|u\|_{W_p^s(\Omega)}^p \leq c_\star |u|_{W_p^s(\Omega)}^p \quad \forall u \in \mathring{W}_p^s(\Omega). \tag{1.23}$$

The same is true with $\|\cdot\|_{W_p^s(\Omega)}$ replaced by $\|\cdot\|_{W_p^s(\Omega)}^*$ on the left-hand side.

For a real number $s > 0$ and $1 < p < \infty$, we consider the linear space of bounded linear functionals on the Sobolev space $\mathring{W}_p^s(\Omega)$, denoted by $W_q^{-s}(\Omega)$, where q is the conjugate of p, i.e. $1/p + 1/q = 1$, equipped with the *dual norm*

$$\|u\|_{W_q^{-s}(\Omega)} := \sup_{\varphi \in \mathring{W}_p^s(\Omega)} \frac{|\langle u, \varphi \rangle|}{\|\varphi\|_{W_p^s(\Omega)}}.$$

Here $\langle u, \varphi\rangle$ denotes the value of the linear functional u at φ; $\langle\cdot,\cdot\rangle$ is called the *duality pairing* between $W_q^{-s}(\Omega)$ and $\mathring{W}_p^s(\Omega)$ and is sometimes denoted by $\langle\cdot,\cdot\rangle_{W_q^{-s}(\Omega)\times\mathring{W}_p^s(\Omega)}$.

We conclude this section with a brief discussion concerning Sobolev spaces on the boundary $\partial\Omega$ of a Lipschitz domain Ω. Recall from Definition 1.31 that for every x on $\partial\Omega$ there exists a Lipschitz continuous function $\varphi : \mathcal{O}' \subset \mathbb{R}^{n-1} \to \mathbb{R}$ such that, using the notation introduced in Definition 1.31,

$$\Gamma \cap \mathcal{O} = \{\zeta = (\zeta', \varphi(\zeta')) : \zeta' \in \mathcal{O}'\}.$$

Thus, locally, $\partial\Omega$ is an $(n-1)$-dimensional hypersurface in $\mathbb{R}^n$. We define the mapping Φ by

$$\Phi(\zeta') := (\zeta', \varphi(\zeta')), \quad \xi' \in \mathcal{O}'.$$

Then, Φ^{-1} exists and it is Lipschitz-continuous on $\Phi(\mathcal{O}')$.

Definition 1.38 Let Ω be a Lipschitz domain in $\mathbb{R}^n$. For $0 \le s \le 1$ and $1 \le p < \infty$ we denote by $W_p^s(\partial\Omega)$ the set of all $u \in L_p(\partial\Omega)$ such that the composition $u \circ \Phi$ belongs to $W_p^s(\mathcal{O}' \cap \Phi^{-1}(\partial\Omega \cap \mathcal{O}))$ for any pair of $\mathcal{O}$ and φ satisfying the conditions of Definition 1.31.

In order to equip $W_p^s(\partial\Omega)$ with a norm, we consider any *atlas* $(\mathcal{O}_j, \Phi_j)_{j=1}^J$ for $\partial\Omega$ such that $\mathcal{O}_j$ and φ_j, $j = 1, \dots, J$, satisfy the conditions of Definition 1.31, and $\Phi_j(\xi') := (\xi', \varphi_j(\xi'))$, with $\xi' \in \mathcal{O}'_j$. We define $\|\cdot\|_{W_p^s(\partial\Omega)}$ by

$$\|u\|_{W_p^s(\partial\Omega)} := \left(\sum_{j=1}^J \|u \circ \Phi_j\|^p_{W_p^s(\mathcal{O}'_j \cap \Phi_j^{-1}(\partial\Omega \cap \mathcal{O}_j))}\right)^{1/p}.$$

In fact, for $0 < s < 1$ it can be shown that this is equivalent to the following norm (which, for the sake of simplicity, is denoted by the same symbol):

$$\|u\|_{W_p^s(\partial\Omega)} := \left(\int_{\partial\Omega} |u|^p \,\mathrm{d}\sigma + \int_{\partial\Omega}\int_{\partial\Omega} \frac{|u(x) - u(y)|^p}{|x - y|^{n-1+sp}} \,\mathrm{d}\sigma(x)\,\mathrm{d}\sigma(y)\right)^{1/p},$$

where $\mathrm{d}\sigma$ denotes the $(n-1)$-dimensional surface measure on $\partial\Omega$.

Suppose that Ω is a domain of class C^m; then, for $0 \le s \le m$ and $1 \le p < \infty$, the Sobolev space $W_p^s(\partial\Omega)$ can be defined analogously as for $0 \le s \le 1$.

The notion of *trace of a function* on a k-dimensional hyperplane intersecting a Lipschitz domain $\Omega \subset \mathbb{R}^n$ has already been considered in the discussion following Theorem 1.34. Now we turn our attention to the concept of trace of a function on the boundary of Ω.

Provided that $\partial\Omega$ is sufficiently smooth, statements analogous to ❶ and ❷ in Theorem 1.34 can be made, using a partition of unity argument, with Ω^k replaced by $\partial\Omega$ and k taken to be equal to $n-1$. Thus, for example, a function $u \in W_p^1(\Omega)$,

$1 < p < \infty$, has a well defined trace, denoted by $\gamma_0(u)$ on the boundary $\partial\Omega$ of a Lipschitz domain Ω; moreover, γ_0 is a bounded linear operator from $W^1_p(\Omega)$ to $W_p^{1-(1/p)}(\partial\Omega)$. If φ belongs to $C^\infty(\overline{\Omega})$ then

$$\gamma_0(\varphi) = \varphi|_{\partial\Omega}. \tag{1.24}$$

More generally, we have the following result (cf. Theorem 7.53 on p. 216 of Adams [1] with $k = 0$).

Theorem 1.39 *Let m be a positive integer and suppose that p is a real number, $1 < p < \infty$. Assume that Ω is a domain of class C^m contained in $\mathbb{R}^n$. Then, there exists a bounded trace operator γ_0 from $W^m_p(\Omega)$ onto $W_p^{m-(1/p)}(\partial\Omega)$.*

When $l < m$, the lth partial derivatives of a function $u \in W^m_p(\Omega)$ have traces in $W_p^{m-l-(1/p)}(\partial\Omega)$. It is standard practice to formulate trace theorems involving higher derivatives in terms of derivatives in the direction of the unit outward normal vector ν to the boundary $\partial\Omega$. Thus, we have the following generalization of Theorem 1.39 (cf. Theorem 7.53 on p. 216 of Adams [1]).

Theorem 1.40 *Let m and l be positive integers such that $m > l$, and let p be a real number such that $1 < p < \infty$. Let Ω be a domain of class C^m contained in $\mathbb{R}^n$. Then, there exists a continuous trace operator*

$$\gamma_l : W^m_p(\Omega) \to \prod_{k=0}^{l} W_p^{m-k-(1/p)}(\partial\Omega)$$

with the property that

$$\gamma_l(\varphi) = \left(\varphi|_{\partial\Omega}, \left.\frac{\partial\varphi}{\partial\nu}\right|_{\partial\Omega}, \ldots, \left.\frac{\partial^l\varphi}{\partial\nu^l}\right|_{\partial\Omega} \right)$$

for every function φ in $C^\infty(\overline{\Omega})$.

We can now describe the space $\mathring{W}^m_p(\Omega)$ in terms of boundary conditions on $\partial\Omega$, expressed as vanishing traces on $\partial\Omega$.

Theorem 1.41 *Let Ω be a domain of class C^m, $m \geq 1$. Then, the space $\mathring{W}^m_p(\Omega)$, $1 < p < \infty$, defined as the closure of $C_0^\infty(\Omega)$ in $W^m_p(\Omega)$, can be characterized as follows:*

$$\mathring{W}^m_p(\Omega) = \left\{ u \in W^m_p(\Omega) : \gamma_{m-1}(u) = 0 \right\}.$$

Trace theorems analogous to these also hold in fractional-order Sobolev spaces. The simplest such result, due to Gagliardo [48], is formulated in the next theorem (see, Theorem 1.5.1.2 on p. 37 of Grisvard [62], and Corollary 1.5.1.6 on p. 39 of Grisvard [62], with $k = 0$ and $l = 0$).

Theorem 1.42 *Suppose that Ω is a Lipschitz domain in $\mathbb{R}^n$, and let $1 < p < \infty$. Assuming that $1/p < s \le 1$, the mapping γ_0 defined on $C^\infty(\overline{\Omega})$ by (1.24) has a unique continuous extension to a linear operator from $W^s_p(\Omega)$ onto $W^{s-(1/p)}_p(\partial\Omega)$; this extension will be still denoted by γ_0. Further, for $1/p < s \le 1$, the space $\mathring{W}^s_p(\Omega)$, defined as the closure of $C^\infty_0(\Omega)$ in $W^s_p(\Omega)$, can be characterized as follows:*

$$\mathring{W}^s_p(\Omega) = \{u \in W^s_p(\Omega) : \gamma_0(u) = 0\}.$$

For $1 < p < \infty$ and letting q be the conjugate of p, i.e. $1/p + 1/q = 1$, the dual space $W^{-s}_q(\partial\Omega)$ of $W^s_p(\partial\Omega)$, $s \ge 0$, is defined as the set of all bounded linear functionals on $W^s_p(\partial\Omega)$, equipped with the *dual norm* $\|\cdot\|_{W^{-s}_q(\partial\Omega)}$ defined by

$$\|u\|_{W^{-s}_q(\partial\Omega)} := \sup_{\varphi \in W^s_p(\partial\Omega)} \frac{|\langle u, \varphi\rangle|}{\|\varphi\|_{W^s_p(\partial\Omega)}}.$$

Here $\langle u, \varphi\rangle$ denotes the value of the linear functional u at φ. It is clear from this definition that for $s = 0$ and $1 < p < \infty$ the dual space of $W^0_p(\partial\Omega) = L_p(\partial\Omega)$ is simply $W^0_q(\partial\Omega) = L_q(\partial\Omega)$.

1.5 Anisotropic Sobolev Spaces

In this section we consider Sobolev spaces that consist of multivariate functions with, potentially, different differentiability properties in the different co-ordinate directions:—hence the attribute *anisotropic*.

Let $\mathbb{R}_+$ be the set of nonnegative real numbers. With a slight abuse of terminology we shall refer to the elements of the set $\mathbb{R}^n_+$ as multi-indices. For $\alpha = (\alpha_1, \dots, \alpha_n) \in \mathbb{R}^n_+$, we define

$$[\alpha] := ([\alpha_1], \dots, [\alpha_n]), \qquad |\alpha| := \alpha_1 + \dots + \alpha_n, \qquad \lfloor\alpha\rfloor := (\lfloor\alpha_1\rfloor, \dots, \lfloor\alpha_n\rfloor),$$

where, for a positive real number x, $[x]$ denotes the integer part of x, and $\lfloor x\rfloor$ denotes the largest integer smaller than x (e.g. $[2.5] = 2$, $[3] = 3$; $\lfloor 2.5\rfloor = 2$, and $\lfloor 3\rfloor = 2$); $[0] := 0$, $\lfloor 0\rfloor := 0$. Let $\mathrm{e}_i = (\delta_{1i}, \dots, \delta_{ni})$, for $i = 1, \dots, n$, where

$$\delta_{ij} := \begin{cases} 1 & \text{for } i = j, \\ 0 & \text{for } i \ne j, \end{cases}$$

is the *Kronecker delta*.

With these notational conventions we then define the following sets:

$$\Omega_i(x) := \{h_i : x + h_i\mathrm{e}_i \in \Omega\},$$
$$\Omega_{ij}(x) := \{(h_i, h_j) : x + c_i h_i \mathrm{e}_i + c_j h_j \mathrm{e}_j \in \Omega;\, c_i, c_j = 0, 1\},$$
$$\vdots$$

$$\Omega_{1\ldots n}(x) := \left\{(h_1,\ldots,h_n) : x + \sum_{k=1}^{n} c_k h_k e_k \in \Omega;\, c_k = 0, 1;\, k = 1,\ldots,n\right\}.$$

Let us also introduce the difference operators Δ_h and Δ_h^k by

$$\Delta_h u(x) := u(x+h) - u(x), \qquad \Delta_h^k u(x) := \Delta_h\big(\Delta_h^{k-1} u(x)\big).$$

For $1 \le p < \infty$, we define the seminorm $|\cdot|_{\alpha,p}$ as follows:

$$|u|_{\alpha,p}^p := \|u\|_{L_p(\Omega)}^p, \quad \text{for } \alpha_1 = \alpha_2 = \cdots = \alpha_n = 0,$$

$$|u|_{\alpha,p}^p := \int_\Omega \int_{\Omega_i(x)} \frac{|\Delta_{h_i e_i} u(x)|^p}{|h_i|^{1+p\alpha_i}}\, dh_i\, dx, \quad \text{for } 0 < \alpha_i < 1;\, \alpha_k = 0,\, k \ne i,$$

$$|u|_{\alpha,p}^p := \int_\Omega \int_{\Omega_{ij}(x)} \frac{|\Delta_{h_i e_i} \Delta_{h_j e_j} u(x)|^p}{|h_i|^{1+p\alpha_i} |h_j|^{1+p\alpha_j}}\, dh_i\, dh_j\, dx,$$
$$\text{for } 0 < \alpha_i, \alpha_j < 1;\, \alpha_k = 0,\, k \ne i, j,$$

$$\vdots$$

$$|u|_{\alpha,p}^p := \int_\Omega \int_{\Omega_{1\ldots n}(x)} \frac{|\Delta_{h_1 e_1} \ldots \Delta_{h_n e_n} u(x)|^p}{|h_1|^{1+p\alpha_1} \ldots |h_n|^{1+p\alpha_n}}\, dh_1 \ldots dh_n\, dx,$$
$$\text{for } 0 < \alpha_1, \ldots, \alpha_n < 1,$$

$$|u|_{\alpha,p}^p := |\partial^{[\alpha]} u|_{\alpha-[\alpha],p}^p, \quad \text{if, for some } k,\, \alpha_k \ge 1,$$

with $\partial^{[\alpha]} u$ understood in the sense of distributions on Ω.

When $p = \infty$, the seminorm $|\cdot|_{\alpha,\infty}$ is defined analogously, as follows:

$$|u|_{\alpha,\infty} := \|u\|_{L_\infty(\Omega)}, \quad \text{for } \alpha_1 = \cdots = \alpha_n = 0,$$

$$|u|_{\alpha,\infty} := \text{ess.sup}_{x\in\Omega, h_i \in \Omega_i(x)} \frac{|\Delta_{h_i e_i} u(x)|}{|h_i|^{\alpha_i}}, \quad \text{for } 0 < \alpha_i < 1,\, \alpha_k = 0,\, k \ne i,$$

and so on.

A finite set of multi-indices $A \subset \mathbb{R}_+^n$ is called *regular* if $0 := (0,\ldots,0) \in A$, and for any $\alpha = (\alpha_1,\ldots,\alpha_n) \in A$ there exist real numbers $\beta_k \ge \alpha_k$, $k = 1,\ldots,n$, such that $\beta_k e_k \in A$ for $k = 1,\ldots,n$. Assuming that A is a regular set of multi-indices, we define the following norms:

$$\|u\|_{W_p^A(\Omega)} := \left(\sum_{\alpha\in A} |u|_{\alpha,p}^p\right)^{1/p}, \quad \text{when } 1 \le p < \infty,$$

$$\|u\|_{W_\infty^A} := \max_{\alpha\in A} |u|_{\alpha,\infty}, \quad \text{when } p = \infty.$$

The set of all $u \in L_{1,\text{loc}}(\Omega)$ such that $\|u\|_{W_p^A(\Omega)} < \infty$ is denoted by $W_p^A(\Omega)$.

Example 1.30 Suppose that $s_1, \ldots, s_n$ are positive real numbers. Let $A := A_0 \cup A_1$, where

$$A_0 := \left\{ \alpha \in \mathbb{N}^n : \sum_{k=1}^{n} \frac{\alpha_k}{s_k} < 1 \right\}, \qquad A_1 := \bigcup_{i=1}^{n} A_{1i},$$

with $\mathbb{N}$ denoting the set of all nonnegative integers and

$$A_{1i} := \left\{ \alpha \in \mathbb{R}^n_+ : \alpha_k \in \mathbb{N} \text{ for } k \neq i;\ \sum_{k=1}^{n} \frac{\alpha_k}{s_k} = 1 \right\}.$$

Then, $W_p^A(\Omega) = W_p^{(s_1,s_2,\ldots,s_n)}(\Omega)$ is an anisotropic Sobolev space equipped with the norm $\|\cdot\|_{W_p^A(\Omega)}$, and the natural seminorm

$$|u|_{W_p^{(s_1,\ldots,s_n)}(\Omega)} := \left(\sum_{\alpha \in A_1} |u|^p_{\alpha,p} \right)^{1/p}, \qquad \text{when } 1 \leq p < \infty,$$

and

$$|u|_{W_\infty^{(s_1,\ldots,s_n)}(\Omega)} := \max_{\alpha \in A_1} |u|_{\alpha,\infty}, \qquad \text{when } p = \infty.$$

Here we have used the attribute *natural* in order to distinguish the seminorm in question from other possible seminorms that can be defined on the space $W_p^A(\Omega)$. When $s_1 = \cdots = s_n = s$, the space $W_p^A(\Omega)$ is the standard (isotropic) Sobolev space $W_p^s(\Omega)$, equipped with a norm that is equivalent to the standard norm of $W_p^s(\Omega)$.

We now consider another class of anisotropic spaces, which we shall make use of in the analysis of time-dependent problems.

Let $\mathcal{U}$ be a Banach space with norm $\|\cdot\|_{\mathcal{U}}$, and let $|\cdot|_{\mathcal{U}}$ be a seminorm on $\mathcal{U}$ such that $|u|_{\mathcal{U}} \leq \|u\|_{\mathcal{U}}$ for all u in $\mathcal{U}$. Suppose that (c, d) is a nonempty open interval of the real line and $1 \leq p \leq \infty$. We consider the set $L_p((c, d); \mathcal{U})$ of all functions $u : (c, d) \to \mathcal{U}$ such that $t \mapsto \|u(t)\|_{\mathcal{U}}$ is measurable on (c, d), with

$$\int_c^d \|u(t)\|^p_{\mathcal{U}}\, \mathrm{d}t < \infty, \qquad \text{when } 1 \leq p < \infty,$$

and

$$\text{ess.sup}_{t\in(c,d)} \|u(t)\|_{\mathcal{U}} < \infty, \qquad \text{when } p = \infty.$$

(As is usual, any two functions that differ from each other only on a subset of zero measure of the interval (c, d) are identified.) It can be shown (see Theorems 2.20.4 and 2.20.8 on pp. 114 and 116 of Kufner et al. [116]) that $L_p((c, d); \mathcal{U})$ is a Banach space with the norm

$$\|u\|_{L_p((c,d);\mathcal{U})} := \left(\int_c^d \|u(t)\|^p_{\mathcal{U}}\, \mathrm{d}t \right)^{1/p}, \qquad \text{when } 1 \leq p < \infty,$$

and

$$\|u\|_{L_\infty((c,d);\mathcal{U})} := \text{ess.sup}_{t\in(c,d)} \|u(t)\|_{\mathcal{U}}, \quad \text{when } p=\infty.$$

The space $L_p((c,d);\mathcal{U})$ has the natural seminorm

$$|u|_{L_p((c,d);\mathcal{U})} := \left(\int_c^d |u(t)|_{\mathcal{U}}^p \, \mathrm{d}t\right)^{1/p}, \quad \text{when } 1\le p<\infty,$$

and

$$|u|_{L_\infty((c,d);\mathcal{U})} := \text{ess.sup}_{\,t\in(c,d)} |u(t)|_{\mathcal{U}}, \quad \text{when } p=\infty.$$

Suppose that $\mathcal{U}$ is a Banach space, with norm $\|\cdot\|_{\mathcal{U}}$. Given a nonempty open interval (c,d) of the real line and a nonnegative integer k, we denote by $C^k((c,d);\mathcal{U})$ the set of all continuous functions

$$u : t \in (c,d) \mapsto u(t) \in \mathcal{U}$$

whose derivatives with respect to t of order $\le k$ are continuous functions of t on (c,d). Instead of $C^0((c,d);\mathcal{U})$, we shall write $C((c,d);\mathcal{U})$.

For a nonnegative integer k and a bounded, nonempty, open interval $(c,d)\subset\mathbb{R}$, we denote by $C^k([c,d];\mathcal{U})$ the set of all u in $C^k((c,d);\mathcal{U})$ such that all derivatives of u with respect to t of order $\le k$ can be continuously extended from the open interval (c,d) to the closed interval $[c,d]$. Instead of $C^0([c,d];\mathcal{U})$, we shall write $C([c,d];\mathcal{U})$. $C([c,d];\mathcal{U})$ is a Banach space equipped with the norm $\|\cdot\|_{C([c,d];\mathcal{U})}$ defined by

$$\|u\|_{C([c,d];\mathcal{U})} := \max_{t\in[c,d]} \|u(t)\|_{\mathcal{U}}.$$

The same is true of $C^k([c,d];\mathcal{U})$ for any nonnegative integer k, when equipped with the norm

$$\|u\|_{C^k([c,d];\mathcal{U})} := \max_{0\le m\le k} \sup_{t\in(c,d)} \|u^{(m)}(t)\|_{\mathcal{U}},$$

where $u^{(m)} = \mathrm{d}^m u/\mathrm{d}t^m$.

Assuming that (c,d) is a nonempty open interval of the real line and $\mathcal{U}$ is a Banach space, we denote by $\mathcal{D}'((c,d);\mathcal{U})$ the linear space of $\mathcal{U}$-valued distributions on (c,d), defined as the set of all continuous linear operators from $\mathcal{D}(c,d)$ into $\mathcal{U}$. For $u\in\mathcal{D}'((c,d);\mathcal{U})$ and $\varphi\in\mathcal{D}(c,d)$ we shall denote the value $u(\varphi)$ of u at φ by $\langle u,\varphi\rangle$. By definition, $\langle u,\varphi\rangle\in\mathcal{U}$.

For a positive integer m, the mth *distributional derivative* $\frac{\mathrm{d}^m u}{\mathrm{d}t^m}$ of $u\in\mathcal{D}'((c,d);\mathcal{U})$ is the continuous linear operator from $\mathcal{D}(c,d)$ into $\mathcal{U}$ defined by

$$\left\langle \frac{\mathrm{d}^m u}{\mathrm{d}t^m}, \varphi\right\rangle = (-1)^m \left\langle u, \frac{\mathrm{d}^m \varphi}{\mathrm{d}t^m}\right\rangle \quad \forall\varphi\in\mathcal{D}(c,d),$$

understood as an equality in $\mathcal{U}$.

To each $u \in L_{1,\text{loc}}((c,d);\mathcal{U})$ we can assign an element $T_u \in \mathcal{D}'((c,d);\mathcal{U})$ by

$$\langle T_u, \varphi\rangle := \int_c^d u(t)\varphi(t)\,\mathrm{d}t \quad (\in \mathcal{U}) \quad \forall \varphi \in \mathcal{D}(c,d).$$

The mapping $u \in L_{1,\text{loc}}((c,d);\mathcal{U}) \mapsto T_u \in \mathcal{D}'((c,d);\mathcal{U})$ is a continuous linear injection, which allows us to identify u with T_u and consider $L_{1,\text{loc}}((c,d);\mathcal{U})$ as a subset of $\mathcal{D}'((c,d);\mathcal{U})$. Thus in particular $L_p((c,d);\mathcal{U})$ will be viewed as a subset of $\mathcal{D}'((c,d);\mathcal{U})$ for all p, $1 \le p \le \infty$.

Let $1 \le p < \infty$, $r > 0$, and write r in the form $r = m + \rho$, $0 \le \rho < 1$, where $m = [r]$ is the integer part of r. We denote by $W_p^r((c,d);\mathcal{U})$ the set of all functions u in $L_p((c,d);\mathcal{U})$ whose mth derivative $u^{(m)} = \mathrm{d}^m u/\mathrm{d}t^m$ on the interval (c,d) in the sense of $\mathcal{U}$-valued distributions is an element of $L_p((c,d);\mathcal{U})$ and

$$\mathcal{N}_{r,p}(u) := \left(\int_c^d \int_c^d \frac{\|u^{(m)}(\tau) - u^{(m)}(\tau')\|_{\mathcal{U}}^p}{|\tau - \tau'|^{1+p\rho}}\,\mathrm{d}\tau\,\mathrm{d}\tau'\right)^{1/p} < \infty$$

when $\rho > 0$; if $\rho = 0$ we define $\mathcal{N}_{r,p}(u) = 0$. The space $W_p^r((c,d);\mathcal{U})$ is a Banach space equipped with the norm

$$\|u\|_{W_p^r((c,d);\mathcal{U})} := \left(\|u\|_{L_p((c,d);\mathcal{U})}^p + \left\|u^{(m)}\right\|_{L_p((c,d);\mathcal{U})}^p + \mathcal{N}_{r,p}(u)^p\right)^{1/p}$$

and the natural seminorm, which is defined for $r = m > 0$ integer and $r > 0$ noninteger, respectively, by

$$|u|_{W_p^r((c,d);\mathcal{U})} := \left\|u^{(m)}\right\|_{L_p((c,d);\mathcal{U})} \quad \text{and} \quad |u|_{W_p^r((c,d);\mathcal{U})} := \mathcal{N}_{r,p}(u).$$

Suppose that Ω is a Lipschitz domain in $\mathbb{R}^n$ and $Q := \Omega \times (c,d)$. We define the anisotropic Sobolev space

$$W_p^{s,r}(Q) := L_p\big((c,d);\, W_p^s(\Omega)\big) \cap W_p^r\big((c,d);\, L_p(\Omega)\big),$$

equipped with the norm

$$\|u\|_{W_p^{s,r}(Q)} := \left(\|u\|_{L_p((c,d);\, W_p^s(\Omega))}^p + \|u\|_{W_p^r((c,d);\, L_p(\Omega))}^p\right)^{1/p}$$

and the natural seminorm

$$|u|_{W_p^{s,r}(Q)} := \left(|u|_{L_p((c,d);\, W_p^s(\Omega))}^p + |u|_{W_p^r((c,d);\, L_p(\Omega))}^p\right)^{1/p}.$$

For $1 \le p < \infty$, the space $W_p^{s,r}(Q)$ can be viewed as a space of type $W_p^A(Q)$; for example, if $s \in \mathbb{N}$, then the corresponding set A is given by

$$\begin{aligned} A = &\big\{(\alpha_1, \dots, \alpha_n, 0) : \alpha_i \in \mathbb{N}, \alpha_1 + \dots + \alpha_n \le s\big\} \\ &\cup \big\{(0, \dots, 0, \beta) : \beta \in \mathbb{N}, \beta < r\big\} \cup \big\{(0, \dots, 0, r)\big\}. \end{aligned}$$

A particularly relevant anisotropic space, $W_2^{s,s/2}(Q)$, $s > 0$, arises in the theory of second-order parabolic partial differential equations on the space-time domain $Q := \Omega \times (0, T)$.

We shall also need the space $\widehat{W}_2^{s,s/2}(Q) := W_2^{(s,\dots,s,s/2)}(Q)$. Clearly,

$$\widehat{W}_2^{s,s/2}(Q) \subset W_2^{s,s/2}(Q). \tag{1.25}$$

We conclude this section by stating two results concerning anisotropic Sobolev spaces.

Theorem 1.43 *Suppose that* $u \in W_2^{s,r}(Q)$, $s, r > 0$, *and let* $\alpha \in \mathbb{N}^n$ *and* $k \in \mathbb{N}$ *be such that* $\frac{|\alpha|}{s} + \frac{k}{r} \leq 1$. *Then,* $\partial_x^\alpha \partial_t^k u$ *belongs to* $W_2^{\mu,\nu}(Q)$, *where* $\frac{\mu}{s} = \frac{\nu}{r} = 1 - (\frac{|\alpha|}{s} + \frac{k}{r})$, *and* ∂_x *and* ∂_t *are the partial derivatives with respect to* $x = (x_1, \dots, x_n)$ *and* t, *respectively.*

For a proof, see Theorem 10.2 on p. 143 and Theorem 18.4 on p. 296 in Besov, Il'in and Nikol'skiĭ [13] and Lemma 7.2 in Grisvard [60]. Theorem 1.43 and (1.25) imply that $\widehat{W}_2^{s,s/2}(Q) = W_2^{s,s/2}(Q)$, with equivalence of norms.

Theorem 1.44 *Suppose that* $u \in W_2^{s,r}(Q)$, $s \geq 0$, $r > 1/2$. *Then, for a nonnegative integer* k, $k < r - 1/2$, *the trace*

$$\frac{\partial^k u}{\partial t^k}(x, 0)$$

is correctly defined as an element of $W_2^q(\Omega)$, *where* $q = \frac{s}{r}(r - k - \frac{1}{2})$.

For a proof of Theorem 1.44, see Theorem 6.7 in Chap. 6 of Nikol'skiĭ [144]. We shall now introduce another important class of function spaces.

1.6 Besov Spaces

For $\delta > 0$ and Ω an open set in $\mathbb{R}^n$, we define

$$\Omega_\delta := \{x \in \Omega : \mathrm{dist}(x, \partial\Omega) > \delta\}.$$

Assuming that $s > 0$, $1 \leq p < \infty$ and $1 \leq q \leq \infty$, the Besov space $B_{p,q}^s(\Omega)$ is defined as follows. Let us write $s = m + \sigma$ where $0 < \sigma \leq 1$ and m is a nonnegative integer. We denote by $B_{p,q}^s(\Omega)$ the set of all u in $L_p(\Omega)$ whose distributional derivatives $\partial^\alpha u$ of order $|\alpha| = m$ satisfy

$$\mathcal{N}_{\alpha,p,q}(u) := \left\{ \int_0^\infty \left[t^{-\sigma} \sup_{|h| \leq t} \left\| \Delta_h^2 \partial^\alpha u \right\|_{L_p(\Omega_{2|h|})} \right]^q \frac{dt}{t} \right\}^{1/q} < \infty$$

if $1 \le q < \infty$, and

$$\mathcal{N}_{\alpha,p,\infty}(u) := \sup_{t>0}\Big[t^{-\sigma} \sup_{|h|\le t} \big\| \Delta_h^2 \partial^\alpha u \big\|_{L_p(\Omega_{2|h|})}\Big] < \infty$$

if $q = \infty$. The norm in $B^s_{p,q}(\Omega)$ is defined by the expression

$$\|u\|_{B^s_{p,q}(\Omega)} := \Big(\|u\|^p_{L_p(\Omega)} + \sum_{|\alpha|=m} \mathcal{N}_{\alpha,p,q}(u)^p \Big)^{1/p}.$$

The space $B^s_{p,q}(\Omega)$ is a Banach space with this norm. If Ω is a Lipschitz domain, then the following relationships hold between Sobolev and Besov spaces:

$$W^s_p(\Omega) = B^s_{p,p}(\Omega)$$

for $s > 0$ noninteger, and for $s > 0$ integer if $p = 2$. For s integer and $p \ne 2$,

$$W^s_p(\Omega) \ne B^s_{p,p}(\Omega).$$

In fact, the following embeddings hold for any $s > 0$ (see, Eq. (3) of Definition 4.2.1 on p. 310 and Eqs. (1), (2) of Theorem 4.6.1 on p. 327 of Triebel [182]):

$$B^s_{p,p}(\Omega) \hookrightarrow W^s_p(\Omega) \hookrightarrow B^s_{p,2}(\Omega) \quad \text{for } 1 < p \le 2,$$

$$B^s_{p,2}(\Omega) \hookrightarrow W^s_p(\Omega) \hookrightarrow B^s_{p,p}(\Omega) \quad \text{for } 2 \le p < \infty.$$

For $s < 0$ and $1 < p < \infty$, we define $B^s_{p,q}(\mathbb{R}^n)$ as the dual space of the Besov space $B^{-s}_{p',q'}(\mathbb{R}^n)$, where $1/p + 1/p' = 1$, $1/q + 1/q' = 1$.

Example 1.31 Consider the Heaviside function H on the interval $\Omega = (-1, 1)$ defined by

$$H(x) = \begin{cases} 1 & \text{if } x \in (0,1), \\ 0 & \text{if } x \in (-1,0]. \end{cases}$$

A simple calculation shows that $H \in B^{1/p}_{p,\infty}(\Omega)$, $p \in [1,\infty)$. Also, $H \in W^s_p(\Omega)$ for all $s < 1/p$, $p \in [1,\infty)$, but $H \notin W^{1/p}_p(0,1)$ for any $p \in [1,\infty)$.

1.7 Interpolation Properties of Sobolev Spaces

For $0 \le s_1, s_2 < \infty$, $s_1 \ne s_2$, $0 < \theta < 1$, $1 \le q \le \infty$, we have (cf. Theorem 2.4.2 on p. 186 and its consequence Eq. (16) on p. 186 in Triebel [182])

$$\big(W^{s_1}_p(\mathbb{R}^n), W^{s_2}_p(\mathbb{R}^n)\big)_{\theta,q} = B^s_{pq}(\mathbb{R}^n), \quad s = (1-\theta)s_1 + \theta s_2.$$

Thus in particular, for $q = p$ and noninteger $s = (1-\theta)s_1 + \theta s_2$, we obtain that

$$\big(W_p^{s_1}(\mathbb{R}^n), W_p^{s_2}(\mathbb{R}^n)\big)_{\theta,p} = W_p^s(\mathbb{R}^n), \quad s = (1-\theta)s_1 + \theta s_2.$$

For $p = 2$ this relation holds without restrictions, i.e.

$$\big(W_2^{s_1}(\mathbb{R}^n), W_2^{s_2}(\mathbb{R}^n)\big)_{\theta,2} = W_2^{(1-\theta)s_1+\theta s_2}(\mathbb{R}^n).$$

Hence, $W_2^s(\mathbb{R}^n)$ are interpolation spaces. Analogous interpolation results hold for Sobolev spaces on a Lipschitz domain $\Omega \subset \mathbb{R}^n$ (cf. Sect. 4.3.1 on p. 317 of Triebel [182] for details).

1.8 Multiplier Spaces

In this section we consider point multipliers (or, simply, multipliers), in Sobolev spaces. These will be extensively used in the remaining chapters to characterize the minimum admissible smoothness of coefficients in differential equations. For proofs and a detailed exposition of the theory we refer to the monographs of Maz'ya and Shaposhnikova [137, 138].

Let Ω be an open set in $\mathbb{R}^n$ and suppose that V and W are two function spaces contained in $\mathcal{D}'(\Omega)$. A function a defined on Ω is called a *multiplier* from V to W if, for every v in V, the product av belongs to W. The set of all multipliers from V to W is denoted by $M(V \to W)$. In particular when $V = W$ we write $M(V)$ instead of $M(V \to V)$. The norm in $M(V \to W)$ is defined as the norm of the operator of multiplication:

$$\|a\|_{M(V\to W)} := \sup\big\{\|av\|_W : \|v\|_V \le 1\big\}.$$

In this section we shall be concerned with multipliers in Sobolev spaces, that is with $M(W_p^t(\Omega) \to W_p^s(\Omega))$ for $p \in (1,\infty)$ and $t \ge s$. Initially, we consider multipliers in pairs of Sobolev spaces on $\mathbb{R}^n$. For the sake of brevity, the symbol $W_p^s(\mathbb{R}^n)$ will be truncated to W_p^s. Motivated by the definition of multiplication of a distribution by a smooth function, for $a \in M(W_p^t \to W_p^s)$ and $u \in W_{p'}^{-s} = (W_p^s)'$, $1/p + 1/p' = 1$, we *define* the product $au \in W_{p'}^{-t} = (W_p^t)'$ by

$$\langle au, \varphi\rangle_{W_{p'}^{-t}\times W_p^t} := \langle u, a\varphi\rangle_{W_{p'}^{-s}\times W_p^s}, \quad \varphi \in W_p^t.$$

Here,

$$\langle\cdot,\cdot\rangle_{W_{p'}^{-t}\times W_p^t} \quad \text{and} \quad \langle\cdot,\cdot\rangle_{W_{p'}^{-s}\times W_p^s}$$

denote the duality pairings between the Sobolev spaces $W_{p'}^{-t}$ and W_p^t, and $W_{p'}^{-s}$ and W_p^s, respectively. This definition implies that

$$M\big(W_{p'}^{-s} \to W_{p'}^{-t}\big) = M\big(W_p^t \to W_p^s\big),$$

and therefore, rather than admitting all real numbers s and t, it suffices to consider $M(W_p^t \to W_p^s)$ for $t \geq s \geq 0$ and $M(W_p^t \to W_p^{-s})$ for $t \geq 0 \geq -s$. We list here a collection of some basic results concerning multipliers.

Lemma 1.45 *Suppose that s and t are either two integers or two nonintegers, $t \geq s \geq 0$ and $p \in (1, \infty)$. If $a \in M(W_p^t \to W_p^s)$, then*

❶ $a \in M(W_p^{t-s} \to L_p)$;

❷ $a \in \begin{cases} M(W_p^{t-\sigma} \to W_p^{s-\sigma}) & \text{for } 0 < \sigma < s \text{ and integer } s, t \text{ and } \sigma, \\ M(B_{p,p}^{t-\sigma} \to B_{p,p}^{s-\sigma}) & \text{for } 0 < \sigma < s \text{ and noninteger } s \text{ and } t; \end{cases}$

❸ $\partial^\alpha a \in M(W_p^t \to W_p^{s-|\alpha|})$, $|\alpha| \leq s$;

❹ $\partial^\alpha a \in M(W_p^{t-s+|\alpha|} \to L_p)$, $|\alpha| \leq s$.

Proof

❶ The result follows from Lemma 2 on p. 40 of Maz'ya and Shaposhnikova [138] (or from Lemma 2.3.4 on p. 40 of [137]) for integer s and t; and from Lemma 4.3.4 on p. 147 in [137] for noninteger s and t.

❷ For integer s, t and σ, the result follows from Proposition 2.7.1 on p. 58 of Maz'ya and Shaposhnikova [137]. For noninteger s and t, the stated result follows from inequality (1) on p. 154 of [138] or from Corollary 4.3.2 on p. 148 in [137].

❸ The result follows from the assumption $a \in M(W_p^t \to W_p^s)$ and part ❶ using Lemma 1 on p. 39 of Maz'ya and Shaposhnikova [138] (or Lemma 2.3.3 on p. 39 of [137]) for integer s and t; and from Lemma 1 on p. 160 in [138] or from Lemma 4.3.5 on p. 149 in [137] for noninteger s and t.

❹ The result follows from part ❶ for integer s and t and $|\alpha| = 0$; from part ❸ for integer s and t and $|\alpha| = s$; and from the assumption $a \in M(W_p^t \to W_p^s)$ and part ❶ using the inequality (11) on p. 42 of Maz'ya and Shaposhnikova [138] (or inequality (2.3.13) on p. 42 in [137]) for s and t integer and $0 < |\alpha| < s$. For noninteger s and t the stated result follows from the corollary of Lemma 1 on p. 160 of [138] or from Corollary 4.3.3 on p. 149 in [137]. □

Lemma 1.46 *Suppose that s and t are either two integers or two nonintegers, $t \geq s \geq 0$, and $p \in (1, \infty)$. Then,*

$$M\big(W_p^t \to W_p^s\big) \subset W_{p,unif}^s,$$

where

$$W_{p,unif}^s := \Big\{u : \sup_{z \in \mathbb{R}^n} \big\|(\tau_z \eta) u\big\|_{W_p^s} < \infty, \forall \eta \in \mathcal{D}\big(\mathbb{R}^n\big) \ s.t. \ \eta = 1 \ on \ B_1\Big\}.$$

If $tp > n$, then $M(W_p^t \to W_p^s) = W_{p,unif}^s$. Here, $(\tau_z \eta)(x) := \eta(x - z)$.

Proof For integer s and t the inclusion $M(W_p^t \to W_p^s) \subset W^s_{p,unif}$ is proved in Sect. 1.3.3 of Maz'ya and Shaposhnikova [138] in the discussion preceding the Theorem on p. 45, while the equality $M(W_p^t \to W_p^s) = W^s_{p,unif}$ for integer s and t and $tp > n$ follows from the Theorem on p. 45 of [138]. For noninteger s and t the result is implied by Proposition 3.2.8 on p. 166 in [138]. □

Lemma 1.47 *Suppose that s and t are either two integers or two nonintegers, $t \geq s \geq 0$, and $a = a(x_1, \dots, x_n)$. If $a \in M(W_2^t(\mathbb{R}^n) \to W_2^s(\mathbb{R}^n))$, then*

$$a \in M\big(W_2^t(\mathbb{R}^{n+k}) \to W_2^s(\mathbb{R}^{n+k})\big)$$

and

$$a \in M\big(W_2^{t,t/2}(\mathbb{R}^n \times \mathbb{R}) \to W_2^{s,s/2}(\mathbb{R}^n \times \mathbb{R})\big).$$

Proof The proof of the result is immediate, using the fact that a is a function of $x_1, \dots, x_n$ only (cf. Proposition 2.7.2 on p. 58 of [137]). □

Lemma 1.48 *Suppose that $s \geq 0$ and $p \in (1, \infty)$; then $M(W_p^s) \subset L_\infty$.*

Proof For integer s the stated result follows from Proposition 2.7.4 on p. 59 of Maz'ya and Shaposhnikova [137]. For noninteger s the result follows from the final inequality stated in part 2 of the proof of Lemma 3.2.2 on p. 159 of [138]. See also (4.3.28) in Lemma 4.3.4 on p. 147 of [137]. □

Lemma 1.49 *Suppose that $p \in (1, \infty)$, and let s and t be nonnegative integers such that $t \geq s$. Further, let*

$$a = \sum_{|\alpha| \leq t} \partial^\alpha a_\alpha,$$

where

$$a_\alpha \in M\big(W_p^t \to W_p^{t-s}\big) \cap M\big(W_{p'}^s \to L_{p'}\big), \quad 1/p + 1/p' = 1;$$

then $a \in M(W_p^t \to W_p^{-s})$.

Proof For $s = 0$ the result follows from ❸ of Lemma 1.45. For $s > 0$ the result follows from part (ii) of Theorem 1 on p. 57 in [138], or from part (ii) of Theorem 2.5.1 on p. 54 of [137]. □

Lemma 1.50

❶ *Let $p \in (1, \infty)$ and assume that s and t are integers, $t > s > 0$ and $tp < n$. If $a \in W^s_{n/t,unif}$, then $a \in M(W_p^t \to W_p^s)$. The result is also true for $t = s$, assuming that $a \in W^s_{n/s,unif} \cap L_\infty$.*

❷ *Let $p \in (1, \infty)$ and assume that s and t are nonintegers, $t > s > 0$. Suppose that either $q \in [n/t, \infty]$ and $tp < n$, or $q \in (p, \infty)$ and $tp = n$. If $a \in B^s_{q,p,unif}$, where*

$$B^s_{q,p,unif} := \left\{u : \sup_{z \in \mathbb{R}^n} \left\|(\tau_z \eta)u\right\|_{B^s_{q,p}} < \infty, \forall \eta \in \mathcal{D}(\mathbb{R}^n) \text{ s.t. } \eta = 1 \text{ on } B_1\right\},$$

then $a \in M(W^t_p \to W^s_p)$. Here, $(\tau_z \eta)(x) := \eta(x - z)$. The result is also true for $t = s$, provided that $a \in B^s_{q,p,unif} \cap L_\infty$.

Proof

❶ The first statement follows from part (i) of Corollary 1 on p. 50 in Sect. 1.3.4 of Maz'ya and Shaposhnikova [138], while the second statement comes from part (ii) of the same result. Alternatively, see part (i) of Corollary 2.3.5 and Proposition 2.3.2 on p. 48 of [137].

❷ The two statements follow, respectively, from part (ii) and part (i) of Theorem 3.3.2 on p. 172 of Maz'ya and Shaposhnikova [138]. Alternatively, see Theorem 4.4.4 on p. 170 of [137]. □

Lemma 1.51 *The linear differential operator L defined by*

$$Lu = \sum_{|\alpha| \le k} a_\alpha(x) \partial^\alpha u, \quad x \in \mathbb{R}^n,$$

is a bounded linear operator from the Sobolev space W^s_p to W^{s-k}_p, $s \ge k$, provided that $a_\alpha \in M(W^{s-|\alpha|}_p \to W^{s-k}_p)$ for every multi-index α, $|\alpha| \le k$.

Proof The result is a direct consequence of part ❸ of Lemma 1.45 and the definition of the multiplier space $M(W^{s-|\alpha|}_p \to W^{s-k}_p)$. □

These results have analogous counterparts in Sobolev spaces on bounded open subsets of $\mathbb{R}^n$. This follows by observing that if Ω is a Lipschitz domain in $\mathbb{R}^n$ and $a \in M(W^t_p(\Omega) \to W^s_p(\Omega))$, then by Theorem 1.33 (and its analogue for fractional-order Sobolev spaces) a can be extended to a function $\tilde{a}$, defined on the whole of $\mathbb{R}^n$, such that $\tilde{a} \in M(W^t_p \to W^s_p) = M(W^t_p(\mathbb{R}^n) \to W^s_p(\mathbb{R}^n))$; the converse is also true: the restriction to Ω of a multiplier $a \in M(W^t_p(\mathbb{R}^n) \to W^s_p(\mathbb{R}^n))$ is an element of $M(W^t_p(\Omega) \to W^s_p(\Omega))$. We note here that for a Lipschitz domain Ω (which is, by definition, bounded), the 'uniform' spaces $W^s_{p,unif}$ and $B^s_{q,p,unif}$ that appear in Lemmas 1.46 and 1.50 are replaced by standard Sobolev and Besov spaces, $W^s_p(\Omega)$ and $B^s_{q,p}(\Omega)$, respectively.

Lemma 1.52 *Let Ω be a Lipschitz domain in $\mathbb{R}^n$ and suppose that $s > 0$ and $p \in (1, \infty)$. If $a \in W^t_q(\Omega)$ then $a \in M(W^s_p(\Omega)) \cap L_\infty(\Omega)$, where*

$$t = s, \qquad q = p \qquad \text{when } sp > n,$$

$$t > s, \qquad q > p \qquad \text{when } sp = n,$$

$$t > s, \qquad q \geq n/s \quad \text{when } sp < n.$$

Proof For $sp > n$, the Sobolev embedding theorem implies that $W_p^s(\Omega) \hookrightarrow L_\infty(\Omega)$. By applying the analogue of Lemma 1.46 in a Lipschitz domain, we deduce that $a \in W_q^t(\Omega) = W_p^s(\Omega) \subset M(W_p^s(\Omega))$.

Now suppose that $t > s$, and either $q > p = n/s$ or $q \geq n/s > p$. Letting $\varepsilon = t - s$ and applying embedding theorems for Sobolev and Besov spaces,

$$a \in W_q^t(\Omega) = W_q^{s+\varepsilon}(\Omega) \hookrightarrow B_{q,p}^s(\Omega) \cap W_{n/s}^s(\Omega),$$

and

$$a \in W_q^{s+\varepsilon}(\Omega) \hookrightarrow L_\infty(\Omega).$$

Thanks to the analogue of Lemma 1.50 in a Lipschitz domain Ω, we have that $a \in M(W_p^s(\Omega))$. □

Lemma 1.53 *Let Ω be a Lipschitz domain in $\mathbb{R}^n$, and suppose that $s > 0$ and $p \in (1, \infty)$. If $a \in L_q(\Omega)$, where*

$$q = p \quad \text{when } sp > n,$$

$$q > p \quad \text{when } sp = n,$$

$$q \geq n/s \quad \text{when } sp < n,$$

then $a \in M(W_p^s(\Omega) \to L_p(\Omega))$.

Proof First suppose that $sp > n$, $q = p$, and let $a \in L_q(\Omega)$. By applying Hölder's inequality and the Sobolev embedding theorem, we obtain

$$\|au\|_{L_p(\Omega)} = \|au\|_{L_q(\Omega)} \leq \|a\|_{L_q(\Omega)}\|u\|_{L_\infty(\Omega)} \leq c\|a\|_{L_q(\Omega)}\|u\|_{W_p^s(\Omega)},$$

and therefore $a \in M(W_p^s(\Omega) \to L_p(\Omega))$.

Now assume that either $sp = n$ and $q > p$, or $sp < n$ and $q \geq n/s$ (and therefore $q > p$). By applying Hölder's inequality and the Sobolev embedding theorem,

$$\|au\|_{L_p(\Omega)} \leq \|a\|_{L_q(\Omega)}\|u\|_{L_{\frac{pq}{q-p}}(\Omega)} \leq c\|a\|_{L_q(\Omega)}\|u\|_{W_p^s(\Omega)}.$$

Hence $a \in M(W_p^s(\Omega) \to L_p(\Omega))$. □

Lemma 1.54 *Let $p \in (1, \infty)$, and suppose that s and t are two real numbers such that either $0 < s \leq n/p < t$ or $s = t > n/p$. Then, $W_p^t(\Omega) \subset M(W_p^s(\Omega))$.*

Proof When $s = t > n/p$, the stated result follows from Lemma 1.52.

Suppose that $0 < s < n/p < t$; then $s < t + s - n/p$, and therefore there exits a real number τ such that $s < \tau < t + s - n/p$; clearly $t > \tau$. Defining $q = n/s$, we deduce that $t - n/p > \tau - n/q$, $t > \tau$ and $q > p$. Thus, by the Sobolev embedding theorem,

$$W_p^t(\Omega) \hookrightarrow W_q^\tau(\Omega),$$

and, because $\tau > s$, $q = n/s$ and $sp < n$, Lemma 1.52 implies that $W_q^\tau(\Omega) \hookrightarrow M(W_p^s(\Omega))$. Hence the desired inclusion.

Finally, let us suppose that $0 < s = n/p < t$; then, there exists a real number $\varepsilon \in (0, 1)$ such that $t > n/(p\varepsilon)$. Let us choose $\tau \in (t - n/(p\varepsilon), t) \cap (s, t)$; then $0 < t - \tau < n/(p\varepsilon)$, and therefore

$$\frac{1}{p} > \frac{1}{p} + \frac{\varepsilon}{n}(\tau - t) > 0.$$

We define

$$q := \left(\frac{1}{p} + \frac{\varepsilon}{n}(\tau - t)\right)^{-1}.$$

Clearly, $q > p$ and

$$t - \frac{n}{p} = \tau - \frac{n}{q} + (1 - \varepsilon)(t - \tau) > \tau - \frac{n}{q},$$

and therefore by the Sobolev embedding theorem we have that

$$W_p^t(\Omega) \hookrightarrow W_q^\tau(\Omega).$$

Because $\tau > s$, $q > p$, $sp = n$, Lemma 1.52 implies that $W_q^\tau(\Omega) \subset M(W_p^s(\Omega))$, whereby $W_p^t(\Omega) \subset M(W_p^s(\Omega))$. □

Our next result is an extension of the familiar Leibniz formula for the differentiation of a product of two smooth functions.

Lemma 1.55 *Let* $p \in (1, \infty)$, *and suppose that* $\mathbf{a} = (a_1, \ldots, a_n)$ *and* u *are two functions such that* $\mathbf{a} \in [L_p(\Omega)]^n$, $u \in M(W_q^1(\Omega))$, *where* $1/p + 1/q = 1$, $1 < p < \infty$. *Then,*

$$\nabla \cdot (\mathbf{a}u) = \mathbf{a} \cdot \nabla u + (\nabla \cdot \mathbf{a})u \quad \text{in } W_p^{-1}(\Omega).$$

Proof Let $w \in L_p(\Omega)$; then, identifying w with the associated regular distribution,

$$\langle \partial_i w, \varphi \rangle = -\langle w, \partial_i \varphi \rangle = -(w, \partial_i \overline{\varphi}), \quad \varphi \in C_0^\infty(\Omega),\ i = 1, \ldots, n.$$

Thus, by Hölder's inequality,

$$\left|\langle \partial_i w, \varphi \rangle\right| \le \|w\|_{L_p(\Omega)} \|\partial_i \varphi\|_{L_q(\Omega)} \le \|w\|_{L_p(\Omega)} \|\varphi\|_{\mathring{W}_q^1(\Omega)}, \quad i = 1, \ldots, n,$$

for every φ in $C_0^\infty(\Omega)$. Since $C_0^\infty(\Omega)$ is dense in $\mathring{W}_q^1(\Omega)$, $1 < q < \infty$, the linear functional $\langle \partial_i w, \cdot \rangle : C_0^\infty(\Omega) \to \mathbb{R}$ can be extended from $C_0^\infty(\Omega)$ to a bounded linear functional on the space $\mathring{W}^1_q(\Omega)$, still denoted by $\partial_i w$, $i = 1, \ldots, n$. Hence, $\partial_i w \in (\mathring{W}^1_q(\Omega))' = W^{-1}_p(\Omega)$, and

$$\langle \partial_i w, \varphi \rangle_{W^{-1}_p(\Omega) \times \mathring{W}^1_q(\Omega)} = -(w, \partial_i \overline{\varphi}), \quad \varphi \in \mathring{W}^1_q(\Omega),\ i = 1, \ldots, n. \tag{1.26}$$

In particular, ∂_i is a bounded linear operator from $L_p(\Omega)$ into $W^{-1}_p(\Omega)$ for each $i \in \{1, \ldots, n\}$.

Now since u belongs to $M(W^1_q(\Omega)) = M(W^1_q(\Omega) \to W^1_q(\Omega))$, it follows from the analogue of Lemma 1.45, part ❸, for a Lipschitz domain Ω that $\partial_i u \in M(W^1_q(\Omega) \to L_q(\Omega))$. Thus also $\partial_i u \in M(\mathring{W}^1_q(\Omega) \to L_q(\Omega)) = M(L_p(\Omega) \to W^{-1}_p(\Omega))$. As a_i, the ith component of $\mathbf{a}$ belongs to $L_p(\Omega)$, we deduce that $a_i(\partial_i u)$ belongs to $W^{-1}_p(\Omega)$.

Next, because $u \in M(W^1_q(\Omega)) \subset W^1_q(\Omega) \cap L_\infty(\Omega)$ by the analogues of Lemmas 1.46 and 1.48 for a Lipschitz domain Ω, we deduce that $a_i u \in L_p(\Omega)$. Therefore, $\partial_i(a_i u)$ belongs to $W^{-1}_p(\Omega)$.

Finally, we show that $(\partial_i a_i)u$ also belongs to $W^{-1}_p(\Omega)$. Indeed, because ∂_i is a bounded linear operator from $L_p(\Omega)$ into the space $W^{-1}_p(\Omega)$ and since

$$u \in M\big(W^1_q(\Omega)\big) \subset M\big(\mathring{W}^1_q(\Omega)\big) = M\big(W^{-1}_p(\Omega)\big),$$

we deduce that $(\partial_i a_i)u \in W^{-1}_p(\Omega)$.

We have thus shown that, under the hypotheses of the lemma, each of the terms $\partial_i(a_i u)$, $a_i(\partial_i u)$, $(\partial_i a_i)u$ belongs to $W^{-1}_p(\Omega)$. It remains to verify that

$$\partial_i(a_i u) = a_i(\partial_i u) + (\partial_i a_i)u \quad \text{in } W^{-1}_p(\Omega),\ i = 1, \ldots, n.$$

Because $a_i \in L_p(\Omega)$ and $u \in M(W^1_q(\Omega)) \subset W^1_q(\Omega) \cap L_\infty(\Omega)$, we deduce that $a_i u \in L_p(\Omega)$ and $a_i(\partial_i u) \in L_1(\Omega) \cap W^{-1}_p(\Omega)$. Hence

$$\begin{aligned}
&\big\langle a_i(\partial_i u), \varphi \big\rangle_{W^{-1}_p(\Omega) \times \mathring{W}^1_q(\Omega)} \\
&= \big(a_i(\partial_i u), \overline{\varphi}\big) = \big(a_i, (\partial_i \overline{u})\overline{\varphi}\big) \\
&= \big(a_i, \partial_i(\overline{u\varphi}) - \overline{u}(\partial_i \overline{\varphi})\big) \\
&= \big(a_i, \partial_i(\overline{u\varphi})\big) - \big(a_i, \overline{u}(\partial_i \overline{\varphi})\big) \\
&= \big(a_i, \partial_i(\overline{u\varphi})\big) - (a_i u, \partial_i \overline{\varphi}) \\
&= -\langle \partial_i a_i, u\varphi \rangle_{W^{-1}_p(\Omega) \times \mathring{W}^1_q(\Omega)} + \big\langle \partial_i(a_i u), \varphi \big\rangle_{W^{-1}_p(\Omega) \times \mathring{W}^1_q(\Omega)},
\end{aligned}$$

for all φ in $C^\infty_0(\Omega)$, where, in the last line, we have used (1.26) and the fact that $u\varphi$ belongs to $\mathring{W}^1_q(\Omega)$ for every φ in $C^\infty_0(\Omega)$. Finally, since u is an element of $M(W^1_q(\Omega)) \subset M(\mathring{W}^1_q(\Omega)) = M(W^{-1}_p(\Omega))$, we have that

$$\big\langle a_i(\partial_i u), \varphi \big\rangle_{W^{-1}_p(\Omega) \times \mathring{W}^1_q(\Omega)} = \big\langle \partial_i(a_i u) - (\partial_i a_i)u, \varphi \big\rangle_{W^{-1}_p(\Omega) \times \mathring{W}^1_q(\Omega)} \tag{1.27}$$

for every φ in $C_0^\infty(\Omega)$. Because $C_0^\infty(\Omega)$ is dense in $\mathring{W}_q^1(\Omega)$, (1.27) implies that

$$\partial_i(a_i u) = a_i(\partial_i u) + (\partial_i a_i)u \quad \text{in } W_p^{-1}(\Omega),\ i = 1, \ldots, n.$$

After summation through $i = 1, \ldots, n$, this yields the desired identity. □

The final result of this section provides a sufficient condition for a function to belong to $M(W_2^t(\Omega) \to W_2^{-s}(\Omega))$.

Lemma 1.56 *Let Ω be a Lipschitz domain in $\mathbb{R}^n$ and let*

$$a = a_0 + \sum_{i=1}^{n} \partial_i a_i,$$

where $a_0 \in M(W_2^t(\Omega) \to L_2(\Omega))$ and

$$a_i \in M\big(W_2^t(\Omega) \to W_2^{1-s}(\Omega)\big) \cap M\big(W_2^{t-1}(\Omega) \to L_2(\Omega)\big), \quad i = 1, \ldots, n,$$

with $0 < s \le 1 \le n/2 < t$, $s \ne 1/2$; then $a \in M(W_2^t(\Omega) \to W_2^{-s}(\Omega))$.

Proof Consider

$$a = a_0 + \sum_{i=1}^{n} \partial_i a_i,$$

where $a_0 \in M(W_2^t(\Omega) \to L_2(\Omega))$, and

$$a_i \in M\big(W_2^t(\Omega) \to W_2^{1-s}(\Omega)\big) \cap M\big(W_2^{t-1}(\Omega) \to L_2(\Omega)\big), \quad i = 1, \ldots, n,$$

and let $u \in W_2^t(\Omega)$, $0 < s \le 1 \le n/2 < t$, where $s \ne 1/2$. Then,

$$a_0 u \in L_2(\Omega), \qquad a_i(\partial_i u) \in L_2(\Omega), \qquad a_i u \in W_2^{1-s}(\Omega), \quad i = 1, \ldots, n.$$

Hence $\partial_i(a_i u) \in W_2^{-s}(\Omega)$, $s \ne 1/2$ (see Remark 12.8 on p. 94 of Lions and Magenes [127]). Now according to the analogue of Lemma 1.46 in a Lipschitz domain Ω we have that a_i belongs to $L_2(\Omega)$, $i = 1, \ldots, n$, and, by Lemma 1.54, $u \in W_2^t(\Omega) \subset M(W_2^1(\Omega))$; thus we deduce from Lemma 1.55 that

$$(\partial_i a_i)u = \partial_i(a_i u) - a_i(\partial_i u) \quad \text{in } W_2^{-1}(\Omega).$$

Since the right-hand side of this equality belongs to $W_2^{-s}(\Omega)$ for $0 < s \le 1$, $s \ne 1/2$, the same is true of the left-hand side. Hence

$$au = a_0 u + \sum_{i=1}^{n} \big(\partial_i(a_i u) - a_i(\partial_i u)\big) \quad \text{in } W_2^{-s}(\Omega), s \ne 1/2.$$

In addition,

$$\begin{aligned}\|au\|_{W_2^{-s}(\Omega)} &\le \|a_0u\|_{W_2^{-s}(\Omega)} + \sum_{i=1}^{n}\left(\left\|\partial_i(a_iu)\right\|_{W_2^{-s}(\Omega)} + \left\|a_i(\partial_iu)\right\|_{W_2^{-s}(\Omega)}\right)\\ &\le \|a_0u\|_{L_2(\Omega)} + \sum_{i=1}^{n}\left(\|a_iu\|_{W_2^{1-s}(\Omega)} + \left\|a_i(\partial_iu)\right\|_{L_2(\Omega)}\right)\\ &\le \left\{\|a_0\|_{M(W_2^t\to L_2)} + \sum_{i=1}^{n}\|a_i\|_{M(W_2^t\to W_2^{1-s})\cap M(W_2^{t-1}\to L_2)}\right\}\|u\|_{W_2^t},\end{aligned}$$

and therefore $au \in M(W_2^t(\Omega) \to W_2^{-s}(\Omega))$, $s \neq 1/2$. □

1.9 Fourier Multipliers and Mollifiers

In this section we consider multipliers in Fourier transform space. We shall then make use of this theory to design mollifiers in various function spaces and to analyze their smoothing properties. The key feature of a mollifier is that it damps the high frequency or high wave-number content in Fourier transform space of the function it is applied to: this property will prove useful in the construction of finite difference schemes for differential equations with data that are nonsmooth with respect to their temporal or spatial variable.

In Sect. 1.9.1 we define Fourier multipliers on Lebesgue spaces; these will be our main tool in the subsequent analysis of mollifiers. In Sect. 1.9.2 we introduce a general definition of mollifier, and in Sect. 1.9.3, using Fourier multipliers on $L_p(\mathbb{R}^n)$, we describe simple sufficient conditions, which ensure that a linear operator is a mollifier on $L_p(\mathbb{R}^n)$. In Sect. 1.9.4 we expand the domains of definition of the mollifiers considered in Sects. 1.9.2 and 1.9.3 to the space of tempered distributions, and in particular to Bessel-potential spaces, Besov spaces and Sobolev spaces of arbitrary real order.

1.9.1 Fourier Multipliers

In this section we outline the main properties of Fourier multipliers. For proofs and further details we refer to the monographs of Bergh and Löfström [9], Grafakos [59], Hörmander [71], Stein [167], and Stein and Weiss [168].

Definition 1.57 Let $1 \le p \le \infty$. We say that $a \in \mathcal{S}'(\mathbb{R}^n)$ is a *Fourier multiplier* in $L_p(\mathbb{R}^n)$ if there exists a positive constant c such that

$$\left\|F^{-1}(a\,Fu)\right\|_{L_p(\mathbb{R}^n)} \le c\|u\|_{L_p(\mathbb{R}^n)} \quad \forall u \in \mathcal{S}\left(\mathbb{R}^n\right);$$

the infimum of the set of all such positive constants c is denoted by $\|a\|_{M_p(\mathbb{R}^n)}$. Equivalently,

$$\|a\|_{M_p(\mathbb{R}^n)} = \sup\bigl\{ \bigl\|F^{-1}(a\,Fu)\bigr\|_{L_p(\mathbb{R}^n)} : u \in \mathcal{S}\bigl(\mathbb{R}^n\bigr),\ \|u\|_{L_p(\mathbb{R}^n)} \le 1 \bigr\}.$$

The linear space of all Fourier multipliers on $L_p(\mathbb{R}^n)$ is denoted by $M_p(\mathbb{R}^n)$ and is equipped with the norm $\|\cdot\|_{M_p(\mathbb{R}^n)}$. For the sake of brevity, we shall sometimes write $F^{-1}a\,Fu$ instead of $F^{-1}(a\,Fu)$.

We observe that since $u \in \mathcal{S}(\mathbb{R}^n)$ and $a \in \mathcal{S}'(\mathbb{R}^n)$, Fu belongs to $\mathcal{S}(\mathbb{R}^n)$ and $a\,Fu$ is a tempered distribution. Thus $F^{-1}(a\,Fu)$ is correctly defined and it belongs to $\mathcal{S}'(\mathbb{R}^n)$. In fact, according to Theorem 1.29,

$$F^{-1}(a\,Fu) = \bigl(F^{-1}a\bigr) * u,$$

and therefore, by E) of Sect. 1.3.3, $F^{-1}(a\,Fu) \in C_M^\infty(\mathbb{R}^n)$.

Since $\mathcal{S}(\mathbb{R}^n)$ is dense in $L_p(\mathbb{R}^n)$, $1 \le p < \infty$, the linear operator

$$F^{-1}a\,F: \quad u \in \mathcal{S}\bigl(\mathbb{R}^n\bigr) \quad \mapsto \quad F^{-1}(a\,Fu) \in L_p\bigl(\mathbb{R}^n\bigr)$$

can be extended, preserving its norm, to a linear operator from $L_p(\mathbb{R}^n)$ to $L_p(\mathbb{R}^n)$. This extension will still be denoted by $F^{-1}a\,F$.

When $p = \infty$, the space $M_p(\mathbb{R}^n)$ can be described explicitly. To do so, we note that the operator $T = F^{-1}a\,F$ commutes with the translation operator on $\mathcal{S}(\mathbb{R}^n)$ in the sense that $\tau_x T = T\tau_x$, $x \in \mathbb{R}^n$. Therefore $a \in M_\infty(\mathbb{R}^n)$ if, and only if, there exists a positive real number c such that

$$\bigl|\bigl(\bigl(F^{-1}a\bigr) * u\bigr)(0)\bigr| \le c\|u\|_{L_\infty(\mathbb{R}^n)}, \quad u \in \mathcal{S}\bigl(\mathbb{R}^n\bigr). \tag{1.28}$$

This inequality implies that a belongs to $M_\infty(\mathbb{R}^n)$ if, and only if, $F^{-1}a$ is an element of the dual space of $L_\infty(\mathbb{R}^n)$, that is, if it is a bounded complex Borel measure on $\mathbb{R}^n$. Hence $M_\infty(\mathbb{R}^n)$ coincides with the set of Fourier transforms of bounded complex Borel measures, and $\|a\|_{M_\infty(\mathbb{R}^n)}$ is equal to the total variation of the measure $F^{-1}a$. In view of (1.28) and the Hahn–Banach theorem, the linear operator $F^{-1}aF : u \in \mathcal{S}(\mathbb{R}^n) \mapsto F^{-1}(a\,Fu) \in L_\infty(\mathbb{R}^n)$ can be extended, without increasing its norm, to a linear operator from $L_\infty(\mathbb{R}^n)$ to $L_\infty(\mathbb{R}^n)$. The extended operator will still be denoted by $F^{-1}aF$.

We recall the following important properties of Fourier multipliers (see, for example, Theorem 6.1.2 on p. 132 in Bergh and Löfström [9], or the monographs by Grafakos [59], Hörmander [71], Stein [167], Stein and Weiss [168]):

$$M_p\bigl(\mathbb{R}^n\bigr) = M_{p'}\bigl(\mathbb{R}^n\bigr), \quad 1 \le p \le \infty,\ 1/p + 1/p' = 1,$$

$$M_p\bigl(\mathbb{R}^n\bigr) \subset M_q\bigl(\mathbb{R}^n\bigr) \subset M_2\bigl(\mathbb{R}^n\bigr) = L_\infty\bigl(\mathbb{R}^n\bigr), \quad 1 \le p \le q \le 2.$$

Furthermore, for $1 \le p_0, p_1 \le \infty$ and $1/p = (1-\theta)/p_0 + \theta/p_1$, with $0 \le \theta \le 1$, we have that

$$\|a\|_{M_p(\mathbb{R}^n)} \le \|a\|^{1-\theta}_{M_{p_0}(\mathbb{R}^n)} \|a\|^{\theta}_{M_{p_1}(\mathbb{R}^n)}, \quad a \in M_{p_0}(\mathbb{R}^n) \cap M_{p_1}(\mathbb{R}^n). \tag{1.29}$$

If a_1 and a_2 belong to $M_p(\mathbb{R}^n)$, then their product $a_1 a_2$ is also contained in $M_p(\mathbb{R}^n)$. In addition, $\|\cdot\|_{M_p(\mathbb{R}^n)}$ is submultiplicative in the sense that

$$\|a_1 a_2\|_{M_p(\mathbb{R}^n)} \le \|a_1\|_{M_p(\mathbb{R}^n)} \|a_2\|_{M_p(\mathbb{R}^n)}$$

for every a_1, a_2 in $M_p(\mathbb{R}^n)$, $1 \le p \le \infty$ (cf. Bergh and Löfström [9], p. 133).

The next two theorems provide convenient tools for verifying that a function is a Fourier multiplier.

Theorem 1.58 (Carlson–Beurling Inequality) *Let $a \in L_2(\mathbb{R}^n)$. Suppose that m is a positive integer such that $m > n/2$ and let $\partial^\alpha a \in L_2(\mathbb{R}^n)$ for every multi-index $\alpha \in \mathbb{N}^n$, $|\alpha| = m$; then, $a \in M_p(\mathbb{R}^n)$ and there exists a constant c, independent of a and p, such that*

$$\|a\|_{M_p(\mathbb{R}^n)} \le c \|a\|^{1-\theta}_{L_2(\mathbb{R}^n)} |a|^{\theta}_{W_2^m(\mathbb{R}^n)}, \quad \theta = n/(2m),$$

for every p, $1 \le p \le \infty$.

Proof Suppose that $t > 0$. Then, by the Cauchy–Schwarz inequality, we have that, for any multi-index $\alpha \in \mathbb{N}^n$, with $|\alpha| = m$,

$$\begin{aligned}
\int_{|\xi|>t} \left|F^{-1}a(\xi)\right| \mathrm{d}\xi &= \int_{|\xi|>t} |\xi|^{-m} |\xi|^m \left|F^{-1}a(\xi)\right| \mathrm{d}\xi \\
&\le \left(\int_{|\xi|>t} |\xi|^{-2m} \,\mathrm{d}\xi \right)^{1/2} \left(\int_{\mathbb{R}^n} |\xi|^{2m} \left|F^{-1}a(\xi)\right|^2 \mathrm{d}\xi \right)^{1/2} \\
&\le c t^{(n/2)-m} |a|_{W_2^m(\mathbb{R}^n)},
\end{aligned}$$

where in the transition to the last line we have used that $F^{-1}a = (2\pi)^{-n}\overline{F\overline{a}}$, whereby $|F^{-1}a|^2 = (2\pi)^{-2n}|F\overline{a}|^2$, together with the identity $|\xi|^{2m}|F\overline{a}|^2 = |F\partial^\alpha \overline{a}|^2$ for $|\alpha| = m$, Parseval's identity, and that $|\overline{a}|_{W_2^m(\mathbb{R}^n)} = |a|_{W_2^m(\mathbb{R}^n)}$.

Similarly,

$$\int_{|\xi|\le t} \left|F^{-1}a(\xi)\right| \mathrm{d}\xi \le c t^{n/2} \|a\|_{L_2(\mathbb{R}^n)}.$$

Now adding the bounds obtained for $|\xi| > t$ and $|\xi| \le t$, choosing

$$t = \left(\frac{|a|_{W_2^m(\mathbb{R}^n)}}{\|a\|_{L_2(\mathbb{R}^n)}} \right)^{1/m}$$

and noting the equalities

$$\|a\|_{M_1(\mathbb{R}^n)} = \|a\|_{M_\infty(\mathbb{R}^n)} = \int_{\mathbb{R}^n} \left|F^{-1}a(\xi)\right| \mathrm{d}\xi,$$

we get

$$\|a\|_{M_p(\mathbb{R}^n)} \le \|a\|_{M_1(\mathbb{R}^n)} = \int_{\mathbb{R}^n} \left|F^{-1}a(\xi)\right| \mathrm{d}\xi \le c\|a\|^{1-\theta}_{L_2(\mathbb{R}^n)} |a|^{\theta}_{W_2^m(\mathbb{R}^n)},$$

with $\theta = n/(2m)$. For $p \in (1, \infty)$ the first inequality here follows from (1.29) with $p_0 = 1$, $p_1 = \infty$ and $\theta = 1 - (1/p)$; for $p = 1, \infty$ it holds trivially. □

When $1 < p < \infty$, a slightly simpler sufficient condition can be given: it is due to Lizorkin [128] and is stated in the next theorem; see also the monograph of Nikol'skiĭ [144], p. 59, where a detailed proof is presented.

Theorem 1.59 (Lizorkin's Multiplier Theorem) *Suppose that $\xi^\alpha \partial^\alpha a$ is a bounded continuous function on the set*

$$\mathbb{R}^n_* = \left\{\xi \in \mathbb{R}^n : \xi_i \neq 0, i = 1, \dots, n\right\},$$

for each multi-index $\alpha \in \{0,1\}^n$; let M_0 be a positive constant such that

$$\left|\xi^\alpha \partial^\alpha a(\xi)\right| \le M_0 \quad \forall \alpha \in \{0,1\}^n, \ \forall \xi \in \mathbb{R}^n_*.$$

Then, $a \in M_p(\mathbb{R}^n)$, $1 < p < \infty$, and there exists a positive constant C_p, depending only on p, such that $\|a\|_{M_p(\mathbb{R}^n)} \le C_p M_0$.

Let Ω be a bounded open set in $\mathbb{R}^n$. We define the set $M_p(\Omega)$ of *local Fourier multipliers* on Ω as the collection of equivalence classes of tempered distributions that are equal to an element of $M_p(\mathbb{R}^n)$ on Ω. As is usual, we identify an equivalence class with any of its elements. The norm in $M_p(\Omega)$ is defined by

$$\|a\|_{M_p(\Omega)} = \inf_{\chi} \|\chi\|_{M_p(\mathbb{R}^n)},$$

where the infimum is taken over all $\chi \in M_p(\mathbb{R}^n)$ such that $a = \chi$ on Ω.

1.9.2 Definition of Mollifier

In order to motivate the general definition of mollifier that will be stated at the end of this section, we begin with a simple example. Given a locally integrable function v defined on $\mathbb{R}$ and a positive real number h, consider the integral average $v \mapsto T_h v$ defined by

$$(T_h v)(x) := \frac{1}{h} \int_{x-h/2}^{x+h/2} v(\xi)\, \mathrm{d}\xi, \quad x \in \mathbb{R},$$

and referred to sometimes as *Steklov mollifier*. In the next chapter we shall assign a specific meaning to the parameter h: it will denote the mesh-size in a finite difference scheme; however, for the time being, the precise interpretation of h is of no significance. Clearly, if v is locally integrable on $\mathbb{R}$, the function $x \mapsto (T_h v)(x)$ is continuous on $\mathbb{R}$, and therefore T_h can be seen as a smoothing operator.

For example, the image of the (discontinuous!) Heaviside function H, defined by

$$H(x) := \begin{cases} 1 & \text{when } x > 0, \\ 0 & \text{otherwise,} \end{cases}$$

under the mapping $v \mapsto T_h v$ is the (continuous!) piecewise linear function

$$(T_h H)(x) = \begin{cases} 0 & \text{when } x < -h/2, \\ (2x+h)/(2h) & \text{when } -h/2 \le x \le h/2, \\ 1 & \text{otherwise.} \end{cases}$$

It is worth noting that, for small values of h, the function $T_h H$ is close to H in the sense that

$$\|T_h H - H\|_{L_p(\mathbb{R})} = \frac{h^{1/p}}{2(1+p)^{1/p}}, \qquad 1 \le p < \infty.$$

The smoothing properties of T_h are best seen in Fourier transform space and, for this purpose, $T_h v$ will be rewritten in the form of a convolution. Denoting by θ the characteristic function of the interval $[-1/2, 1/2]$ and defining $\theta_h(x) := h^{-1}\theta(h^{-1}x)$, we can write

$$T_h v = \theta_h * v.$$

It is clear from this representation that T_h is a translation-invariant linear operator. Moreover, by Young's inequality (1.17),

$$\|T_h v\|_{L_p(\mathbb{R})} \le \|\theta_h\|_{L_1(\mathbb{R})} \|v\|_{L_p(\mathbb{R})} = \|\theta\|_{L_1(\mathbb{R})} \|v\|_{L_p(\mathbb{R})}, \qquad v \in L_p(\mathbb{R}).$$

Hence, T_h is a bounded linear operator on $L_p(\mathbb{R})$, uniformly in h; or, in other words, the family of operators $\{T_h : h > 0\}$ is *uniformly bounded* on $L_p(\mathbb{R})$.

In order to clarify the effect of mollification on smooth functions, we note that, since $F\theta_h(\xi) = F\theta(h\xi)$, $T_h v$ can be rewritten in the form

$$T_h v = v + h^2 \partial^2 B_h^{(0)} v,$$

where

$$B_h^{(0)} v := F^{-1} \frac{1 - F\theta(h\xi)}{h^2 \xi^2} F v$$

and

$$F\theta(\xi) = \frac{\sin(\xi/2)}{\xi/2}.$$

Here ∂^2 denotes the second derivative with respect to x. Similarly to T_h, $B_h^{(0)}$ is a translation-invariant linear operator on $L_p(\mathbb{R})$.

Now let us suppose that the Fourier transform of v contains 'low frequencies (wave-numbers)' only, i.e. $Fv(\xi) = 0$ for $|h\xi| > 2\delta$ for some constant δ, $0 < \delta \leq \pi/2$; then, according to the Paley–Wiener theorem (Theorem 1.30), v is infinitely many times continuously differentiable on $\mathbb{R}$. Thanks to Theorem 1.58, the function

$$\frac{1 - F\theta(h\xi)}{h^2\xi^2},$$

featuring in the definition of $B_h^{(0)}v$, is a local Fourier multiplier on $L_p(-\delta, \delta)$. Thus

$$\left\|\partial^2 B_h^{(0)} v\right\|_{L_p(\mathbb{R})} = \left\|B_h^{(0)} \partial^2 v\right\|_{L_p(\mathbb{R})} \leq C\left\|\partial^2 v\right\|_{L_p(\mathbb{R})},$$

and therefore

$$\|T_h v - v\|_{L_p(\mathbb{R})} \leq Ch^2\left\|\partial^2 v\right\|_{L_p(\mathbb{R})}.$$

In other words, if v is sufficiently smooth (that is, if $\partial^2 v \in L_p(\mathbb{R})$), then $T_h v$ approximates v with $\mathcal{O}(h^2)$ error as $h \to 0$.

Let us now turn our attention to understanding the effect of mollification on the 'high frequency (wave number)' content of a function. By considering the Fourier transform of $T_h v$ and applying Theorem 1.29 we obtain

$$F(T_h v)(\xi) = F\theta(h\xi)Fv(\xi) = \frac{\sin(\xi h/2)}{\xi h/2} Fv(\xi).$$

This shows that, for any fixed $h \in (0, 1]$ and $|\xi| \gg 1$, the magnitude of $F(T_h v)(\xi)$ is smaller than that of $Fv(\xi)$, by a factor that is bounded by a multiple of $1/\xi$. A more detailed picture emerges by writing

$$T_h v = h D_h B_h^{(1)} v,$$

where $D_h w := (w(x + h) - w(x))/h$ is the first-order divided difference of the function w on the uniform mesh $h\mathbb{Z}$, and

$$B_h^{(1)} v = F^{-1} \frac{1}{\imath h\xi} \exp\left(-\frac{\imath h\xi}{2}\right) Fv.$$

Similarly to $B_h^{(0)}$, $B_h^{(1)}$ is a translation-invariant linear operator on $L_p(\mathbb{R})$. Suppose that the Fourier transform of v is supported on 'high frequencies (wave-numbers)' only, i.e. $Fv(\xi) = 0$ for $|h\xi| \leq \delta$ for some constant δ. Because

$$\frac{1}{\imath h\xi} \exp\left(-\frac{\imath h\xi}{2}\right)$$

is a local Fourier multiplier on $L_p((-\infty, \delta) \cup (\delta, \infty))$, it follows that

$$\|D_h B_h^{(1)} v\|_{L_p(\mathbb{R})} = \|B_h^{(1)} D_h v\|_{L_p(\mathbb{R})} \leq C \|D_h v\|_{L_p(\mathbb{R})},$$

and therefore,

$$\|T_h v\|_{L_p(\mathbb{R})} \leq C h \|D_h v\|_{L_p(\mathbb{R})}.$$

In other words, if v is a 'nonsmooth' function in the sense that the only requirement on the regularity of v is that $\sup_{h \in (0,1]} \|D_h v\|_{L_p(\mathbb{R})} \leq \text{Const.}$, then $T_h v$ approximates zero with $\mathcal{O}(h)$ error as $h \to 0$. For example, in the case of the function $x \in \mathbb{R} \mapsto v(x) := H(x) - H(x-1)$, with compact support $[0, 1]$ and jump discontinuities at $x = 0$ and $x = 1$, $\sup_{h \in (0,1]} \|D_h v\|_{L_1(\mathbb{R})} = 2$, and therefore $\|T_h v\|_{L_1(\mathbb{R})} = \mathcal{O}(h)$ as $h \to 0$.

Thus, to summarize our findings, T_h is a translation-invariant linear operator, uniformly bounded on $L_p(\mathbb{R})$; further, if v is smooth, in the sense that the Fourier transform of v has compact support in the interval $[-2\delta/h, 2\delta/h]$, then $T_h v - v$ is $\mathcal{O}(h^\mu)$ with $\mu = 2$ in the $L_p(\mathbb{R})$ norm, and finally if v is nonsmooth, in the sense that the Fourier transform of v is supported in the complement of the interval $[-\delta/h, \delta/h]$, then $T_h v$ is of size $\mathcal{O}(h^\nu)$, with $\nu = 1$, in the $L_p(\mathbb{R})$ norm. Motivated by this example, and following Kreiss, Thomée and Widlund [115] and Thomée and Wahlbin [178], we adopt the following definition of *mollifier*.

Definition 1.60 A family of linear translation-invariant operators $\{T_h : 0 < h \leq h_0\}$, each of which is uniformly bounded on $L_p(\mathbb{R}^n)$, is called a *family of mollifiers of order* (μ, ν) if, for some real number δ with $0 < \delta \leq \pi/2$, there exist translation-invariant linear operators $B_{h,\alpha}^{(0)}$, $|\alpha| = \mu$, and $B_{h,\alpha}^{(1)}$, $|\alpha| = \nu$, and positive constants $C_\alpha^{(0)}$, $|\alpha| = \mu$, and $C_\alpha^{(1)}$, $|\alpha| = \nu$, independent of h, such that:

(i) for every $v \in L_p(\mathbb{R}^n)$ with $Fv(\xi) = 0$ for $|h\xi| > 2\delta$,

$$T_h v = v + h^\mu \sum_{|\alpha|=\mu} \partial^\alpha B_{h,\alpha}^{(0)} v,$$

$$\|B_{h,\alpha}^{(0)} v\|_{L_p(\mathbb{R}^n)} \leq C_\alpha^{(0)} \|v\|_{L_p(\mathbb{R}^n)};$$

(ii) for every $v \in L_p(\mathbb{R}^n)$ with $Fv(\xi) = 0$ for $|h\xi| < \delta$,

$$T_h v = h^\nu \sum_{|\alpha|=\nu} D_h^\alpha B_{h,\alpha}^{(1)} v,$$

$$\|B_{h,\alpha}^{(1)} v\|_{L_p(\mathbb{R}^n)} \leq C_\alpha^{(1)} \|v\|_{L_p(\mathbb{R}^n)},$$

where, for a multi-index $\alpha = (\alpha_1, \ldots, \alpha_n)$, D_h^α denotes the forward divided difference of order $|\alpha|$ on the uniform mesh $h\mathbb{Z}^n$.

Given a particular h, (i) requires that $T_h v$ approximates v to order $\mathcal{O}(h^\mu)$ whenever the Fourier transform of v contains low 'frequencies (wave-numbers)' only (with respect to the given resolution h). Similarly, (ii) requires that $T_h v$ approximates 0 to order $\mathcal{O}(h^\nu)$ whenever the Fourier transform of v contains 'high frequencies (wave-numbers)' only. The integers μ and ν will be called the *order of approximation* and the *order of mollification* of T_h, respectively. In the next section we consider a general class of mollifiers of order (μ, ν); thereafter, we shall confine ourselves to mollifiers with $\mu = 2$.

1.9.3 An Admissible Class of Mollifiers

We shall generalize our simple example of the Steklov mollifier introduced in the previous section by considering mollifiers of the form

$$T_h v := \theta_h * v,\ \theta_h(x) = h^{-n}\theta\big(h^{-1}x\big), \tag{1.30}$$

with $\theta \in L_1(\mathbb{R}^n)$. The next theorem gives a precise characterization of the admissible class of θ in terms of Fourier multipliers (cf. Kreiss, Thomée and Widlund [115] and Thomée and Wahlbin [178]).

Theorem 1.61 *Let $p \in [1, \infty)$, and assume that θ belongs to $L_1(\mathbb{R}^n)$ and its Fourier transform $F\theta$ can be expressed as follows:*

$$F\theta(\xi) = 1 + \sum_{|\alpha|=\mu} \xi^\alpha b_\alpha^{(0)}(\xi), \quad b_\alpha^{(0)} \in M_p(B_{2\delta}), \tag{1.31}$$

$$F\theta(\xi) = \sum_{|\alpha|=\nu} \left(\sin\frac{\xi}{2}\right)^\alpha b_\alpha^{(1)}(\xi), \quad b_\alpha^{(1)} \in M_p\big(\mathbb{R}^n \setminus \bar{B}_\delta\big). \tag{1.32}$$

Then, $F\theta \in M_p$ and (1.30) *defines a family of mollifiers T_h on $L_p(\mathbb{R}^n)$ of order (μ, ν) in the sense of Definition* 1.60.

Proof Clearly, for each $h > 0$, the operator T_h is linear and translation-invariant on $L_p(\mathbb{R}^n)$. Further, thanks to Young's inequality (1.17),

$$\big\|F^{-1}(F\theta \cdot Fv)\big\|_{L_p(\mathbb{R}^n)} = \|\theta * v\|_{L_p(\Omega)} \le \|\theta\|_{L_1(\mathbb{R}^n)}\|v\|_{L_p(\mathbb{R}^n)} \tag{1.33}$$

for each $v \in \mathcal{S}(\mathbb{R}^n)$, and therefore $F\theta \in M_p$, $\|F\theta\|_{M_p} \le \|\theta\|_{L_1(\mathbb{R}^n)}$. Since the $L_1(\mathbb{R}^n)$ norms of θ and θ_h are equal, inequality (1.33) also implies that the family $\{T_h : h > 0\}$ is uniformly bounded on $L_p(\mathbb{R}^n)$. In order to verify that $T_h v = \theta_h * v$ is a mollifier of order (μ, ν), we define the linear operators $B_{h,\alpha}^{(0)}$ and $B_{h,\alpha}^{(1)}$ by

$$B_{h,\alpha}^{(0)} v := F^{-1}\big(\imath^{-|\alpha|} b_\alpha^{(0)}(h\xi) Fv\big),$$

$$B_{h,\alpha}^{(1)} v := F^{-1}\left((2\imath)^{-|\alpha|} \exp\left(-\frac{\imath h}{2}\xi \cdot \alpha\right) b_\alpha^{(1)}(h\xi) Fv\right).$$

Thus, by definition, $B_{h,\alpha}^{(0)}$, $|\alpha| = \mu$, and $B_{h,\alpha}^{(1)}$, $|\alpha| = \nu$, are translation-invariant linear operators on $L_p(\mathbb{R}^n)$. Moreover,

$$\|B_{h,\alpha}^{(0)} v\|_{L_p(\mathbb{R}^n)} \le \|b_\alpha^{(0)}\|_{M_p(B_{2\delta})} \|v\|_{L_p(\mathbb{R}^n)}$$

for all $v \in L_p(\mathbb{R}^n)$ such that $Fv(\xi) = 0$, $|h\xi| > 2\delta$, and

$$\|B_{h,\alpha}^{(1)} v\|_{L_p(\mathbb{R}^n)} \le \|b_\alpha^{(1)}\|_{M_p(\mathbb{R}^n \setminus \bar{B}_\delta)} \|v\|_{L_p(\mathbb{R}^n)}$$

for all $v \in L_p(\mathbb{R}^n)$ such that $Fv(\xi) = 0$, $|h\xi| < \delta$. Since $F\theta_h(\xi) = (F\theta)(h\xi)$ and since in Fourier transform space ∂^α and D_h^α correspond to multiplication by $(\imath\xi)^\alpha$ and by

$$h^{-|\alpha|} \prod_{j=1}^n (\mathrm{e}^{\imath h\xi_j} - 1)^{\alpha_j} = (2\imath)^{|\alpha|} \exp\left(\frac{\imath h}{2} \xi \cdot \alpha\right) \left(h^{-1} \sin \frac{h\xi}{2}\right)^\alpha,$$

respectively, the required representations of $T_h v$ in terms of $B_{h,\alpha}^{(0)}$ and $B_{h,\alpha}^{(1)}$ follow from (1.31) and (1.32). □

In order to illustrate the significance of this theorem, we construct a family of mollifiers of order (μ, ν) in one dimension ($n = 1$), which generalizes our simple example of the Steklov mollifier discussed in Sect. 1.9.2. For two integers, $\mu \ge 1$ and $\nu \ge 1$, let $p_{\mu,\nu}$ be a polynomial of degree $k \ge \nu$ such that

$$p_{\mu,\nu}(\sin \xi) = \xi^\nu + \xi^{\mu+\nu} \sum_{m=0}^\infty c_m \xi^m$$

where c_m, $m = 0, 1, \ldots$, are suitable constants and $c_0 \ne 0$, with the infinite series converging absolutely and uniformly for $|\xi| \le 2\delta$ and some δ in the interval $(0, 2\pi]$.

Let us consider the function

$$S_{\mu,\nu}(\xi) := \frac{p_{\mu,\nu}(\sin(\xi/2))}{(\xi/2)^\nu}, \quad \xi \in \mathbb{R}.$$

The extension of $S_{\mu,\nu}$ from the real line to the complex plane is an entire function $U(\zeta)$ satisfying the hypotheses of the Paley–Wiener theorem (Theorem 1.30) with $\rho = k/2$; therefore, there exists a distribution $\theta_{\mu,\nu}$, with compact support $\operatorname{supp} \theta_{\mu,\nu} \subset [-\frac{1}{2}k, \frac{1}{2}k]$, such that

$$F\theta_{\mu,\nu} = S_{\mu,\nu}.$$

Moreover because $S_{\mu,\nu} \in L_2(\mathbb{R})$, Plancherel's theorem implies that $\theta_{\mu,\nu} \in L_2(\mathbb{R})$; further, since $\theta_{\mu,\nu}$ has compact support, it follows by Hölder's inequality that $\theta_{\mu,\nu} \in L_1(\mathbb{R})$. Thus, the linear operator $T_h^{\mu,\nu}$ defined by

$$T_h^{\mu,\nu} u = \theta_h^{\mu,\nu} * u, \quad \theta_h^{\mu,\nu} = h^{-1} \theta_{\mu,\nu}(h^{-1}x), \tag{1.34}$$

belongs to the class of mollifiers considered. Now we show that $\{T_h^{\mu,\nu} : h > 0\}$ is a family of mollifiers of order (μ, ν).

Theorem 1.62 *The operator* $T_h^{\mu,\nu}$ *defined by* (1.34) *is a mollifier of order* (μ, ν) *on* $L_p(\mathbb{R}^n)$, $1 \le p < \infty$.

Proof We shall apply Theorem 1.61 in tandem with the Carlson–Beurling inequality stated in Theorem 1.58. Let us define the function $b^{(0)}$ by

$$b^{(0)}(\xi) := \frac{F\theta_{\mu,\nu}(\xi) - 1}{(\xi/2)^{\mu}}, \qquad |\xi| \le 2\delta.$$

Then,

$$F\theta_{\mu,\nu}(\xi) = 1 + \left(\frac{\xi}{2}\right)^{\mu} b^{(0)}(\xi), \quad |\xi| \le 2\delta.$$

In order to show that $b^{(0)} \in M_p(B_{2\delta})$, let us observe that since $F\theta_{\mu,\nu} = S_{\mu,\nu}$, we have that

$$b^{(0)}(\xi) = \sum_{m=0}^{\infty} \frac{c_m}{2^m} \xi^m, \quad |\xi| \le 2\delta,$$

$$\frac{\mathrm{d}}{\mathrm{d}\xi} b^{(0)}(\xi) = \sum_{m=0}^{\infty} \frac{(m+1)c_{m+1}}{2^{m+1}} \xi^m, \quad |\xi| \le 2\delta,$$

and therefore $b^{(0)} \in C^1(\overline{B}_{2\delta})$. Let φ belong to $C_0^\infty(\mathbb{R})$ and suppose that $\varphi(\xi) = 1$, $|\xi| \le 2\delta$; the existence of such a function is ensured by Lemma 1.15. By applying Theorem 1.58 with $n = 1$ and $m = 1$, we deduce that $\varphi b^{(0)} \in M_p$. Therefore $b^{(0)} \in M_p(B_{2\delta})$, and hence we have (1.31).

In order to verify (1.32), we write

$$F\theta_{\mu,\nu}(\xi) = S_{\mu,\nu}(\xi) = \left(\sin\frac{\xi}{2}\right)^{\nu} b^{(1)}(\xi), \quad |\xi| \ge \delta,$$

where

$$\begin{aligned} b^{(1)}(\xi) &= \left(\frac{\xi}{2}\right)^{-\nu} \frac{p_{\mu,\nu}(\sin(\xi/2))}{(\sin(\xi/2))^{\nu}} \\ &= \left(\frac{\xi}{2}\right)^{-\nu} q_{\mu,\nu}\big(\sin(\xi/2)\big), \quad |\xi| \ge \delta, \end{aligned}$$

and $q_{\mu,\nu}$ is a polynomial of degree $k - \nu$. Clearly $q_{\mu,\nu}(\sin(\xi/2)) \in M_p$. Let $\psi \in C^\infty(\mathbb{R})$ be such that $\psi(\xi) = 1$, $|\xi| \ge \delta$, and $\psi(\xi) = 0$ in the neighbourhood of $\xi = 0$. Then, according to Theorem 1.58, $(\xi/2)^{-\nu}\psi \in M_p$. Since the product of two elements in M_p is also contained in M_p (see Sect. 1.9.1), it follows that $\psi b^{(1)} \in M_p$; thus, $b^{(1)}$ belongs to $M_p(\mathbb{R} \setminus \bar{B}_\delta)$, and hence (1.32). □

1.9.4 Mollifiers of Tempered Distributions

In the remainder of this section we focus on a family of mollifiers with fixed order of approximation, $\mu = 2$, which will be used extensively throughout the rest of the book. They arise from the polynomials $p_{2,\nu}(\xi) = \xi^\nu$, $\nu \geq 1$. The Taylor series of $p_{2,\nu}(\sin \xi)$ in the neighbourhood of $\xi = 0$ has the form

$$p_{2,\nu}(\sin \xi) = \xi^\nu + \xi^{\nu+2} \sum_{m=0}^{\infty} c_m \xi^m,$$

where c_m are suitable constants, $c_0 \neq 0$, and the infinite series converges absolutely and uniformly for all $|\xi| < 2\delta$, where $\delta \in (0, 2\pi]$. This expansion indicates that the order of approximation is indeed $\mu = 2$. Clearly,

$$S_{2,\nu}(\xi) = \frac{p_{2,\nu}(\sin(\xi/2))}{(\xi/2)^\nu} = \left(\frac{\sin(\xi/2)}{\xi/2}\right)^\nu, \quad \nu = 1, 2, \ldots.$$

In fact, we shall extend the range of values for ν by allowing $\nu = 0$. In this case, $S_{2,0}(\xi) \equiv 1$.

Let θ_ν denote the inverse Fourier transform of $S_{2,\nu}$, $\nu \geq 0$. For $\nu = 0$, $\theta_0 = \delta$, the Dirac distribution concentrated at zero. When $\nu \geq 1$, a simple calculation reveals that θ_ν is a B-spline of degree $\nu - 1$ supported on the interval $[-\nu/2, \nu/2]$. For example,

$$\theta_1(x) = \begin{cases} 1 & \text{if } |x| \leq 1/2, \\ 0 & \text{otherwise}, \end{cases}$$

$$\theta_2(x) = \begin{cases} 1 - |x| & \text{if } |x| \leq 1, \\ 0 & \text{otherwise}, \end{cases}$$

$$\theta_3(x) = \begin{cases} (3 - 4x^2)/4 & \text{if } |x| \leq 1/2, \\ (3 - 2|x|)^2/8 & \text{if } 1/2 < |x| \leq 3/2, \\ 0 & \text{otherwise}. \end{cases}$$

Letting $\theta_h^\nu(x) := h^{-1}\theta_\nu(h^{-1}x)$ for $\nu \geq 1$ and $\theta_h^0 := \theta_0$, we consider the associated family of linear operators:

$$T_h^\nu v := \theta_h^\nu * v.$$

For $\nu = 0$, $T_h^0 : L_p(\mathbb{R}) \to L_p(\mathbb{R})$ is simply the identity operator, whereas for $\nu \geq 1$ it follows from Theorem 1.62 that T_h^ν is a family of mollifiers on $L_p(\mathbb{R})$, $1 \leq p < \infty$, of order $(2, \nu)$.

Our definition of T_h^ν is easily generalized to the case of n dimensions. Let $\nu = (\nu_1, \ldots, \nu_n)$, $\nu_i \geq 0$, $i = 1, \ldots, n$, and let $\theta_h^\nu \in \mathcal{E}'(\mathbb{R}^n)$ denote the tensor product of the univariate distributions $\theta_h^{\nu_i} \in \mathcal{E}'(\mathbb{R})$, $i = 1, \ldots, n$. We define

$$T_h^\nu v := \theta_h^\nu * v. \tag{1.35}$$

This mollifier exhibits different orders of smoothing in the various co-ordinate directions: in the ith direction the order of mollification is ν_i, with $\nu_i = 0$ signifying *no mollification* in the ith direction. Since θ_h^ν is a distribution with compact support, $T_h^\nu v$ is correctly defined for any distribution $v \in \mathcal{D}'(\mathbb{R}^n)$, and therefore T_h^ν can be seen as a mollifier on $\mathcal{D}'(\mathbb{R}^n)$. When $\nu = (0, \dots, 0)$, T_h^ν is the identity operator in $\mathcal{D}'(\mathbb{R}^n)$, which has no smoothing properties.

Next, we shall investigate, for a tempered distribution v, the amount of smoothing required to ensure that $T_h^\nu v$ is a continuous function. To this end, we shall first establish a set of preliminary results, the first of which characterizes the smoothness of θ_ν in one dimension ($n = 1$). For this purpose we require a further class of function spaces, called Bessel-potential spaces.

Given a real number s and a real number $p > 1$, we consider the *Bessel-potential space* of tempered distributions:

$$H_p^s(\mathbb{R}^n) = \left\{u \in \mathcal{S}'(\mathbb{R}^n) : F^{-1}\left(\left(1 + |\xi|^2\right)^{s/2} Fu\right) \in L_p(\mathbb{R}^n)\right\}.$$

$H_p^s(\mathbb{R}^n)$ is a Banach space with the norm

$$\|u\|_{H_p^s(\mathbb{R}^n)} = \left\|F^{-1}\left(\left(1 + |\xi|^2\right)^{s/2} Fu\right)\right\|_{L_p(\mathbb{R}^n)}.$$

Concerning the relationship between the Sobolev spaces $W_p^s(\mathbb{R}^n)$ and Bessel-potential spaces $H_p^s(\mathbb{R}^n)$, for $s \in \mathbb{R}$ and $p \in (1, \infty)$, we have that

$$W_p^s(\mathbb{R}^n) = H_p^s(\mathbb{R}^n) \quad \text{for } s = 0, \pm 1, \pm 2, \dots.$$

Before we state and prove the next lemma, which provides a characterization of the smoothness of θ_ν in terms of Bessel-potential spaces, we recall the following embeddings:

❶ assuming that $p \in (1, \infty)$ and $s > n/p$,

$$H_p^s(\mathbb{R}^n) \hookrightarrow BC(\mathbb{R}^n); \tag{1.36}$$

❷ assuming that $p \in (1, \infty)$ and $s > 0$,

$$W_p^s(\mathbb{R}^n) \hookrightarrow H_{p(\varepsilon)}^{s-\varepsilon}(\mathbb{R}^n), \quad p(\varepsilon) := \frac{p}{1 - \frac{\varepsilon p}{n}}, \tag{1.37}$$

for integer $s > 0$, $1 < p < \infty$ and $\varepsilon = 0$; for noninteger $s > 0$, $1 < p \le 2$ and $\varepsilon = 0$; and for noninteger $s > 0$, $2 < p < \infty$ and $0 < \varepsilon < n/p$.

The first of these continuous embeddings, (1.36), follows from Eq. (16) on p. 206 of Triebel [181]. When $s > 0$ is an integer, $1 < p < \infty$ and $\varepsilon = 0$, the second continuous embedding, (1.37), is an immediate consequence of the equality $W_p^s(\mathbb{R}^n) = H_p^s(\mathbb{R}^n)$. When $s > 0$ is noninteger, $1 < p \le 2$ and $\varepsilon = 0$, (1.37) is a trivial consequence of the continuous embedding $W_p^s(\mathbb{R}^n) = B_{p,p}^s(\mathbb{R}^n) \hookrightarrow H_p^s(\mathbb{R}^n)$ (cf. Theorem 5(C) on p. 155 of Stein [167]). Finally, when $s > 0$ is noninteger and

$2 < p < \infty$, suppose that $0 < \varepsilon < n/p$; then, (1.37) follows from the continuous embedding $W_p^s(\mathbb{R}^n) = B_{p,p}^s(\mathbb{R}^n) \hookrightarrow H_{p(\varepsilon)}^{s-\varepsilon}(\mathbb{R}^n)$ (cf. (17) on p. 206 of Triebel [181] with $t = s - \varepsilon$ and $q = p(\varepsilon)(> p)$).

Lemma 1.63 *Suppose that $\nu \in \mathbb{N}$, $\sigma \in \mathbb{R}$ and $p \in (1, \infty)$ are such that $\sigma + (1/p) < \nu$. Then, $\theta_\nu \in H_{p'}^\sigma(\mathbb{R})$, where $1/p + 1/p' = 1$.*

Proof Let $\nu = 0$. For $-\sigma > 1/p$, $H_p^{-\sigma}(\mathbb{R})$ is embedded in $BC(\mathbb{R})$ (cf. (1.36) with $n = 1$). Therefore $\theta_0 = \delta$ belongs to $(H_p^{-\sigma}(\mathbb{R}))' = H_{p'}^\sigma(\mathbb{R})$.

Let $\nu = 1$. It is easily seen that $\theta_1 \in W_q^\tau(\mathbb{R})$ for $\tau < 1/q$ and every $q \in (1, \infty)$. We recall that $W_q^\tau(\mathbb{R})$ is embedded in $H_{q(\varepsilon)}^{\tau-\varepsilon}(\mathbb{R})$ with $q(\varepsilon) = q/(1 - \varepsilon q)$ (cf. (1.37) with $n = 1$), where $\varepsilon = 0$ if $1 < q \le 2$ and $0 < \varepsilon < 1/q$ if $2 < q < \infty$. Thus, if $1 < p' \le 2$, by taking $q = p'$, $\tau = \sigma$ and $\varepsilon = 0$, we directly deduce that $\theta_1 \in H_q^\tau(\mathbb{R}) = H_{p'}^\sigma(\mathbb{R})$. If on the other hand $2 < p' < \infty$, then we choose $q = p'/(1 + \varepsilon p')$ and $\tau = \sigma + \varepsilon$, with $0 < \varepsilon < 1/2 - 1/p'$; hence $2 < q < p'$ and $\tau < 1/q$, and so, again, we have that $\theta_1 \in W_q^\tau(\mathbb{R}) \hookrightarrow H_{q(\varepsilon)}^{\tau-\varepsilon}(\mathbb{R}) = H_{p'}^\sigma(\mathbb{R})$.

For $\nu \ge 2$, the proof can be reduced to the case of $\nu = 1$ by noting that $\theta_\nu = \theta_1 * \cdots * \theta_1$ (ν-fold convolution). Indeed, let us choose α_j and p_j', $-\infty < \alpha_j < 1/p_j'$, $1 < p_j' < \infty$, $j = 1, \ldots, \nu$, such that

$$\sigma = \alpha_1 + \cdots + \alpha_\nu \quad \text{and} \quad \frac{1}{p'} = \frac{1}{p_1'} + \cdots + \frac{1}{p_\nu'} - (\nu - 1).$$

Since

$$F^{-1}\big((1 + |\xi|^2)^{\sigma/2} F\theta_\nu\big) = F^{-1}(1 + |\xi|^2)^{\alpha_1/2} F\theta_1 * \cdots * F^{-1}(1 + |\xi|^2)^{\alpha_\nu/2} F\theta_1$$

and $\theta_1 \in H_{p_j'}^{\alpha_j}(\mathbb{R})$ for $\alpha_j < 1/p_j'$, $j = 1, \ldots, \nu$, Young's inequality (1.17) yields

$$\|F^{-1}\big((1 + |\xi|^2)^{\sigma/2} F\theta_\nu\big)\|_{L_{p'}(\mathbb{R})} \le \prod_{j=1}^{\nu} \|F^{-1}(1 + |\xi|^2)^{\alpha_j/2} F\theta_1\|_{L_{p_j'}(\mathbb{R})},$$

and therefore $\theta_\nu \in H_{p'}^\sigma(\mathbb{R})$. □

In order to extend this result to n dimensions, we need the following lemma.

Lemma 1.64 *For $s \in \mathbb{R}$ and $\sigma_j \in \mathbb{R}$, $j = 1, \ldots, n$, consider the function a_n defined by*

$$a_n(\xi) := (1 + |\xi|^2)^{-s/2} \prod_{j=1}^{n} (1 + \xi_j^2)^{-\sigma_j/2}, \quad \xi = (\xi_1, \ldots, \xi_n) \in \mathbb{R}^n.$$

Let $\mathcal{P}_n$ denote the collection of all nonempty subsets of $I_n := \{1, \dots, n\}$. If

$$s + \sum_{j \in \mathcal{P}} \sigma_j \geq 0 \quad \forall \mathcal{P} \in \mathcal{P}_n,$$

then a_n is a Fourier multiplier on $L_p(\mathbb{R}^n)$, $1 < p < \infty$.

Proof The proof is based on Lizorkin's multiplier theorem (Theorem 1.59). Clearly a_n is a continuous function on $\mathbb{R}^n_*$; in order to show that it is bounded on $\mathbb{R}^n_*$ we note that every $\xi = (\xi_1, \dots, \xi_n) \in \mathbb{R}^n$ can be written as

$$\xi = r(\cos\gamma_1, \dots, \cos\gamma_n), \quad \text{where } r = |\xi|, \ \cos^2\gamma_1 + \cdots + \cos^2\gamma_n = 1,$$

and $\gamma_i \in [0, \pi]$. Let $Q := \{j \in I_n : \sigma_j < 0\}$ and, for each $\xi \in \mathbb{R}^n$, let $P(\xi) := \{j \in I_n \setminus Q : \cos^2\gamma_j \geq 1/n\}$. If $Q = \emptyset$, then $P(\xi) \neq \emptyset$ for all $\xi \in \mathbb{R}^n$. Therefore $P(\xi) \cup Q \neq \emptyset$ for all $\xi \in \mathbb{R}^n$. For $\sigma \in \mathbb{R}$ shall write $(\sigma)_+ := \max\{\sigma, 0\}$. Hence, for each $\xi \in \mathbb{R}^n$ we have that

$$a_n(\xi) \leq n^{\frac{1}{2}\sum_{j=1}^n (\sigma_j)_+} \left(1 + |\xi|^2\right)^{-\frac{1}{2}s_0(\xi)}, \quad \text{where } s_0(\xi) = s + \textstyle\sum_{j \in P(\xi) \cup Q} \sigma_j.$$

Since $s_0(\xi) \geq 0$ for each $\xi \in \mathbb{R}^n$, we deduce that a_n is bounded on $\mathbb{R}^n_*$. Thus we have shown that $\xi^\alpha \partial^\alpha a_n \in BC(\mathbb{R}^n_*)$ for $\alpha = (0, \dots, 0)$. We shall prove by induction that $\xi^\alpha \partial^\alpha a_n \in BC(\mathbb{R}^n_*)$ for all $\alpha \in \{0, 1\}^n$. Let $\alpha \in \{0, 1\}^n$ and assume that $\xi^\beta \partial^\beta a_n \in BC(\mathbb{R}^n_*)$ for all $\beta \leq \alpha$. Fix $j \in I_n$ such that $\alpha_j = 0$ (if there is no such j, the proof is complete). By the Leibniz formula

$$\xi_j \xi^\alpha \partial_j \partial^\alpha a_n(\xi) = \sum_{\beta \leq \alpha} \binom{\alpha}{\beta} \left(\xi^\beta \partial^\beta g_j(\xi)\right)\left(\xi^{\alpha-\beta} \partial^{\alpha-\beta} a_n(\xi)\right),$$

where

$$g_j(\xi) := -s\xi_j^2\left(1 + |\xi|^2\right)^{-1} - \sigma_j \xi_j^2\left(1 + \xi_j^2\right)^{-1}.$$

As $\xi^\beta \partial^\beta g_j$ is a bounded continuous function on $\mathbb{R}^n_*$, by recalling the inductive hypothesis we deduce that $\xi_j \xi^\alpha \partial_j \partial^\alpha a_n$ belongs to $BC(\mathbb{R}^n_*)$. Hence, by induction, $\xi^\alpha \partial^\alpha a_n \in BC(\mathbb{R}^n_*)$ for all $\alpha \in \{0, 1\}^n$. Thanks to Lizorkin's multiplier theorem a_n is therefore a Fourier multiplier on $L_p(\mathbb{R}^n)$, $1 < p < \infty$. □

By applying Lemma 1.64, we obtain the following extension of Lemma 1.63 to n dimensions.

Lemma 1.65 *Suppose that $s \in \mathbb{R}$, $p \in (1, \infty)$, $\nu = (\nu_1, \dots, \nu_n) \in \mathbb{N}^n$, and assume that there exist n real numbers σ_j, $j = 1, \dots, n$, such that*

$$\sigma_j + (1/p) < \nu_j \quad \forall j \in I_n, \tag{1.38}$$

$$s + \sum_{j \in \mathcal{P}} \sigma_j \geq 0 \quad \forall \mathcal{P} \in \mathcal{P}_n, \tag{1.39}$$

where $I_n := \{1, \dots, n\}$ *and* $\mathcal{P}_n$ *is the collection of all nonempty subsets of* I_n. *Then,* $\theta_\nu \in H_{p'}^{-s}(\mathbb{R}^n)$.

Proof Let us define the *partial Fourier transform* F_j by

$$(F_j u)(x_1, \dots, \xi_j, \dots, x_n) := \int_{-\infty}^{\infty} u(x) e^{-\imath x_j \xi_j}\, dx_j, \quad u \in \mathcal{S}(\mathbb{R}^n),$$

and extend the definition of F_j to $\mathcal{S}'(\mathbb{R}^n)$ in the usual way. Similarly, we define F_j^{-1}. With these definitions, we can write $F = F_1 \cdots F_n$. Since θ_ν is the tensor product of θ_{ν_j}, $j = 1, \dots, n$, it follows that

$$F\theta_\nu(\xi) = \prod_{j=1}^{n} F_j \theta_{\nu_j}(\xi_j).$$

By recalling the definition of a_n from Lemma 1.64, we have that

$$F^{-1}\big((1+|\xi|^2)^{-s/2} F\theta_\nu\big) = F^{-1} a_n * \left(\prod_{j=1}^{n} F_j^{-1} (1+|\xi_j|^2)^{\sigma_j/2} F_j \theta_{\nu_j} \right).$$

Under the hypotheses of the lemma a_n is a Fourier multiplier on $L_{p'}(\mathbb{R}^n)$ by Lemma 1.64, and $\theta_{\nu_j} \in H_{p'}^{\sigma_j}(\mathbb{R})$, $j = 1, \dots, n$, by Lemma 1.63. Hence,

$$\|\theta_\nu\|_{H_{p'}^{-s}(\mathbb{R}^n)} = \big\| F^{-1}\big((1+|\xi|^2)^{-s/2} F\theta_\nu\big) \big\|_{L_{p'}(\mathbb{R}^n)} \le \|a_n\|_{M_p} \prod_{j=1}^{n} \|\theta_{\nu_j}\|_{H_{p'}^{\sigma_j}(\mathbb{R})},$$

and therefore $\theta_\nu \in H_{p'}^{-s}(\mathbb{R}^n)$. □

Thus we have characterized the smoothness of the function θ_ν in terms of Bessel-potential spaces. Next we show that if $\theta_\nu \in H_{p'}^{-s}(\mathbb{R}^n)$ and $u \in H_p^s(\mathbb{R}^n)$ then $T_h^\nu u = \theta_h^\nu * u$ is a continuous function on $\mathbb{R}^n$. In fact, we shall establish a more general result from which the continuity of $T_h^\nu u$ easily follows.

Lemma 1.66 *Suppose that* $u \in \mathcal{S}'(\mathbb{R}^n)$, $v \in \mathcal{E}'(\mathbb{R}^n)$ *and let*

$$U := F^{-1}\big((1+|\xi|^2)^{s/2} Fu\big), V := F^{-1}\big((1+|\xi|^2)^{-s/2} Fv\big).$$

If $U \in L_p(\mathbb{R}^n)$, $V \in L_{p'}(\mathbb{R}^n)$, $1/p + 1/p' \ge 1$, $p \in (1, \infty)$, *then* $U * V$ *belongs to* $L_r(\mathbb{R}^n)$, $1/r = 1/p + 1/p' - 1$, *and* $u * v = U * V$. *In particular, if* $1/p + 1/p' = 1$, *then* $u * v = U * V$ *is a bounded uniformly continuous function on* $\mathbb{R}^n$.

Proof The fact that $U * V$ belongs to $L_r(\mathbb{R}^n)$ is the consequence of Young's inequality. Since $u * v = F^{-1}(Fu \cdot Fv)$ in $\mathcal{S}'(\mathbb{R}^n)$, it suffices to show that $Fu \cdot Fv = F(U * V)$ in $\mathcal{S}'(\mathbb{R}^n)$ to deduce that $u * v = U * V$ in $\mathcal{S}'(\mathbb{R}^n)$.

Because $Fu \in \mathcal{S}'(\mathbb{R}^n)$, $Fv \in C_M^\infty(\mathbb{R}^n)$ and $(1+|\xi|^2)^{s/2} \in C_M^\infty(\mathbb{R}^n)$, by recalling the definition of multiplication of a tempered distribution by a function from $C_M^\infty(\mathbb{R}^n)$, we obtain

$$\begin{aligned}
\langle Fu \cdot Fv, \varphi \rangle &= \langle Fu, Fv \cdot \varphi \rangle \\
&= \big\langle (1+|\xi|^2)^{s/2} Fu, (1+|\xi|^2)^{-s/2} Fv \cdot \varphi \big\rangle \\
&= \big\langle (1+|\xi|^2)^{s/2} Fu \cdot (1+|\xi|^2)^{-s/2} Fv, \varphi \big\rangle = \langle FU \cdot FV, \varphi \rangle,
\end{aligned}$$

for all $\varphi \in \mathcal{S}(\mathbb{R}^n)$. As $FV \in C_M^\infty(\mathbb{R}^n)$, it follows that $F(\varphi FV) \in \mathcal{S}(\mathbb{R}^n)$, and therefore

$$\langle FU \cdot FV, \varphi \rangle = \langle FU, \varphi \cdot FV \rangle = \big\langle U, F(\varphi FV) \big\rangle = \langle U, V_- * F\varphi \rangle, \quad \varphi \in \mathcal{S}\big(\mathbb{R}^n\big).$$

Since U is a regular distribution and $V_- * F\varphi = F(\varphi FV) \in \mathcal{S}(\mathbb{R}^n)$, by applying Fubini's theorem we obtain

$$\begin{aligned}
\langle U, V_- * F\varphi \rangle &= \int_{\mathbb{R}^n} U(x)(V_- * F\varphi)(x)\,\mathrm{d}x \\
&= \int_{\mathbb{R}^n} U(x) \int_{\mathbb{R}^n} F\varphi(y) V(y-x)\,\mathrm{d}y\,\mathrm{d}x \\
&= \int_{\mathbb{R}^n} \left(\int_{\mathbb{R}^n} U(x) V(y-x)\,\mathrm{d}x \right) F\varphi(y)\,\mathrm{d}y \\
&= \int_{\mathbb{R}^n} (U * V)(y) F\varphi(y)\,\mathrm{d}y \\
&= \langle U * V, F\varphi \rangle = \big\langle F(U * V), \varphi \big\rangle, \quad \varphi \in \mathcal{S}\big(\mathbb{R}^n\big).
\end{aligned}$$

Thus $Fu \cdot Fv = FU \cdot FV = F(U * V)$, and therefore $u * v = U * V$ in $\mathcal{S}'(\mathbb{R}^n)$.

Finally, if $1/p + 1/p' = 1$ then, by Young's inequality, $U * V \in L_\infty(\mathbb{R}^n)$ and, since $\tau_h(U * V) - (U * V) = (\tau_h U - U) * V$, also

$$\big\| \tau_h(U * V) - (U * V) \big\|_{L_\infty(\mathbb{R}^n)} \le \|\tau_h U - U\|_{L_p(\mathbb{R}^n)} \|V\|_{L_{p'}(\mathbb{R}^n)}.$$

As $\|\tau_h U - U\|_{L_p(\mathbb{R}^n)} \to 0$ when $|h| \to 0$, it follows that $U * V$ is a bounded uniformly continuous function on $\mathbb{R}^n$. □

Now we are ready to prove the main result of this section.

Theorem 1.67 *Suppose that $u \in H_p^s(\mathbb{R}^n)$, $s \in \mathbb{R}$, $p \in (1, \infty)$, $\nu \in \mathbb{N}^n$, and assume that there exist n real numbers σ_j, $j = 1, \ldots, n$, such that inequalities* (1.38) *and* (1.39) *hold. Then, $T_h^\nu u$ is a bounded uniformly continuous function on $\mathbb{R}^n$.*

Proof Thanks to Lemma 1.65, $\theta_\nu \in H^{-s}_{p'}(\mathbb{R}^n)$, $1/p + 1/p' = 1$, and therefore also $\theta^\nu_h \in H^{-s}_{p'}(\mathbb{R}^n)$. By recalling Lemma 1.66 with

$$U = F^{-1}\big(1+|\xi|^2\big)^{s/2} Fu \in L_p\big(\mathbb{R}^n\big) \quad \text{and} \quad V = F^{-1}\big(1+|\xi|^2\big)^{-s/2} F\theta^\nu_h \in L_{p'}\big(\mathbb{R}^n\big),$$

and noting that convolution is commutative, it follows that $T^\nu_h u = \theta^\nu_h * u = u * \theta^\nu_h = U * V$ is bounded and uniformly continuous on $\mathbb{R}^n$. □

Analogous results hold in Besov and Sobolev spaces.

Theorem 1.68 *Suppose that* $u \in B^s_{p,p}(\mathbb{R}^n)$, $s \in \mathbb{R}$, $p \in (1,\infty)$, $\nu \in \mathbb{N}^n$, *and assume that there exist* n *real numbers* σ_j, $j = 1,\dots,n$, *such that* (1.38) *and* (1.39) *hold. Then,* $T^\nu_h u$ *is a uniformly continuous function on* $\mathbb{R}^n$.

Proof Let us observe that if (1.38) and (1.39) hold, then (1.39) holds with strict inequality: indeed, if (1.38) and (1.39) are satisfied for some set of σ_j, $j = 1,\dots,n$, then there is a $\delta > 0$ such that $\sigma_j < \nu_j - \frac{1}{p} - \delta$, $j = 1,\dots,n$. Letting $\sigma'_j = \sigma_j + \delta$, $j = 1,\dots,n$, we deduce that (1.38) and (1.39) hold with σ_j replaced by σ'_j, and $\geq$ replaced by $>$ in (1.39). Now let

$$s_* := s + \min_{\mathcal{P}\in\mathcal{P}_n} \sum_{j\in\mathcal{P}} \sigma'_j;$$

clearly, $s_* > 0$.

Let us note the continuous embedding $B^s_{p,p}(\mathbb{R}^n) \hookrightarrow H^{s-\varepsilon}_{p(\varepsilon)}(\mathbb{R}^n)$ for $s \in \mathbb{R}$, $p \in (1,\infty)$, with $0 < \varepsilon < n/p$ and $p(\varepsilon) := \frac{p}{1-(\varepsilon p/n)}$ (cf. (17) on p. 206 of Triebel [181]). By choosing a sufficiently small ε in the interval $(0, \min(s_*, n/p))$, we can thus ensure that the strict versions of the inequalities (1.38) and (1.39) hold with s and p replaced by $s-\varepsilon$ and $p(\varepsilon)$, respectively, and the stated result follows from Theorem 1.67. □

Theorem 1.69 *Suppose that* $u \in W^s_p(\mathbb{R}^n)$, $s \in \mathbb{R}$, $p \in (1,\infty)$, $\nu \in \mathbb{N}^n$, *and assume that there exist* n *real numbers* σ_j, $j = 1,\dots,n$, *such that the inequalities* (1.38) *and* (1.39) *hold. Then,* $T^\nu_h u$ *is a bounded uniformly continuous function on* $\mathbb{R}^n$.

Proof By noting that, for $p \in (1,\infty)$,

$$W^s_p\big(\mathbb{R}^n\big) = \begin{cases} H^s_p(\mathbb{R}^n) & \text{if } s = 0, \pm 1, \pm 2, \dots, \\ B^s_{p,p}(\mathbb{R}^n) & \text{if } s \neq \text{integer}, \end{cases}$$

the result follows from Theorems 1.67 and 1.68. □

1.9.5 Multipliers and Mollifiers on Periodic Spaces

Many of the results discussed in previous sections have natural counterparts in spaces of periodic functions. Here we give a brief summary of some simple facts, which we require in the construction and error analysis of finite difference schemes.

1.9.5.1 Distributions on a Torus

Let $\mathbb{T}^n$ denote the *n-torus*; in the n-dimensional Euclidean space $\mathbb{R}^n$ it can be represented by the cube

$$\mathbb{T}^n := \big\{x = (x_1, \dots, x_n) \in \mathbb{R}^n : |x_j| \le \pi,\, j = 1, \dots, n\big\},$$

where 'opposite points' are identified. In other words, $x \in \mathbb{T}^n$ and $y \in \mathbb{T}^n$ are identified whenever $x - y = 2k\pi$ for some $k = (k_1, \dots, k_n) \in \mathbb{Z}^n$.

We denote by $C^\infty(\mathbb{T}^n)$ the set of all infinitely many times continuously differentiable complex-valued functions defined on $\mathbb{T}^n$. For any $\varphi \in C^\infty(\mathbb{T}^n)$, $\varphi(x) = \varphi(y)$ for all x and y in $\mathbb{T}^n$ such that $x - y = 2k\pi$ for some $k \in \mathbb{Z}^n$.

Definition 1.70 A sequence $\{\varphi_m\}_{m=1}^\infty \subset C^\infty(\mathbb{T}^n)$ is said to converge to φ in $C^\infty(\mathbb{T}^n)$ if $\partial^\alpha \varphi_m$ converges to $\partial^\alpha \varphi$, uniformly on $\mathbb{T}^n$, as $m \to \infty$, for every multi-index α. When equipped with convergence in this sense, the set $C^\infty(\mathbb{T}^n)$ will be denoted by $\mathcal{D}(\mathbb{T}^n)$.

Suppose that u is a linear functional on $\mathcal{D}(\mathbb{T}^n)$, whose value at $\varphi \in \mathcal{D}(\mathbb{T}^n)$ is denoted by $\langle u, \varphi\rangle$. We shall say that u is a continuous linear functional on $\mathcal{D}(\mathbb{T}^n)$ if $\langle u, \varphi_m\rangle \to \langle u, \varphi\rangle$ as $m \to \infty$, whenever $\varphi_m \to \varphi$ in $\mathcal{D}(\mathbb{T}^n)$.

Definition 1.71 A continuous linear functional on $\mathcal{D}(\mathbb{T}^n)$ is called a distribution on $\mathbb{T}^n$. The set of all distributions on $\mathbb{T}^n$ is denoted by $\mathcal{D}'(\mathbb{T}^n)$.

We define addition in $\mathcal{D}'(\mathbb{T}^n)$, multiplication by a complex number, differentiation, tensor product, translation, and multiplication by a function from $\mathcal{D}(\mathbb{T}^n)$ analogously as in the case of $\mathcal{D}'(\Omega)$, $\Omega \subset \mathbb{R}^n$.

Let $1 \le p \le \infty$; then $L_p(\mathbb{T}^n)$ is defined as the set of all Lebesgue-measurable functions v on $\mathbb{T}^n$ such that $v(x) = v(y)$ for a.e. x and y in $\mathbb{T}^n$ such that $x - y = 2k\pi$ for some $k \in \mathbb{Z}^n$, and

$$\|v\|_{L_p(\mathbb{T}^n)} := \left(\int_{\mathbb{T}^n} |v(x)|^p \,\mathrm{d}x\right)^{1/p} < \infty \quad \text{if } 1 \le p < \infty, \qquad \text{and}$$

$$\|v\|_{L_\infty(\mathbb{T}^n)} := \text{ess.sup}_{x\in\mathbb{T}^n} |v(x)| < \infty \quad \text{if } p = \infty.$$

Any function u in $L_p(\mathbb{T}^n)$, $1 \le p \le \infty$, can be identified with an element of $\mathcal{D}'(\mathbb{T}^n)$ via

$$\langle u, \varphi \rangle = \int_{\mathbb{T}^n} u(x)\varphi(x)\,\mathrm{d}x.$$

Thus, $\mathcal{D}(\mathbb{T}^n) \subset L_p(\mathbb{T}^n) \subset \mathcal{D}'(\mathbb{T}^n)$. In particular, every trigonometric polynomial

$$T(x) = \sum_{k\in\Lambda} a_k \mathrm{e}^{\imath k\cdot x}, \quad x \in \mathbb{T}^n,$$

where Λ is a finite subset of $\mathbb{Z}^n$ and the a_k are complex numbers, is an element of $\mathcal{D}'(\mathbb{T}^n)$.

For $u \in \mathcal{D}'(\mathbb{T}^n)$, we define the Fourier coefficients of u by

$$\hat{u}(k) := \langle u, \mathrm{e}^{-\imath kx} \rangle, \quad k \in \mathbb{Z}^n.$$

In particular if $u \in L_p(\mathbb{T}^n)$, $1 \le p \le \infty$, then by identifying it with an element of $\mathcal{D}'(\mathbb{T}^n)$, as indicated above, we have that

$$\hat{u}(k) = \int_{\mathbb{T}^n} u(x)\mathrm{e}^{-\imath x\cdot k}\,\mathrm{d}x, \quad k \in \mathbb{Z}^n.$$

We recall the following results concerning Fourier series (see, for example, Edwards [43], Chap. 12).

(i) Any function φ in $\mathcal{D}(\mathbb{T}^n)$ can be expanded into an infinite series

$$\varphi(x) = \frac{1}{(2\pi)^n} \sum_{k\in\mathbb{Z}^n} a_k \mathrm{e}^{\imath k\cdot x}, \tag{1.40}$$

which converges in $\mathcal{D}(\mathbb{T}^n)$, where $\{a_k\}_{k\in\mathbb{Z}^n}$ is a sequence of complex numbers such that

$$|a_k| \le c_m(1+|k|)^{-m}, \quad k \in \mathbb{Z}^n, \tag{1.41}$$

for all $m = 0, 1, 2, \dots$, and where c_m are appropriate positive constants. In fact, $a_k = \hat{\varphi}(k)$, $k \in \mathbb{Z}^n$. The converse of this statement is also true: if $\{a_k\}_{k\in\mathbb{Z}^n}$ satisfies the condition (1.41) then the series $(2\pi)^{-n}\sum_{k\in\mathbb{Z}^n} a_k \mathrm{e}^{\imath k\cdot x}$ converges in $\mathcal{D}(\mathbb{T}^n)$; denoting by $\varphi(x)$ the limiting function, we have that $\hat{\varphi}(k) = a_k$ for $k \in \mathbb{Z}^n$. The expansion (1.40) is called the Fourier series of φ. A sequence $\{a_k\}_{k\in\mathbb{Z}^n}$ satisfying (1.41) is said to be *rapidly decreasing*. For a rapidly decreasing sequence $a = \{a_k\}_{k\in\mathbb{Z}^n}$, the function defined by the right-hand side of (1.40) will be denoted by $a^\vee$. Thus,

$$a^\vee(x) := \frac{1}{(2\pi)^n} \sum_{k\in\mathbb{Z}^n} a_k \mathrm{e}^{\imath k\cdot x},$$

where the infinite series on the right-hand side converges in $\mathcal{D}(\mathbb{T}^n)$.

(ii) Any distribution $u \in \mathcal{D}'(\mathbb{T}^n)$ can be represented as an infinite series

$$u = \frac{1}{(2\pi)^n} \sum_{k\in\mathbb{Z}^n} a_k \mathrm{e}_k, \quad \text{with } \mathrm{e}_k(x) := \mathrm{e}^{\imath k\cdot x}, \tag{1.42}$$

which converges in $\mathcal{D}'(\mathbb{T}^n)$, where $\{a_k\}_{k\in\mathbb{Z}^n}$ is a sequence of complex numbers such that

$$|a_k| \le c_m \big(1+|k|\big)^m, \quad k \in \mathbb{Z}^n, \tag{1.43}$$

for some $m \in \mathbb{N}$, where c_m is an appropriate positive constant. In fact, $a_k = \hat{\varphi}(k)$, $k \in \mathbb{Z}^n$. The converse of this statement is also true: if $\{a_k\}_{k\in\mathbb{Z}^n}$ satisfies the condition (1.43) then the series $(2\pi)^{-n}\sum_{k\in\mathbb{Z}^n} a_k \mathrm{e}_k$ converges in $\mathcal{D}'(\mathbb{T}^n)$; denoting by $u \in \mathcal{D}'(\mathbb{T}^n)$ its limit, we have that $\hat{u}(k) = a_k$ for $k \in \mathbb{Z}^n$; see Theorem 1.72 and the subsequent discussion for a proof of this result. The expansion (1.42) is called the Fourier series of u. A sequence $\{a_k\}_{k\in\mathbb{Z}^n}$ satisfying (1.43) is said have *at most polynomial growth*, or that $\{a_k\}_{k\in\mathbb{Z}^n}$ is a *tempered sequence*. For a sequence $a = \{a_k\}_{k\in\mathbb{Z}^n}$ that has at most polynomial growth, the distribution defined by the right-hand side of (1.42) will be, again, denoted by $a^\vee$.

1.9.5.2 Periodic Distributions

Let us consider the space $\mathcal{S}'(\mathbb{R}^n)$ of tempered distributions on $\mathbb{R}^n$. An element $u \in \mathcal{S}'(\mathbb{R}^n)$ is called a *periodic distribution* if

$$u = \tau_{-2k\pi} u \tag{1.44}$$

holds for all k in $\mathbb{Z}^n$; in other words,

$$\langle u, \varphi\rangle = \langle u, \tau_{2k\pi}\varphi\rangle \tag{1.45}$$

for all φ in $\mathcal{S}(\mathbb{R}^n)$ and all $k \in \mathbb{Z}^n$, where $\langle\cdot,\cdot\rangle$ denotes the duality pairing between $\mathcal{S}'(\mathbb{R}^n)$ and $\mathcal{S}(\mathbb{R}^n)$. The set of all periodic distributions on $\mathbb{R}^n$ will be denoted by $\mathcal{S}'_\pi(\mathbb{R}^n)$.

Any complex-valued function u defined on $[-\pi,\pi)^n$ can be extended 2π periodically to the whole of $\mathbb{R}^n$; the extended function will be denoted by the same symbol. Thus, for example, $x \in \mathbb{R}^n \mapsto \mathrm{e}_k(x) := \mathrm{e}^{\imath k\cdot x}$ belongs to $\mathcal{S}'_\pi(\mathbb{R}^n)$ for all k in $\mathbb{Z}^n$. Furthermore, if $\{\alpha_k\}_{k\in\mathbb{Z}^n}$ is a sequence of complex numbers of at most polynomial growth (i.e. (1.43) holds), then $\sum_{k\in\mathbb{Z}^n} a_k \mathrm{e}_k$ belongs to $\mathcal{S}'_\pi(\mathbb{R}^n)$, where convergence takes place in $\mathcal{S}'(\mathbb{R}^n)$. Thus, by (1.43), any distribution on $\mathbb{T}^n$ can be thought of as a periodic distribution on $\mathbb{R}^n$. The next result (cf. Triebel [183], Sect. 9.1.2) shows that the converse is also true. The proof is simple, and for the sake of completeness it is included here; we recall that F denotes the Fourier transform on $\mathcal{S}'(\mathbb{R}^n)$.

Theorem 1.72 *$u \in S'(\mathbb{R}^n)$ is a periodic distribution (i.e. $u \in S'_\pi(\mathbb{R}^n)$) if, and only if, there exists a sequence $\{a_k\}_{k\in\mathbb{Z}^n}$ of at most polynomial growth such that*

$$u = \frac{1}{(2\pi)^n} \sum_{k\in\mathbb{Z}^n} a_k \mathrm{e}_k, \quad \text{where } \mathrm{e}_k(x) := \mathrm{e}^{\imath k\cdot x}, \tag{1.46}$$

and the infinite series on the right-hand side converges in $\mathcal{S}'(\mathbb{R}^n)$.

Proof Suppose that $u \in S'_\pi(\mathbb{R}^n)$; then, by applying (1.44) (or (1.45)) we deduce that

$$\langle Fu, \varphi\rangle = \langle u, F\varphi\rangle = \big\langle u, \tau_{-2k\pi}(F\varphi)\big\rangle = \big\langle Fu, \mathrm{e}^{\imath 2\pi k\cdot x}\varphi\big\rangle \quad \forall k \in \mathbb{Z}^n,\ \forall \varphi \in S\big(\mathbb{R}^n\big).$$

Therefore,

$$Fu = \mathrm{e}^{\imath 2\pi k\cdot x}\, Fu \quad \forall k \in \mathbb{Z}^n.$$

Now if x belongs to the interior of $\operatorname{supp} Fu$ it follows from this equality that $\mathrm{e}^{\imath 2\pi k\cdot x} = 1$ for all $k \in \mathbb{Z}^n$, and hence $x \in \mathbb{Z}^n$. This implies that $\operatorname{supp} Fu \subset \mathbb{Z}^n$, and there exists a sequence of complex numbers $\{b_k\}_{k\in\mathbb{Z}^n}$ such that

$$Fu = \sum_{k\in\mathbb{Z}^n} b_k \delta_k,$$

where δ_k is the Dirac distribution concentrated at $k \in \mathbb{Z}^n$. Suppose that $\varphi \in \mathcal{S}(\mathbb{R}^n)$ with $\varphi(0) = 1$ and $\operatorname{supp}\varphi \subset \{x \in \mathbb{R}^n : |x| \le 1\}$. Then,

$$\langle Fu, \tau_m\varphi\rangle = \sum_{k\in\mathbb{Z}^n} b_k \langle \delta_k, \tau_m\varphi\rangle = \sum_{k\in\mathbb{Z}^n} b_k \delta_{km} = b_m, \quad m \in \mathbb{Z}^n,$$

where δ_{km} is the Kronecker delta. As Fu is an element of $\mathcal{S}'(\mathbb{R}^n)$, it follows from this equality that the sequence $\{b_k\}_{k\in\mathbb{Z}^n}$ is of at most polynomial growth. Now,

$$u = F^{-1}(Fu) = \sum_{k\in\mathbb{Z}^n} b_k F^{-1}\delta_k = (2\pi)^{-n} \sum_{k\in\mathbb{Z}^n} b_k \mathrm{e}^{\imath k\cdot x},$$

which implies (1.46) with $a_k = b_k$, $k \in \mathbb{Z}^n$.

The proof of the converse statement is straightforward. □

We note that the mapping $u \mapsto \{a_k\}_{k\in\mathbb{Z}^n}$ in this theorem is an injection. This allows one to identify $u \in \mathcal{S}'_\pi(\mathbb{R}^n)$ with $u \in \mathcal{D}'(\mathbb{T}^n)$ through (1.42). Consequently, distributions on the torus $\mathbb{T}^n$ and periodic distributions on $\mathbb{R}^n$ can be identified.

1.9.5.3 Mollifiers on Function Spaces of Periodic Functions

We consider the following spaces of periodic distributions.

Periodic Sobolev Spaces Suppose that $1 < p < \infty$ and $m = 1, 2, \ldots$; then

$$W_p^m(\mathbb{T}^n) := \left\{ u \in L_p(\mathbb{T}^n) : \|u\|_{W_p^m(\mathbb{T}^n)} = \left(\sum_{|\alpha|\le m} \|\partial^\alpha u\|^p_{L_p(\mathbb{T}^n)} \right)^{1/p} < \infty \right\}.$$

Further, we put $W_p^0(\mathbb{T}^n) := L_p(\mathbb{T}^n)$, $1 < p < \infty$. For noninteger $s > 0$ and $1 < p < \infty$, $W_p^s(\mathbb{T}^n)$ is defined analogously (cf. the discussion following Theorem 1.35). For any positive real number s and $1 < p < \infty$, $W_p^{-s}(\mathbb{T}^n)$ denotes the dual space of $W_q^s(\mathbb{T}^n)$, equipped with the associated dual norm:

$$\|u\|_{W_p^{-s}(\mathbb{T}^n)} := \sup_{0\ne\varphi\in W_q^s(\mathbb{T}^n)} \frac{|\langle u, \varphi\rangle|}{\|\varphi\|_{W_q^s(\mathbb{T}^n)}}, \qquad 1/p + 1/q = 1.$$

Periodic Bessel-Potential Spaces Suppose that $1 < p < \infty$ and $-\infty < s < \infty$; then

$$H_p^s(\mathbb{T}^n) := \left\{ u \in \mathcal{D}'(\mathbb{T}^n) : \left\| \left((1 + |k|^2)^{s/2} \hat{u} \right)^\vee \right\|_{L_p(\mathbb{T}^n)} < \infty \right\}.$$

Next we consider mollifiers on these function spaces. Assuming that T_1 and T_2 are trigonometric polynomials, their convolution $T_1 * T_2$ is defined by

$$(T_1 * T_2)(x) = \int_{\mathbb{T}^n} T_1(x - y) T_2(y)\, \mathrm{d}y.$$

This definition can be extended to periodic distributions following the same route as in the nonperiodic case discussed earlier in this chapter.

Let $\theta_\nu(x)$ denote the B-spline of degree $\nu - 1$, $\nu \ge 1$ (see Sect. 1.9.4) supported on the interval $[-\nu/2, \nu/2]$ of the real line. When $\nu = 0$ we define θ_0 as the Dirac distribution concentrated at 0. For $h > 0$, let $\theta_h^\nu(x) := h^{-1}\theta(h^{-1}x)$ when $\nu \ge 1$ and $\theta_h^0 := \theta_0$. We shall suppose that h has been chosen small enough to ensure that the support of θ_h^ν, the closed interval $[-\nu h/2, \nu h/2]$, is contained in the open interval $(-\pi, \pi)$. Let us assume that θ_h^ν has been extended 2π periodically to the whole real line, and consider the family of mollifiers T_h^ν defined by

$$T_h^\nu u = \theta_h^\nu * u, \quad u \in \mathcal{D}'(\mathbb{T}^1).$$

The multidimensional counterpart of this mollifier is defined in the same way as in the nonperiodic case: assuming that $\nu = (\nu_1, \ldots, \nu_n)$, where ν_i are nonnegative integers, let θ_h^ν denote the tensor product of the univariate distributions $\theta_h^{\nu_i}$, $i = 1, \ldots, n$. We define

$$T_h^\nu u = \theta_h^\nu * u, \quad u \in \mathcal{D}'(\mathbb{T}^n).$$

The next two results are the 'periodic analogues' of Theorems 1.67 and 1.69.

Theorem 1.73 *Suppose that $u \in H_p^s(\mathbb{T}^n)$, with $s \in \mathbb{R}$, $p \in (1, \infty)$, let $\nu \in \mathbb{N}^n$, and assume that there exist n real numbers σ_j, $j = 1, \ldots, n$, such that the inequalities* (1.38) *and* (1.39) *hold. Then, $T_h^\nu u$ is a bounded uniformly continuous function on $\mathbb{T}^n$.*

Theorem 1.74 *Suppose that $u \in W_p^s(\mathbb{T}^n)$, with $s \in \mathbb{R}$, $p \in (1, \infty)$, let $\nu \in \mathbb{N}^n$, and assume that there exist n real numbers σ_j, $j = 1, \ldots, n$, such that the inequalities* (1.38) *and* (1.39) *hold. Then, $T_h^\nu u$ is a bounded uniformly continuous function on $\mathbb{T}^n$.*

To prove these results one proceeds in the same way as in the nonperiodic case, except that Lizorkin's multiplier theorem is replaced in the proof by a multiplier theorem, stated in Theorem 1.75 below, due to Marcinkiewicz.

1.9.5.4 Fourier Multipliers on Periodic Spaces

A sequence $\{a(k)\}_{k\in\mathbb{Z}^n}$ is called a *Fourier multiplier* on $L_p(\mathbb{T}^n)$, $1 \le p \le \infty$, if there exists a positive constant C_p such that

$$\|(a\hat{u})^\vee\|_{L_p(\mathbb{T}^n)} \le C_p \|u\|_{L_p(\mathbb{T}^n)} \quad \forall u \in L_p(\mathbb{T}^n).$$

The smallest constant C_p for which this inequality holds will be denoted by $\|a\|_{m_p(\mathbb{T}^n)}$. The set of all Fourier multipliers on $L_p(\mathbb{T}^n)$ will be labelled by $m_p(\mathbb{T}^n)$. It can be shown that $\|\cdot\|_{m_p(\mathbb{T}^n)}$ is a norm on $m_p(\mathbb{T}^n)$. The next theorem, due to Marcinkiewicz, provides a characterization of Fourier multipliers on $L_p(\mathbb{T}^n)$. Before stating it, let us introduce some notation.

Assuming that $\{a_k\}_{k\in\mathbb{Z}^n}$ is a sequence of complex numbers and

$$\mathrm{e}_j = (\delta_{1j}, \ldots, \delta_{nj}), \quad j = 1, \ldots, n,$$

where δ_{ij} is the Kronecker delta, we define the *partial undivided difference operator* Δ_j in the jth co-ordinate direction by

$$\Delta_j a(k) = a(k + \mathrm{e}_j) - a(k), \quad j = 1, \ldots, n, \ k \in \mathbb{Z}^n.$$

We also require the notion of *total variation* of a sequence $a = \{a_k\}_{k\in\mathbb{Z}^n}$, defined by

$$\mathrm{Var}(a) := \sup_{k\in\mathbb{Z}^n} \max_{0\neq\alpha\in\{0,1\}^n} \sum\nolimits_{\nu}^{\alpha} |\Delta^\alpha a_\nu|;$$

here $\Delta^\alpha := \Delta_1^{\alpha_1} \ldots \Delta_n^{\alpha_n}$, and, for $\alpha \in \{0, 1\}^n$, we have used the multi-index notation

$$\sum\nolimits_{\nu}^{\alpha} := \sum\nolimits_{\nu_1}^{\alpha_1} \cdots \sum\nolimits_{\nu_n}^{\alpha_n}$$

with

$$\sum_{\nu_j}^{\alpha_j} := \begin{cases} \max_{\nu_j=\pm 2^{|k_j|-1},\ldots,\pm 2^{|k_j|}-1} & \text{if } \alpha_j = 0, \\ \sum_{\nu_j=\pm 2^{|k_j|-1},\ldots,\pm 2^{|k_j|}-1} & \text{if } \alpha_j = 1. \end{cases}$$

When $k_j = 0$ for some j, then it is assumed that the corresponding maximization (when $\alpha_j = 0$) or sum (when $\alpha_j = 1$) is only through $\nu_j = 0$; the $+$ or the $-$ sign is chosen in $\pm$ depending on whether $k_j > 0$ or $k_j < 0$.

Theorem 1.75 (Marcinkiewicz Multiplier Theorem) *Let $a = \{a_k\}_{k\in\mathbb{Z}^n}$ be a sequence of complex numbers such that the following conditions hold:*

$$\sup_{k\in\mathbb{Z}^n} |a_k| \le M_0, \quad \mathrm{Var}(a) \le M_0.$$

Then, a is a Fourier multiplier on $L_p(\mathbb{T}^n)$, $1 < p < \infty$; that is, there exists a constant C_p, depending only on p, such that $\|a\|_{m_p(\mathbb{T}^n)} \le C_p M_0$.

For a proof of this result, we refer to Zygmund [206], Vol. II, p. 232, in the case of $n = 1$, and Nikol'skiĭ [144], p. 57, in the general case. Quite apart from its relevance in the analysis of smoothing operators, the Marcinkiewicz multiplier theorem will be one of our main tools in the error analysis of finite difference schemes in discrete L_p norms.

When $n = 1$ it is usually simple to show by direct calculations that a sequence $a = \{a_k\}_{k\in\mathbb{Z}^n}$ has bounded total variation; when $n > 1$, however, because of the complicated structure of $\mathrm{Var}(a)$, this can be a tedious exercise. It is therefore useful to seek a simpler criterion, under slightly stronger hypotheses on a. The next two theorems indicate how this can be achieved.

We define the concept of *Lebesgue point* for a locally integrable function. Suppose that $x_0 \in \mathbb{R}^n$ and f is a function defined and locally integrable in an open neighbourhood of x_0. We say that x_0 is a Lebesgue point of f provided that

$$\lim_{\varepsilon\to 0} \frac{1}{|B(x_0, \varepsilon)|} \int_{B(x_0,\varepsilon)} \big|f(y) - f(x_0)\big| \, dy = 0.$$

Clearly, each point of continuity of a function f is a Lebesgue point of f. The following example shows that the converse statement is false.

Example 1.32 Consider the function A defined on $\mathbb{R}^2$ by

$$A(x, y) = \begin{cases} \frac{x^2}{x^2+y^2} & \text{when } (x, y) \neq (0, 0), \\ \frac{1}{2} & \text{when } (x, y) = (0, 0). \end{cases}$$

Then, A is continuous on $\mathbb{R}^2 \setminus \{(0, 0)\}$, but not at $(0, 0)$; nevertheless, each point $(x, y) \in \mathbb{R}^2$ is a Lebesgue point of A.

In fact, according to Lebesgue's differentiation theorem (cf. (1.11)), for any function f that is defined and locally integrable on an open set Ω in $\mathbb{R}^n$, almost every $x_0 \in \Omega$ is a Lebesgue point of f.

For the proof of the next result we refer to Theorem 3.4.2 and Remark 3.4.4 in the monograph of Schmeisser and Triebel [162].

Theorem 1.76 *Let $1 < p < \infty$ and let $A \in L_\infty(\mathbb{R}^n)$ be an element of $M_p(\mathbb{R}^n)$. Suppose additionally that each point $k \in \mathbb{Z}^n$ is a Lebesgue point of A. Then, the sequence $a = \{a_k\}_{k\in\mathbb{Z}^n}$, defined by $a_k = A(k)$ is a Fourier multiplier in $L_p(\mathbb{T}^n)$, and $\|a\|_{m_p(\mathbb{T}^n)} \le \|A\|_{M_p(\mathbb{R}^n)}$.*

As an application of this powerful result, we state the following analogue of Lizorkin's theorem for $L_p(\mathbb{T}^n)$.

Theorem 1.77 *Let $1 < p < \infty$. Suppose that $A \in L_\infty(\mathbb{R}^n)$ is such that $\xi^\alpha \partial^\alpha A(\xi)$ is a bounded continuous function of ξ on the set*

$$\mathbb{R}^n_* := \{\xi \in \mathbb{R}^n : \xi_i \neq 0, i = 1, \dots, n\}$$

for every multi-index $\alpha \in \{0, 1\}^n$, and each $k \in \mathbb{Z}^n$ is a Lebesgue point of A. Suppose further that $a = \{a_k\}_{k \in \mathbb{Z}^n}$ is a sequence of complex numbers defined by $a_k := A(k)$, $k \in \mathbb{Z}^n$. Then, a is a Fourier multiplier on $L_p(\mathbb{T}^n)$, and $\|a\|_{m_p(\mathbb{T}^n)} \leq \|A\|_{M_p(\mathbb{R}^n)}$.

Proof The stated result is an immediate consequence of Theorems 1.59 and 1.76. □

Example 1.33 Consider the sequence of real numbers $a = \{a_k\}_{k \in \mathbb{Z}^2}$, with $k = (k_1, k_2)$, defined by

$$a_k = \begin{cases} \frac{k_1^2}{k_1^2 + k_2^2} & \text{when } k \neq (0, 0), \\ \frac{1}{2} & \text{when } k = (0, 0). \end{cases}$$

Then, a is a Fourier multiplier on $L_p(\mathbb{T}^2)$, $1 < p < \infty$. This follows from Theorem 1.77 by noting the following: the function A defined in Example 1.32 is a Fourier multiplier on $L_p(\mathbb{R}^2)$ thanks to Theorem 1.59; each $\xi \in \mathbb{R}^2$ is a Lebesgue point of A, whereby each $k \in \mathbb{Z}^2$ is a Lebesgue point of A; and $a_k = A(k)$, $k \in \mathbb{Z}^n$.

With all the prerequisites now in place, we are ready to embark on the numerical approximation of partial differential equations.

The remaining chapters are devoted to the construction and analysis of finite difference methods for the approximate solution of elliptic, parabolic and hyperbolic equations. As we have already emphasized in the Introduction, our key concern are instances when the data and the solution to the problem under the consideration are not smooth enough to allow the use of conventional tools from the theory of finite difference methods. In particular, since neither the coefficients in the differential equations under consideration nor the initial or boundary data will be required to be continuous functions, sampling the data at the points of a finite difference grid, as is usual in the classical theory of finite difference methods, in generally infeasible. We shall therefore mollify the data in the process of constructing various finite difference schemes, so as to ensure that the mollified data are continuous and can be, thereby, meaningfully sampled at the points of the finite difference grid.

Chapter 2
Elliptic Boundary-Value Problems

In the first part of this chapter we focus on the question of well-posedness of boundary-value problems for linear partial differential equations of elliptic type. The second part is devoted to the construction and the error analysis of finite difference schemes for these problems. It will be assumed throughout that the coefficients in the equation, the boundary data and the resulting solution are real-valued functions.

2.1 Existence and Uniqueness of Solutions

Suppose that Ω is a bounded open set in $\mathbb{R}^n$, k is a positive integer and $a_{\alpha\beta}$, $0 \leq |\alpha|, |\beta| \leq k$, with $\alpha, \beta \in \mathbb{N}^n$, are real-valued-functions defined on Ω. We consider the linear partial differential operator $P(x, \partial)$ of order $2k$ defined by

$$P(x,\partial)u := \sum_{0\leq|\alpha|,|\beta|\leq k} (-1)^{|\alpha|}\partial^\alpha\big(a_{\alpha\beta}(x)\partial^\beta u\big), \quad x \in \Omega. \tag{2.1}$$

The *principal part* $P_0(x, \partial)$ of the differential operator $P(x, \partial)$ is defined by

$$P_0(x,\partial)u := \sum_{|\alpha|,|\beta|=k} (-1)^{|\alpha|}\partial^\alpha\big(a_{\alpha\beta}(x)\partial^\beta u\big), \quad x \in \Omega.$$

$P(x, \partial)$ is said to be an *elliptic operator* on Ω if, and only if,

$$\sum_{|\alpha|,|\beta|=k} a_{\alpha\beta}(x)\xi^\alpha\xi^\beta > 0 \quad \forall x \in \Omega,\ \forall \xi \in \mathbb{R}^n \setminus \{0\}.$$

$P(x, \partial)$ is called *uniformly elliptic* on Ω if, and only if, there exists a positive real number $\tilde{c}$ such that

$$\sum_{|\alpha|,|\beta|=k} a_{\alpha\beta}(x)\xi^\alpha\xi^\beta \geq \tilde{c}|\xi|^{2k} \quad \forall x \in \Omega,\ \forall \xi \in \mathbb{R}^n. \tag{2.2}$$

B.S. Jovanović, E. Süli, *Analysis of Finite Difference Schemes*,
Springer Series in Computational Mathematics 46,
DOI 10.1007/978-1-4471-5460-0_2,

Example 2.1 Consider the second-order partial differential operator, corresponding to $k = 1$ above, defined by

$$P(x, \partial)u := -\sum_{i,j=1}^{n} \frac{\partial}{\partial x_j}\left(a_{ij}(x)\frac{\partial u}{\partial x_i}\right) + \sum_{i=1}^{n}\left[-\frac{\partial}{\partial x_i}\big(a_i(x)u\big) + b_i(x)\frac{\partial u}{\partial x_i}\right] + c(x)u, \tag{2.3}$$

with a_{ij}, $i, j = 1, \ldots, n$; a_i, b_i, $i = 1, \ldots, n$; and c being real-valued functions defined on an open set $\Omega \subset \mathbb{R}^n$, and such that

$$\sum_{i,j=1}^{n} a_{ij}(x)\xi_i\xi_j \geq \tilde{c}\sum_{i=1}^{n}\xi_i^2 \quad \forall x \in \Omega,\ \forall \xi = (\xi_1, \ldots, \xi_n) \in \mathbb{R}^n, \tag{2.4}$$

for a positive real number $\tilde{c}$, independent of x and ξ; then $P(x, \partial)$ is a second-order uniformly elliptic operator on Ω.

Example 2.2 Consider the partial differential operator $P(x, \partial)$, defined by

$$P(x, \partial)u := \partial_1^2 M_1(u) + 2\partial_1\partial_2 M_3(u) + \partial_2^2 M_2(u),$$

where $\partial_i := \partial/\partial x_i$ and $\partial_i^2 := \partial^2/\partial x_i^2$ for $i = 1, 2$,

$$M_1(u) := a_1(x)\partial_1^2 u + a_0(x)\partial_2^2 u,$$

$$M_2(u) := a_0(x)\partial_1^2 u + a_2(x)\partial_2^2 u,$$

$$M_3(u) := a_3(x)\partial_1\partial_2 u,$$

and a_i, $i = 0, 1, 2, 3$, are four real-valued functions defined on a bounded open set $\Omega \subset \mathbb{R}^2$ such that there exist positive real numbers c_1 and c_2 for which

$$a_i(x) \geq c_1, \quad i = 1, 2, 3, \qquad a_1(x)a_2(x) - a_0^2(x) \geq c_2 \quad \forall x \in \Omega.$$

Under these hypotheses $P(x, \partial)$ is a fourth-order uniformly elliptic operator on Ω. The same is true if the above inequalities satisfied by the coefficients a_i are replaced by

$$a_i(x) \geq c_1, \quad i = 1, 2, \qquad a_1(x)a_2(x) - \big(a_0(x) + a_3(x)\big)^2 \geq c_2 \quad \forall x \in \Omega.$$

A partial differential equation on Ω is usually supplemented with *boundary conditions* on $\partial\Omega$. The differential equation in tandem with the boundary conditions imposed forms a *boundary-value problem*.

Example 2.3 For the second-order partial differential equation considered in Example 2.1 the following boundary conditions are the most common, with g denoting a given real-valued function defined on the boundary $\partial\Omega$ in each case:

❶ Dirichlet boundary condition: $u = g$ on $\partial\Omega$;
❷ Oblique derivative boundary condition:

$$\sum_{i,j=1}^{n} a_{ij}(x)\frac{\partial u}{\partial x_i}\nu_j + \sum_{i=1}^{n} a_i(x)u\nu_i + \sigma(x)u = g \quad \text{on } \partial\Omega,$$

where ν_j is the jth component of the unit outward normal vector ν to $\partial\Omega$ and σ is a given real-valued function defined on $\partial\Omega$ such that

$$\sigma + \frac{1}{2}\sum_{i=1}^{n}(a_i + b_i)\nu_i \geq 0 \quad \text{on } \partial\Omega.$$

The differential operator

$$u \mapsto \sum_{i,j=1}^{n} a_{ij}(x)\frac{\partial u}{\partial x_i}\nu_j + \sum_{i=1}^{n} a_i(x)u\nu_i, \quad x \in \partial\Omega,$$

is called the *co-normal derivative* corresponding to the partial differential operator from Example 2.1. A particularly important special case arises when $a_{ij} = \delta_{ij}$, $i, j = 1, \dots, n$, and $a_i = 0$, $i = 1, \dots, n$. Then, the oblique derivative boundary condition becomes:

$$\partial_\nu u + \sigma u = g \quad \text{on } \partial\Omega,$$

and is referred to as *Robin boundary condition*. Here,

$$\partial_\nu = \frac{\partial}{\partial \nu} := \sum_{i=1}^{n} \nu_i \frac{\partial}{\partial x_i}$$

denotes the (outward) normal derivative on $\partial\Omega$; it is assumed that

$$\sigma + \frac{1}{2}\sum_{i=1}^{n} b_i\nu_i \geq 0 \quad \text{on } \partial\Omega.$$

In particular, when $\sigma = 0$ on $\partial\Omega$, the resulting boundary condition

$$\partial_\nu u = g \quad \text{on } \partial\Omega$$

is called a *Neumann boundary condition*.

In many problems that arise in applications boundary conditions of different kind are enforced on different parts of the boundary; for example, $\partial\Omega$ may be the union of two disjoint subsets $\partial\Omega_1$ and $\partial\Omega_2$, with Dirichlet boundary condition imposed on $\partial\Omega_1$ and an oblique derivative boundary condition imposed on $\partial\Omega_2$. In most of what follows we shall, for simplicity, confine ourselves to the study of elliptic boundary-value problems subject to homogeneous Dirichlet boundary conditions (corresponding, in the case of a second-order elliptic equation, to $g \equiv 0$ in Example 2.3, part ❶).

Returning to the general elliptic equation of order $2k$, we formulate the classical homogeneous Dirichlet boundary-value problem.

Definition 2.1 Let $\Omega \subset \mathbb{R}^n$ be a bounded open set and suppose that $f \in C(\Omega)$ and $a_{\alpha\beta} \in C^{|\alpha|}(\Omega)$, $|\alpha|, |\beta| \le k$. A function

$$u \in C^{2k}(\Omega) \cap C^{k-1}(\overline{\Omega})$$

is a *classical solution* of the homogeneous Dirichlet problem if

$$P(x, \partial)u := \sum_{0\le|\alpha|,|\beta|\le k} (-1)^{|\alpha|}\partial^\alpha \big(a_{\alpha\beta}(x)\partial^\beta u\big) = f(x)$$

for every x in Ω, and

$$\partial_\nu^m u = 0 \quad \text{on } \partial\Omega, \text{ for } 0 \le m \le k-1.$$

It is assumed here that the differential operator $P(x, \partial)$, with $x \in \Omega$, is elliptic or uniformly elliptic on Ω. Frequently, the smoothness requirements on the data stated in this definition are not satisfied. As is demonstrated by the next example, in such instances the corresponding homogeneous Dirichlet boundary-value problem has no classical solution.

Example 2.4 Let $\Omega = (-1, 1)^n \subset \mathbb{R}^n$ and consider *Poisson's equation*

$$-\Delta u := -\sum_{i=1}^{n} \frac{\partial^2 u}{\partial x_i^2} = f \quad \text{in } \Omega,$$

subject to the homogeneous Dirichlet boundary condition

$$u = 0 \quad \text{on } \partial\Omega.$$

Suppose further that $f(x) = \operatorname{sgn}(\frac{1}{2} - |x|)$, $x \in \Omega$.

Clearly, this problem has no classical solution, $u \in C^2(\Omega) \cap C(\overline{\Omega})$, for otherwise Δu would be a continuous function on Ω, which is impossible as $\operatorname{sgn}(\frac{1}{2} - |x|)$ is not continuous on Ω.

In order to overcome the limitations of Definition 2.1 highlighted by this example, we generalize the notion of classical solution by weakening the differentiability requirements on both the data and the corresponding solution.

Definition 2.2 Let $\Omega \subset \mathbb{R}^n$ be a bounded open set and suppose that $f \in L_2(\Omega)$ and $a_{\alpha\beta} \in M\big(W_2^{2k-|\beta|}(\Omega) \to W_2^{|\alpha|}(\Omega)\big)$, $|\alpha|, |\beta| \leq k$. A function

$$u \in W_2^{2k}(\Omega) \cap \mathring{W}_2^k(\Omega)$$

is a *strong solution* of the homogeneous Dirichlet problem if

$$P(x, \partial)u := \sum_{0 \leq |\alpha|, |\beta| \leq k} (-1)^{|\alpha|} \partial^\alpha \big(a_{\alpha\beta}(x)\partial^\beta u\big) = f(x)$$

for almost every x in Ω.

While for classical solutions both the partial differential equation and the boundary condition are assumed to hold in the pointwise sense, for strong solutions the partial differential equation is to be understood in terms of equivalence classes consisting of functions that are equal almost everywhere on Ω; also, instead of being imposed explicitly, the boundary condition has been incorporated into the function space $W_2^{2k}(\Omega) \cap \mathring{W}_2^k(\Omega)$ in which a solution is sought. Unfortunately, it is not easy to show that the homogeneous Dirichlet problem for the partial differential equation (2.1) possesses a strong solution; in fact, as is illustrated by Example 2.5 below a strong solution will not exist unless $\partial\Omega$ and the data are sufficiently smooth. Thus we shall further relax the differentiability requirements on u and weaken the concept of solution by converting the boundary-value problem into a variational problem. The first step in this process is to create a bilinear functional associated with the differential operator $P(x, \partial)$ using integration by parts. Suppose that $u \in W_2^{2k}(\Omega)$, $f \in L_2(\Omega)$, and $v \in C_0^\infty(\Omega)$; then

$$\begin{aligned}\int_\Omega v(x) f(x)\,\mathrm{d}x &= \int_\Omega v P(x, \partial) u\,\mathrm{d}x \\ &= \sum_{0 \leq |\alpha|, |\beta| \leq k} (-1)^{|\alpha|} \int_\Omega v \partial^\alpha \big(a_{\alpha\beta}(x)\partial^\beta u\big)\,\mathrm{d}x \\ &= \sum_{0 \leq |\alpha|, |\beta| \leq k} \int_\Omega a_{\alpha\beta}(x)\partial^\beta u \partial^\alpha v\,\mathrm{d}x.\end{aligned}$$

In the transition to the last expression, by partial integration, we made use of the fact that $\operatorname{supp} v \subset\subset \Omega$. Motivated by this identity we introduce the following notation:

$$a(u, v) := \sum_{0 \leq |\alpha|, |\beta| \leq k} \int_\Omega a_{\alpha\beta}(x)\partial^\beta u \partial^\alpha v\,\mathrm{d}x,$$

$$(f, v) := \int_\Omega f(x) v(x)\,\mathrm{d}x.$$

Clearly $a(\cdot,\cdot)$ is correctly defined for u that is merely in $\mathring{W}_2^k(\Omega)$ and for v in the same space; in fact, $a(\cdot,\cdot)$ is a bilinear functional on the product space $\mathring{W}_2^k(\Omega) \times \mathring{W}_2^k(\Omega)$; similarly, $v \mapsto (f, v)$ is a linear functional on $\mathring{W}_2^k(\Omega)$.

These considerations motivate the following definition.

Definition 2.3 Let $\Omega \subset \mathbb{R}^n$ be a bounded open set and suppose that $f \in W_2^{-k}(\Omega)$ and $a_{\alpha\beta} \in L_\infty(\Omega)$, $|\alpha|, |\beta| \le k$. A function

$$u \in \mathring{W}_2^k(\Omega)$$

is a *weak solution* of the homogeneous Dirichlet problem if

$$a(u, v) = \langle f, v\rangle$$

for every $v \in \mathring{W}_2^k(\Omega)$, where now $\langle \cdot,\cdot\rangle$ denotes the duality pairing between $W_2^{-k}(\Omega)$ and $\mathring{W}_2^k(\Omega)$, i.e. $\langle f, v\rangle$ signifies the value of the linear functional $f \in W_2^{-k}(\Omega) = [\mathring{W}_2^k(\Omega)]'$ at $v \in \mathring{W}_2^k(\Omega)$.

Remark 2.1 By applying the Sobolev embedding theorem, it is easily seen that the bilinear functional $a(\cdot,\cdot)$ is well defined under even weaker regularity hypotheses on the coefficients $a_{\alpha\beta}$. Indeed, it suffices to assume in Definition 2.3 that

$$a_{\alpha\beta} \in M\big(W_2^{k-|\alpha|} \to L_{p_\beta}(\Omega)\big), \quad |\alpha|, |\beta| \le k,$$

where $p_\beta = 2$ when $|\beta| = k$, $p_\beta = 2n/(n + 2(k - |\beta|))$ when $0 < k - |\beta| < n/2$; $p_\beta > 1$ (but arbitrarily close to 1) when $k - |\beta| = n/2$; and $p_\beta = 1$ when $k - |\beta| > n/2$.

Next we show that the homogeneous Dirichlet boundary-value problem has a unique weak solution. The proof is based on a simple application of the Lax–Milgram theorem (Theorem 1.13) and the following result.

Theorem 2.4 (Gårding's Inequality) *Suppose that $\Omega \subset \mathbb{R}^n$ is a Lipschitz domain. Let $P(x, \partial)$ be a linear partial differential operator of order 2k of the form* (2.1) *such that, for some $\tilde{c} > 0$, the uniform ellipticity condition* (2.2) *holds. Suppose also that*

$$a_{\alpha\beta} \in C(\Omega) \quad \textit{for } |\alpha| = |\beta| = k$$

and

$$a_{\alpha\beta} \in L_\infty(\Omega) \quad \textit{for } |\alpha|, |\beta| \le k.$$

Then, there exist constants $c_0 > 0$ and $\lambda_0 \ge 0$ such that

$$a(v, v) + \lambda_0 \|v\|^2_{L_2(\Omega)} \ge c_0 \|v\|^2_{W_2^k(\Omega)} \quad \textit{for all } v \in \mathring{W}_2^k(\Omega). \tag{2.5}$$

The proof of this results is long and technical, and will not be presented here; the interested reader is referred to Theorem 9.17 on p. 292 of Renardy and Rogers [155], for example.

For second-order uniformly elliptic operators of the form (2.3) the proof of Gårding's inequality is much simpler, and we shall confine ourselves to this case; in fact, as will be seen below, in the case of a second-order uniformly elliptic operator the smoothness hypotheses on the coefficients in the principal part of the operator can be slightly relaxed: they need not be continuous functions, as long as they belong to $L_\infty(\Omega)$. We note that the bilinear functional corresponding to the operator (2.3) is given by

$$a(u,v) = \sum_{i,j=1}^{n} \int_\Omega a_{ij}(x) \frac{\partial u}{\partial x_i} \frac{\partial v}{\partial x_j} \, \mathrm{d}x + \sum_{i=1}^{n} a_i(x) u \frac{\partial v}{\partial x_i} \, \mathrm{d}x$$

$$+ \int_\Omega b_i(x) \frac{\partial u}{\partial x_i} v \, \mathrm{d}x + \int_\Omega c(x) u v \, \mathrm{d}x, \quad u, v \in \mathring{W}_2^1(\Omega).$$

Theorem 2.5 *Suppose that $\Omega \subset \mathbb{R}^n$ is a Lipschitz domain. Let $P(x,\partial)$ be the second-order linear partial differential operator defined by* (2.3) *where a_{ij}, a_i, $b_j \in L_\infty(\Omega)$, $i, j = 1, \dots, n$, and $c \in L_\infty(\Omega)$ are such that, for some $\tilde{c} > 0$, the uniform ellipticity condition* (2.4) *holds. Then, there exist real numbers $c_0 > 0$ and $\lambda_0 \geq 0$ such that*

$$a(v,v) + \lambda_0 \|v\|^2_{L_2(\Omega)} \geq c_0 \|v\|^2_{W_2^1(\Omega)} \quad \forall v \in \mathring{W}_2^1(\Omega).$$

Proof Thanks to (2.4) and the Cauchy–Schwarz inequality we have that

$$a(v,v) = \sum_{i,j=1}^{n} \int_\Omega a_{ij}(x) \frac{\partial v}{\partial x_i} \frac{\partial v}{\partial x_j} \, \mathrm{d}x + \sum_{i=1}^{n} \int_\Omega a_i(x) v \frac{\partial v}{\partial x_i} \, \mathrm{d}x$$

$$+ \sum_{i=1}^{n} \int_\Omega b_i(x) \frac{\partial v}{\partial x_i} v \, \mathrm{d}x + \int_\Omega c(x) v^2 \, \mathrm{d}x$$

$$\geq \tilde{c} \int_\Omega |\nabla v|^2 \, \mathrm{d}x - \int_\Omega \left[2 \sum_{i=1}^{n} \left(a_i^2 + b_i^2\right) \right]^{1/2} |\nabla v| |v| \, \mathrm{d}x$$

$$- \|c\|_{L_\infty(\Omega)} \int_\Omega |v|^2 \, \mathrm{d}x,$$

where, as usual $|\nabla u| = \left[(\frac{\partial u}{\partial x_1})^2 + \cdots + (\frac{\partial u}{\partial x_n})^2 \right]^{1/2}$. By applying the elementary inequality

$$ab \leq \varepsilon a^2 + \frac{1}{4\varepsilon} b^2$$

with $\varepsilon = \tilde{c}/2$, we obtain

$$a(v, v) \geq \frac{\tilde{c}}{2} \int_\Omega |\nabla v|^2 \, dx - C\|v\|^2_{L_2(\Omega)},$$

where

$$C = \frac{1}{\tilde{c}} \left\| \sum_{i=1}^{n} (a_i^2 + b_i^2) \right\|_{L_\infty(\Omega)} + \|c\|_{L_\infty(\Omega)}.$$

Equivalently,

$$a(v, v) \geq \frac{\tilde{c}}{2} \|v\|^2_{W_2^1(\Omega)} - \left(C + \frac{\tilde{c}}{2}\right) \|v\|^2_{L_2(\Omega)},$$

which proves Gårding's inequality with $c_0 = \tilde{c}/2$ and $\lambda_0 = C + (\tilde{c}/2)$. □

Remark 2.2 We note that Theorem 2.5 can be proved under even weaker hypotheses on a_{ij}, a_i and b_i. Indeed, it suffices to assume that

$$\begin{aligned} a_{ij} &\in M\big(L_2(\Omega) \to L_2(\Omega)\big), \quad i, j = 1, \ldots, n, \\ a_i, b_i &\in M\big(W_2^1(\Omega) \to L_2(\Omega)\big), \quad i = 1, \ldots, n, \\ c &\in M\big(W_2^1(\Omega) \to L_p(\Omega)\big), \end{aligned}$$

where $p = 2n/(n+2)$ if $n > 2$; $p > 1$ (but arbitrarily close to 1) if $n = 2$; and $p = 1$ if $n = 1$.

We now state the main result of this section, which concerns the existence of a weak solution to a homogeneous Dirichlet boundary-value problem.

Theorem 2.6 *Let $P(x, \partial)$ be a linear partial differential operator of order $2k$ of the form* (2.1), *satisfying the conditions of Theorem* 2.4 *on a Lipschitz domain $\Omega \subset \mathbb{R}^n$. Then, there exists a $\lambda_0 \geq 0$ such that, for any $\lambda \geq \lambda_0$ and any $f \in W_2^{-k}(\Omega)$, the homogeneous Dirichlet problem for the operator*

$$\tilde{P}(x, \partial) = P(x, \partial) + \lambda$$

has a unique weak solution $u \in \mathring{W}_2^k(\Omega)$. Furthermore, this solution satisfies

$$\|u\|_{W_2^k(\Omega)} \leq C\|f\|_{W_2^{-k}(\Omega)}.$$

Proof According to Theorem 2.4 there exists a $\lambda_0 \geq 0$ such that the Gårding inequality (2.5) holds. For $\lambda \geq \lambda_0$ we consider the bilinear functional

$$\tilde{a}(u, v) = a(u, v) + \lambda(u, v), \quad u, v \in \mathring{W}_2^k(\Omega),$$

associated with the operator $\tilde{P}$. We shall prove that $\tilde{a}(\cdot,\cdot)$ satisfies the conditions of the Lax–Milgram theorem (Theorem 1.13) on $\mathring{W}_2^k(\Omega) \times \mathring{W}_2^k(\Omega)$. Let us take $\mathcal{U} = \mathring{W}_2^k(\Omega)$ in Theorem 1.13 and recall that $\mathring{W}_2^k(\Omega)$ is a real Hilbert space. The $\mathcal{U}$-coercivity of $\tilde{a}(\cdot,\cdot)$ is a straightforward consequence of (2.5):

$$\tilde{a}(v,v) = a(v,v) + \lambda\|v\|^2_{L_2(\Omega)} \geq c_0\|v\|^2_{W_2^k(\Omega)} \quad \forall v \in \mathring{W}_2^k(\Omega).$$

We shall now verify that $\tilde{a}(\cdot,\cdot)$ is bounded on $\mathring{W}_2^k(\Omega) \times \mathring{W}_2^k(\Omega)$. Given $v, w \in \mathring{W}_2^k(\Omega)$, using the Cauchy–Schwarz inequality repeatedly we obtain the following chain of inequalities, which ultimately lead to the conclusion that $\tilde{a}(\cdot,\cdot)$ is a bounded bilinear functional on $\mathring{W}_2^k(\Omega) \times \mathring{W}_2^k(\Omega)$:

$$\begin{aligned}
|\tilde{a}(v,w)| &\leq |a(v,w)| + \lambda|(v,w)| \\
&\leq \sum_{0\leq|\alpha|,|\beta|\leq k} \int_\Omega |a_{\alpha\beta}(x)||\partial^\beta v||\partial^\alpha w|\,\mathrm{d}x + \lambda|(v,w)| \\
&\leq \max_{0\leq|\alpha|,|\beta|\leq k} \|a_{\alpha\beta}\|_{L_\infty(\Omega)} \sum_{0\leq|\alpha|,|\beta|\leq k} \int_\Omega |\partial^\beta v||\partial^\alpha w|\,\mathrm{d}x + \lambda|(v,w)| \\
&\leq c_1\|v\|_{\mathcal{U}}\|w\|_{\mathcal{U}}.
\end{aligned}$$

Thus, by the Lax–Milgram theorem (Theorem 1.13), for each $f \in W_2^{-k}(\Omega) = \mathcal{U}'$, there exists a unique weak solution $u \in \mathring{W}_2^k(\Omega)$ to the homogeneous Dirichlet problem. □

In the case of second-order elliptic equations we have an analogous result.

Theorem 2.7 *Let $P(x,\partial)$ be a linear second-order partial differential operator of the form* (2.3)*, satisfying the conditions of Theorem* 2.5 *on a Lipschitz domain $\Omega \subset \mathbb{R}^n$. Then, there exists a $\lambda_0 \geq 0$ such that, for any $\lambda \geq \lambda_0$ and any $f \in W_2^{-1}(\Omega)$, the homogeneous Dirichlet problem for the operator*

$$\tilde{P}(x,\partial) = P(x,\partial) + \lambda$$

has a unique weak solution $u \in \mathring{W}_2^1(\Omega)$, and this solution satisfies

$$\|u\|_{W_2^1(\Omega)} \leq C\|f\|_{W_2^{-1}(\Omega)}.$$

Furthermore, if $a_i, b_i \in W_p^1(\Omega)$, $i = 1,\ldots,n$, where $p = n/2$ when $n > 2$; $p > 1$ is arbitrary when $n = 2$; and $p = 1$ when $n = 1$; and

$$c(x) - \frac{1}{2}\sum_{i=1}^n \frac{\partial}{\partial x_i}\big(a_i(x) + b_i(x)\big) \geq 0$$

for almost every $x \in \Omega$, then $\lambda_0 = 0$. In other words, the homogeneous Dirichlet problem corresponding to the operator $P(x, \partial)$ has a unique weak solution $u \in \mathring{W}_2^1(\Omega)$ under these hypotheses.

Proof The first part of the theorem is proved in exactly the same way as the corresponding statement in Theorem 2.6. In order to prove the second part let us observe that, by the divergence theorem,

$$\int_\Omega [a_i(x) + b_i(x)] \frac{\partial v}{\partial x_i} v \,\mathrm{d}x = -\frac{1}{2} \int_\Omega \frac{\partial}{\partial x_i} \big(a_i(x) + b_i(x)\big) v^2 \,\mathrm{d}x \quad \forall v \in \mathring{W}_2^1(\Omega);$$

we note that because $a_i, b_i \in W_p^1(\Omega)$, $i = 1, \ldots, n$, where p is as assumed, Hölder's inequality, followed by the application of Sobolev's embedding theorem, implies that the function appearing as the integrand on the right-hand side is an element of $L_1(\Omega)$. Therefore the right-hand side of this equality is meaningful.

Consequently,

$$a(v, v) \geq \tilde{c} \sum_{i=1}^{n} \int_\Omega \left| \frac{\partial v}{\partial x_i} \right|^2 \mathrm{d}x. \tag{2.6}$$

By applying the Friedrichs inequality (1.23) with $s = 1$ and $p = 2$, the right-hand side of (2.6) can be further bounded below to obtain

$$a(v, v) \geq c_0 \|v\|^2_{W_2^1(\Omega)}, \tag{2.7}$$

where $c_0 = \tilde{c}/c_\star$, and hence the $\mathring{W}_2^1(\Omega)$-coercivity of the bilinear functional $a(\cdot, \cdot)$. The boundedness of $a(\cdot, \cdot)$ on the space $\mathring{W}_2^1(\Omega) \times \mathring{W}_2^1(\Omega)$ follows from the boundedness of $\tilde{a}(\cdot, \cdot) = a(\cdot, \cdot) + \lambda(\cdot, \cdot)$ by setting $\lambda = 0$. The required result is now obtained from the Lax–Milgram theorem (Theorem 1.13). □

Remark 2.3 We note that Theorem 2.7 continues to hold when the regularity hypotheses of Theorem 2.5 are replaced by the weaker ones from Remark 2.2.

Having developed relatively straightforward sufficient conditions for the existence of a unique weak solution to an elliptic boundary-value problem, the question that we now need to address is whether a weak solution might possess additional regularity to qualify as a strong solution. The answer to this question very much depends on additional regularity of the data (i.e. the coefficients, the right-hand side of the partial differential equation, and the boundary $\partial\Omega$). Since a general discussion of regularity properties of weak solutions to elliptic boundary-value problems is beyond the scope of this book, we shall confine ourselves to Poisson's equation subject to a homogeneous Dirichlet boundary condition, which is sufficiently illustrative of the key ideas. We begin with a simple example, which shows that a weak solution to an elliptic boundary-value problem need not be a strong solution to the problem, and that a strong solution may not even exist.

Example 2.5 Suppose that $\Omega = \{(x, y) \in \mathbb{R}^2 : x^2 + y^2 < \mathrm{e}^{-1}\}$ and let $f(x, y) := -\Delta(\log|\log(x^2 + y^2)|)$, with $\log := \log_{\mathrm{e}}$ and the differential operator Δ understood in the sense of distributions on Ω. It is easily seen by changing from Cartesian co-ordinates to polar co-ordinates that the function $u : (x, y) \mapsto \log|\log(x^2 + y^2)|$ belongs to $\mathring{W}_2^1(\Omega)$ and that, therefore, $f \in W_2^{-1}(\Omega)$. Thus, u is the unique weak solution to the boundary-value problem: $-\Delta u = f$ on Ω (with the equality understood as being between two elements of $W_2^{-1}(\Omega)$), subject to the boundary condition $u = 0$ on $\partial\Omega$. However, the function u is *not* a strong solution and, as a matter of fact, the boundary-value problem has no strong solution, since $f \notin L_2(\Omega)$.

In fact, even if f belongs to $W_2^{s-2}(\Omega)$, $s \geq 2$, it does not automatically follow that the weak solution to Poisson's equation $-\Delta u = f$, with a homogeneous Dirichlet boundary condition on $\partial\Omega$, belongs to $W_2^s(\Omega) \cap \mathring{W}_2^1(\Omega)$. Whether or not this is the case depends on the smoothness of $\partial\Omega$. In particular if Ω is a bounded polygonal domain in $\mathbb{R}^2$, the regularity of the solution is ultimately limited by the size of the maximum internal angle of Ω; the next theorem is a special case of a more general result, due to Grisvard [61].

Theorem 2.8 *Suppose that* $f \in W_2^{s-2}(\Omega)$, $1 \leq s < 3$, $s \neq 3/2, 5/2$, *with* $\Omega = (0, 1)^2$, *and consider the homogeneous Dirichlet boundary-value problem for Poisson's equation*:

$$\begin{aligned} -\Delta u &= f \quad \text{on } \Omega, \\ u &= 0 \quad \text{on } \partial\Omega. \end{aligned}$$

Then, the unique weak solution u in $\mathring{W}_2^1(\Omega)$ belongs to $W_2^s(\Omega) \cap \mathring{W}_2^1(\Omega)$.

The limitation $s < 3$ on the Sobolev exponent in Theorem 2.8 is sharp in the sense that the stated regularity result is invalid for $s \geq 3$ unless f satisfies certain *compatibility conditions* at the four corners of the square. More precisely, u belongs to the space $W_2^s(\Omega) \cap \mathring{W}_2^1(\Omega)$ for $s \in \mathbb{N}$, $s \geq 3$, provided that $f \in W_2^{s-2}(\Omega)$ and the following conditions hold *at the four corners*:

$$\begin{aligned} &f = 0, \\ &\partial_1^2 f - \partial_2^2 f = 0, \\ &\cdots\cdots\cdots\cdots\cdots\cdots \\ &\partial_1^{2k} f - \partial_1^{2k-2}\partial_2^2 f + \cdots + (-1)^k \partial_2^{2k} f = 0, \quad \text{with } k = \left[\frac{s-2}{2}\right]. \end{aligned} \tag{2.8}$$

The proof proceeds similarly to the one in Volkov [193], where an analogous regularity result was shown for classical solutions. For details we refer to the work of Hell [70].

Next we formulate a result that concerns the existence of weak solutions to the homogeneous Dirichlet problem for the fourth-order uniformly elliptic equation considered in Example 2.2.

Theorem 2.9 *Let $\Omega \subset \mathbb{R}^2$ be a Lipschitz domain. Consider the partial differential operator $P(x, \partial)$, defined by*

$$P(x,\partial)u := \partial_1^2 M_1(u) + 2\partial_1\partial_2 M_3(u) + \partial_2^2 M_2(u),$$

where

$$\begin{aligned} M_1(u) &:= a_1(x)\partial_1^2(u) + a_0(x)\partial_2^2 u, \\ M_2(u) &:= a_0(x)\partial_1^2 u + a_2(x)\partial_2^2 u, \\ M_3(u) &:= a_3(x)\partial_1\partial_2 u, \end{aligned}$$

and $a_i \in L_\infty(\Omega)$, $i = 0, 1, 2, 3$, are such that there exist positive constants c_1 and c_2 for which

$$a_i(x) \geq c_1, \quad i = 1, 2, 3, \qquad a_1(x)a_2(x) - a_0^2(x) \geq c_2, \quad x \in \Omega.$$

Then, for any $f \in W_2^{-2}(\Omega)$, the homogeneous Dirichlet boundary-value problem for $P(x, \partial)$ has a unique weak solution u in $\mathring{W}_2^2(\Omega)$.

Proof The proof is, again, based on the Lax–Milgram theorem (Theorem 1.13); its nontrivial part is to verify that the bilinear functional

$$a(u,v) = \big(M_1(u), \partial_1^2 v\big) + 2\big(M_3(u), \partial_1\partial_2 v\big) + \big(M_2(u), \partial_2^2 v\big), \quad u, v \in \mathring{W}_2^2(\Omega),$$

is $\mathring{W}_2^2(\Omega)$-coercive. Clearly,

$$\begin{aligned} a(v,v) = \int_\Omega &\big[a_1(x)|\partial_1^2 v|^2 + 2a_3(x)|\partial_1\partial_2 v|^2 \\ &+ a_2(x)|\partial_2^2 v|^2 + 2a_0(x)\partial_1^2 v\,\partial_2^2 v\big]\,\mathrm{d}x \quad \forall v \in \mathring{W}_2^2(\Omega). \end{aligned}$$

As v is real-valued (by the convention stated at the beginning of the chapter), we have the following identity:

$$\begin{aligned} a(v,v) = &\frac{1}{2}\int_\Omega a_1(x)\left(\partial_1^2 v + \frac{a_0(x)}{a_1(x)}\partial_2^2 v\right)^2 \mathrm{d}x \\ &+ \frac{1}{2}\int_\Omega a_2(x)\left(\partial_2^2 v + \frac{a_0(x)}{a_2(x)}\partial_1^2 v\right)^2 \mathrm{d}x \\ &+ \frac{1}{2}\int_\Omega \left(a_1(x) - \frac{a_0^2(x)}{a_2(x)}\right)|\partial_1^2 v|^2\,\mathrm{d}x \\ &+ \frac{1}{2}\int_\Omega \left(a_2(x) - \frac{a_0^2(x)}{a_1(x)}\right)|\partial_2^2 v|^2\,\mathrm{d}x \end{aligned}$$

$$+2\int_{\Omega} a_3(x)|\partial_1\partial_2 v|^2\,\mathrm{d}x \quad \forall v \in \mathring{W}_2^2(\Omega).$$

Therefore,

$$\begin{aligned} a(v,v) \geq \frac{1}{2}\int_{\Omega}\left(a_1(x) - \frac{a_0^2(x)}{a_2(x)}\right)|\partial_1^2 v|^2\,\mathrm{d}x \\ + \frac{1}{2}\int_{\Omega}\left(a_2(x) - \frac{a_0^2(x)}{a_1(x)}\right)|\partial_2^2 v|^2\,\mathrm{d}x \\ + 2\int_{\Omega} a_3(x)|\partial_1\partial_2 v|^2\,\mathrm{d}x \quad \forall v \in \mathring{W}_2^2(\Omega). \end{aligned}$$

By noting the assumptions on the coefficients a_i, $i = 0, 1, 2, 3$, it follows that there exists a positive constant $\tilde{c}$ such that

$$a(v,v) \geq \tilde{c}|v|^2_{W_2^2(\Omega)} \quad \forall v \in \mathring{W}_2^2(\Omega).$$

Finally, by the Friedrichs inequality (1.23) with $s = p = n = 2$,

$$\|v\|^2_{W_2^2(\Omega)} \leq c_\star |v|^2_{W_2^2(\Omega)} \quad \forall v \in \mathring{W}_2^2(\Omega),$$

and hence

$$a(v,v) \geq c_0\|v\|^2_{W_2^2(\Omega)} \quad \forall v \in \mathring{W}_2^2(\Omega),$$

where $c_0 = \tilde{c}/c_\star$. □

Remark 2.4 Suppose that the homogeneous Dirichlet boundary condition

$$\partial_\nu^m u = 0 \quad \text{on } \partial\Omega \text{ for } m = 0, 1,$$

for the partial differential operator $P(x, \partial)$ defined in Theorem 2.9 has been replaced by the following set of boundary conditions:

$$u = 0, \quad M_1(u)\nu_1 + M_3(u)\nu_2 = 0, \qquad M_3(u)\nu_1 + M_2(u)\nu_2 = 0 \quad \text{on } \partial\Omega.$$

The weak formulation of the corresponding boundary-value problem is: find $u \in W_2^2(\Omega) \cap \mathring{W}_2^1(\Omega)$ such that

$$a(u,v) = \langle f, v\rangle$$

for every $v \in W_2^2(\Omega) \cap \mathring{W}_2^1(\Omega)$. Again, by using the Lax–Milgram theorem (Theorem 1.13), it is easy to prove that, under the same conditions on a_i, $i = 0, 1, 2, 3$, as in Theorem 2.9, this problem too has a unique weak solution, now in the function space $W_2^2(\Omega) \cap \mathring{W}_2^1(\Omega)$.

Finally, we return to the boundary-value problem considered in Example 2.4, which has been shown to have no classical solution. By applying Theorem 2.7 with $a_{ij}(x) \equiv 1$, $i = j$, $a_{ij}(x) \equiv 0$, $i \neq j$, $1 \leq i, j \leq n$, $b_i(x) \equiv 0$, $c(x) \equiv 0$, $f(x) = \operatorname{sgn}(\frac{1}{2} - |x|)$, and $\Omega = (-1, 1)^n$, we see that there is a unique weak solution $u \in \mathring{W}_2^1(\Omega)$ to this problem. In fact, it can be shown that this weak solution belongs to $W_2^2(\Omega) \cap \mathring{W}_2^1(\Omega)$ and it is, therefore, a strong solution to the boundary-value problem (see Grisvard [62, 63]).

Remark 2.5 The existence and uniqueness of a weak solution to a Neumann, Robin, or oblique derivative boundary-value problem for a second-order uniformly elliptic equation can be established in a similar fashion, using the Lax–Milgram theorem (Theorem 1.13).

Remark 2.6 Theorems 2.6 and 2.7 imply that the weak formulation of the Dirichlet boundary-value problem for the operator $\tilde{P}(x, \partial) = P(x, \partial) + \lambda$, $\lambda \geq \lambda_0 \geq 0$, is *well-posed in the sense of Hadamard*; that is, for each $f \in W_2^{-k}(\Omega)$, there exists a unique (weak) solution $u \in \mathring{W}_2^k(\Omega)$; moreover, "small" changes in f give rise to "small" changes in the corresponding solution u. The latter property follows by noting that if u_1 and u_2 are weak solutions in $\mathring{W}_2^k(\Omega)$ of the homogeneous Dirichlet problem for $\tilde{P}(x, \partial)$ corresponding to right-hand sides f_1 and f_2 in $W_2^{-k}(\Omega)$, respectively, then $u_1 - u_2$ is the unique weak solution in $\mathring{W}_2^k(\Omega)$ of the homogeneous Dirichlet boundary-value problem for the operator $\tilde{P}(x, \partial)$ corresponding to the right-hand side $f_1 - f_2$ in $W_2^{-k}(\Omega)$. It thus follows from Theorems 2.6 and 2.7 that

$$\|u_1 - u_2\|_{W_2^k(\Omega)} \leq C \|f_1 - f_2\|_{W_2^{-k}(\Omega)},$$

where C is a positive constant, independent of u_1, u_2, f_1 and f_2; this implies the continuous dependence of the solution to the homogeneous Dirichlet boundary-value problem on the right-hand side of the equation.

2.2 Approximation of Elliptic Problems

We begin this section by outlining the general approach to the construction of finite difference schemes for elliptic boundary-value problems; we then introduce basic results from the theory of finite difference schemes and present some classical tools for the error analysis of finite difference schemes for partial differential equations with smooth solutions. The limitations of the classical theory will lead us to consider finite difference schemes with mollified data, and we shall develop a theoretical framework for the error analysis of such nonstandard schemes. We conclude by considering finite difference approximations of second- and fourth-order elliptic equations with variable coefficients, and derive sharp error bounds in various mesh-dependent (discrete) norms, under minimal smoothness requirements on the data and the associated solution.

2.2.1 Introduction to the Theory of Finite Difference Schemes

Assuming that Ω is a bounded open set in $\mathbb{R}^n$, we consider a boundary-value problem on Ω of the general form

$$\mathcal{L}u = f \quad \text{in } \Omega, \tag{2.9}$$

$$lu = g \quad \text{on } \Gamma = \partial\Omega, \tag{2.10}$$

where $\mathcal{L}$ is a linear partial differential operator, and l is a linear operator that specifies the boundary condition. For example, we may have

$$\mathcal{L}u := -\sum_{i,j=1}^{n} \frac{\partial}{\partial x_j}\left(a_{ij}(x)\frac{\partial u}{\partial x_i}\right) + \sum_{i=1}^{n} b_i(x)\frac{\partial u}{\partial x_i} + c(x)u,$$

where the $a_{ij}(x)$, $i, j = 1, \ldots, n$, satisfy (2.4), with one of the following choices of the boundary operator l (Dirichlet, Neumann or oblique derivative):

$$lu := u,$$

or

$$lu := \frac{\partial u}{\partial \nu},$$

or

$$lu := \sum_{i,j=1}^{n} a_{ij}(x)\frac{\partial u}{\partial x_i}\nu_j + \sigma(x)u,$$

where ν is the unit outward normal vector to Γ, ν_j is the jth component of ν, $j = 1, \ldots, n$, and σ is a bounded, nonnegative function defined on Γ.

The construction of a finite difference scheme for the boundary-value problem (2.9), (2.10) consists of two basic steps: first, the domain $\overline{\Omega}$ is replaced by a finite set of points, called the *mesh* or *grid*, and second, the derivatives in the differential equation and in the boundary condition are replaced by divided differences. To describe the first of these two steps more precisely, suppose that we have approximated $\overline{\Omega} = \Omega \cup \Gamma$ by the mesh

$$\overline{\Omega}^h := \Omega^h \cup \Gamma^h,$$

where $\Omega^h \subset \Omega$ is the set of *interior mesh-points*, and $\Gamma^h \subset \Gamma$ is the set of *boundary mesh-points*. Typically the mesh consist of a finite set of points obtained by considering the intersections of n families of parallel hyperplanes, each element of each family being perpendicular to one of the co-ordinate axes. If the domain Ω is not axiparallel, adjustments may need to be made to the mesh near the boundary $\partial\Omega$, which may be curved. The parameter $h = (h_1, \ldots, h_n)$ measures the spacing of the mesh; in particular, $h_i > 0$ denotes the mesh-size in the ith co-ordinate direction. Once the mesh has been constructed, we proceed by replacing the derivatives featuring in $\mathcal{L}$ by divided differences, and approximate the boundary condition in a similar fashion. This yields a *finite difference scheme* of the form

$$\mathcal{L}_h U(x) = f_h(x), \quad x \in \Omega^h, \tag{2.11}$$

$$l_h U(x) = g_h(x), \quad x \in \Gamma^h, \tag{2.12}$$

where $\mathcal{L}_h$ and l_h are linear difference operators, representing discrete counterparts of $\mathcal{L}$ and l, while f_h and g_h are suitable approximations of f and g, respectively. In algebraic terms, (2.11), (2.12) is a system of linear equations involving the values of the approximate solution U at the mesh-points.

Assuming that (2.11), (2.12) has a unique solution U, when the mesh spacing is small the sequence of values of the approximate solution at the mesh-points, $\{U(x) : x \in \overline{\Omega}^h\}$, is expected to resemble $\{u(x) : x \in \overline{\Omega}^h\}$, the set of values of the exact solution u at the mesh-points. However the closeness of $U(x)$ to $u(x)$ at $x \in \overline{\Omega}^h$ is by no means obvious, and the proof of such approximation results represents the central theme of this book. We shall consider a range of problems of the form (2.9), (2.10), and derive sharp bounds on the error between the analytical solution u (typically a weak solution) and its finite difference approximation U in terms of positive powers of the discretization parameter h. Bounds of this kind imply, in particular, that the error between the analytical solution u and its finite difference approximation U converges to zero with a certain rate, in a certain norm, as $h \to 0$.

2.2.2 *Finite Difference Approximation in One Space Dimension*

In this section we shall focus on the finite difference approximation of a two-point boundary-value problem. We begin by developing some basic results about mesh-functions (i.e. functions that are defined on the finite difference mesh), finite difference operators and mesh-dependent (discrete) norms.

2.2.2.1 Meshes, Mesh-Functions and Mesh-Dependent Norms

Meshes Suppose that N is a positive integer, $N \geq 2$, let $h := 1/N$, and consider the *uniform mesh* on the unit interval $(0, 1)$ of the real line, defined by

$$\Omega^h := \{x_i : x_i = ih, i = 1, \ldots, N-1\}.$$

We further define

$$\overline{\Omega}^h := \Omega^h \cup \{0, 1\}.$$

Let S^h denote the linear space of real-valued functions defined on the mesh $\overline{\Omega}^h$, and let S_0^h be the linear space of all real-valued functions defined on the mesh $\overline{\Omega}^h$ that are equal to zero on $\Gamma^h := \overline{\Omega}^h \setminus \Omega^h$. Any element of the set S^h (or of S_0^h) will be referred to as a *mesh-function*.

For a mesh-function $V \in S^h$ we define $V_i := V(x_i) = V(ih)$. We equip the linear space S_0^h with the inner product

$$(V, W)_h = (V, W)_{L_2(\Omega^h)} := \sum_{x \in \Omega^h} hV(x)W(x) = \sum_{i=1}^{N-1} hV_i W_i, \tag{2.13}$$

which closely resembles the inner product

$$(v, w) = \int_0^1 v(x)w(x)\,\mathrm{d}x,$$

of the Hilbert space $L_2(\Omega)$. The inner product $(\cdot, \cdot)_h$ induces the norm $\|\cdot\|_h$ on S_0^h defined by

$$\|V\|_h = \|V\|_{L_2(\Omega^h)} := (V, V)_h^{1/2}. \tag{2.14}$$

Analogously, we equip the linear space S^h with the inner product

$$[V, W]_h = (V, W)_{L_2(\overline{\Omega}^h)} := \frac{h}{2}\big[V(0)W(0) + V(1)W(1)\big] + (V, W)_h$$

and the induced norm

$$[\![V]\!]_h = \|V\|_{L_2(\overline{\Omega}^h)} := [V, V]_h^{1/2}.$$

We shall also need the meshes

$$\Omega_-^h := \Omega^h \cup \{0\}, \qquad \Omega_+^h := \Omega^h \cup \{1\}.$$

On the linear space of real-valued functions defined on the mesh Ω_-^h, we consider the inner product

$$[V, W)_h = (V, W)_{L_2(\Omega_-^h)} := \sum_{x \in \Omega_-^h} hV(x)W(x) = \sum_{i=0}^{N-1} hV_i W_i$$

and the associated norm

$$[\![V\|_h = \|V\|_{L_2(\Omega_-^h)} := [V, V)_h^{1/2},$$

with an analogous definition of the inner product $(V, W]_h = (V, W)_{L_2(\Omega_+^h)}$ and the corresponding norm $\|V]\!]_h = \|V\|_{L_2(\Omega_+^h)}$ on the linear space of real-valued mesh-functions defined on Ω_+^h.

Finite Difference Operators The forward, backward and central divided difference operators D_x^+, D_x^- and D_x^0 on the mesh $\overline{\Omega}_h$ are defined, respectively, by

$$D_x^+ V := \frac{V^+ - V}{h}, \qquad D_x^- V := \frac{V - V^-}{h}, \qquad D_x^0 V := \frac{1}{2}\big(D_x^+ V + D_x^- V\big),$$

where we have used the notation

$$V^\pm = V^\pm(x) := V(x \pm h).$$

With these definitions, we have the following discrete Leibniz formulae:

$$D_x^+(VW) = \big(D_x^+ V\big)W^+ + V\big(D_x^+ W\big) = \big(D_x^+ V\big)W + V^+\big(D_x^+ W\big),$$
$$D_x^-(VW) = \big(D_x^- V\big)W^- + V\big(D_x^- W\big) = \big(D_x^- V\big)W + V^-\big(D_x^- W\big),$$

and the summation-by-parts formula:

$$\big[D_x^+ V, W)_h = -(V, D_x^- W\big]_h + V(1)W(1) - V(0)W(0), \tag{2.15}$$

which immediately yields the following result.

Lemma 2.10 *Suppose that $V \in S_0^h$; then,*

$$\big(-D_x^+ D_x^- V, V\big)_h = \sum_{i=1}^{N} h\big|D_x^- V_i\big|^2 = \sum_{i=0}^{N-1} h\big|D_x^+ V_i\big|^2. \tag{2.16}$$

Proof Let us write $U_i = D_x^- V_i$, $i = 1, \dots, N$, and note that

$$\big(-D_x^+ D_x^- V, V\big)_h = -\big(D_x^+ U, V\big)_h = -\big[D_x^+ U, V\big)_h = \big(U, D_x^- V\big]_h = \big\|D_x^- V\big]\!\big|_h^2,$$

thanks to our assumption that $V \in S_0^h$, which implies that $V_0 = V(0) = 0$ and $V_N = V(1) = 0$, and using the identity (2.15). The second equality in (2.16) follows simply by noting that $D_x^- V_i = D_x^+ V_{i-1}$, $i = 1, \dots, N$, and shifting the index i in the summation. □

The Discrete Laplace Operator on S_0^h On the set S_0^h, we define the linear operator $\Lambda : S_0^h \to S_0^h$ by

$$(\Lambda V)(x) := \begin{cases} -D_x^+ D_x^- V(x) & \text{if } x \in \Omega^h, \\ 0 & \text{if } x \in \Gamma^h = \overline{\Omega}^h \setminus \Omega^h. \end{cases}$$

Since

$$\begin{aligned}(\Lambda V, W)_h &= -\big(D_x^+ D_x^- V, W\big)_h = \big(D_x^- V, D_x^- W\big]_h = \big(D_x^- W, D_x^- V\big]_h \\ &= \big[D_x^+ V, D_x^+ W\big)_h = \big[D_x^+ W, D_x^+ V\big)_h = (\Lambda W, V)_h,\end{aligned}$$

Λ is a symmetric linear operator on S_0^h. Moreover, thanks to (2.16),

$$(\Lambda V, V) = \| D_x^- V]|_h^2 = |[D_x^+ V \|_h^2 > 0 \quad \text{for all } V \in S_0^h \setminus \{0\},$$

and therefore Λ is positive definite on S_0^h. Thus Λ has $N-1$ distinct positive eigenvalues, which are easily shown to be (see Samarskiĭ [159], Sect. 2.4.2)

$$\lambda_k = \frac{4}{h^2} \sin^2 \frac{k\pi h}{2}, \quad k = 1, 2, \ldots, N-1; \tag{2.17}$$

these eigenvalues satisfy the inequalities

$$8 < \lambda_k < \frac{4}{h^2}, \quad k = 1, 2, \ldots, N-1. \tag{2.18}$$

The corresponding $N-1$ eigenfunctions V^k, $k = 1, \ldots, N-1$, satisfying $\Lambda V^k = \lambda_k V^k$, are

$$V^k(x) = \sin k\pi x, \quad x \in \overline{\Omega}^h, \ k = 1, 2, \ldots, N-1.$$

The set of eigenfunctions $\{V^1, \ldots, V^{N-1}\}$ is an orthogonal system in S_0^h with respect to the inner product $(\cdot, \cdot)_h$; that is,

$$\left(V^k, V^l\right)_h = \frac{1}{2}\delta_{kl}, \quad k, l = 1, 2, \ldots, N-1, \tag{2.19}$$

where δ_{kl} is the Kronecker delta; in fact, $\{V^1, \ldots, V^{N-1}\}$ forms a basis of the linear space S_0^h. Consequently an arbitrary mesh-function $V \in S_0^h$ can be expressed as a linear combination of these eigenfunctions:

$$V(x) = \sum_{k=1}^{N-1} b_k \sin k\pi x, \quad x \in \overline{\Omega}^h, \tag{2.20}$$

where

$$b_k = 2\left(V, V^k\right)_h.$$

By noting the orthogonality of the eigenfunctions we deduce the following *discrete Parseval identity*:

$$\|V\|_h^2 = \frac{1}{2} \sum_{k=1}^{N-1} b_k^2. \tag{2.21}$$

Analogously,

$$\| D_x^- V]|_h^2 = |[D_x^+ V \|_h^2 = (\Lambda V, V)_h = \frac{1}{2} \sum_{k=1}^{N-1} \lambda_k b_k^2, \tag{2.22}$$

$$\left\|D_x^+ D_x^- V\right\|_h^2 = (\Lambda V, \Lambda V)_h = \frac{1}{2}\sum_{k=1}^{N-1} \lambda_k^2 b_k^2. \tag{2.23}$$

It follows from (2.18) and (2.21)–(2.23) that

$$\left\|D_x^+ D_x^- V\right\|_h \geq 2\sqrt{2}\left\|D_x^- V\right]\!\!\right]_h = 2\sqrt{2}\left[\!\!\left[D_x^+ V\right\|_h \geq 8\|V\|_h \tag{2.24}$$

for each $V \in S_0^h$.

Discrete Sobolev Norms on S_0^h The discrete analogues of Sobolev seminorms and norms are defined similarly to their 'continuous' counterparts introduced in Chap. 1. In particular, we define

$$\begin{aligned}
|V|_{1,h} &= |V|_{W_2^1(\Omega^h)} := \left\|D_x^- V\right]\!\!\right]_h = \left[\!\!\left[D_x^+ V\right\|_h, \\
|V|_{2,h} &= |V|_{W_2^2(\Omega^h)} := \left\|D_x^+ D_x^- V\right\|_h, \\
\|V\|_{k,h} &= \|V\|_{W_2^k(\Omega^h)} := \left(\|V\|^2_{W_2^{k-1}(\Omega^h)} + |V|^2_{W_2^k(\Omega^h)}\right)^{1/2},
\end{aligned} \tag{2.25}$$

where $k = 1, 2$, with the convention that $W_2^0(\Omega^h) = L_2(\Omega^h)$. The inequalities (2.24) imply that the seminorms $|\cdot|_{W_2^1(\Omega^h)}$ and $|\cdot|_{W_2^2(\Omega^h)}$ are equivalent to the norms $\|\cdot\|_{W_2^1(\Omega^h)}$ and $\|\cdot\|_{W_2^2(\Omega^h)}$, respectively, on S_0^h.

Lemma 2.11 (Discrete Friedrichs Inequality) *There exists a positive constant $c_\star$ such that*

$$\|V\|^2_{W_2^1(\Omega^h)} \leq c_\star \left\|D_x^- V\right\|^2_{L_2(\Omega_+^h)} \tag{2.26}$$

for all $V \in S_0^h$.

Proof The last inequality in (2.24) implies (2.26) with $c_\star = 9/8$. □

Lemma 2.12 (Discrete Sobolev Embedding) *For all $V \in S_0^h$ the following inequality holds*

$$\|V\|_{\infty,h} := \max_{x\in\overline{\Omega}^h} |V(x)| \leq \frac{1}{2}\left\|D_x^- V\right\|_{L_2(\Omega_+^h)} \tag{2.27}$$

Proof Using the Cauchy–Schwarz inequality, we obtain from the identity

$$|V_i|^2 = (1-ih)|V_i|^2 + ih|V_i|^2 = (1-ih)\left|\sum_{j=1}^{i}(D_x^- V_j)h\right|^2 + ih\left|\sum_{j=i+1}^{N}(D_x^- V_j)h\right|^2$$

that

$$|V_i|^2 \le (1-ih)\left(\sum_{j=1}^{i} h\right)\sum_{j=1}^{i}(D_x^- V_j)^2 h + ih\left(\sum_{j=i+1}^{N} h\right)\sum_{j=i+1}^{N}(D_x^- V_j)^2 h$$

$$= ih(1-ih)\sum_{j=1}^{N}(D_x^- V_j)^2 h.$$

The required inequality then follows by taking the maximum over the index $i \in \{0, 1, \ldots, N\}$ and noting that, for all such i, $0 \le ih(1-ih) \le 1/4$. □

The Discrete Laplace Operator on S^h We define the linear operator $\overline{\Lambda} : S^h \to S^h$ by

$$(\overline{\Lambda} V)(x) := \begin{cases} -\frac{2}{h} D_x^+ V(0) & \text{if } x = 0, \\ -D_x^+ D_x^- V(x) & \text{if } x \in \Omega^h, \\ \frac{2}{h} D_x^- V(1) & \text{if } x = 1. \end{cases}$$

Assuming that each $V \in S^h$ is extended outside $\overline{\Omega}^h$ as an even function, we have that

$$(\overline{\Lambda} V)(x) = \left(-D_x^+ D_x^- V\right)(x) \quad \text{for } x \in \overline{\Omega}^h.$$

The linear operator $\overline{\Lambda}$ is symmetric with respect to the inner product $[\cdot, \cdot]_h$. The eigenvalues of $\overline{\Lambda}$ are given by the formula (2.17), but now for $k = 0, 1, 2, \ldots, N$. In fact, since $\lambda_0 = 0$ is an eigenvalue, $\overline{\Lambda} : S^h \to S^h$ is only nonnegative (positive semidefinite) rather than positive definite; that is,

$$[\overline{\Lambda} V, V]_h \ge 0 \quad \text{for all } V \in S^h \setminus \{0\}.$$

The eigenfunctions of $\overline{\Lambda}$ corresponding to the eigenvalues λ_k, $k = 0, \ldots, N$, are:

$$W^0(x) = 1, \qquad W^k(x) = \cos k\pi x, \quad k = 1, 2, \ldots, N;$$

these form an orthogonal system in the sense that

$$\left[W^k, W^l\right]_h = \begin{cases} 1 & \text{if } k = l = 0, N, \\ \frac{1}{2} & \text{if } k = l = 1, 2, \ldots, N-1, \\ 0 & \text{if } k \ne l, \end{cases}$$

and they span the linear space S^h; hence each mesh-function $V \in S^h$ can be expressed as

$$V(x) = \frac{1}{2} a_0 + \sum_{k=1}^{N-1} a_k \cos k\pi x + \frac{1}{2} a_N \cos N\pi x, \tag{2.28}$$

where

$$a_k = 2[V, \cos k\pi x]_h, \quad \text{for } k = 0, 1, \ldots, N.$$

When $V \in S_0^h$, the expansions (2.20) and (2.28) coincide at all points of the mesh $\overline{\Omega}^h$.

By noting the orthogonality of the eigenfunctions W^k, $k = 0, \ldots, N$, it is easily seen that for any mesh-function V contained in S^h the following identities hold:

$$[\![V]\!]_h^2 = \frac{1}{4}a_0^2 + \frac{1}{2}\sum_{k=1}^{N-1} a_k^2 + \frac{1}{4}a_N^2,$$

$$\|D_x^- V]\!]_h^2 = [\overline{\Lambda}V, V]_h = \frac{1}{2}\sum_{k=1}^{N-1} \lambda_k a_k^2 + \frac{1}{4}\lambda_N a_N^2,$$

$$[\![\overline{\Lambda}V]\!]_h^2 = \frac{1}{2}\sum_{k=1}^{N-1} \lambda_k^2 a_k^2 + \frac{1}{4}\lambda_N^2 a_N^2.$$

Next, we introduce analogous discrete Sobolev norms on the linear space S^h, consisting of all real-valued functions defined on the mesh $\overline{\Omega}^h$.

Discrete Sobolev Norms on S^h Similarly as on S_0^h, we introduce on S^h the following discrete analogues of the Sobolev norms $\|\cdot\|_{W_2^k(\Omega)}$, $k = 1, 2$:

$$[\![V]\!]_{1,h} = \|V\|_{W_2^1(\overline{\Omega}^h)} := \left([\![V]\!]_h^2 + \|D_x^- V]\!]_h^2\right)^{1/2},$$

$$[\![V]\!]_{2,h} = \|V\|_{W_2^2(\overline{\Omega}^h)} := \left([\![V]\!]_h^2 + \|D_x^- V]\!]_h^2 + [\![\overline{\Lambda}V]\!]_h^2\right)^{1/2}.$$

Fractional-Order Discrete Sobolev Norms Next we shall define fractional-order Sobolev norms on S_0^h and derive an interpolation inequality that relates these to the integer-order discrete Sobolev norms defined earlier. We shall limit ourselves to the case when the Sobolev index r is in the range $(0, 1) \cup (1, 2)$. We define the seminorm $|\cdot|_{W_2^r(\Omega^h)}$ by

$$|V|_{W_2^r(\Omega^h)} := \begin{cases} \left(h^2 \sum_{x,y\in\overline{\Omega}^h, x\neq y} \frac{[V(x)-V(y)]^2}{|x-y|^{1+2r}}\right)^{1/2} & \text{if } 0 < r < 1, \\ \left(h^2 \sum_{x,y\in\Omega_-^h, x\neq y} \frac{[D_x^+ V(x)-D_x^+ V(y)]^2}{|x-y|^{1+2r}}\right)^{1/2} & \text{if } 1 < r < 2, \end{cases}$$

and we introduce the corresponding fractional-order discrete Sobolev norm

$$\|V\|_{W_2^r(\Omega^h)} := \left(\|V\|_{W_2^{[r]}(\Omega^h)}^2 + |V|_{W_2^r(\Omega^h)}^2\right)^{1/2}, \quad 0 < r < 2,\ r \neq 1.$$

Higher order fractional-order discrete Sobolev norms can be defined similarly.

Next we state an interpolation inequality that establishes a relationship between fractional-order discrete Sobolev norms and the integer-order norms defined earlier.

Lemma 2.13 *Suppose that* $r \in (0, 1)$. *Then, there exists a positive real number* $C(r)$ *such that, for each mesh-function* $V \in S_0^h$,

$$\|V\|_{W_2^r(\Omega^h)} \leq C(r)\|V\|_{L_2(\Omega^h)}^{1-r}\|V\|_{W_2^1(\Omega^h)}^r, \quad 0 < r < 1.$$

Proof Given a mesh-function $V \in S_0^h$, we decompose it as a finite linear combination of sine functions, as in (2.20), and define the norm $B_r(\cdot)$ on S_0^h in terms of the corresponding expansion coefficients b_k, $k = 1, \dots, N-1$, by

$$B_r(V) := \left(\frac{1}{2}\sum_{k=1}^{N-1} k^{2r} b_k^2\right)^{1/2}.$$

It is left to the reader to verify that $B_r(\cdot)$ is indeed a norm on S_0^h. By noting (2.17), the elementary inequality

$$\sin x \geq \frac{2}{\pi}x, \quad 0 \leq x \leq \pi/2,$$

and Hölder's inequality with exponents $p := 1/(1-r)$ and $p' := 1/r$ we obtain

$$B_r(V) \leq \left[\frac{1}{2}\sum_{k=1}^{N-1}\left(\frac{\lambda_k}{4}\right)^r b_k^2\right]^{1/2} = 2^{-r}\left[\frac{1}{2}\sum_{k=1}^{N-1} b_k^{2(1-r)}\left(\lambda_k b_k^2\right)^r\right]^{1/2}$$

$$\leq 2^{-r}\left(\frac{1}{2}\sum_{k=1}^{N-1} b_k^2\right)^{(1-r)/2}\left(\frac{1}{2}\sum_{k=1}^{N-1}\lambda_k b_k^2\right)^{r/2},$$

and hence, by the discrete Parseval identities (2.21) and (2.22),

$$B_r(V) \leq 2^{-r}\|V\|_{L_2(\Omega^h)}^{1-r}|V|_{W_2^1(\Omega^h)}^r. \tag{2.29}$$

The rest of the proof is devoted to showing that the norm $B_r(\cdot)$ is equivalent to $\|\cdot\|_{W_2^r(\Omega^h)}$. For this purpose, we extend the function $V \in S_0^h$ from

$$\overline{\Omega}^h = \{kh : k = 0, \dots, N\}$$

to the mesh

$$\{kh : k = 0, \pm1, \pm2, \dots, \pm N\}$$

as an odd function; that is, $V(-x) := -V(x)$ for each x in $\overline{\Omega}^h$. The resulting function is then further extended to the infinite lattice

$$h\mathbb{Z} = \{kh : k = 0, \pm1, \pm2, \dots\}$$

as a 2-periodic function; as before, $h := 1/N$ and $N \geq 2$. Let $\omega^h := (-1, 1) \cap h\mathbb{Z}$ and $\overline{\omega}^h := [-1, 1] \cap h\mathbb{Z}$. For mesh-functions V defined on $\overline{\omega}^h$ we consider

$$N_r(V) := \left\{ h^2 \sum_{x\in\overline{\omega}^h}^{\star} \sum_{t\in\overline{\omega}^h, t\neq 0}^{\star} \frac{[V(x) - V(x-t)]^2}{|t|^{1+2r}} \right\}^{1/2},$$

where

$$h \sum_{x\in\overline{\omega}^h}^{\star} W(x) := \frac{1}{2}h\big[W(-1) + W(1)\big] + h \sum_{x\in\omega^h} W(x) = [W, 1]_{L_2(\overline{\omega}^h)}.$$

By noting the periodicity of the extended function (still denoted by V) and the expansion (2.20), we obtain

$$\begin{aligned}
N_r(V)^2 &= h^2 \sum_{x\in\overline{\omega}^h}^{\star} \sum_{t\in\overline{\omega}^h, t\neq 0}^{\star} |t|^{-1-2r} V(x)\big[-V(x-t) + 2V(x) - V(x+t)\big] \\
&= h^2 \sum_{x\in\overline{\omega}^h}^{\star} \sum_{t\in\overline{\omega}^h, t\neq 0}^{\star} |t|^{-1-2r} \sum_{l=1}^{N-1} b_l \sin l\pi x \sum_{k=1}^{N-1} 4b_k \sin^2 \frac{k\pi t}{2} \sin k\pi x \\
&= \sum_{l=1}^{N-1} \sum_{k=1}^{N-1} b_l b_k h \sum_{x\in\overline{\omega}^h}^{\star} \sin l\pi x \sin k\pi x h \sum_{t\in\overline{\omega}^h, t\neq 0}^{\star} |t|^{-1-2r} 4 \sin^2 \frac{k\pi t}{2} \\
&= 8 \sum_{k=1}^{N-1} b_k^2 h \sum_{t\in\Omega_+^h}^{\star\star} t^{-1-2r} \sin^2 \frac{k\pi t}{2}.
\end{aligned}$$

Here we have used the notation

$$h \sum_{t\in\Omega_+^h}^{\star\star} W(t) := h \sum_{t\in\Omega^h} W(t) + \frac{1}{2}hW(1) = (W, 1)_h + \frac{1}{2}hW(1).$$

After further transformation, we obtain

$$N_r(V)^2 = 16\left(\frac{\pi}{2}\right)^{2r} \frac{1}{2} \sum_{k=1}^{N-1} k^{2r} b_k^2 C(k, r),$$

where

$$C(k, r) := \frac{k\pi h}{2} \sum_{t\in\Omega_+^h}^{\star\star} \left(\frac{k\pi t}{2}\right)^{-1-2r} \sin^2 \frac{k\pi t}{2}.$$

It is easily seen that $C(k,r)$ is the Riemann sum for the integral

$$\int_0^{k\pi/2} x^{-1-2r} \sin^2 x \, dx$$

and can be therefore bounded from below and above as follows:

$$\frac{1}{8}\left(\frac{2}{\pi}\right)^{2r} \le C(k,r) \le \pi^{2-2r}\left(1+\frac{1}{2-2r}\right)+\frac{1}{2r}\left(\frac{2}{\pi}\right)^{2r}.$$

Thus we deduce that $N_r(\cdot)$ and $B_r(\cdot)$ are equivalent norms on S_0^h.

By noting inequality (2.29), the equivalence of the seminorm $|\cdot|_{W_2^1(\Omega^h)}$ and the norm $\|\cdot\|_{W_2^1(\Omega^h)}$ on the linear space S_0^h, in conjunction with the obvious inequality $|V|_{W_2^r(\Omega^h)} \le N_r(V)$, we then arrive at the desired inequality. That completes the proof. □

Remark 2.7 The lemma can also be proved by using the cosine expansion (2.28) and the norm

$$A_r(V) := \left(\frac{1}{2}\sum_{k=1}^{N-1} k^{2r} a_k^2 + \frac{1}{4} N^{2r} a_N^2\right)^{1/2}.$$

It can be shown that this norm is equivalent to $N_r(\cdot)$, provided that V has been extended periodically outside $\overline{\Omega}^h$ as an even function.

Remark 2.8 A similar argument shows, for $r \in (1,2)$, that there exists a positive real number $C_1(r)$ such that

$$\|V\|_{W_2^r(\Omega^h)} \le C_1(r)\|V\|_{W_2^1(\Omega^h)}^{2-r}\|V\|_{W_2^2(\Omega^h)}^{r-1}, \quad 1 < r < 2.$$

Remark 2.9 Finally we note that, similarly as on S_0^h, one can define a fractional-order discrete Sobolev norm on S^h as follows:

$$\|V\|_{W_2^r(\overline{\Omega}^h)}^2 := \left(\|V\|_{W_2^{[r]}(\overline{\Omega}^h)}^2 + |V|_{W_2^r(\Omega^h)}^2\right)^{1/2}, \quad 0 < r < 2,\ r \ne 1.$$

After this brief summary of notational conventions in one dimension, we consider a simple one-dimensional model problem, construct its finite difference approximation and derive bounds on the error, in the discrete norms defined above, between the analytical solution and its finite difference approximation.

2.2.3 Finite Difference Scheme for a Univariate Problem

We give a simple illustration of the general framework of finite difference approximation by considering the following two-point boundary-value problem for a second-order linear (ordinary) differential equation:

$$-u'' + c(x)u = f(x), \quad x \in (0, 1), \tag{2.30}$$

$$u(0) = 0, \qquad u(1) = 0. \tag{2.31}$$

We shall assume that $c \geq 0$ almost everywhere on $(0, 1)$, $c \in L_\infty(0, 1)$ and $f \in W_2^{-1}(0, 1)$.

The first step in the construction of a finite difference scheme for this boundary-value problem is to define the mesh. Let N be an integer, $N \geq 2$, and let $h := 1/N$ be the mesh-size; the mesh-points are $x_i := ih$, $i = 0, \ldots, N$. We then define

$$\Omega^h := \{x_i : i = 1, \ldots, N-1\},$$
$$\Gamma^h := \{x_0, x_N\} \quad \text{and} \quad \overline{\Omega}^h := \Omega^h \cup \Gamma^h.$$

Let us suppose that the unique weak solution $u \in \overset{\circ}{W}{}_2^1(0, 1)$ to this boundary-value problem is sufficiently smooth (e.g. $u \in C^4([0, 1])$). Then, by Taylor series expansion of u about the mesh-point x_i, $1 \leq i \leq N-1$, we deduce that, as $h \to 0$,

$$\begin{aligned} u(x_{i\pm1}) &= u(x_i \pm h) \\ &= u(x_i) \pm hu'(x_i) + \frac{h^2}{2}u''(x_i) \pm \frac{h^3}{6}u'''(x_i) + \mathcal{O}(h^4), \end{aligned}$$

so that

$$D_x^+ u(x_i) := \frac{u(x_{i+1}) - u(x_i)}{h} = u'(x_i) + \mathcal{O}(h),$$

$$D_x^- u(x_i) := \frac{u(x_i) - u(x_{i-1})}{h} = u'(x_i) + \mathcal{O}(h),$$

$$D_x^0 u(x_i) := \frac{u(x_{i+1}) - u(x_{i-1})}{2h} = u'(x_i) + \mathcal{O}(h^2)$$

and

$$\begin{aligned} D_x^+ D_x^- u(x_i) &= D_x^- D_x^+ u(x_i) \\ &= \frac{u(x_{i+1}) - 2u(x_i) + u(x_{i-1})}{h^2} \\ &= u''(x_i) + \mathcal{O}(h^2). \end{aligned}$$

Recall that D_x^+ and D_x^- are called the *forward* and *backward* divided difference operator, respectively, D_x^0 is referred to as the *central-difference operator*, while $D_x^+ D_x^-$ is the (symmetric) *second divided difference* operator. It follows from these Taylor series expansions that, for a sufficiently smooth function u (e.g. for $u \in C^2([0, 1])$), $D_x^+ u(x_i)$ and $D_x^+ u(x_i)$ approximate $u'(x_i)$ to $\mathcal{O}(h)$ for $i = 0, \ldots, N-1$

and $i = 1, \ldots, N$, respectively, while the central difference approximation $D_x^0 u(x_i)$ is more accurate: it approximates $u'(x_i)$ to $\mathcal{O}(h^2)$ for $i = 1, \ldots, N-1$ (provided that $u \in C^3([0, 1])$. Similarly, the *second divided difference* $D_x^+ D_x^- u(x_i)$ is an $\mathcal{O}(h^2)$ approximation to $u''(x_i)$, $i = 1, \ldots, N-1$, (as long as $u \in C^4([0, 1])$. Thus we replace the second derivative u'' in (2.30) by the second divided difference to obtain

$$-D_x^+ D_x^- u(x_i) + c(x_i)u(x_i) \approx f(x_i), \quad i = 1, \ldots, N-1, \tag{2.32}$$

$$u(x_0) = 0, \qquad u(x_N) = 0. \tag{2.33}$$

Here we have implicitly assumed that both c and f are continuous functions on the interval $(0, 1)$; thus, $c(x_i)$ and $f(x_i)$ are correctly defined for all $i = 1, \ldots, N-1$. We shall also suppose that

$$c(x) \geq 0 \quad \forall x \in (0, 1). \tag{2.34}$$

Now (2.32) and (2.33) indicate that we should seek our approximation U to u by solving the system of difference equations:

$$-D_x^+ D_x^- U_i + c(x_i)U_i = f(x_i), \quad i = 1, \ldots, N-1, \tag{2.35}$$

$$U_0 = 0, \qquad U_N = 0. \tag{2.36}$$

Using matrix notation, this can be written as

$$\mathcal{A}U = F,$$

where

$$\mathcal{A} := \begin{bmatrix} \frac{2}{h^2} + c(x_1) & -\frac{1}{h^2} & & & 0 \\ -\frac{1}{h^2} & \frac{2}{h^2} + c(x_2) & -\frac{1}{h^2} & & \\ & \ddots & \ddots & \ddots & \\ & & -\frac{1}{h^2} & \frac{2}{h^2} + c(x_{N-2}) & -\frac{1}{h^2} \\ 0 & & & -\frac{1}{h^2} & \frac{2}{h^2} + c(x_{N-1}) \end{bmatrix},$$

$$U := (U_1, U_2, \ldots, U_{N-1})^{\mathrm{T}}$$

and

$$F := \big(f(x_1), f(x_2), \ldots, f(x_{N-1})\big)^{\mathrm{T}}.$$

Thus $\mathcal{A}$ is a symmetric tridiagonal $(N-1) \times (N-1)$ matrix, and U and F are column vectors of size $N-1$.

We begin the analysis of the finite difference scheme (2.35), (2.36) by showing that it has a unique solution; this will be achieved by proving that the matrix $\mathcal{A}$ is nonsingular. For this purpose, we introduce the inner product (2.13). Let S_0^h denote the set of all real-valued functions V defined at the mesh-points x_i, $i = 0, \ldots, N$, such that $V_0 = V_N = 0$.

We define the linear operator $A : S_0^h \to S_0^h$ by

$$(AV)_i := -D_x^+ D_x^- V_i + c(x_i) V_i, \quad i = 1, \ldots, N-1,$$
$$(AV)_0 = (AV)_N := 0.$$

Returning to the finite difference scheme (2.35), (2.36) and using Lemma 2.10 and (2.34), we see that, for $V \in S_0^h$,

$$\begin{aligned}(AV, V)_h &= \left(-D_x^+ D_x^- V + cV, V\right)_h \\ &= \left(-D_x^+ D_x^- V, V\right)_h + (cV, V)_h \\ &\geq \sum_{i=1}^{N} h \left|D_x^- V_i\right|^2 = \|D_x^- V\|^2_{L_2(\Omega_+^h)},\end{aligned} \tag{2.37}$$

where the norm $\|\cdot\|_{L_2(\Omega_+^h)}$ has been defined in the previous section. Thus, if $AV = 0$ for some V, then $D_x^- V_i = 0$, $i = 1, \ldots, N$; because $V_0 = V_N = 0$, this implies that $V_i = 0$, $i = 0, \ldots, N$. Hence $AV = 0$ if, and only if, $V = 0$. We deduce that $A : S_0^h \to S_0^h$ is invertible and, consequently, $\mathcal{A}$ is a nonsingular matrix; thus (2.35), (2.36) has a unique solution, $U = \mathcal{A}^{-1} F$. We summarize our findings in the next theorem.

Theorem 2.14 *Suppose that* c *and* f *are continuous functions on the interval* $(0, 1)$, *and* $c(x) \geq 0$ *for* $x \in (0, 1)$; *then, the finite difference scheme* (2.35), (2.36) *possesses a unique solution* U *in* S_0^h.

We note that by Theorem 2.7, for $c \in C([0, 1])$ satisfying (2.34) and $f \in C([0, 1])$, the boundary-value problem (2.30), (2.31) has a unique weak solution $u \in \overset{\circ}{W}{}_2^1(0, 1)$; in fact, by Sobolev's embedding theorem u belongs to $C([0, 1])$ and therefore $u'' = f - cu \in C([0, 1])$. However to derive an error bound between u and its finite difference approximation U we shall have to assume that u is even more regular (the precise regularity hypothesis required in the analysis will be stated below). A key ingredient in our error analysis will be the fact that the scheme (2.35), (2.36) is stable (or discretely well-posed) in the sense that "small" perturbations in the data result in "small" perturbations in the corresponding finite difference solution. Actually, we shall prove the discrete version of the inequality appearing in Remark 2.6. For this purpose, we shall consider the *discrete* L_2 *norm* (2.14) and the *discrete Sobolev norm* (2.25). From (2.37) and the discrete Friedrichs inequality (2.26) we deduce, with $c_0 = 1/c_\star = 8/9$, that

$$(AV, V)_h \geq c_0 \|V\|^2_{W_2^1(\Omega^h)}. \tag{2.38}$$

Now the stability of the finite difference scheme (2.35), (2.36) easily follows.

Theorem 2.15 *The scheme* (2.35), (2.36) *is stable in the sense that*

$$\|U\|_{W_2^1(\Omega^h)} \leq \frac{1}{c_0}\|f\|_{L_2(\Omega^h)}, \tag{2.39}$$

where $c_0 = 8/9$.

Proof From (2.38) and (2.35) we have that

$$\begin{aligned} c_0\|U\|^2_{W_2^1(\Omega^h)} &\leq (AU, U)_h = (f, U)_h \\ &\leq \|f\|_{L_2(\Omega^h)}\|U\|_{L_2(\Omega^h)} \leq \|f\|_{L_2(\Omega^h)}\|U\|_{W_2^1(\Omega^h)}, \end{aligned}$$

and hence we deduce (2.39). □

Theorem 2.15 implies that if U_1 and U_2 are solutions of the problem (2.35), (2.36) corresponding to right-hand sides f_1 and f_2, respectively, then

$$\|U_1 - U_2\|_{W_2^1(\Omega^h)} \leq \frac{1}{c_0}\|f_1 - f_2\|_{L_2(\Omega^h)}.$$

Therefore, in analogy with the boundary-value problem (2.30), (2.31), the difference scheme (2.35), (2.36) is well-posed in the sense of Remark 2.6. It is important to note that the 'stability constant' $1/c_0$ is independent of the discretization parameter h: the spacing of the finite difference mesh.

By exploiting this stability result it is easy to derive a bound on the error between the analytical solution u, and its finite difference approximation U. We define the *global error*, e, by

$$e_i := u(x_i) - U_i, \quad i = 0, \ldots, N.$$

Obviously $e_0 = 0$, $e_N = 0$, and

$$Ae_i = \varphi_i, \quad i = 1, \ldots, N-1, \tag{2.40}$$

where the mesh-function φ, defined by

$$\varphi_i := Au(x_i) - f(x_i), \quad i = 1, \ldots, N-1,$$

is called the *truncation error* of the finite difference scheme. A simple calculation using (2.30) reveals that

$$\varphi_i = u''(x_i) - D_x^+ D_x^- u(x_i), \quad i = 1, \ldots, N-1.$$

Since the global error satisfies (2.40), we can apply (2.39) to deduce that

$$\|u - U\|_{W_2^1(\Omega^h)} = \|e\|_{W_2^1(\Omega^h)} \leq \frac{1}{c_0}\|\varphi\|_{L_2(\Omega^h)}. \tag{2.41}$$

It remains to bound $\|\varphi\|_{L_2(\Omega^h)}$.

Assuming now that $u \in C^4([0,1])$, the Taylor series expansions stated at the beginning of this section imply that

$$\varphi_i = u''(x_i) - D_x^+ D_x^- u(x_i) = \mathcal{O}(h^2);$$

thus, there exists a positive constant C, independent of h, such that

$$|\varphi_i| \leq Ch^2.$$

Consequently,

$$\|\varphi\|_{L_2(\Omega^h)} = \left(\sum_{i=1}^{N-1} h|\varphi_i|^2\right)^{1/2} \leq Ch^2. \tag{2.42}$$

Combining (2.41) and (2.42), it follows that

$$\|u - U\|_{W_2^1(\Omega^h)} \leq \frac{C}{c_0} h^2. \tag{2.43}$$

In fact, a more careful treatment of the remainder term in the Taylor series expansion of u reveals that, for $i = 1, \ldots, N-1$,

$$\varphi_i = u''(x_i) - D_x^+ D_x^- u(x_i) = -\frac{1}{12} h^2 u''''(\xi_i), \quad \xi_i \in (x_{i-1}, x_{i+1}).$$

Thus

$$|\varphi_i| \leq \frac{1}{12} h^2 \max_{x \in [0,1]} \left|u''''(x)\right|, \tag{2.44}$$

and hence

$$C = \frac{1}{12} \max_{x \in [0,1]} \left|u''''(x)\right|$$

in (2.42). As $c_0 = 1/c_\star$ and $c_\star = 9/8$, we deduce that $c_0 = 8/9$. Substituting the values of the constants C and c_0 into (2.43), it follows that

$$\|u - U\|_{W_2^1(\Omega^h)} \leq \frac{3}{32} h^2 \left\|u''''\right\|_{C([0,1])}.$$

Thus we have proved the following result.

Theorem 2.16 *Let $f \in C([0,1])$, $c \in C([0,1])$, with $c(x) \geq 0$ for all $x \in [0,1]$, and suppose that the corresponding solution of the boundary-value problem* (2.30), (2.31) *belongs to $C^4([0,1])$; then,*

$$\|u - U\|_{W_2^1(\Omega^h)} \leq \frac{3}{32} h^2 \left\|u''''\right\|_{C([0,1])}. \tag{2.45}$$

We note that by the argument following Theorem 2.14 the hypotheses $f \in C([0,1])$, $c \in C([0,1])$, $c \geq 0$ imply that the unique weak solution of the boundary-value problem (2.30), (2.31) belongs to $C^2([0,1])$, and it is therefore a classical solution. Thus, the word *solution* in this theorem means *classical solution*.

It follows from (2.37) with $V = e$, (2.40), the Cauchy–Schwarz inequality, the last inequality in (2.24), (2.27) and (2.44) that

$$\|u - U\|_{\infty,h} \leq \frac{1}{48\sqrt{2}} h^2 \|u''''\|_{C([0,1])}. \tag{2.46}$$

We thus deduce the following result.

Theorem 2.17 *Suppose that the assumptions of Theorem* 2.16 *are satisfied; then, the error bound* (2.46) *holds.*

This simple stability and error analysis of the finite difference scheme (2.35), (2.36) already contains the key ingredients of a general error analysis of finite difference approximations, and it is instructive to highlight them here.

(1) The first step is to prove the stability of the scheme in an appropriate mesh-dependent norm (cf. (2.39), for example). A typical stability result for the abstract finite difference scheme (2.11), (2.12) considered at the beginning of the section is of the form

$$c_0 |||U|||_{\Omega^h} \leq \|f_h\|_{\Omega^h} + \|g_h\|_{\Gamma^h}, \tag{2.47}$$

where $|||\cdot|||_{\Omega^h}$, $\|\cdot\|_{\Omega^h}$ and $\|\cdot\|_{\Gamma^h}$ are mesh-dependent norms involving mesh-points of Ω^h (or $\overline{\Omega}^h$) and Γ^h, respectively, and c_0 is a positive constant, independent of h.

(2) The second step is to estimate the size of the *truncation error*,

$$\varphi_{\Omega^h} := L_h u - f_h \quad \text{in } \Omega^h,$$

$$\varphi_{\Gamma^h} := l_h u - g_h \quad \text{on } \Gamma^h.$$

In the case of the finite difference scheme (2.11), (2.12), $\varphi_{\Gamma^h} = 0$, and therefore φ_{Γ^h} did not appear explicitly in our error analysis. If

$$\|\varphi_{\Omega^h}\|_{\Omega^h} + \|\varphi_{\Gamma^h}\|_{\Gamma^h} \to 0 \quad \text{as } h \to 0,$$

for a sufficiently smooth solution u of (2.9), (2.10), we say that the scheme (2.11), (2.12) is *consistent.* If p is the largest positive real number such that

$$\|\varphi_{\Omega^h}\|_{\Omega^h} + \|\varphi_{\Gamma^h}\|_{\Gamma^h} \leq C h^p \quad \text{as } h \to 0,$$

(where C is a positive constant independent of h) for all sufficiently smooth u, then the scheme is said to have *order of accuracy p*.

The finite difference scheme (2.11), (2.12) is said to *converge* to (2.9), (2.10) (and U is said to converge to u) in the norm $|||\cdot|||_{\Omega^h}$, if

$$|||u - U|||_{\Omega^h} \to 0 \quad \text{as } h \to 0.$$

If q is the largest positive real number such that, for all u sufficiently smooth,

$$|||u - U|||_{\Omega^h} \leq Ch^q \quad \text{as } h \to 0$$

(where C is a positive constant independent of h), then the scheme is said to have *order of convergence* q.

From these definitions we deduce the following fundamental theorem.

Theorem 2.18 *Suppose that the finite difference scheme* (2.11), (2.12) *for problem* (2.9), (2.10) *is stable* (*i.e.* (2.47) *holds for all* f_h *and* g_h *and corresponding solution* U, *with* c_0 *independent of* h) *and that the scheme is consistent*; *then* (2.11), (2.12) *is a convergent approximation of* (2.9), (2.10) *and the order of convergence is not less than the order of accuracy.*

Proof We define the *global error* $e := u - U$; then,

$$L_h e = L_h(u - U) = L_h u - L_h U = L_h u - f_h.$$

Thus,

$$L_h e = \varphi_{\Omega^h}$$

and similarly

$$l_h e = \varphi_{\Gamma^h}.$$

By stability,

$$c_0 |||u - U|||_{\Omega^h} = c_0 |||e|||_{\Omega^h} \leq \|\varphi_{\Omega^h}\|_{\Omega^h} + \|\varphi_{\Gamma^h}\|_{\Gamma^h},$$

and hence we arrive at the stated result. □

Paraphrasing Theorem 2.18, *stability* and *consistency* of the scheme imply its *convergence*. This abstract result is at the heart of the error analysis of finite difference approximations of differential equations.

2.2.4 The Multi-dimensional Case

Since the two-dimensional case is sufficiently representative, for the sake of notational simplicity we shall confine our attention to elliptic boundary-value problems in the plane.

Meshes and Divided Difference Operators Assuming that N is an integer, $N \geq 2$, we shall use a uniform square mesh Ω^h with mesh-size $h := 1/N$ over the unit square $\Omega := (0,1)^2$, defined by

$$\Omega^h := \{x = (x_1, x_2) = (ih, jh) : i, j = 1, \ldots, N-1\},$$

and the square mesh

$$\overline{\Omega}^h := \{(ih, jh) : i, j = 0, \ldots, N\}.$$

Let $\Gamma := \partial\Omega$ be the boundary of Ω and define

$$\Gamma^h := h\mathbb{Z}^2 \cap \Gamma = \overline{\Omega}^h \setminus \Omega^h.$$

Analogously, let

$$\Gamma_{ik} := \{x \in \Gamma : x_i = k,\ 0 < x_{3-i} < 1\}, \quad i = 1, 2,\ k = 0, 1,$$

and define

$$\Gamma_{ik}^h := \Gamma_{ik} \cap h\mathbb{Z}^2, \qquad \overline{\Gamma}_{ik}^h := \overline{\Gamma}_{ik} \cap h\mathbb{Z}^2, \qquad \Gamma_*^h := \Gamma^h \setminus \left(\cup_{i,k}\Gamma_{ik}^h\right).$$

Let us also introduce

$$\Omega_i^h := \Omega^h \cup \Gamma_{i0}^h, \qquad \Omega_{i+2}^h := \Omega^h \cup \Gamma_{i1}^h, \quad i = 1, 2,$$
$$\Omega_{kl}^h := \Omega^h \cup \Gamma_{1k}^h \cup \Gamma_{2l}^h \cup \{(k,l)\}, \quad k, l = 0, 1.$$

Let S^h be the set of all real-valued functions defined on the mesh $\overline{\Omega}^h$. We shall use the notation $V_{ij} := V(ih, jh)$. By S_0^h we denote the set of all real-valued functions defined on the mesh $\overline{\Omega}^h$ that vanish at all points of Γ^h. The set S_0^h is equipped with the inner product

$$(V, W)_h = (V, W)_{L_2(\Omega^h)} := h^2 \sum_{x \in \Omega^h} V(x) W(x) = h^2 \sum_{i,j=1}^{N-1} V_{ij} W_{ij}, \quad (2.48)$$

and the norm

$$\|V\|_h = \|V\|_{L_2(\Omega^h)} := (V, V)_h^{1/2}.$$

The norms $\|\cdot\|_{L_2(\Omega_i^h)}$ and $\|\cdot\|_{L_2(\Omega_{kl}^h)}$ are defined analogously to $\|\cdot\|_{L_2(\Omega^h)}$.

The forward, backward and central divided difference operators on the mesh $\overline{\Omega}_h$ are defined analogously as in the one-dimensional case:

$$D_{x_i}^+ V := \frac{V^{+i} - V}{h}, \qquad D_{x_i}^- V := \frac{V - V^{-i}}{h}, \qquad D_{x_i}^0 V := \frac{1}{2}\left(D_{x_i}^+ V + D_{x_i}^- V\right),$$

where

$$V^{\pm i} := V^{\pm i}(x) = V(x \pm h\mathrm{e}_i), \quad \mathrm{e}_i := (\delta_{i1}, \delta_{i2}), \quad i = 1, 2,$$

and δ_{ik} is the Kronecker delta.

Discrete Sobolev Norms Analogously as in the one-dimensional case, we define the following discrete Sobolev seminorms on S^h:

$$\begin{aligned} |V|_{W_2^1(\Omega^h)} &:= \left(\|D_{x_1}^+ V\|^2_{L_2(\Omega_1^h)} + \|D_{x_2}^+ V\|^2_{L_2(\Omega_2^h)}\right)^{1/2} \\ &= \left(\|D_{x_1}^- V\|^2_{L_2(\Omega_3^h)} + \|D_{x_2}^- V\|^2_{L_2(\Omega_4^h)}\right)^{1/2}, \\ |V|_{W_2^2(\Omega^h)} &:= \left(\|D_{x_1}^+ D_{x_1}^- V\|^2_{L_2(\Omega^h)} + \|D_{x_1}^+ D_{x_2}^+ V\|^2_{L_2(\Omega_{00}^h)}\right. \\ &\quad \left. + \|D_{x_2}^+ D_{x_2}^- V\|^2_{L_2(\Omega^h)}\right)^{1/2} \end{aligned} \tag{2.49}$$

and the corresponding discrete Sobolev norms

$$\|V\|_{W_2^k(\Omega^h)} := \left(\|V\|^2_{W_2^{k-1}(\Omega^h)} + |V|^2_{W_2^k(\Omega^h)}\right)^{1/2}, \quad k = 1, 2, \tag{2.50}$$

with the notational convention $W_2^0(\Omega^h) := L_2(\Omega^h)$.

Let us also introduce the following inner products

$$[V, W]_h := h^2 \sum_{x \in \Omega^h} V(x)W(x) + \frac{h^2}{2} \sum_{x \in \Gamma^h \setminus \Gamma_*^h} V(x)W(x) + \frac{h^2}{4} \sum_{x \in \Gamma_*^h} V(x)W(x),$$

$$[V, W]_{i,h} := h^2 \sum_{x \in \Omega_i^h} V(x)W(x) + \frac{h^2}{2} \sum_{x \in \Gamma^h \setminus (\Gamma_{i0}^h \cup \overline{\Gamma}_{i1}^h)} V(x)W(x), \quad i = 1, 2,$$

and the associated norms

$$\begin{aligned} [\![V]\!]_h &= [\![V]\!]_{L_2(\Omega^h)} := [V, V]_h^{1/2}, \\ [\![V]\!]_i &= [\![V]\!]_{i,h} := [V, V]_{i,h}^{1/2}. \end{aligned}$$

In analogy with the one-dimensional case, we define the following discrete Sobolev seminorms and norms on S^h:

$$\begin{aligned} [V]_{W_2^1(\Omega^h)} &:= \left([\![D_{x_1}^+ V]\!]_1^2 + [\![D_{x_2}^+ V]\!]_2^2\right)^{1/2}, \\ [V]_{W_2^2(\Omega^h)} &:= \left([\![\overline{\Lambda}_1 V]\!]^2_{L_2(\Omega^h)} + \|D_{x_1}^+ D_{x_2}^+ V\|^2_{L_2(\Omega_{00}^h)} + [\![\overline{\Lambda}_2 V]\!]^2_{L_2(\Omega^h)}\right)^{1/2}, \\ [\![V]\!]_{W_2^k(\Omega^h)} &:= \left([\![V]\!]^2_{W_2^{k-1}(\Omega^h)} + [V]^2_{W_2^k(\Omega^h)}\right)^{1/2}, \quad k = 1, 2, \end{aligned}$$

where

$$(\overline{\Lambda}_i V)(x) := \begin{cases} -\frac{2}{h} D_{x_i}^+ V & \text{if } x \in \overline{\Gamma}_{i0}^h, \\ -D_{x_i}^+ D_{x_i}^- V(x) & \text{if } x \in \Omega^h \cup \Gamma_{3-i,0}^h \cup \Gamma_{3-i,1}^h, \\ \frac{2}{h} D_{x_i}^- V & \text{if } x \in \overline{\Gamma}_{i1}^h. \end{cases}$$

The Discrete Laplace Operator on S_0^h We consider the discrete analogue of the Laplace operator in two space dimensions, defined on $h\mathbb{Z}^2$ by

$$\Delta_h V := D_{x_1}^+ D_{x_1}^- V + D_{x_2}^+ D_{x_2}^- V.$$

The mapping $\Lambda : S_0^h \to S_0^h$ defined, for $V \in S_0^h$, by

$$(\Lambda V)(x) = \begin{cases} -(\Delta_h V)(x) & \text{if } x \in \Omega^h, \\ 0 & \text{if } x \in \Gamma^h, \end{cases}$$

positive definite operator with respect to the inner product $(\cdot, \cdot)_h$. In particular, for $V \in S_0^h$ we have that

$$(\Lambda V, V)_h = (-\Delta_h V, V)_h = |V|^2_{W_2^1(\Omega^h)}. \tag{2.51}$$

Furthermore,

$$\|\Delta_h V\|_h^2 = \|D_{x_1}^+ D_{x_1}^- V\|^2_{L_2(\Omega^h)} + 2\|D_{x_1}^+ D_{x_2}^+ V\|^2_{L_2(\Omega_{00}^h)} + \|D_{x_2}^+ D_{x_2}^- V\|^2_{L_2(\Omega^h)},$$

and therefore,

$$\|\Delta_h V\|_h^2 \geq |V|^2_{W_2^2(\Omega^h)}.$$

Similarly,

$$\|\Delta_h V\|_h^2 \geq 16(-\Delta_h V, V)_h \geq 16^2 \|V\|_h^2 = 16^2 \|V\|^2_{L_2(\Omega^h)}$$

and

$$|V|_{W_2^2(\Omega^h)} \geq 2\sqrt{2}|V|_{W_2^1(\Omega^h)} \geq 8\sqrt{2}\|V\|_{L_2(\Omega^h)}, \quad V \in S_0^h. \tag{2.52}$$

Consequently, on the linear space S_0^h the seminorms $|\cdot|_{W_2^1(\Omega^h)}$ and $|\cdot|_{W_2^2(\Omega^h)}$ are equivalent to the norms $\|\cdot\|_{W_2^1(\Omega^h)}$ and $\|\cdot\|_{W_2^2(\Omega^h)}$, respectively.

Lemma 2.19 (Discrete Friedrichs Inequality) *There exists a positive real number $c_\star$, independent of h, such that*

$$\|V\|^2_{W_2^1(\Omega^h)} \leq c_\star \left(\|D_{x_1}^+ V\|^2_{L_2(\Omega_1^h)} + \|D_{x_2}^+ V\|^2_{L_2(\Omega_2^h)}\right) \tag{2.53}$$

for all V in S_0^h.

Proof Inequality (2.53) with $c_\star = 17/16$ follows directly from the definition (2.49) of the seminorm $|\cdot|_{W_2^1(\Omega^h)}$ and the second inequality in (2.52). □

Fractional-Order Discrete Sobolev Norms We define the fractional-order discrete Sobolev seminorm $|\cdot|_{W_2^r(\Omega^h)}$ by

$$|V|^2_{W_2^r(\Omega^h)} := \sum_{i=1}^{2} h^3 \sum_{\substack{x_i,t_i=0\\ x_i\neq t_i}}^{Nh} \sum_{x_{3-i}=h}^{(N-1)h} \frac{[V(x)-V(t_i e_i + x_{3-i}e_{3-i})]^2}{|x_i-t_i|^{1+2r}}$$

if $0<r<1$, and by

$$|V|^2_{W_2^r(\Omega^h)} := \sum_{i=1}^{2} h^3 \sum_{\substack{x_i,t_i=0\\ x_i\neq t_i}}^{(N-1)h} \sum_{x_{3-i}=0}^{(N-1)h} \frac{[D^+_{x_i}V(x)-D^+_{x_i}V(t_i e_i + x_{3-i}e_{3-i})]^2}{|x_i-t_i|^{1+2(r-1)}}$$

$$+ \sum_{i=1}^{2} h^3 \sum_{\substack{x_{3-i},t_{3-i}=0\\ x_{3-i}\neq t_{3-i}}}^{Nh} \sum_{x_i=0}^{(N-1)h} \frac{[D^+_{x_i}V(x)-D^+_{x_i}V(x_i e_i + t_{3-i}e_{3-i})]^2}{|x_{3-i}-t_{3-i}|^{1+2(r-1)}}$$

if $1<r<2$. We also introduce the associated fractional-order discrete Sobolev norm by

$$\|V\|_{W_2^r(\Omega^h)} := \left(\|V\|^2_{W_2^{[r]}(\Omega^h)} + |V|^2_{W_2^r(\Omega^h)}\right)^{1/2}, \quad 0<r<2,\ r\neq 1.$$

Similarly as in one dimension, we have the interpolation inequalities

$$\begin{aligned} \|V\|_{W_2^r(\Omega^h)} &\le C(r)\|V\|^{1-r}_{L_2(\Omega^h)}\|V\|^{r}_{W_2^1(\Omega^h)}, \quad 0<r<1,\\ \|V\|_{W_2^r(\Omega^h)} &\le C(r)\|V\|^{2-r}_{W_2^1(\Omega^h)}\|V\|^{r-1}_{W_2^2(\Omega^h)}, \quad 1<r<2, \end{aligned} \tag{2.54}$$

which follow directly from their one-dimensional counterparts.

2.2.5 Approximation of a Generalized Poisson Problem

In Sect. 2.2.3 we presented a detailed error analysis for a finite difference approximation of a simple two-point boundary-value problem. Here we shall undertake a similar study for the generalized Poisson equation in two space dimensions subject to a homogeneous Dirichlet boundary condition:

$$-\Delta u + c(x,y)u = f(x,y) \quad \text{in } \Omega, \tag{2.55}$$

$$u = 0 \quad \text{on } \Gamma = \partial\Omega, \tag{2.56}$$

where $\Omega := (0, 1) \times (0, 1)$, c is a continuous function on $\overline{\Omega}$ and $c(x, y) \geq 0$. For the sake of notational simplicity we have denoted the two independent variables by x and y, instead of x_1 and x_2. As far as the smoothness of the function f is concerned, we shall consider two distinct cases:

(a) First we shall assume that f is continuous on $\overline{\Omega}$. In this case, the error analysis proceeds along the same lines as in Sect. 2.2.3.
(b) We shall then consider the case when f is in $L_2(\Omega)$ only; then the boundary-value problem (2.55), (2.56) does not necessarily have a classical solution; nevertheless, a weak solution still exists. This lack of smoothness gives rise to some technical difficulties both in the formulation of an adequate finite difference scheme and its error analysis. Since the point values of f need not be meaningful at the mesh-points (after all, f can be changed on a subset of Ω of zero Lebesgue measure without altering it as an element of $L_2(\Omega)$), instead of sampling the function f at the mesh-points we shall sample a mollified right-hand side $T_h f$. Also, since the analytical solution may not have a Taylor expansion with the required number of terms, we shall apply a different technique, based on integral representation theorems, to estimate the size of the truncation error.

We begin by considering the first of these two cases.

(a) ($f \in C(\overline{\Omega})$) The first step in the construction of the finite difference approximation to (2.55), (2.56) is to define the mesh. Let N be an integer, $N \geq 2$, and let $h := 1/N$; the mesh-points are (x_i, y_j), $i, j = 0, \dots, N$, where $x_i := ih$, $y_j := jh$. These mesh-points form the mesh

$$\overline{\Omega}^h := \{(x_i, y_j) : i, j = 0, \dots, N\}.$$

Similarly as in Sect. 2.2.2, we consider the set of interior mesh-points

$$\Omega^h := \{(x_i, y_j) : i, j = 1, \dots, N-1\}$$

and the set of boundary mesh-points

$$\Gamma^h := \overline{\Omega}^h \setminus \Omega^h.$$

In analogy with (2.35), (2.36), the finite difference approximation of (2.55), (2.56) is:

$$-\left(D_x^+ D_x^- U_{ij} + D_y^+ D_y^- U_{ij}\right) + c(x_i, y_j) U_{ij} = f(x_i, y_j), \quad (x_i, y_j) \in \Omega^h, \tag{2.57}$$

$$U = 0 \quad \text{on } \Gamma^h. \tag{2.58}$$

In expanded form, this can be written as follows:

$$-\left(\frac{U_{i+1,j} - 2U_{ij} + U_{i-1,j}}{h^2} + \frac{U_{i,j+1} - 2U_{ij} + U_{i,j-1}}{h^2}\right) + c(x_i, y_j) U_{ij}$$
$$= f(x_i, y_j) \quad \text{if } (x_i, y_j) \in \Omega^h, \tag{2.59}$$

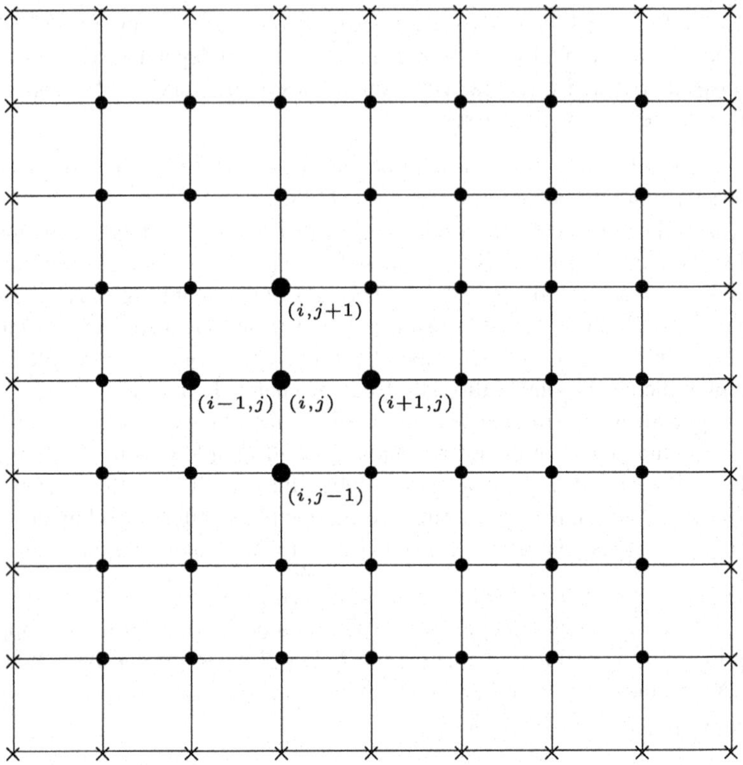

Fig. 2.1 The set of interior mesh-points Ω^h, denoted by •, the set of boundary mesh-points Γ^h, denoted by $\times$, and a typical five-point difference stencil

$$U_{ij} = 0 \quad \text{if } (x_i, y_j) \in \Gamma^h, \tag{2.60}$$

where the divided difference operators $D_x^\pm = D_{x_1}^\pm$ and $D_y^\pm = D_{x_2}^\pm$ have been defined in Sect. 2.2.2.

For each i and j, $1 \le i, j \le N-1$, the finite difference equation (2.59) involves five values of the approximate solution U: $U_{i,j}$, $U_{i-1,j}$, $U_{i+1,j}$, $U_{i,j-1}$, $U_{i,j+1}$, as indicated in Fig. 2.1; hence its name: *five-point difference scheme*. It is possible to write (2.59), (2.60) as a system of linear equations

$$\mathcal{A}U = F, \tag{2.61}$$

where

$$\begin{aligned} U &:= (U_{11}, U_{12}, \dots, U_{1,N-1}, U_{21}, U_{22}, \dots, U_{2,N-1}, \dots, \\ &\quad \dots, U_{i1}, U_{i2}, \dots, U_{i,N-1}, \dots, U_{N-1,1}, U_{N-1,2}, \dots, U_{N-1,N-1})^{\mathrm{T}}, \\ F &:= (F_{11}, F_{12}, \dots, F_{1,N-1}, F_{21}, F_{22}, \dots, F_{2,N-1}, \dots, \\ &\quad \dots, F_{i1}, F_{i2}, \dots, F_{i,N-1}, \dots, F_{N-1,1}, F_{N-1,2}, \dots, F_{N-1,N-1})^{\mathrm{T}}, \end{aligned}$$

Fig. 2.2 The sparsity structure of the banded matrix $\mathcal{A}$: K is an $(N-1)\times(N-1)$ symmetric tridiagonal matrix, $J=(-1/h^2)I$, I is the $(N-1)\times(N-1)$ identity matrix, and O is the $(N-1)\times(N-1)$ zero matrix

$$\mathcal{A}=\begin{pmatrix} K & J & O & O & \cdots & O & O \\ J & K & J & O & \cdots & O & O \\ O & J & K & J & \cdots & O & O \\ O & O & J & K & \cdots & O & O \\ \cdots & \cdots & \cdots & \cdots & \cdots & \cdots & \cdots \\ O & O & O & O & \cdots & K & J \\ O & O & O & O & \cdots & J & K \end{pmatrix}$$

and $\mathcal{A}$ is an $(N-1)^2 \times (N-1)^2$ sparse, banded matrix.

A typical row of the matrix contains five nonzero entries, corresponding to the five values of U in the finite difference stencil shown in Fig. 2.1, while the sparsity structure of $\mathcal{A}$ is indicated in Fig. 2.2.

Next we show that (2.57), (2.58) has a unique solution. We proceed in the same way as in the previous section for the finite difference approximation of the two-point boundary-value problem. For two functions, V and W, defined on Ω^h, we introduce the discrete L_2-inner product (2.48):

$$(V,W)_h := \sum_{i=1}^{N-1}\sum_{j=1}^{N-1} h^2 V_{ij} W_{ij}.$$

Again, let S_0^h denote the set of functions V defined on $\overline{\Omega}^h$ such that $V=0$ on Γ^h.

We define the linear operator

$$A : S_0^h \to S_0^h$$

at mesh-points of Ω^h and Γ^h, respectively, as follows:

$$\begin{aligned}(AV)_{ij} &:= -\left(D_x^+ D_x^- V_{ij} + D_y^+ D_y^- V_{ij}\right) + c(x_i)V_i, \quad i,j=1,\dots,N-1,\\ (AV)_{i0} &= (AV)_{iN} = (AV)_{0j} = (AV)_{Nj} := 0, \quad i,j=0,\dots,N.\end{aligned}$$

Returning to the analysis of the finite difference scheme (2.57), (2.58), we note that, since $c(x,y)\geq 0$ on $\overline{\Omega}$, by (2.51) and (2.49) we have that

$$\begin{aligned}(AV,V)_h &= \left(-D_x^+D_x^-V - D_y^+D_y^-V + cV, V\right)_h \\ &= \left(-D_x^+D_x^-V, V\right)_h + \left(-D_y^+D_y^-V, V\right)_h + (cV,V)_h \\ &\geq \sum_{i=1}^{N}\sum_{j=1}^{N-1} h^2 \left|D_x^- V_{ij}\right|^2 + \sum_{i=1}^{N-1}\sum_{j=1}^{N} h^2\left|D_y^- V_{ij}\right|^2, \end{aligned} \tag{2.62}$$

for any V in S_0^h. This implies, just as in the one-dimensional analysis presented in the previous section, that $\mathcal{A}$ is a nonsingular matrix. Indeed if $AV=0$, then (2.62) yields:

$$D_x^- V_{ij} = \frac{V_{ij} - V_{i-1,j}}{h} = 0, \qquad \begin{array}{l} i = 1, \dots, N, \\ j = 1, \dots, N-1; \end{array}$$

$$D_y^- V_{ij} = \frac{V_{ij} - V_{i,j-1}}{h} = 0, \qquad \begin{array}{l} i = 1, \dots, N-1, \\ j = 1, \dots, N. \end{array}$$

Since $V = 0$ on Γ^h, these imply that $V = 0$ on $\overline{\Omega}^h$. Thus $AV = 0$ if, and only if, $V = 0$. Hence $\mathcal{A}$ is nonsingular, and $U = \mathcal{A}^{-1}F$ is the unique solution of (2.57), (2.59); the solution may be found by solving the system of linear equations (2.61).

In order to prove the stability of the finite difference scheme (2.57), (2.58), we consider, similarly as in the one-dimensional case, the discrete L_2 norm

$$\|V\|_{L_2(\Omega^h)} := (V, V)_h^{1/2},$$

and the discrete W_2^1 norm (see (2.50))

$$\|V\|_{W_2^1(\Omega^h)} := \left(\|V\|_{L_2(\Omega^h)}^2 + \|D_x^- V\|_{L_2(\Omega_x^h)}^2 + \|D_y^- V\|_{L_2(\Omega_y^h)}^2\right)^{1/2},$$

where

$$\Omega_x^h := \Omega_3^h = \{(x_i, y_j) : i = 1, \dots, N, j = 1, \dots, N-1\},$$

$$\Omega_y^h := \Omega_4^h = \{(x_i, y_j) : i = 1, \dots, N-1, j = 1, \dots, N\}.$$

The norm $\|\cdot\|_{W_2^1(\Omega^h)}$ is the discrete analogue of the Sobolev norm $\|\cdot\|_{W_2^1(\Omega)}$ defined by

$$\|u\|_{W_2^1(\Omega)} := \left(\|u\|_{L_2(\Omega)}^2 + \left\|\frac{\partial u}{\partial x}\right\|_{L_2(\Omega)}^2 + \left\|\frac{\partial u}{\partial y}\right\|_{L_2(\Omega)}^2\right)^{1/2}.$$

In terms of this notation the inequality (2.62) has the following form:

$$(AV, V)_h \geq \|D_x^- V\|_{L_2(\Omega_x^h)}^2 + \|D_y^- V\|_{L_2(\Omega_y^h)}^2. \tag{2.63}$$

The discrete Friedrichs inequality (2.53) and inequality (2.63) imply that

$$(AV, V)_h \geq c_0 \|V\|_{W_2^1(\Omega^h)}^2, \tag{2.64}$$

where $c_0 = 1/c_\star = 16/17$.

Theorem 2.20 *The scheme* (2.57), (2.58) *is stable in the sense that*

$$\|U\|_{W_2^1(\Omega^h)} \leq \frac{1}{c_0} \|f\|_{L_2(\Omega^h)}, \tag{2.65}$$

where $c_0 = 16/17$.

Proof The proof of this stability result is completely analogous to that of its one-dimensional counterpart (2.39), now using (2.64) and the Cauchy–Schwarz inequality. □

Having established the stability of the difference scheme (2.57), (2.58), we turn to the question of its accuracy. We define the *global error* e by

$$e_{ij} := u(x_i, y_j) - U_{ij}, \quad i, j = 0, \ldots, N,$$

and the *truncation error* φ by

$$\varphi_{ij} := Au(x_i, y_j) - f(x_i, y_j), \quad i, j = 1, \ldots, N-1.$$

Then,

$$Ae_{ij} = \varphi_{ij}, \quad i, j = 1, \ldots, N-1,$$
$$e = 0 \quad \text{on } \Gamma^h.$$

By noting (2.65) we have

$$\begin{aligned} \|u - U\|_{W_2^1(\Omega^h)} &= \|e\|_{W_2^1(\Omega^h)} \\ &\le \frac{1}{c_0} \|\varphi\|_{L_2(\Omega^h)}. \end{aligned} \tag{2.66}$$

Thus, in order to obtain a bound on the global error, it suffices to estimate the size of the truncation error in the $\|\cdot\|_{L_2(\Omega^h)}$ norm. To do so, let us assume that $u \in C^4(\overline{\Omega})$; then, by expanding each term in φ in a Taylor series about the point (x_i, y_j), we obtain

$$\begin{aligned} \varphi_{ij} &= \Delta u(x_i, y_j) - \left(D_x^+ D_x^- u(x_i, y_j) + D_y^+ D_y^- u(x_i, y_j)\right) \\ &= \left[\frac{\partial^2 u}{\partial x^2}(x_i, y_j) - D_x^+ D_x^- u(x_i, y_j)\right] + \left[\frac{\partial^2 u}{\partial y^2}(x_i, y_j) - D_y^+ D_y^- u(x_i, y_j)\right] \\ &= -\frac{h^2}{12}\left(\frac{\partial^4 u}{\partial x^4}(\xi_i, y_j) + \frac{\partial^4 u}{\partial y^4}(x_i, \eta_j)\right), \quad i, j = 1, \ldots, N-1, \end{aligned}$$

where $\xi_i \in (x_{i-1}, x_{i+1})$, $\eta_j \in (y_{j-1}, y_{j+1})$.

Thus,

$$|\varphi_{ij}| \le \frac{h^2}{12}\left(\left\|\frac{\partial^4 u}{\partial x^4}\right\|_{C(\overline{\Omega})} + \left\|\frac{\partial^4 u}{\partial y^4}\right\|_{C(\overline{\Omega})}\right),$$

and we deduce that the truncation error φ satisfies the bound

$$\|\varphi\|_{L_2(\Omega^h)} \le \frac{h^2}{12}\left(\left\|\frac{\partial^4 u}{\partial x^4}\right\|_{C(\overline{\Omega})} + \left\|\frac{\partial^4 u}{\partial y^4}\right\|_{C(\overline{\Omega})}\right). \tag{2.67}$$

Finally (2.66) and (2.67) yield the following error bound.

Theorem 2.21 *Let* $f \in C(\overline{\Omega})$, $c \in C(\overline{\Omega})$, *with* $c(x, y) \geq 0$, $(x, y) \in \overline{\Omega}$, *and suppose that the corresponding weak solution of the boundary-value problem* (2.55), (2.56) *belongs to* $C^4(\overline{\Omega})$; *then*

$$\|u - U\|_{W_2^1(\Omega^h)} \leq \frac{17h^2}{192}\left(\left\|\frac{\partial^4 u}{\partial x^4}\right\|_{C(\overline{\Omega})} + \left\|\frac{\partial^4 u}{\partial y^4}\right\|_{C(\overline{\Omega})}\right). \tag{2.68}$$

Proof Recall that $1/c_0 = c_\star = 17/16$, and combine (2.66) and (2.67). □

According to this result, the five-point difference scheme (2.57), (2.58) for the boundary-value problem (2.55), (2.56) is second-order convergent, provided that u is sufficiently smooth; i.e. $u \in C^4(\overline{\Omega})$.

Elliptic regularity theory tells us (see, for example, Ladyzhenskaya and Ural'tseva [118], Gilbarg and Trudinger [53] or Renardy and Rogers [155]) that if the right-hand side and the coefficients are "sufficiently smooth", then the associated classical solution of the elliptic problem is "as smooth as one would expect" in the *interior* of the domain on which the problem is posed; e.g. in the case of a second-order elliptic boundary-value problem, if $f \in C^{k,\alpha}(\Omega)$, $k \geq 0$, $0 < \alpha < 1$, then $u \in C^{k+2,\alpha}(\Omega)$. Unfortunately, in general, the solution will not be smooth up to the boundary if the boundary is not of class $C^{k+2,\alpha}$, as is the case when Ω is a square. For a simple illustration, we refer to Example 9.52 on p. 325 of Renardy and Rogers [155]; a more detailed account of regularity theory for elliptic equations in domains with nonsmooth boundaries is given in Grisvard [62, 63] and Dauge [28]. Thus, in general, the solution of our simple model problem (2.55), (2.56), will not belong to $C^4(\overline{\Omega})$ even if f and c are smooth functions, because the boundary $\Gamma = \partial\Omega$ is only of class $C^{0,1}$. Consequently, the hypothesis $u \in C^4(\overline{\Omega})$ that was made in the statement of Theorem 2.21 is unrealistic (unless f satisfies suitable compatibility conditions at the four corners of Ω (cf. (2.8))).

Our analysis has another limitation: it was performed under the assumption that $f \in C(\overline{\Omega})$, which was necessary in order to ensure that the values of f are meaningfully defined at the mesh-points. However, in applications one often encounters differential equations where f is a lot less smooth (e.g. f is piecewise continuous, or $f \in L_2(\Omega)$, or f is a Borel measure). When $f \in L_2(\Omega)$, for example, we know that the homogeneous Dirichlet boundary-value problem for the partial differential equation $-\Delta u + cu = f$, with c bounded and nonnegative, still has a unique weak solution in $H_0^1(\Omega)$, so it is natural to ask whether one can construct a second-order accurate finite difference approximation of the weak solution. This brings us to case (b), formulated at the beginning of the section.

(b) ($f \in L_2(\Omega)$). We shall use the same finite difference mesh as in case (a), but we shall modify the difference scheme (2.57), (2.58) to cater for the fact that f is not continuous on $\overline{\Omega}$. The idea is to replace $f(x_i, y_j)$ in (2.57) by a cell-average

Fig. 2.3 The cell K_{ij}

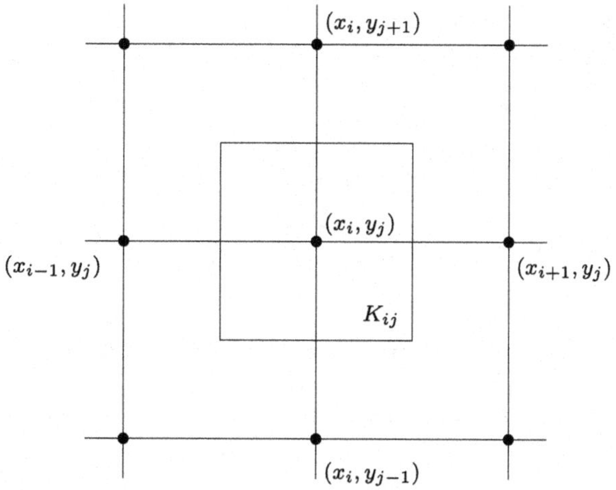

of f:

$$(T_h^{11} f)_{ij} := \frac{1}{h^2} \int_{K_{ij}} f(x, y)\,\mathrm{d}x\,\mathrm{d}y,$$

where the 'cell' K_{ij} is defined by

$$K_{ij} := \left(x_i - \frac{h}{2}, x_i + \frac{h}{2}\right) \times \left(y_j - \frac{h}{2}, y_j + \frac{h}{2}\right),$$

with $i, j = 1, \dots, N-1$.

This seemingly ad hoc approach has the following justification. Integrating the partial differential equation $-\Delta u + cu = f$ over the cell K_{ij} and using the divergence theorem we have that

$$-\int_{\partial K_{ij}} \frac{\partial u}{\partial \nu}\,\mathrm{d}s + \int_{K_{ij}} cu\,\mathrm{d}x\,\mathrm{d}y = \int_{K_{ij}} f\,\mathrm{d}x\,\mathrm{d}y, \tag{2.69}$$

where ∂K_{ij} is the boundary of K_{ij}, and ν is the unit outward normal to ∂K_{ij}.

The normal vectors to ∂K_{ij} point in the co-ordinate directions, so the normal derivative $\partial u/\partial \nu$ can be approximated by divided differences using the values of u at the five mesh-points marked by • in Fig. 2.3, in conjunction with a midpoint quadrature rule along each edge of K_{ij} to approximate the contour integral featuring in the first term of (2.69) (cf. Examples 2.6 and 2.7).

Approximating the second integral on the left by a midpoint quadrature rule, now in two dimensions, on K_{ij}, and dividing both sides by $\mathrm{meas}(K_{ij}) = h^2$, we obtain

$$\begin{aligned}&-\big(D_x^+ D_x^- u(x_i, y_j) + D_y^+ D_y^- u(x_i, y_j)\big) + c(x_i, y_j)u(x_i, y_j)\\ &\quad\approx \frac{1}{h^2} \int_{K_{ij}} f(x, y)\,\mathrm{d}x\,\mathrm{d}y.\end{aligned}$$

We note here that $(T_h^{11} f)_{ij}$ is correctly defined for $f \in L_2(\Omega)$; indeed,

$$
\begin{aligned}
\left|(T_h^{11} f)_{ij}\right| &= \frac{1}{h^2}\left|\int_{K_{ij}} f(x, y)\,\mathrm{d}x\,\mathrm{d}y\right| \\
&\le \frac{1}{h^2}\left(\int_{K_{ij}} 1^2\,\mathrm{d}x\,\mathrm{d}y\right)^{1/2}\left(\int_{K_{ij}} |f(x, y)|^2\,\mathrm{d}x\,\mathrm{d}y\right)^{1/2} \\
&= h^{-1}\|f\|_{L_2(K_{ij})} (< \infty).
\end{aligned} \tag{2.70}
$$

Thus we define our finite difference scheme for (2.55), (2.56) by

$$
-(D_x^+ D_x^- + D_y^+ D_y^-)U_{ij} + c(x_i, y_j)U_{ij} = \left(T_h^{11} f\right)_{ij}, \quad (x_i, y_j) \in \Omega^h, \tag{2.71}
$$

$$
U = 0 \quad \text{on } \Gamma^h. \tag{2.72}
$$

Remark 2.10 Finite difference schemes that arise from integral formulations of a differential equation, such as (2.69), are called *finite volume methods.*

Since we have not changed the difference operator on the left-hand side, the argument presented in (a) concerning the existence and uniqueness of a solution to the difference scheme (2.57), (2.58) still applies to (2.71), (2.72); therefore, (2.71), (2.72) has a unique solution U in S_0^h. Moreover, we have the following stability result.

Theorem 2.22 *The scheme* (2.71), (2.72) *is stable in the sense that*

$$
\|U\|_{W_2^1(\Omega^h)} \le \frac{1}{c_0}\|f\|_{L_2(\Omega)}, \tag{2.73}
$$

where $c_0 = 16/17$.

Proof From (2.64) and (2.70) we have

$$
\begin{aligned}
c_0\|U\|^2_{W_2^1(\Omega^h)} &\le (AU, U)_h = \left(T_h^{11} f, U\right)_h \\
&\le \left\|T_h^{11} f\right\|_{L_2(\Omega^h)}\|U\|_{L_2(\Omega^h)} \le \left\|T_h^{11} f\right\|_{L_2(\Omega^h)}\|U\|_{W_2^1(\Omega^h)} \\
&\le \|f\|_{L_2(\Omega)}\|U\|_{W_2^1(\Omega^h)},
\end{aligned}
$$

and hence (2.73). □

Having established the stability of the scheme (2.71), (2.72) we consider the question of its accuracy. Let us define the global error, e, as before:

$$
e_{ij} := u(x_i, y_j) - U_{ij}, \quad i, j = 0, \ldots, N.
$$

Clearly, for $i, j = 1, \dots, N-1$ we have

$$\begin{aligned}
Ae_{ij} &= Au(x_i, y_j) - AU_{ij} \\
&= Au(x_i, y_j) - \left(T_h^{11} f\right)_{ij} \\
&= -\left(D_x^+ D_x^- u(x_i, y_j) + D_y^+ D_y^- u(x_i, y_j)\right) + c(x_i, y_j) u(x_i, y_j) \\
&\quad + \left[T_h^{11}\left(\frac{\partial^2 u}{\partial x^2}\right)(x_i, y_j) + T_h^{11}\left(\frac{\partial^2 u}{\partial y^2}\right)(x_i, y_j) - T_h^{11}(cu)(x_i, y_j)\right].
\end{aligned} \tag{2.74}$$

By noting that

$$\begin{aligned}
T_h^{11}\left(\frac{\partial^2 u}{\partial x^2}\right)(x_i, y_j) &= \frac{1}{h}\int_{y_j-h/2}^{y_j+h/2} \frac{\frac{\partial u}{\partial x}(x_i + h/2, y) - \frac{\partial u}{\partial x}(x_i - h/2, y)}{h}\,\mathrm{d}y \\
&= \frac{1}{h}\int_{y_j-h/2}^{y_j+h/2} D_x^+ \frac{\partial u}{\partial x}(x_i - h/2, y)\,\mathrm{d}y \\
&= D_x^+\left[\frac{1}{h}\int_{y_j-h/2}^{y_j+h/2} \frac{\partial u}{\partial x}(x_i - h/2, y)\,\mathrm{d}y\right],
\end{aligned}$$

and that, similarly,

$$T_h^{11}\left(\frac{\partial^2 u}{\partial y^2}\right)(x_i, y_j) = D_y^+\left[\frac{1}{h}\int_{x_i-h/2}^{x_i+h/2} \frac{\partial u}{\partial y}(x, y_j - h/2)\,\mathrm{d}x\right],$$

equality (2.74) can be rewritten as

$$Ae = D_x^+ \varphi_1 + D_y^+ \varphi_2 + \psi,$$

where

$$\begin{aligned}
\varphi_1(x_i, y_j) &:= \frac{1}{h}\int_{y_j-h/2}^{y_j+h/2} \frac{\partial u}{\partial x}(x_i - h/2, y)\,\mathrm{d}y - D_x^- u(x_i, y_j), \\
\varphi_2(x_i, y_j) &:= \frac{1}{h}\int_{x_i-h/2}^{x_i+h/2} \frac{\partial u}{\partial y}(x, y_j - h/2)\,\mathrm{d}x - D_y^- u(x_i, y_j), \\
\psi(x_i, y_j) &:= (cu)(x_i, y_j) - T_h^{11}(cu)(x_i, y_j).
\end{aligned}$$

Thus,

$$Ae = D_x^+ \varphi_1 + D_y^+ \varphi_2 + \psi \quad \text{in } \Omega^h, \tag{2.75}$$

$$e = 0 \quad \text{on } \Gamma^h. \tag{2.76}$$

As the stability result (2.73) implies only the crude bound

$$\|e\|_{W_2^1(\Omega^h)} \le \frac{1}{c_0}\|D_x^+\varphi_1 + D_y^+\varphi_2 + \psi\|_{L_2(\Omega^h)},$$

which does not exploit the special form of the truncation error,

$$\varphi := D_x^+\varphi_1 + D_y^+\varphi_2 + \psi,$$

we shall proceed in a different way. The idea is to sharpen (2.73) by proving a discrete analogue of the well-posedness result from Theorem 2.7; we recall that this states that the following bound holds for the boundary-value problem (2.55), (2.56):

$$\|u\|_{\mathring{W}_2^1(\Omega)} \le \frac{1}{c_0}\|f\|_{W_2^{-1}(\Omega)}.$$

In order to obtain a discrete counterpart of this inequality, we consider the discrete negative Sobolev norm $\|\cdot\|_{W_2^{-1}(\Omega^h)}$, defined by

$$\|V\|_{W_2^{-1}(\Omega^h)} := \sup_{V \in S_0^h \setminus \{0\}} \frac{|(V, W)_h|}{\|W\|_{W_2^1(\Omega^h)}}.$$

Theorem 2.23 *The scheme* (2.71), (2.72) *is stable in the sense that*

$$\|U\|_{W_2^1(\Omega^h)} \le \frac{1}{c_0}\|T_h^{11} f\|_{W_2^{-1}(\Omega^h)}, \tag{2.77}$$

where $c_0 = 16/17$.

Proof From (2.64), by noting the definition of the $\|\cdot\|_{W_2^{-1}(\Omega^h)}$ norm, we have that

$$\begin{aligned} c_0\|U\|^2_{W_2^1(\Omega^h)} &\le (AU, U)_h = \left(T_h^{11} f, U\right)_h \\ &\le \|T_h^{11} f\|_{W_2^{-1}(\Omega^h)}\|U\|_{W_2^1(\Omega^h)}, \end{aligned}$$

and hence (2.77). □

Now we apply Theorem 2.23 to (2.75), (2.76) to deduce that

$$\|e\|_{W_2^1(\Omega^h)} \le \frac{1}{c_0}\|D_x^+\varphi_1 + D_y^+\varphi_2 + \psi\|_{W_2^{-1}(\Omega^h)}. \tag{2.78}$$

In order to bound the right-hand side of (2.78) let us consider the expression

$$\left(D_x^+\varphi_1 + D_y^+\varphi_2 + \psi, W\right)_h$$

for $W \in S_0^h \setminus \{0\}$. Using summation by parts, we shall pass the difference operators D_x^+ and D_y^+ from φ_1 and φ_2, respectively, onto W. As $W = 0$ on the set Γ^h, we have that

$$\begin{aligned}
\left(D_x^+\varphi_1, W\right)_h &= \sum_{j=1}^{N-1} h\left(\sum_{i=1}^{N-1} h\frac{\varphi_1(x_{i+1}, y_j) - \varphi_1(x_i, y_j)}{h} W_{ij}\right) \\
&= -\sum_{j=1}^{N-1} h\left(\sum_{i=1}^{N} h\varphi_1(x_i, y_j)\frac{W_{ij} - W_{i-1,j}}{h}\right) \\
&= -\sum_{j=1}^{N-1} h\left(\sum_{i=1}^{N} h\varphi_1(x_i, y_j) D_x^- W_{ij}\right) \\
&= -\sum_{i=1}^{N}\sum_{j=1}^{N-1} h^2\varphi_1(x_i, y_j) D_x^- W_{ij} \\
&\le \left(\sum_{i=1}^{N}\sum_{j=1}^{N-1} h^2\left|\varphi_1(x_i, y_j)\right|^2\right)^{1/2} \left(\sum_{i=1}^{N}\sum_{j=1}^{N-1} h^2\left|D_x^- W_{ij}\right|^2\right)^{1/2}.
\end{aligned}$$

We thus deduce that

$$\left|\left(D_x^+\varphi_1, W\right)_h\right| \le \|\varphi_1\|_{L_2(\Omega_x^h)} \left\|D_x^- W\right\|_{L_2(\Omega_x^h)}. \tag{2.79}$$

Similarly,

$$\left|\left(D_y^+\varphi_2, W\right)_h\right| \le \|\varphi_2\|_{L_2(\Omega_y^h)} \left\|D_y^- W\right\|_{L_2(\Omega_y^h)}. \tag{2.80}$$

By the Cauchy–Schwarz inequality we also have that

$$\left|(\psi, W)_h\right| \le \|\psi\|_{L_2(\Omega^h)} \|W\|_{L_2(\Omega^h)}. \tag{2.81}$$

Now, by combining (2.79)–(2.81) and noting the elementary inequality

$$|a_1 b_1 + a_2 b_2 + a_3 b_3| \le \left(a_1^2 + a_2^2 + a_3^2\right)^{1/2}\left(b_1^2 + b_2^2 + b_3^2\right)^{1/2},$$

we arrive at the bound

$$\begin{aligned}
&\left|\left(D_x^+\varphi_1 + D_y^+\varphi_2 + \psi, W\right)_h\right| \\
&\quad\le \left(\|\varphi_1\|^2_{L_2(\Omega_x^h)} + \|\varphi_2\|^2_{L_2(\Omega_y^h)} + \|\psi\|^2_{L_2(\Omega^h)}\right)^{1/2} \\
&\qquad\times \left(\left\|D_x^- W\right\|^2_{L_2(\Omega_x^h)} + \left\|D_y^- W\right\|^2_{L_2(\Omega_y^h)} + \|W\|^2_{L_2(\Omega^h)}\right)^{1/2} \\
&\quad= \left(\|\varphi_1\|^2_{L_2(\Omega_x^h)} + \|\varphi_2\|^2_{L_2(\Omega_y^h)} + \|\psi\|^2_{L_2(\Omega^h)}\right)^{1/2} \|W\|_{W_2^1(\Omega^h)}.
\end{aligned}$$

Dividing both sides by $\|W\|_{W_2^1(\Omega^h)}$ and taking the supremum over all $W \in S_0^h \setminus \{0\}$ yields the following inequality:

$$\left\| D_x^+\varphi_1 + D_y^+\varphi_2 + \psi \right\|_{W_2^{-1}(\Omega^h)} \le \left(\|\varphi_1\|^2_{L_2(\Omega_x^h)} + \|\varphi_2\|^2_{L_2(\Omega_y^h)} + \|\psi\|^2_{L_2(\Omega^h)} \right)^{1/2}. \tag{2.82}$$

Inserting (2.82) into (2.78) we obtain the following bound on the global error in terms of the truncation error of the scheme.

Lemma 2.24 *The global error, $e := u - U$, of the finite difference scheme* (2.71), (2.72) *satisfies the bound*

$$\|e\|_{W_2^1(\Omega^h)} \le \frac{1}{c_0} \left(\|\varphi_1\|^2_{L_2(\Omega_x^h)} + \|\varphi_2\|^2_{L_2(\Omega_y^h)} + \|\psi\|^2_{L_2(\Omega^h)} \right)^{1/2}, \tag{2.83}$$

where $c_0 = 16/17$, and φ_1, φ_2 and ψ are defined by

$$\varphi_1(x_i, y_j) := \frac{1}{h} \int_{y_j-h/2}^{y_j+h/2} \frac{\partial u}{\partial x}(x_i - h/2, y)\mathrm{d}y - D_x^- u(x_i, y_j), \tag{2.84}$$

$$\varphi_2(x_i, y_j) := \frac{1}{h} \int_{x_i-h/2}^{x_i+h/2} \frac{\partial u}{\partial y}(x, y_j - h/2)\,\mathrm{d}x - D_y^- u(x_i, y_j), \tag{2.85}$$

$$\psi(x_i, y_j) := (cu)(x_i, y_j) - \frac{1}{h^2} \int_{x_i-h/2}^{x_i+h/2} \int_{y_j-h/2}^{y_j+h/2} (cu)(x, y)\,\mathrm{d}x\,\mathrm{d}y, \tag{2.86}$$

with $i = 1, \dots, N$ and $j = 1, \dots, N-1$ in (2.84); *$i = 1, \dots, N-1$ and $j = 1, \dots, N$ in* (2.85); *and $i, j = 1, \dots N-1$ in* (2.86).

To complete the error analysis, it remains to bound φ_1, φ_2 and ψ. Using Taylor series expansions it is easily seen that

$$\left|\varphi_1(x_i, y_j)\right| \le \frac{h^2}{24} \left(\left\| \frac{\partial^3 u}{\partial x \partial y^2} \right\|_{C(\overline{\Omega})} + \left\| \frac{\partial^3 u}{\partial x^3} \right\|_{C(\overline{\Omega})} \right), \tag{2.87}$$

$$\left|\varphi_2(x_i, y_j)\right| \le \frac{h^2}{24} \left(\left\| \frac{\partial^3 u}{\partial x^2 \partial y} \right\|_{C(\overline{\Omega})} + \left\| \frac{\partial^3 u}{\partial y^3} \right\|_{C(\overline{\Omega})} \right), \tag{2.88}$$

$$\left|\psi(x_i, y_j)\right| \le \frac{h^2}{24} \left(\left\| \frac{\partial^2 (cu)}{\partial x^2} \right\|_{C(\overline{\Omega})} + \left\| \frac{\partial^2 (cu)}{\partial y^2} \right\|_{C(\overline{\Omega})} \right), \tag{2.89}$$

which yield the required bounds on $\|\varphi_1\|_{L_2(\Omega_x^h)}$, $\|\varphi_2\|_{L_2(\Omega_y^h)}$ and $\|\psi\|_{L_2(\Omega^h)}$. We thus arrive at the following theorem.

Theorem 2.25 *Let $f \in L_2(\Omega)$, $c \in C^2(\overline{\Omega})$ with $c(x, y) \ge 0$, $(x, y) \in \overline{\Omega}$, and suppose that the corresponding weak solution of the boundary-value problem* (2.55),

(2.56) *belongs to* $C^3(\overline{\Omega})$. *Then,*

$$\|u - U\|_{W_2^1(\Omega^h)} \le \frac{17}{384} h^2 M_3, \tag{2.90}$$

where

$$\begin{aligned} M_3 = \Bigg\{ & \left(\left\| \frac{\partial^3 u}{\partial x \partial y^2} \right\|_{C(\overline{\Omega})} + \left\| \frac{\partial^3 u}{\partial x^3} \right\|_{C(\overline{\Omega})} \right)^2 \\ & + \left(\left\| \frac{\partial^3 u}{\partial x^2 y} \right\|_{C(\overline{\Omega})} + \left\| \frac{\partial^3 u}{\partial y^3} \right\|_{C(\overline{\Omega})} \right)^2 \\ & + \left(\left\| \frac{\partial^2 (cu)}{\partial x^2} \right\|_{C(\overline{\Omega})} + \left\| \frac{\partial^2 (cu)}{\partial y^2} \right\|_{C(\overline{\Omega})} \right)^2 \Bigg\}^{1/2}. \end{aligned}$$

Proof As $1/c_0 = 17/16$, by substituting (2.87)–(2.89) into the right-hand side of (2.83) the estimate (2.90) immediately follows. □

By comparing (2.90) with (2.68) we see that while the smoothness requirement on the solution has been relaxed from $u \in C^4(\overline{\Omega})$ to $u \in C^3(\overline{\Omega})$, second-order convergence has been retained.

The hypothesis $u \in C^3(\overline{\Omega})$ can be further relaxed by using integral representations of φ_1, φ_2 and ψ instead of Taylor series expansions. We show how this is done for φ_1 and ψ; φ_2 is handled analogously to φ_1. The argument is based on repeated use the Newton–Leibniz formula

$$w(b) - w(a) = \int_a^b w'(x)\,\mathrm{d}x.$$

In order to simplify the notation, let us write $x_{i\pm1/2} := x_i \pm h/2$ and $y_{j\pm1/2} := y_j \pm h/2$; we then have that

$$\begin{aligned} \varphi_1(x_i, y_j) &= \frac{1}{h^2} \int_{x_{i-1}}^{x_i} \int_{y_{j-1/2}}^{y_{j+1/2}} \left[\frac{\partial u}{\partial x}(x_{i-1/2}, y) - \frac{\partial u}{\partial x}(x, y_j) \right] \mathrm{d}x\,\mathrm{d}y \\ &= \frac{1}{h^2} \int_{x_{i-1}}^{x_i} \int_{y_{j-1/2}}^{y_{j+1/2}} \left[\frac{\partial u}{\partial x}(x_{i-1/2}, y) - \frac{\partial u}{\partial x}(x, y) \right] \mathrm{d}x\,\mathrm{d}y \\ &\quad + \frac{1}{h^2} \int_{x_{i-1}}^{x_i} \int_{y_{j-1/2}}^{y_{j+1/2}} \left[\frac{\partial u}{\partial x}(x, y) - \frac{\partial u}{\partial x}(x, y_j) \right] \mathrm{d}x\,\mathrm{d}y \\ &= \frac{1}{h^2} \int_{y_{j-1/2}}^{y_{j+1/2}} \left[\int_{x_{i-1}}^{x_i} \left(\int_x^{x_{i-1/2}} \frac{\partial^2 u}{\partial x^2}(\xi, y)\,\mathrm{d}\xi \right) \mathrm{d}x \right] \mathrm{d}y \\ &\quad + \frac{1}{h^2} \int_{x_{i-1}}^{x_i} \left[\int_{y_{j-1/2}}^{y_{j+1/2}} \left(\int_{y_j}^{y} \frac{\partial^2 u}{\partial x \partial y}(x, \eta)\,\mathrm{d}\eta \right) \mathrm{d}x \right] \mathrm{d}y. \end{aligned}$$

We thus deduce by partial integration that

$$\begin{aligned}
\varphi_1(x_i, y_j) = \frac{1}{h^2} \int_{y_{j-1/2}}^{y_{j+1/2}} \Bigg[& x \int_x^{x_{i-1/2}} \frac{\partial^2 u}{\partial x^2}(\xi, y)\,\mathrm{d}\xi \Bigg|_{x=x_{i-1}}^{x=x_i} \\
& + \int_{x_{i-1}}^{x_i} x \frac{\partial^2 u}{\partial x^2}(x, y)\,\mathrm{d}x \Bigg] \mathrm{d}y \\
+ \frac{1}{h^2} \int_{x_{i-1}}^{x_i} \Bigg[& y \int_{y_j}^{y} \frac{\partial^2 u}{\partial x \partial y}(x, \eta)\,\mathrm{d}\eta \Bigg|_{y=y_{j-1/2}}^{y=y_{j+1/2}} \\
& - \int_{y_{j-1/2}}^{y_{j+1/2}} y \frac{\partial^2 u}{\partial x \partial y}(x, y)\,\mathrm{d}y \Bigg] \mathrm{d}x \\
= \frac{1}{h^2} \int_{y_{j-1/2}}^{y_{j+1/2}} \Bigg[& \int_{x_{i-1}}^{x_{i-1/2}} (x - x_{i-1}) \frac{\partial^2 u}{\partial x^2}(x, y)\,\mathrm{d}x \\
& + \int_{x_{i-1/2}}^{x_i} (x - x_i) \frac{\partial^2 u}{\partial x^2}(x, y)\,\mathrm{d}x \Bigg] \mathrm{d}y \\
- \frac{1}{h^2} \int_{x_{i-1}}^{x_i} \Bigg[& \int_{y_{j-1/2}}^{y_j} (y - y_{j-1/2}) \frac{\partial^2 u}{\partial x \partial y}(x, y)\,\mathrm{d}y \\
& + \int_{y_j}^{y_{j+1/2}} (y - y_{j+1/2}) \frac{\partial^2 u}{\partial x \partial y}(x, y)\,\mathrm{d}y \Bigg] \mathrm{d}x.
\end{aligned}$$

We define the piecewise quadratic functions

$$A_i(x) = \begin{cases} \frac{1}{2}(x - x_{i-1})^2 & \text{if } x \in [x_{i-1}, x_{i-1/2}], \\ \frac{1}{2}(x - x_i)^2 & \text{if } x \in [x_{i-1/2}, x_i], \end{cases}$$

$$B_j(y) = \begin{cases} \frac{1}{2}(y - y_{j-1/2})^2 & \text{if } y \in [y_{j-1/2}, y_j], \\ \frac{1}{2}(y - y_{j+1/2})^2 & \text{if } y \in [y_j, y_{j+1/2}], \end{cases}$$

and note that A_i and B_j are continuous functions of their respective arguments; furthermore,

$$A_i(x_{i-1}) = A_i(x_i) = 0 \quad \text{and} \quad B_j(y_{j-1/2}) = B_j(y_{j+1/2}) = 0.$$

Integration by parts then yields

$$\begin{aligned}
\varphi_1(x_i, y_j) = & \frac{1}{h^2} \int_{y_{j-1/2}}^{y_{j+1/2}} \left[\int_{x_{i-1}}^{x_i} A_i'(x) \frac{\partial^2 u}{\partial x^2}(x, y)\,\mathrm{d}x \right] \mathrm{d}y \\
& - \frac{1}{h^2} \int_{x_{i-1}}^{x_i} \left[\int_{y_{j-1/2}}^{y_{j+1/2}} B_j'(y) \frac{\partial^2 u}{\partial x \partial y}(x, y)\,\mathrm{d}y \right] \mathrm{d}x
\end{aligned}$$

$$= -\frac{1}{h^2}\int_{y_{j-1/2}}^{y_{j+1/2}}\left[\int_{x_{i-1}}^{x_i} A_i(x)\frac{\partial^3 u}{\partial x^3}(x,y)\,\mathrm{d}x\right]\mathrm{d}y$$

$$+\frac{1}{h^2}\int_{x_{i-1}}^{x_i}\left[\int_{y_{j-1/2}}^{y_{j+1/2}} B_j(y)\frac{\partial^3 u}{\partial x\partial y^2}(x,y)\,\mathrm{d}y\right]\mathrm{d}x. \tag{2.91}$$

Now

$$\left|A_i(x)\right| \le \frac{1}{8}h^2, \quad x\in[x_{i-1},x_i] \quad \text{and} \quad \left|B_j(y)\right| \le \frac{1}{8}h^2, \quad y\in[y_{j-1/2},y_{j+1/2}],$$

and therefore,

$$\left|\varphi_1(x_i,y_j)\right| \le \frac{1}{8}\int_{x_{i-1}}^{x_i}\int_{y_{j-1/2}}^{y_{j+1/2}}\left|\frac{\partial^3 u}{\partial x^3}(x,y)\right|\mathrm{d}x\,\mathrm{d}y$$

$$+\frac{1}{8}\int_{x_{i-1}}^{x_i}\int_{y_{j-1/2}}^{y_{j+1/2}}\left|\frac{\partial^3 u}{\partial x\partial y^2}(x,y)\right|\mathrm{d}x\,\mathrm{d}y.$$

Consequently,

$$\|\varphi_1\|^2_{L_2(\Omega_x^h)} \le \frac{h^4}{32}\left(\left\|\frac{\partial^3 u}{\partial x^3}\right\|^2_{L_2(\Omega)}+\left\|\frac{\partial^3 u}{\partial x\partial y^2}\right\|^2_{L_2(\Omega)}\right). \tag{2.92}$$

Analogously,

$$\|\varphi_2\|^2_{L_2(\Omega_y^h)} \le \frac{h^4}{32}\left(\left\|\frac{\partial^3 u}{\partial y^3}\right\|^2_{L_2(\Omega)}+\left\|\frac{\partial^3 u}{\partial x^2\partial y}\right\|^2_{L_2(\Omega)}\right). \tag{2.93}$$

In order to estimate ψ, we note that

$$\psi(x_i,y_j) = \frac{1}{h^2}\int_{x_{i-1/2}}^{x_{i+1/2}}\int_{y_{j-1/2}}^{y_{j+1/2}}\left(\int_x^{x_i}\frac{\partial w}{\partial x}(s,y)\,\mathrm{d}s\right.$$

$$\left.+\int_y^{y_j}\frac{\partial w}{\partial y}(x,t)\,\mathrm{d}t+\int_x^{x_i}\int_y^{y_j}\frac{\partial^2 w}{\partial x\partial y}(s,t)\,\mathrm{d}s\,\mathrm{d}t\right)\mathrm{d}x\,\mathrm{d}y$$

$$= -\frac{1}{h^2}\int_{x_{i-1/2}}^{x_{i+1/2}}\int_{y_{j-1/2}}^{y_{j+1/2}} C_i(x)\frac{\partial^2 w}{\partial x^2}(x,y)\,\mathrm{d}x\,\mathrm{d}y$$

$$-\frac{1}{h^2}\int_{x_{i-1/2}}^{x_{i+1/2}}\int_{y_{j-1/2}}^{y_{j+1/2}} B_j(y)\frac{\partial^2 w}{\partial y^2}(x,y)\,\mathrm{d}x\,\mathrm{d}y$$

$$+\frac{1}{h^2}\int_{x_{i-1/2}}^{x_{i+1/2}}\int_{y_{j-1/2}}^{y_{j+1/2}}\left(\int_x^{x_i}\int_y^{y_j}\frac{\partial^2 w}{\partial x\partial y}(s,t)\,\mathrm{d}s\,\mathrm{d}t\right)\mathrm{d}x\,\mathrm{d}y,$$

where $w(x, y) = c(x, y)u(x, y)$ and

$$C_i(x) = \begin{cases} \frac{1}{2}(x - x_{i-1/2})^2 & \text{if } x \in [x_{i-1/2}, x_i], \\ \frac{1}{2}(x - x_{i+1/2})^2 & \text{if } x \in [x_i, x_{i+1/2}]. \end{cases}$$

Hence,

$$\begin{aligned} |\psi(x_i, y_j)| \le \frac{1}{8}\bigg(& \int_{x_{i-1/2}}^{x_{i+1/2}} \int_{y_{j-1/2}}^{y_{j+1/2}} \left| \frac{\partial^2 w}{\partial x^2}(x, y) \right| \mathrm{d}x\, \mathrm{d}y \\ & + \int_{x_{i-1/2}}^{x_{i+1/2}} \int_{y_{j-1/2}}^{y_{j+1/2}} \left| \frac{\partial^2 w}{\partial y^2}(x, y) \right| \mathrm{d}x\, \mathrm{d}y \\ & + 2 \int_{x_{i-1/2}}^{x_{i+1/2}} \int_{y_{j-1/2}}^{y_{j+1/2}} \left| \frac{\partial^2 w}{\partial x \partial y} \right| \mathrm{d}x\, \mathrm{d}y \bigg), \end{aligned}$$

so that, with $w := cu$, we have

$$\|\psi\|^2_{L_2(\Omega^h)} \le \frac{3h^4}{64} \left(\left\| \frac{\partial^2 w}{\partial x^2} \right\|^2_{L_2(\Omega)} + \left\| \frac{\partial^2 w}{\partial y^2} \right\|^2_{L_2(\Omega)} + 4 \left\| \frac{\partial^2 w}{\partial x \partial y} \right\|^2_{L_2(\Omega)} \right). \quad (2.94)$$

By substituting (2.92)–(2.94) into the right-hand side of (2.83) and noting that $1/c_0 = 16/17$, we obtain the following result.

Theorem 2.26 *Let $f \in L_2(\Omega)$, $c \in M(W_2^2(\Omega))$, with $c(x, y) \ge 0$ for all (x, y) in $\overline{\Omega}$, and suppose that the corresponding weak solution of the boundary-value problem* (2.55), (2.56) *belongs to $W_2^3(\Omega) \cap \mathring{W}_2^1(\Omega)$. Then,*

$$\|u - U\|_{W_2^1(\Omega^h)} \le Ch^2 \|u\|_{W_2^3(\Omega)}, \quad (2.95)$$

where C is a positive constant (computable from (2.83) *and* (2.92)–(2.94)), *independent of h and u.*

We note that, by the analogue of Lemma 1.46 on a Lipschitz domain, $M(W_2^2(\Omega)) \subset W_2^2(\Omega)$, and therefore, by Sobolev embedding $c \in M(W_2^2(\Omega))$ is a continuous function with well-defined values at the mesh-points.

It can be verified by numerical experiments that the error bound (2.95) is best possible in the sense that further weakening of the regularity hypothesis on u leads to a loss of second-order convergence. Error bounds of the type (2.95), where the highest possible order of convergence is attained under the weakest hypothesis on the smoothness of the solution, are called *optimal* or *compatible with the smoothness of the solution*. Thus, for example, (2.95) is an optimal error bound for the difference scheme (2.71), (2.72), but (2.90) is not. At this point it does not concern us whether the smoothness requirements on the coefficients in the equation are the weakest possible: that issue will be addressed later, in our discussion of optimal error bounds under minimal smoothness hypotheses on the coefficients and the source term f.

We shall now explore the convergence rate of the finite difference scheme in the norm $\|\cdot\|_{W_2^1(\Omega^h)}$ under even weaker regularity hypotheses on the solution, resulting in a loss of second-order convergence established above for $u \in W_2^3(\Omega) \cap \mathring{W}_2^1(\Omega)$. Suppose, for example, that $u \in W_2^2(\Omega) \cap \mathring{W}_2^1(\Omega)$. From (2.91), by noting that

$$|A_i'(x)| \leq \frac{1}{2}h, \quad x \in [x_{i-1}, x_i] \quad \text{and} \quad |B_j'(y)| \leq \frac{1}{2}h, \quad y \in [y_{j-1/2}, y_{j+1/2}],$$

we have that

$$|\varphi_1(x_i, y_j)| \leq \frac{1}{2h}\int_{x_{i-1}}^{x_i}\int_{y_{j-1/2}}^{y_{j+1/2}} \left|\frac{\partial^2 u}{\partial x^2}(x,y)\right| \mathrm{d}x\,\mathrm{d}y + \frac{1}{2h}\int_{x_{i-1}}^{x_i}\int_{y_{j-1/2}}^{y_{j+1/2}} \left|\frac{\partial^2 u}{\partial x \partial y}(x,y)\right| \mathrm{d}x\,\mathrm{d}y.$$

Consequently,

$$\|\varphi_1\|^2_{L_2(\Omega_x^h)} \leq \frac{h^2}{2}\left(\left\|\frac{\partial^2 u}{\partial x^2}\right\|^2_{L_2(\Omega)} + \left\|\frac{\partial^2 u}{\partial x \partial y}\right\|^2_{L_2(\Omega)}\right). \tag{2.96}$$

Analogously,

$$\|\varphi_2\|^2_{L_2(\Omega_y^h)} \leq \frac{h^2}{2}\left(\left\|\frac{\partial^2 u}{\partial y^2}\right\|^2_{L_2(\Omega)} + \left\|\frac{\partial^2 u}{\partial x \partial y}\right\|^2_{L_2(\Omega)}\right). \tag{2.97}$$

From (2.83), (2.96), (2.97) and (2.94), under the assumptions that $c \in M(W_2^2(\Omega))$, $c \geq 0$ on $\overline{\Omega}$ and $u \in W_2^2(\Omega) \cap \mathring{W}_2^1(\Omega)$, we deduce that:

$$\|u - U\|_{W_2^1(\Omega^h)} \leq Ch\|u\|_{W_2^2(\Omega)}, \tag{2.98}$$

where C is a positive constant, independent of h and u.

Application of Function Space Interpolation When $u \in W_2^s(\Omega)$, $2 < s < 3$, an error bound can be obtained from (2.95) and (2.98) by function space interpolation. For the sake of simplicity we shall confine ourselves to Poisson's equation (i.e. $c(x, y) \equiv 0$). In that case the constant C featuring in (2.95) and (2.98) represents an absolute constant (i.e. it is independent of $c(x, y)$). Let us consider the mapping $L : u \mapsto u - U$, with U understood as a linear function of $f = -\Delta u$. Evidently, L is a linear operator. It follows from (2.95) that the operator L, considered as a linear mapping $L : W_2^3(\Omega) \to W_2^1(\Omega^h)$, is bounded and

$$\|L\|_{W_2^3(\Omega)\to W_2^1(\Omega^h)} \leq Ch^2.$$

In the same way, it follows from (2.98) that the operator L, considered as a linear mapping $L : W_2^2(\Omega) \to W_2^1(\Omega^h)$, is bounded and

$$\|L\|_{W_2^2(\Omega)\to W_2^1(\Omega^h)} \leq Ch.$$

By the results of Sect. 1.1.5, the operator L, considered as a linear mapping $L : (W_2^3(\Omega), W_2^2(\Omega))_{\theta,q} \to (W_2^1(\Omega^h), W_2^1(\Omega^h))_{\theta,q}$, is also bounded and, thanks to (1.8),

$$\|L\|_{(W_2^3(\Omega),W_2^2(\Omega))_{\theta,q}\to(W_2^1(\Omega^h),W_2^1(\Omega^h))_{\theta,q}} \leq \left(Ch^2\right)^{1-\theta}(Ch)^\theta = Ch^{2-\theta}.$$

Furthermore,

$$\left(W_2^1(\Omega^h), W_2^1(\Omega^h)\right)_{\theta,q} = W_2^1(\Omega^h),$$
$$\left(W_2^3(\Omega), W_2^2(\Omega)\right)_{\theta,q} = W_2^{3-\theta}(\Omega).$$

Thus we obtain the following error bound:

$$\|u - U\|_{W_2^1(\Omega^h)} \leq Ch^{2-\theta}\|u\|_{W_2^{3-\theta}(\Omega)}, \quad 0 < \theta < 1.$$

By writing $3 - \theta = s$ here and supplementing the resulting bounds with the ones corresponding to the limiting cases $s = 2$ and $s = 3$, we deduce that

$$\|u - U\|_{W_2^1(\Omega^h)} \leq Ch^{s-1}\|u\|_{W_2^s(\Omega)}, \quad 2 \leq s \leq 3,$$

where C is a positive real number, independent of h and u.

In the next section we shall show how the tedious use of integral representation theorems can be avoided in the error analysis of finite difference methods by appealing to the Bramble–Hilbert lemma and its variants.

2.3 Convergence Analysis on Uniform Meshes

In the previous section we derived an optimal bound on the global error between the unique weak solution u to a homogeneous Dirichlet boundary-value problem for the generalized Poisson equation and its finite difference approximation U, under the hypothesis that $u \in W_2^s(\Omega) \cap \mathring{W}_2^1(\Omega)$, $s \in [2, 3]$. We used integral representations for $s = 2, 3$ in conjunction with function space interpolation for $s \in (2, 3)$. Here we shall consider the same problem by using a different technique; our main tool will be the Bramble–Hilbert lemma.

2.3.1 The Bramble–Hilbert Lemma

We begin by stating the Bramble–Hilbert lemma in its simplest form, in the case of integer-order Sobolev spaces (cf. [20]). We shall then illustrate its use in the error analysis of simple discretization methods and describe its generalizations to fractional-order and anisotropic Sobolev spaces. We shall also formulate a multilinear version of the Bramble–Hilbert lemma.

Theorem 2.27 (Bramble–Hilbert Lemma) *Let $\Omega \subset \mathbb{R}^n$ be a Lipschitz domain and, for a positive integer m and a real number $p \in [1, \infty]$, let η be a bounded linear functional on the Sobolev space $W_p^m(\Omega)$ such that*

$$\mathcal{P}_{m-1} \subset \operatorname{Ker}(\eta),$$

where $\mathcal{P}_{m-1}$ denotes the set of all polynomials of degree $m-1$ in n variables. Then, there exists a positive real number $C = C(m, p, n, \Omega)$ such that

$$|\eta(v)| \le C\|\eta\||v|_{W_p^m(\Omega)} \quad \forall v \in W_p^m(\Omega).$$

The proof of this result will be presented below in a more general context. First, however, we consider a series of examples that illustrate the application of Theorem 2.27.

Example 2.6 In this example we apply the Bramble–Hilbert lemma to provide a bound on the error in the numerical quadrature rule

$$\int_{-1}^{1} v(t)\,\mathrm{d}t \approx 2v(0),$$

called the *midpoint rule*. We shall assume that $v \in W_p^2(-1, 1)$, $1 \le p \le \infty$. In order to estimate the error committed, let us consider the linear functional

$$\eta(v) := \int_{-1}^{1} v(t)\,\mathrm{d}t - 2v(0)$$

defined on $W_p^2(-1, 1)$. Clearly, $\mathcal{P}_1 \subset \operatorname{Ker}(\eta)$ and

$$\begin{aligned}
|\eta(v)| &\le \int_{-1}^{1} |v(t)|\,\mathrm{d}t + 2|v(0)| \\
&= \int_{-1}^{1} |v(t)|\,\mathrm{d}t + \left| \int_{-1}^{1}\int_{t}^{0} v'(s)\,\mathrm{d}s\,\mathrm{d}t + \int_{-1}^{1} v(t)\,\mathrm{d}t \right| \\
&\le 2\int_{-1}^{1} |v(t)|\,\mathrm{d}t + 2\int_{-1}^{1} |v'(t)|\,\mathrm{d}t \\
&\le 2 \cdot 2^{1-\frac{1}{p}} \left(\|v\|_{L_p(-1,1)} + \|v'\|_{L_p(-1,1)} \right) \\
&\le 2 \cdot 4^{1-\frac{1}{p}} \|v\|_{W_p^1(-1,1)} \le 2 \cdot 4^{1-\frac{1}{p}} \|v\|_{W_p^2(-1,1)}.
\end{aligned}$$

From the Bramble–Hilbert lemma we deduce that there exists a positive constant $C = C(p)$ such that

$$|\eta(v)| \le C|v|_{W_p^2(-1,1)}.$$

In the next example we consider a similar analysis on the interval $[-h,h]$. Using a scaling argument we shall reduce the problem to the one considered in Example 2.6.

Example 2.7 Let us suppose that we are required to estimate the size of the error in the midpoint rule on the interval $[-h,h]$, for $h>0$:

$$\int_{-h}^{h} u(x)\,\mathrm{d}x \approx 2hu(0).$$

To do so, we consider the linear functional

$$\eta_h(u) := \int_{-h}^{h} u(x)\,\mathrm{d}x - 2hu(0),$$

and introduce the following change of variable, in order to map $[-h,h]$ on the 'canonical interval' $[-1,1]$:

$$x = ht, \quad t \in [-1,1], \qquad v(t) := u(x).$$

Then, with η as in the previous example,

$$\eta_h(u) = h\eta_1(v) = h\eta(v).$$

Therefore, according to the final inequality in Example 2.6, and returning from the interval $[-1,1]$ to $[-h,h]$,

$$\left|\eta_h(u)\right| \le Ch|v|_{W_p^2(-1,1)} = Ch \cdot h^{2-\frac{1}{p}}|u|_{W_p^2(-h,h)}.$$

In particular, for $p=2$ we have that

$$\left|\eta_h(u)\right| \le Ch^{5/2}|u|_{W_2^2(-h,h)}.$$

Using the error bound for the midpoint rule on the interval $[-h,h]$ established in this last example by means of the Bramble–Hilbert lemma it is possible to obtain an optimal-order bound on the global error in a finite difference approximation of a two-point boundary-value problem. We shall explain how this is done. In the next section we shall then extend the technique to multiple space dimensions.

Let us consider the two-point boundary-value problem

$$-u'' = f(x), \quad x \in (0,1),$$
$$u(0) = 0, \qquad u(1) = 0.$$

Given the *nonuniform* finite difference mesh $0 = x_0 < x_1 < \cdots < x_N = 1$ with spacing $h_i := x_i - x_{i-1}$, $i = 1,\dots,N$, we define $\hbar_i := (h_{i+1}+h_i)/2$, $i = 1,\dots,N-1$, and introduce the backward and forward divided difference operators

$$D_x^- V_i := \frac{V_i - V_{i-1}}{h_i}, \qquad D_x^+ V_i := \frac{V_{i+1} - V_i}{\hbar_i},$$

and the following inner products and norms:

$$(V, W)_h := \sum_{i=1}^{N-1} \hbar_i V_i W_i, \quad \|V\|_{L_2(\Omega^h)} := (V, V)_h^{1/2},$$
$$(V, W]_h := \sum_{i=1}^{N} h_i V_i W_i, \quad \|V\|_{L_2(\Omega^h_+)} := (V, V]_h^{1/2},$$

where $\Omega^h := \{x_1, \dots, x_{N-1}\}$ and $\Omega^h_+ := \{x_1, \dots, x_N\}$. Let us consider the following finite difference approximation of the two-point boundary-value problem:

$$-D_x^+ D_x^- U_i = T_h^1 f_i, \quad i = 1, \dots, N-1,$$
$$U_0 = 0, \qquad U_N = 0,$$

where $T_h^1 f$ denotes the mollification of f defined by

$$T_h^1 f_i := \frac{1}{\hbar_i} \int_{x_{i-1/2}}^{x_{i+1/2}} f(x) \mathrm{d}x, \quad i = 1, \dots, N-1.$$

In order to derive a bound on the global error $e := u - U$ at the mesh-points, we note that

$$-D_x^+ D_x^- e_i = -D_x^+ \eta_i, \quad i = 1, \dots, N-1,$$
$$e_0 = 0, \qquad e_N = 0,$$

where

$$\begin{aligned}\eta_i &:= D_x^- u(x_i) - u'(x_{i-1/2}) \\ &= \frac{1}{2h_i}\left[\int_{-h_i}^{h_i} u'\left(x_{i-1/2} + \frac{1}{2}x\right) \mathrm{d}x - 2h_i u'(x_{i-1/2})\right] \\ &= \frac{1}{2h_i} \eta_{h_i}\left(u'\left(x_{i-1/2} + \frac{1}{2}\cdot\right)\right), \quad i = 1, \dots, N,\end{aligned}$$

where η_{h_i} is as in Example 2.7. We thus deduce that

$$|\eta_i| \le C h_i^{3/2} |u'|_{W_2^2(x_{i-1}, x_i)},$$

where C is a positive constant, independent of h_i. Consequently,

$$\|\eta\|^2_{L_2(\Omega^h_+)} = \sum_{i=1}^{N} h_i |\eta_i|^2 \le C^2 \sum_{i=1}^{N} h_i h_i^3 |u'|^2_{W_2^2(x_{i-1}, x_i)}$$

$$= C^2 \sum_{i=1}^{N} h_i^4 |u|^2_{W_2^3(x_{i-1},x_i)} \le C^2 h^4 |u|^2_{W_2^3(0,1)},$$

where $h = \max_i h_i$. We complete the error analysis by showing that the quantity $\|D_x^- e\|_{L_2(\Omega_+^h)}$ can be bounded in terms of $\|\eta\|_{L_2(\Omega_+^h)}$. Indeed, by summation by parts and using the Cauchy–Schwarz inequality we obtain

$$\begin{aligned}\|D_x^- e\|^2_{L_2(\Omega_+^h)} &= (-D_x^+ D_x^- e, e)_h = (-D_x^+ \eta, e)_h \\ &= (\eta, D_x^- e]_h \le \|\eta\|_{L_2(\Omega_+^h)} \|D_x^- e\|_{L_2(\Omega_+^h)}.\end{aligned}$$

Hence,

$$\|D_x^- e\|_{L_2(\Omega_+^h)} \le \|\eta\|_{L_2(\Omega_+^h)},$$

and therefore

$$|u - U|_{W_2^1(\Omega^h)} := \|D_x^- (u - U)\|_{L_2(\Omega_+^h)} \le Ch^2 |u|_{W_2^3(0,1)},$$

where C is a positive constant, independent of h and u. We note that we did not have to impose any regularity requirements on the nonuniform mesh to prove this error bound; in the next section, we shall develop a similar analysis in two dimensions.

First, however, we formulate a generalization of the Bramble–Hilbert lemma to Sobolev spaces of any positive (not necessarily integer) order.

Theorem 2.28 *Let $\Omega \subset \mathbb{R}^n$ be a Lipschitz domain and, for real numbers $s > 0$ and $p \in [1, \infty]$, let η be a bounded linear functional on the Sobolev space $W_p^s(\Omega)$ such that, by writing $s = m + \alpha$ with m a nonnegative integer and $0 < \alpha \le 1$,*

$$\mathcal{P}_m \subset \mathrm{Ker}(\eta).$$

Then, there exists a positive real number $C = C(s, p, n, \Omega)$ such that

$$|\eta(v)| \le C\|\eta\| |v|_{W_p^s(\Omega)} \quad \forall v \in W_p^s(\Omega).$$

Proof This result is a simple consequence of Theorem 1.13 with $\mathcal{U}_0 = L_p(\Omega)$, $\mathcal{U}_1 = W_p^s(\Omega)$, $S_0(u) = \|u\|_{L_p(\Omega)}$, $S_1(u) = \|u\|_{W_p^s(\Omega)}$, $S(u) = |\eta(u)|$, by noting that, according to the Theorem 1.36, $W_p^s(\Omega)$ is compactly embedded in $L_p(\Omega)$ for any $s > 0$. □

One can apply this result to the midpoint rule to deduce, in the same manner as in the integer-order case considered earlier, that the linear functional η defined on $W_p^s(-1, 1)$, $1/p < s \le 2$, $1 \le p \le \infty$, by

$$\eta(v) = \int_{-1}^{1} v(t)\,dt - 2v(0)$$

satisfies the bound

$$\left|\eta(v)\right| \le C|v|_{W_p^s(-1,1)},$$

for any v in $W_p^s(\Omega)$, $1/p < s \le 2$, $1 \le p \le \infty$. Thus in particular, with $p = 2$, we obtain the following bound on the global error in the finite difference approximation of the two-point boundary-value problem considered:

$$|u - U|_{W_2^1(\Omega^h)} \le Ch^s|u'|_{W_2^s(0,1)},$$

where $h = \max_i h_i$, provided that $u \in W_2^{s+1}(0,1)$ (whereby $u' \in W_2^s(0,1)$), $1/2 < s \le 2$. In the next section we extend this result to Poisson's equation on the unit square. First we shall however formulate a generalization of the Bramble–Hilbert lemma to anisotropic Sobolev spaces of the type $W_p^A(\Omega)$.

Let $A \subset \mathbb{R}_+^n$ be a regular set of nonnegative real multi-indices (cf. Sect. 1.5). We denote the convex hull in $\mathbb{R}^n$ of the set A by $\kappa(A)$. Let $\partial_0\kappa(A)$ be the part of the boundary of $\kappa(A)$ that has empty intersection with the co-ordinate hyperplanes, and let $A_\partial = A \cap \overline{\partial_0\kappa(A)}$. Let B be a nonempty subset of A_∂ such that $B \cup \{0\}$ is a regular set of multi-indices, and define

$$\nu(B) := \left\{\beta \in \mathbb{N}_+^n : \partial^{\lfloor\alpha\rfloor}x^\beta \equiv 0 \ \ \forall\alpha \in B\right\}.$$

Let $\mathcal{P}_B$ denote the set of all polynomials in n variables of the form

$$P(x) = \sum_{\alpha\in\nu(B)} p_\alpha x^\alpha.$$

Theorem 2.29 *Suppose that Ω is a Lipschitz domain in $\mathbb{R}^n$ and let the sets A and B of real nonnegative multi-indices satisfy the conditions formulated in the previous paragraph. Then, there exists a positive real number $C = C(A, B, p, n, \Omega)$ such that*

$$\inf_{P\in\mathcal{P}_B} \|v - P\|_{W_p^A(\Omega)} \le C\sum_{\alpha\in B} |v|_{\alpha,p} \quad \forall v \in W_p^A(\Omega).$$

Moreover, if η is a bounded linear functional on $W_p^A(\Omega)$, with norm $\|\eta\|$, such that

$$\mathcal{P}_B \subset \mathrm{Ker}(\eta),$$

then

$$\left|\eta(v)\right| \le C\|\eta\|\sum_{\alpha\in B} |v|_{\alpha,p} \quad \forall v \in W_p^A(\Omega).$$

Proof This result is a simple consequence of Theorem 1.13 with $\mathcal{U}_0 = L_p(\Omega)$, $\mathcal{U}_1 = W_p^A(\Omega)$, $S_0(u) = \|u\|_{L_p(\Omega)}$,

$$S_1(u) = \|u\|_{L_p(\Omega)} + \sum_{\alpha\in B} |v|_{\alpha,p},$$

and $S(u) = |\eta(u)|$, and noting that $W_p^A(\Omega)$, equipped with the norm $S_1(\cdot)$ is compactly embedded in $L_p(\Omega)$. □

As a further generalization, we state the following multilinear version of the Bramble–Hilbert lemma: this will be used extensively in the bilinear case throughout the book.

Lemma 2.30 *Suppose that A_k, B_k and Ω_k satisfy the same conditions in $\mathbb{R}^{n_k}$, $k = 1, \ldots, m$, as A, B and Ω did in the previous theorem. Let $(v_1, \ldots, v_m) \mapsto \eta(v_1, \ldots, v_m)$ be a bounded multilinear functional on the function space*

$$W_{p_1}^{A_1}(\Omega_1) \times \cdots \times W_{p_m}^{A_m}(\Omega_m),$$

which vanishes whenever one of its entries has the form $v_k = x^\alpha$, $x \in \Omega_k$, $\alpha \in \nu(B_k)$. Then, there exists a real number

$$C = C(A_1, B_1, p_1, \Omega_1, n_1, \ldots, A_m, B_m, p_m, \Omega_m, n_m)$$

such that

$$\left|\eta(v_1, \ldots, v_m)\right| \leq C\|\eta\| \prod_{k=1}^{m} \sum_{\alpha \in B_k} |v_k|_{\alpha, p_k}$$

for every $(v_1, \ldots, v_m)$ in $W_{p_1}^{A_1}(\Omega_1) \times \cdots \times W_{p_m}^{A_m}(\Omega_m)$.

When $m = 2$, this result will be referred to as the *bilinear version of the Bramble–Hilbert lemma*. In the case of standard, integer-order isotropic Sobolev spaces, the bilinear version of the Bramble–Hilbert lemma can be found in Ciarlet [26], Theorem 4.2.5. In the general case the proof is analogous, and is once again a simple consequence of Theorem 2.29.

2.3.2 Optimal Error Bounds on Uniform Meshes

In this section we shall use the Bramble–Hilbert lemma to derive an optimal bound on the global error of the finite difference (or, more precisely, finite volume) approximation (2.71), (2.72) of the homogeneous Dirichlet boundary-value problem (2.55), (2.56) on a uniform mesh of size h; in the next section we shall extend this analysis to nonuniform meshes. Thus, we consider the following finite difference scheme:

$$-(D_x^+ D_x^- + D_y^+ D_y^-)U_{ij} + c(x_i, y_j)U_{ij} = \left(T_h^{11} f\right)_{ij}, \quad (x_i, y_j) \in \Omega^h, \tag{2.99}$$

$$U = 0 \quad \text{on } \Gamma^h. \tag{2.100}$$

Let $e := u - U$ denote the global error of the scheme; then, according to Lemma 2.24,

$$\|e\|_{W_2^1(\Omega^h)} \leq \frac{1}{c_0}\left(\|\varphi_1\|^2_{L_2(\Omega^h_x)} + \|\varphi_2\|^2_{L_2(\Omega^h_y)} + \|\psi\|^2_{L_2(\Omega^h)}\right)^{1/2}, \qquad (2.101)$$

where φ_1, φ_2, and ψ are defined by

$$\varphi_1(x_i, y_j) := \frac{1}{h}\int_{y_{j-1/2}}^{y_{j+1/2}} \frac{\partial u}{\partial x}(x_{i-1/2}, y)\,\mathrm{d}y - D_x^- u(x_i, y_j),$$

$$\varphi_2(x_i, y_j) := \frac{1}{h}\int_{x_{i-1/2}}^{x_{i+1/2}} \frac{\partial u}{\partial y}(x, y_{j-1/2})\,\mathrm{d}x - D_y^- u(x_i, y_j),$$

$$\psi(x_i, y_j) := (cu)(x_i, y_j) - \frac{1}{h^2}\int_{x_{i-1/2}}^{x_{i+1/2}}\int_{y_{j-1/2}}^{y_{j+1/2}} (cu)(x, y)\,\mathrm{d}x\,\mathrm{d}y,$$

with $x_{i\pm 1/2} = x_i \pm h/2$ and $y_{j\pm 1/2} = y_j \pm h/2$.

We shall use the Bramble–Hilbert lemma to estimate φ_1, φ_2 and ψ in terms of appropriate powers of the discretization parameter h and suitable Sobolev seminorms of the analytical solution u. We begin by considering φ_1. Let us introduce the change of variables

$$x = x_{i-1/2} + \tilde{x}h, \quad -\frac{1}{2} \leq \tilde{x} \leq \frac{1}{2}; \qquad y = y_j + \tilde{y}h, \quad -\frac{1}{2} \leq \tilde{y} \leq \frac{1}{2},$$

and define $\tilde{v}(\tilde{x}, \tilde{y}) := h\frac{\partial u}{\partial x}(x, y)$. Then,

$$\varphi_1(x_i, y_j) = \frac{1}{h}\tilde{\varphi}_1(\tilde{v}),$$

where

$$\tilde{\varphi}_1(\tilde{v}) := \int_{-1/2}^{1/2}\int_{-1/2}^{1/2} \left[\tilde{v}(0, \tilde{y}) - \tilde{v}(\tilde{x}, 0)\right]\mathrm{d}\tilde{x}\,\mathrm{d}\tilde{y}.$$

Thanks to the trace theorem (Theorem 1.42),

$$\left|\tilde{\varphi}_1(\tilde{v})\right| \leq C_s \|\tilde{v}\|_{W_2^s(\tilde{K})}, \quad s > 1/2,$$

where

$$\tilde{K} := \left(-\frac{1}{2}, \frac{1}{2}\right) \times \left(-\frac{1}{2}, \frac{1}{2}\right),$$

and $C_s = C(s)$ is a positive constant. Thus $\tilde{\varphi}_1$ is a bounded linear functional (of the argument $\tilde{v}$) on $W_2^s(\tilde{K})$ for $s > 1/2$.

Moreover, $\tilde{\varphi}_1 = 0$ when $\tilde{v}(\tilde{x}, \tilde{y}) = \tilde{x}^k \tilde{y}^l$, $k, l \in \{0, 1\}$. According to Theorem 2.28, there exists a positive constant $C = C(s)$ such that

$$\left|\tilde{\varphi}_1(\tilde{v})\right| \leq C|\tilde{v}|_{W_2^s(\tilde{K})}, \quad 1/2 < s \leq 2.$$

Hence, by defining

$$K_{ij} := (x_{i-1/2}, x_{i+1/2}) \times (y_{j-1/2}, y_{j+1/2})$$

and returning from $\tilde{x}$ and $\tilde{y}$ to the original variables x and y, we deduce that

$$|\tilde{\varphi}_1(\tilde{v})| \leq Ch^s \left|\frac{\partial u}{\partial x}\right|_{W_2^s(K_{ij})}, \quad 1/2 < s \leq 2,$$

so that

$$|\varphi_1(x_i, y_j)| \leq Ch^{s-1} \left|\frac{\partial u}{\partial x}\right|_{W_2^s(K_{ij})}, \quad 1/2 < s \leq 2.$$

By noting that the Sobolev seminorm on the unit square is *superadditive* on the family $\{K_{ij}\}$ of mutually disjoint Lebesgue-measurable subsets K_{ij} of Ω, i.e. for $w \in W_2^s(\Omega)$ one has

$$\left(\sum_{i=1}^{N-1}\sum_{j=1}^{N-1} |w|^2_{W_2^s(K_{ij})}\right)^{1/2} \leq |w|_{W_2^s(\cup_{i,j=1}^{N-1} K_{ij})},$$

it follows with $w = \partial u/\partial x$ that

$$\|\varphi_1\|_{L_2(\Omega_x^h)} \leq Ch^s \left|\frac{\partial u}{\partial x}\right|_{W_2^s(\Omega)}, \quad 1/2 < s \leq 2, \tag{2.102}$$

where C is a positive constant, dependent only on s. Analogously,

$$\|\varphi_2\|_{L_2(\Omega_y^h)} \leq Ch^s \left|\frac{\partial u}{\partial y}\right|_{W_2^s(\Omega)}, \quad 1/2 < s \leq 2. \tag{2.103}$$

To complete the error analysis it remains to estimate $\psi(x_i, y_j)$. For this purpose we shall write $w := cu$ and note that

$$\psi(x_i, y_j) = w(x_i, y_j) - \frac{1}{h^2}\int_{x_{i-1/2}}^{x_{i+1/2}}\int_{y_{j-1/2}}^{y_{j+1/2}} w(x, y)\,dx\,dy.$$

Let us also consider the following change of variables:

$$x = x_i + \tilde{x}h, \quad -\frac{1}{2} \leq \tilde{x} \leq \frac{1}{2}; \qquad y = y_j + \tilde{y}h, \quad -\frac{1}{2} \leq \tilde{y} \leq \frac{1}{2},$$

and define $\tilde{w}(\tilde{x}, \tilde{y}) := w(x, y)$. Then,

$$\psi(x_i, y_j) = \tilde{\psi}(\tilde{w}),$$

where

$$\tilde{\psi}(\tilde{w}) := \tilde{w}(0, 0) - \int_{-1/2}^{1/2}\int_{-1/2}^{1/2} \tilde{w}(\tilde{x}, \tilde{y})\,d\tilde{x}\,d\tilde{y}.$$

By Sobolev's embedding theorem $\tilde{\psi}$ is a bounded linear functional (of $\tilde{w}$) on $W_2^s(\tilde{K})$ for $s > 1$, where, as before, $\tilde{K} := (-1/2, 1/2) \times (-1/2, 1/2)$. Furthermore, $\tilde{\psi}(\tilde{w}) = 0$ whenever $\tilde{w} = \tilde{x}^k \tilde{y}^l$ with $k, l \in \{0, 1\}$. Thus, by the Bramble–Hilbert lemma,

$$|\tilde{\psi}(\tilde{w})| \leq C|\tilde{w}|_{W_2^s(\tilde{K})}, \quad 1 < s \leq 2,$$

and consequently, after returning from the $(\tilde{x}, \tilde{y})$ co-ordinate system to the original variables x and y, we obtain the bound

$$|\psi(x_i, y_j)| \leq Ch^{s-1}|w|_{W_2^s(K_{ij})}, \quad 1 < s \leq 2,$$

and finally, after squaring and summing over $i, j = 1, \ldots, N-1$,

$$\|\psi\|_{L_2(\Omega^h)} \leq Ch^s|cu|_{W_2^s(\Omega)}, \quad 1 < s \leq 2. \tag{2.104}$$

Thus, by assuming that the weak solution $u \in W_2^s(\Omega) \cap \overset{\circ}{W}{}_2^1(\Omega)$ and that $c \in M(W_2^s(\Omega))$, for $1 < s \leq 2$, after substituting (2.102), (2.103) and (2.104) into (2.101), we arrive at the following bound on the global error:

$$\|u - U\|_{W_2^1(\Omega^h)} \leq Ch^s \left(\left|\frac{\partial u}{\partial x}\right|_{W_2^s(\Omega)} + \left|\frac{\partial u}{\partial y}\right|_{W_2^s(\Omega)} + \|c\|_{M(W_2^s(\Omega))}\|u\|_{W_2^s(\Omega)} \right),$$

where C is a positive constant depending on s, but independent of h; or, more crudely, after bounding $|\partial u/\partial x|_{W_2^s(\Omega)} + |\partial u/\partial y|_{W_2^s(\Omega)}$ by $\|u\|_{W_2^{s+1}(\Omega)}$, and writing $s-1$ instead s, we obtain

$$\|u - U\|_{W_2^1(\Omega^h)} \leq Ch^{s-1}\|u\|_{W_2^s(\Omega)}, \quad 2 < s \leq 3.$$

This should be compared with the error bound derived in the previous section using integral representations based on the Newton–Leibniz formula for $s = 2$ and $s = 3$ and by function space interpolation for $2 < s < 3$.

2.4 Convergence Analysis on Nonuniform Meshes

Our objective in this section is to develop the error analysis of finite difference (or, more precisely, finite volume) approximations on nonuniform meshes for the model Poisson equation with homogeneous Dirichlet boundary condition:

$$-\Delta u = f \quad \text{in } \Omega, \tag{2.105}$$

$$u = 0 \quad \text{on } \Gamma = \partial\Omega, \tag{2.106}$$

where $\Omega := (0, 1) \times (0, 1)$. When $f \in W_2^{-1}(\Omega)$, this boundary-value problem has a unique weak solution u in $\overset{\circ}{W}{}_2^1(\Omega)$; furthermore, if $f \in W_2^s(\Omega)$ then u belongs to $W_2^{s+2}(\Omega)$, $-1 \leq s < 1$, $s \neq \pm 1/2$ (see, Theorem 2.8).

As has already been indicated earlier, the key idea behind the construction of a finite volume method for (2.105), (2.106) is to make use of the divergence form of the differential operator $\Delta = \nabla \cdot \nabla$ appearing in the equation $-\Delta u = f$ by integrating both sides over mutually disjoint 'cells' $K_{ij} \subset \Omega$, and use the divergence theorem to convert integrals over the cells K_{ij} into contour integrals along their boundaries, which are then discretized by means of numerical quadrature rules. This construction gives rise to a finite difference scheme whose right-hand side involves the integral average of f over individual cells, the particular form of the difference scheme being dependent on the shapes of the cells and the numerical quadrature formula used. For example, if Ω has been partitioned by a uniform square mesh of mesh-size h, then the resulting scheme coincides with (2.71), (2.72) (with $c \equiv 0$).

2.4.1 Cartesian-Product Nonuniform Meshes

We begin by considering Cartesian-product nonuniform meshes. For the purposes of the error analysis it is helpful to reformulate the finite volume scheme as a Petrov–Galerkin finite element method based on bilinear or piecewise linear trial functions on the underlying mesh and piecewise constant test functions on the dual 'box mesh'. We shall prove that, as in the case of uniform meshes considered in the previous section, the scheme is stable in the discrete W_2^1 norm. This stability result will then, similarly to the arguments in the previous section, lead to an optimal-order error bound in the discrete W_2^1 norm under minimal smoothness requirements on the exact solution and without any additional assumptions on the spacing of the mesh. In particular, the mesh is not required to be quasi-uniform (in a sense that will be made precise). If quasi-uniformity is assumed, then an additional error bound holds, in the discrete maximum norm. In the next section similar results will be derived for a general one-parameter family of schemes.

The problem (2.105), (2.106) is approximated on the nonuniform mesh $\overline{\Omega}^h$, which is the Cartesian product of the one-dimensional meshes

$$\{x_i,\ i=0,\dots,M : x_0=0,\ x_i - x_{i-1} = h_i, x_M = 1\},$$
$$\{y_j,\ j=0,\dots,N : y_0=0,\ y_j - y_{j-1} = k_j, y_N = 1\}.$$

We then define

$$\Omega^h := \Omega \cap \overline{\Omega}^h, \qquad \Gamma^h := \Gamma \cap \overline{\Omega}^h,$$
$$\Omega_x^h := \overline{\Omega}^h \cap \big((0,1] \times (0,1)\big), \qquad \Omega_y^h := \overline{\Omega}^h \cap \big((0,1) \times (0,1]\big),$$
$$\Gamma_x^h := \overline{\Omega}^h \cap \big(\{0,1\} \times (0,1)\big), \qquad \Gamma_y^h := \overline{\Omega}^h \cap \big((0,1) \times \{0,1\}\big).$$

To each mesh-point (x_i, y_j) in Ω^h we assign a cell

$$K_{ij} := (x_{i-1/2}, x_{i+1/2}) \times (y_{j-1/2}, y_{j+1/2}),$$

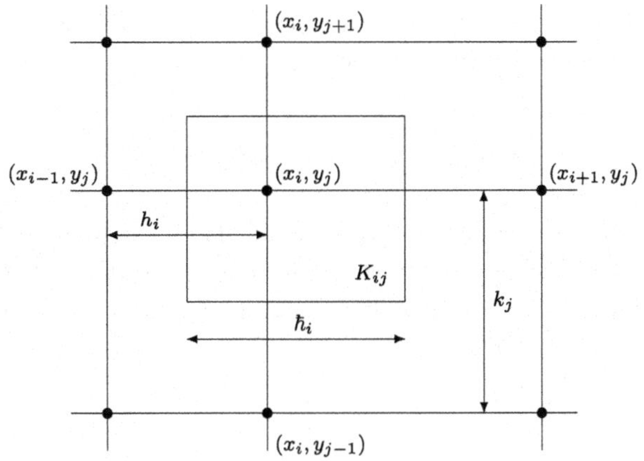

Fig. 2.4 Section of the Cartesian-product nonuniform mesh $\overline{\Omega}^h$, showing nine mesh-points and the cell K_{ij} associated with the mesh-point (x_i, y_j)

as shown in Fig. 2.4, where

$$x_{i-1/2} := x_i - \frac{1}{2}h_i, \qquad x_{i+1/2} := x_i + \frac{1}{2}h_{i+1},$$

$$y_{j-1/2} := y_j - \frac{1}{2}k_j, \qquad y_{j+1/2} := y_j + \frac{1}{2}k_{j+1},$$

and we denote the edge-lengths of the cell K_{ij} by

$$\hbar_i := \frac{1}{2}(h_i + h_{i+1}) \quad \text{and} \quad \bar{k}_j := \frac{1}{2}(k_j + k_{j+1}).$$

A simple calculation based on the definition of the fractional-order Sobolev norm shows that χ_{ij}, the characteristic function of the set $(-h_{i+1}/2, h_i/2) \times (-k_{j+1}/2, k_j/2)$, belongs to $W_2^\tau(\mathbb{R}^2)$ for all $\tau < 1/2$. Assuming that f belongs to $W_2^s(\Omega)$ for some $s > -1/2$, and extending f from Ω onto $\mathbb{R}^2$ by preserving its Sobolev class, we deduce from Theorem 1.69 that the convolution $\chi_{ij} * f$ is a continuous function on $\mathbb{R}^2$ (whose values on Ω^h are independent of the particular form of the extension). Convolution of (2.105) with χ_{ij} then yields

$$-\frac{1}{\text{meas } K_{ij}} \int_{\partial K_{ij}} \frac{\partial u}{\partial \nu} \,\mathrm{d}s = \frac{1}{\text{meas } K_{ij}} (\chi_{ij} * f)(x_i, y_j), \tag{2.107}$$

where ν denotes the unit outward normal vector to ∂K_{ij}.

We remark that if f is a locally integrable function on Ω then, similarly as in the case of uniform meshes considered earlier, the right-hand side of (2.107) is simply

$$\left(T_h^{11} f\right)_{ij} = \frac{1}{\hbar_i \bar{k}_j} \int_{x_{i-1/2}}^{x_{i+1/2}} \int_{y_{j-1/2}}^{y_{j+1/2}} f(x, y) \,\mathrm{d}x \,\mathrm{d}y.$$

Let $\mathcal{S}^h$ signify the set of all real-valued continuous piecewise bilinear functions defined on the rectangular partition of $\overline{\Omega}$ induced by $\overline{\Omega}^h$, and let $\mathcal{S}_0^h$ be the subset of

$\mathcal{S}^h$ consisting of those functions that vanish on Γ. Motivated by the form of (2.107), we define the finite volume approximation of u as $U \in \mathcal{S}_0^h$ satisfying

$$-\frac{1}{\hbar_i \hbar_j}\int_{\partial K_{ij}} \frac{\partial U}{\partial \nu}\,\mathrm{d}s = \frac{1}{\hbar_i \hbar_j}(\chi_{ij} * f)(x_i, y_j) \quad \text{for } (x_i, y_j) \in \Omega^h. \tag{2.108}$$

First, we shall prove that this method is stable by proceeding in the same way as in the case of uniform meshes considered in the previous section. To this end, we shall rewrite (2.108) as a finite difference scheme on $\overline{\Omega}^h$ by using the *averaging operator* μ_x defined by

$$\mu_x V_{ij} := \frac{1}{8\hbar_i}(h_i V_{i-1,j} + 6\hbar_i V_{ij} + h_{i+1} V_{i+1,j}) \tag{2.109}$$

and the divided differences

$$D_x^- V_{ij} := \frac{V_{ij} - V_{i-1,j}}{h_i} \quad \text{and} \quad D_x^+ V_{ij} := \frac{V_{i+1,j} - V_{ij}}{\hbar_i},$$

with analogous definitions for μ_y, D_y^- and D_y^+. With these notational conventions,

$$-\int_{\partial K_{ij}} \frac{\partial U}{\partial \nu}\,\mathrm{d}s = -\hbar_i \hbar_j \left(D_x^+ D_x^- \mu_y + D_y^+ D_y^- \mu_x\right) U_{ij}. \tag{2.110}$$

By inserting (2.110) into (2.108), the finite volume method (2.108) can be restated as the finite difference scheme

$$-(D_x^+ D_x^- \mu_y + D_y^+ D_y^- \mu_x) U = T_h^{11} f \quad \text{in } \Omega^h, \tag{2.111}$$

$$U = 0 \quad \text{on } \Gamma^h, \tag{2.112}$$

where

$$\left(T_h^{11} f\right)_{ij} := \frac{1}{\hbar_i \hbar_j}(\chi_{ij} * f)(x_i, y_j).$$

We begin the analysis of the scheme (2.111), (2.112) by investigating its stability in the discrete W_2^1 norm, $\|\cdot\|_{W_2^1(\Omega^h)}$, defined by

$$\|V\|_{W_2^1(\Omega^h)} := \left(\|V\|^2_{L_2(\Omega^h)} + |V|^2_{W_2^1(\Omega^h)}\right)^{1/2},$$

where $\|\cdot\|_{L_2(\Omega^h)}$ is the discrete L_2 norm on the linear space of real-valued mesh-functions defined on Ω^h:

$$\|V\|_{L_2(\Omega^h)} := (V, V)_h^{1/2}, \qquad (V, W)_h := \sum_{i=1}^{M-1} \sum_{j=1}^{N-1} \hbar_i \hbar_j V_{ij} W_{ij},$$

and $|\cdot|_{W_2^1(\Omega^h)}$ is the discrete W_2^1 seminorm defined by

$$|V|_{W_2^1(\Omega^h)} := \left(\|D_x^- V\|^2_{L_2(\Omega_x^h)} + \|D_y^- V\|^2_{L_2(\Omega_y^h)}\right)^{1/2},$$

with

$$\|V\|^2_{L_2(\Omega_x^h)} := (V, V]_x^2, \qquad (V, W]_x := \sum_{i=1}^{M}\sum_{j=1}^{N-1} h_i \hbar k_j V_{ij} W_{ij},$$

$$\|V\|^2_{L_2(\Omega_y^h)} := (V, V]_y^2, \qquad (V, W]_y := \sum_{i=1}^{M-1}\sum_{j=1}^{N} \hbar_i k_j V_{ij} W_{ij}.$$

The associated discrete W_2^{-1} norm is then defined by

$$\|V\|_{W_2^{-1}(\Omega^h)} := \sup_{W \in S_0^h \setminus \{0\}} \frac{|(V, W)_h|}{\|W\|_{W_2^1(\Omega^h)}}.$$

Lemma 2.31 *Suppose that V is a mesh-function defined on $\overline{\Omega}_h$.*

(a) *If $V = 0$ on Γ_x^h, then*

$$(\mu_x V, V]_y \geq \frac{1}{2}\|V\|^2_{L_2(\Omega_y^h)}. \tag{2.113}$$

(b) *If $V = 0$ on Γ_y^h, then*

$$(\mu_y V, V]_x \geq \frac{1}{2}\|V\|^2_{L_2(\Omega_x^h)}. \tag{2.114}$$

Proof We shall only prove inequality (2.113), the proof of (2.114) being analogous. Let us assume for a moment that j is fixed, $1 \leq j \leq N$. Then,

$$\begin{aligned}
\sum_{i=1}^{M-1} \hbar_i (\mu_x V_{ij}) V_{ij} &= \frac{1}{8}\sum_{i=1}^{M-1}\left(h_i V_{i-1,j} V_{ij} + 6\hbar_i V_{ij}^2 + h_{i+1} V_{i+1,j} V_{ij}\right)\\
&\geq \frac{1}{8}\left(\sum_{i=1}^{M-1} 5\hbar_i V_{ij}^2 - \frac{1}{2}\sum_{i=2}^{M} h_i V_{ij}^2 - \frac{1}{2}\sum_{i=0}^{M-2} h_{i+1} V_{ij}^2\right)\\
&\geq \frac{1}{2}\sum_{i=1}^{M-1} \hbar_i V_{ij}^2.
\end{aligned}$$

We then multiply this by k_j and sum through the index $j \in \{1, \ldots, N\}$ to deduce the desired inequality. □

We shall also require the following discrete analogue of the Friedrichs inequality on Cartesian-product nonuniform meshes.

Lemma 2.32 *Suppose that V is a mesh-function defined on $\overline{\Omega}_h$ such that $V = 0$ on Γ_h. Then,*

$$\|V\|^2_{W^1_2(\Omega^h)} \leq \frac{3}{2}|V|^2_{W^1_2(\Omega^h)}. \tag{2.115}$$

Proof Let V be a mesh-function defined on $\overline{\Omega}_h$ such that $V = 0$ on Γ_h. Then, the expression

$$\|V\|^2_{L_2(\Omega^h)} = \sum_{i=1}^{M-1}\sum_{j=1}^{N-1} \hbar_i k\!\!\!^{-}_j V_{ij}^2$$

can be bounded as follows:

$$\begin{aligned}
\|V\|^2_{L_2(\Omega^h)} &= \frac{1}{2}\sum_{i=1}^{M-1}\sum_{j=1}^{N-1} \hbar_i k\!\!\!^{-}_j V_{ij}^2 + \frac{1}{2}\sum_{i=1}^{M-1}\sum_{j=1}^{N-1} \hbar_i k\!\!\!^{-}_j V_{ij}^2 \\
&= \frac{1}{2}\left(\sum_{i=1}^{M-1}\sum_{j=1}^{N-1} \hbar_i k\!\!\!^{-}_j \left|\sum_{m=1}^{i} h_m D_x^- V_{mj}\right|^2 + \sum_{i=1}^{M-1}\sum_{j=1}^{N-1} \hbar_i k\!\!\!^{-}_j \left|\sum_{n=1}^{j} k_n D_y^- V_{in}\right|^2\right) \\
&\leq \frac{1}{2}\sum_{i=1}^{M-1}\sum_{j=1}^{N-1} \hbar_i k\!\!\!^{-}_j \left(\sum_{m=1}^{i} h_m\right)\left(\sum_{m=1}^{i} h_m |D_x^- V_{mj}|^2\right) \\
&\quad + \frac{1}{2}\sum_{i=1}^{M-1}\sum_{j=1}^{N-1} \hbar_i k\!\!\!^{-}_j \left(\sum_{n=1}^{j} k_n\right)\left(\sum_{n=1}^{j} k_n |D_y^- V_{in}|^2\right) \\
&\leq \frac{1}{2}\left(\sum_{m=1}^{M}\sum_{j=1}^{N-1} h_m k\!\!\!^{-}_j |D_x^- V_{mj}|^2\right)\left(\sum_{i=1}^{M-1} \hbar_i \sum_{m=1}^{i} h_m\right) \\
&\quad + \frac{1}{2}\left(\sum_{i=1}^{M-1}\sum_{n=1}^{N} \hbar_i k_n |D_y^- V_{in}|^2\right)\left(\sum_{j=1}^{N-1} k\!\!\!^{-}_j \sum_{n=1}^{j} k_n\right) \\
&\leq \frac{1}{2}\left(\|D_x^- V\|^2_{L_2(\Omega_x^h)} + \|D_x^- V\|^2_{L_2(\Omega_y^h)}\right) = \frac{1}{2}|V|^2_{W_2^1(\Omega^h)}.
\end{aligned}$$

Adding $|V|^2_{W^1_2(\Omega^h)}$ to both sides completes the proof of the lemma. □

By using this discrete Friedrichs inequality we shall now prove that the finite difference scheme is stable; the key to the proof is the following result.

Theorem 2.33 *Let* $L^h V := -(D_x^+ D_x^- \mu_y + D_y^+ D_y^- \mu_x)V$. *Then,*

$$\|V\|_{W_2^1(\Omega^h)} \leq 3\|L^h V\|_{W_2^{-1}(\Omega^h)} \tag{2.116}$$

for any mesh-function V *defined on* $\overline{\Omega}^h$ *and such that* $V = 0$ *on* Γ^h.

Proof By taking the $(\cdot, \cdot)_h$ inner product of $L^h V$ with V we obtain

$$\left(-(D_x^+ D_x^- \mu_y)V, V\right)_h + \left(-(D_y^+ D_y^- \mu_x)V, V\right)_h = \left(L^h V, V\right)_h.$$

By performing summations by parts in the two terms on the left-hand side we get

$$\left(D_x^- \mu_y V, D_x^- V\right]_x + \left(D_y^- \mu_x V, D_y^- V\right]_y = \left(L^h V, V\right)_h.$$

Since D_x^- commutes with μ_y and D_y^- commutes with μ_x, we can apply (2.113) and (2.114) to obtain

$$\frac{1}{2}\left(\|D_x^- V\|^2_{L_2(\Omega_x^h)} + \|D_y^- V\|^2_{L_2(\Omega_y^h)}\right) \leq \left(L^h V, V\right)_h.$$

By recalling (2.115) and the definition of $\|\cdot\|_{W_2^{-1}(\Omega^h)}$ we get (2.116). □

Theorem 2.33 now implies the stability of the scheme.

Theorem 2.34 *For any* $f \in W_2^s(\Omega)$, $s > -1/2$, *the scheme* (2.108) (*or, equivalently,* (2.111), (2.112)) *has a unique solution* U. *Moreover,*

$$\|U\|_{W_2^1(\Omega^h)} \leq 3\|T_h^{11} f\|_{W_2^{-1}(\Omega^h)}.$$

Having proved stability, we are now ready to embark on the error analysis of the scheme. We shall derive an optimal-order error bound for the finite difference method (2.111), (2.112), which can also be seen as a superconvergence result for the finite volume method (2.108) considered as a Petrov–Galerkin finite element method, on a family of Cartesian-product nonuniform meshes. By *superconvergence* we mean that $\mathcal{O}(h^2)$ convergence of the error between u and its continuous piecewise bilinear approximation U is observed in the *discrete* W_2^1 norm while only $\mathcal{O}(h)$ convergence will be seen if $u - U$ is measured in the norm of the Sobolev space $W_2^1(\Omega)$. The result will be shown to hold without any additional assumptions on the spacing of the mesh: in particular the mesh is not required to be quasi-uniform (the definition of quasi-uniform mesh will be given in the statement of Theorem 2.38).

Theorem 2.35 *Suppose that* $u \in W_2^{s+1}(\Omega) \cap \mathring{W}_2^1(\Omega)$, $1/2 < s \leq 2$. *Then,*

$$\|u - U\|_{W_2^1(\Omega^h)} \leq C h^s |u|_{W_2^{s+1}(\Omega)}, \tag{2.117}$$

where $h = \max_{i,j}(h_i, k_j)$, *and* $C = C(s)$ *is a positive constant independent of* u *and the discretization parameters.*

In the proof of Theorem 2.35 we shall make use of anisotropic Sobolev spaces on rectangular subdomains of $\mathbb{R}^2$. For $\omega = (a,b) \times (c,d)$ and a pair (r,s) of nonnegative real numbers, we denote by $W_2^{r,s}(\omega)$ the anisotropic Sobolev space consisting of all functions $u \in L_2(\omega)$ such that

$$|u|_{W_2^{r,0}(\omega)} := \left(\int_c^d |u(\cdot, y)|^2_{W_2^r(a,b)}\, \mathrm{d}y \right)^{1/2} < \infty,$$

$$|u|_{W_2^{0,s}(\omega)} := \left(\int_a^b |u(x, \cdot)|^2_{W_2^s(c,d)}\, \mathrm{d}x \right)^{1/2} < \infty.$$

The linear space $W_2^{r,s}(\omega)$ is a Banach space equipped with the norm

$$\|u\|_{W_2^{r,s}(\omega)} := \left(\|u\|^2_{L_2(\omega)} + |u|^2_{W_2^{r,0}(\omega)} + |u|^2_{W_2^{0,s}(\omega)} \right)^{1/2}.$$

For $s \geq 0$, $W_2^{s,s}(\omega)$ coincides with the standard (isotropic) Sobolev space $W_2^s(\omega)$, and the norm $\|\cdot\|_{W_2^{s,s}(\omega)}$ is equivalent to the Sobolev norm $\|\cdot\|_{W_2^s(\omega)}$ (cf. Sect. 18 of Besov, Il'in and Nikol'skiĭ [13]).

Proof of Theorem 2.35 Let us define the global error as $e := u - U$. Then, by applying the difference operator L^h defined in Theorem 2.33 to e and noting the definition of the finite difference scheme, we deduce that

$$L^h e = \left(T_h^{11} \frac{\partial^2 u}{\partial x^2} - D_x^+ D_x^- \mu_y u \right) + \left(T_h^{11} \frac{\partial^2 u}{\partial y^2} - D_y^+ D_y^- \mu_x u \right).$$

However,

$$\left(T_h^{11} \frac{\partial^2 u}{\partial x^2} \right)_{ij} = \frac{1}{\hbar_i \hbar_j} \int_{y_{j-1/2}}^{y_{j+1/2}} \left[\frac{\partial u}{\partial x}(x_{i+1/2}, y) - \frac{\partial u}{\partial x}(x_{i-1/2}, y) \right] \mathrm{d}y$$
$$= D_x^+ \left(T_-^{01} \frac{\partial u}{\partial x} \right)_{ij},$$

where

$$\left(T_-^{01} w\right)_{ij} = \frac{1}{\hbar_j} \int_{y_{j-1/2}}^{y_{j+1/2}} w(x_{i-1/2}, y)\, \mathrm{d}y.$$

Consequently,

$$\begin{aligned} L^h e &= D_x^+ \eta_1 + D_y^+ \eta_2 \quad \text{in } \Omega^h, \\ e &= 0 \quad \text{on } \Gamma^h, \end{aligned} \tag{2.118}$$

where

$$\eta_1 := T_-^{01} \frac{\partial u}{\partial x} - D_x^- \mu_y u, \qquad \eta_2 := T_-^{10} \frac{\partial u}{\partial y} - D_y^- \mu_x u,$$

and T_{-}^{10} is defined analogously to T_{-}^{01} above. By applying Theorem 2.33 to the finite difference equations (2.118) we have that

$$\|e\|_{W_2^1(\Omega^h)} \leq 3\big\|D_x^+\eta_1 + D_y^+\eta_2\big\|_{W_2^{-1}(\Omega^h)}.$$

It remains to bound the right-hand side of this inequality. We observe to this end that, for any mesh-function V defined on $\overline{\Omega}^h$ and vanishing on Γ^h,

$$-\big(D_x^+\eta_1 + D_y^+\eta_2, V\big)_h = \big(\eta_1, D_x^- V\big]_x + \big(\eta_2, D_y^- V\big]_y.$$

By noting the definition of the norm $\|\cdot\|_{W_2^{-1}(\Omega^h)}$ we thus deduce that

$$\big\|D_x^+\eta_1 + D_y^+\eta_2\big\|_{W_2^{-1}(\Omega^h)} \leq \|\eta_1\|_{L_2(\Omega_x^h)} + \|\eta_2\|_{L_2(\Omega_y^h)}.$$

Hence,

$$\|u - U\|_{W_2^1(\Omega^h)} \leq 3\big(\|\eta_1\|_{L_2(\Omega_x^h)} + \|\eta_2\|_{L_2(\Omega_y^h)}\big). \tag{2.119}$$

It remains to bound the right-hand side of (2.119). We only consider the term involving η_1; the norm of η_2 is bounded analogously.

To this end, we first define

$$(\mu_y u)(x, y_j) := \frac{1}{8\hbar_j}\big[k_j u(x, y_{j-1}) + 6\hbar_j u(x, y_j) + k_{j+1} u(x, y_{j+1})\big],$$

and for fixed x, $0 \leq x \leq 1$, we let $I_y w(x,\cdot)$ denote the univariate continuous piecewise linear interpolant of $w(x,\cdot)$ on the mesh $\overline{\Omega}_y^h$. Then,

$$(\mu_y w)(x, y_j) = \frac{1}{\hbar_j}\int_{y_{j-1/2}}^{y_{j+1/2}} (I_y w)(x, y)\,\mathrm{d}y,$$

and therefore,

$$\begin{aligned}
(\mu_y u)_{ij} - (\mu_y u)_{i-1,j} &= \int_{x_{i-1}}^{x_i} \frac{\partial}{\partial x}(\mu_y u)(x, y_j)\,\mathrm{d}x \\
&= \int_{x_{i-1}}^{x_i} \frac{\partial}{\partial x}\frac{1}{\hbar_j}\int_{y_{j-1/2}}^{y_{j+1/2}} (I_y u)(x, y)\,\mathrm{d}x\,\mathrm{d}y \\
&= \frac{1}{\hbar_j}\int_{x_{i-1}}^{x_i}\int_{y_{j-1/2}}^{y_{j+1/2}} \frac{\partial}{\partial x}(I_y u)(x, y)\,\mathrm{d}x\,\mathrm{d}y \\
&= \frac{1}{\hbar_j}\int_{x_{i-1}}^{x_i}\int_{y_{j-1/2}}^{y_{j+1/2}} I_y\Big(\frac{\partial u}{\partial x}\Big)(x, y)\,\mathrm{d}x\,\mathrm{d}y.
\end{aligned}$$

Thus we find that $(\eta_1)_{ij}$ can be expressed as

$$(\eta_1)_{ij} = \frac{1}{h_i\hbar_j}\int_{x_{i-1}}^{x_i}\int_{y_{j-1/2}}^{y_{j+1/2}} \Big[\frac{\partial u}{\partial x}(x_{i-1/2}, y) - \Big(I_y\frac{\partial u}{\partial x}\Big)(x, y)\Big]\,\mathrm{d}x\,\mathrm{d}y.$$

By splitting η_1 as the sum of η_{11} and η_{12}, where

$$(\eta_{11})_{ij} := \frac{1}{h_i \hbar_j} \int_{x_{i-1}}^{x_i} \int_{y_j}^{y_{j+1/2}} \left[\frac{\partial u}{\partial x}(x_{i-1/2}, y) - \left(I_y \frac{\partial u}{\partial x} \right)(x, y) \right] \mathrm{d}x\, \mathrm{d}y,$$

$$(\eta_{12})_{ij} := \frac{1}{h_i \hbar_j} \int_{x_{i-1}}^{x_i} \int_{y_{j-1/2}}^{y_j} \left[\frac{\partial u}{\partial x}(x_{i-1/2}, y) - \left(I_y \frac{\partial u}{\partial x} \right)(x, y) \right] \mathrm{d}x\, \mathrm{d}y,$$

the task of estimating η_1 is reduced to bounding η_{11} and η_{12}.

Let us first consider η_{11}. By introducing the change of variables

$$x = x_{i-1/2} + \tilde{x} h_i, \quad -\frac{1}{2} \le \tilde{x} \le \frac{1}{2}; \qquad y = y_j + \tilde{y} k_{j+1}, \quad 0 \le \tilde{y} \le 1,$$

and defining $\tilde{v}(\tilde{x}, \tilde{y}) := h_i \frac{\partial u}{\partial x}(x, y)$, we can write

$$(\eta_{11})_{ij} = \frac{k_{j+1}}{h_i \hbar_j} \tilde{\eta}_{11}(\tilde{v}),$$

where

$$\tilde{\eta}_{11}(\tilde{v}) := \int_{-1/2}^{1/2} \int_0^{1/2} \left[\tilde{v}(0, \tilde{y}) - \tilde{v}(\tilde{x}, 0)(1 - \tilde{y}) - \tilde{v}(\tilde{x}, 1)\tilde{y} \right] \mathrm{d}\tilde{x}\, \mathrm{d}\tilde{y}.$$

Now $\tilde{\eta}_{11}$ can be regarded as a linear functional (with the argument $\tilde{v}$) defined on $W_2^s(\tilde{K}^*)$, where $s > 1/2$ and

$$\tilde{K}^* := \left(-\frac{1}{2}, \frac{1}{2} \right) \times (0, 1).$$

Thanks to the trace theorem (Theorem 1.42),

$$\left| \tilde{\eta}_{11}(\tilde{v}) \right| \le C \|\tilde{v}\|_{W_2^s(\tilde{K}^*)}, \quad s > 1/2,$$

and therefore $|\tilde{\eta}_{11}(\cdot)|$ is a bounded sublinear functional on $W_2^s(\tilde{K}^*)$. Moreover, if $\tilde{v}(\tilde{x}, \tilde{y}) = \tilde{x}^k \tilde{y}^l$, $k, l \in \{0, 1\}$, then $\tilde{\eta}_{11}(\tilde{v}) = 0$. By applying Theorem 1.9 with

$$\mathcal{U}_1 = W_2^s(\tilde{K}^*), \qquad \mathcal{U}_0 = L_2(\tilde{K}^*),$$

$$S = |\tilde{\eta}_{11}|, \qquad S_1 = \left(|\cdot|^2_{W_2^{0,s}(\tilde{K}^*)} + |\cdot|^2_{W_2^{s,0}(\tilde{K}^*)} \right)^{1/2}, \qquad S_0 = \|\cdot\|_{L_2(\tilde{K}^*)},$$

and noting that for $s > 0$ the Sobolev space $W_2^s(\tilde{K}^*)$ is compactly embedded in $L_2(\tilde{K}^*)$, we deduce that

$$\left| \tilde{\eta}_{11}(\tilde{v}) \right| \le C \left(|\tilde{v}|^2_{W_2^{0,s}(\tilde{K}^*)} + |\tilde{v}|^2_{W_2^{s,0}(\tilde{K}^*)} \right)^{1/2}$$

for $1/2 < s \le 2$. By defining $K_{ij}^- := (x_{i-1}, x_i) \times (y_{j-1}, y_j)$, $K_{i,j+1}^- := (x_{i-1}, x_i) \times (y_j, y_{j+1})$ and returning from the $(\tilde{x}, \tilde{y})$–variables to the original (x, y) coordinates, we thus have that

$$|\tilde{\eta}_{11}(\tilde{v})|^2 \le C\left(\frac{h_i^2 k_{j+1}^{2s}}{h_i k_{j+1}}\left|\frac{\partial u}{\partial x}\right|^2_{W_2^{0,s}(K_{i,j+1}^-)} + \frac{h_i^{2(s+1)}}{h_i k_{j+1}}\left|\frac{\partial u}{\partial x}\right|^2_{W_2^{s,0}(K_{i,j+1}^-)}\right),$$

and therefore

$$|(\eta_{11})_{ij}|^2 \le C\left(\frac{k_{j+1}^{2s+1}}{h_i \hbar_j^2}\left|\frac{\partial u}{\partial x}\right|^2_{W_2^{0,s}(K_{i,j+1}^-)} + \frac{h_i^{2s-1} k_{j+1}}{\hbar_j^2}\left|\frac{\partial u}{\partial x}\right|^2_{W_2^{s,0}(K_{i,j+1}^-)}\right).$$

Analogously,

$$|(\eta_{12})_{ij}|^2 \le C\left(\frac{k_j^{2s+1}}{h_i \hbar_j^2}\left|\frac{\partial u}{\partial x}\right|^2_{W_2^{0,s}(K_{ij}^-)} + \frac{h_i^{2s-1} k_j}{\hbar_j^2}\left|\frac{\partial u}{\partial x}\right|^2_{W_2^{s,0}(K_{ij}^-)}\right).$$

By noting the superadditivity of the Sobolev seminorm on a family of mutually disjoint Lebesgue-measurable subsets of Ω, we thus have that

$$\|\eta_1\|^2_{L_2(\Omega_x^h)} \le C h^{2s}\left(\left|\frac{\partial u}{\partial x}\right|^2_{W_2^{0,s}(\Omega)} + \left|\frac{\partial u}{\partial x}\right|^2_{W_2^{s,0}(\Omega)}\right), \tag{2.120}$$

where $h = \max_{i,j}(h_i, k_j)$. Analogously,

$$\|\eta_2\|^2_{L_2(\Omega_y^h)} \le C h^{2s}\left(\left|\frac{\partial u}{\partial y}\right|^2_{W_2^{0,s}(\Omega)} + \left|\frac{\partial u}{\partial y}\right|^2_{W_2^{s,0}(\Omega)}\right). \tag{2.121}$$

By substituting (2.120) and (2.121) into (2.119) we thus obtain the desired error bound

$$\|u - U\|_{W_2^1(\Omega^h)} \le C h^s |u|_{W_2^{s+1}(\Omega)}, \quad 1/2 < s \le 2.$$

That completes the proof of the theorem. □

On a quasi-uniform mesh, the finite volume method (2.108) can be shown to be (almost) optimally accurate in the *discrete maximum norm* $\|\cdot\|_{\infty,h}$ defined by

$$\|V\|_{\infty,h} := \max_{(x,y)\in\overline{\Omega}^h} |V(x, y)|.$$

We shall say that $\{\overline{\Omega}^h\}$ is a family of *quasi-uniform Cartesian-product meshes* on $\overline{\Omega} = [0, 1] \times [0, 1]$ if there exists a positive constant $C_\star$ such that

$$h := \max_{i,j}(h_i, k_j) \le C_\star \min_{i,j}(h_i, k_j).$$

Some auxiliary results are required to prove an error bound in the discrete maximum norm; these are formulated in the next two lemmas, the first of which states a version of the *inverse inequality* (see, for example, Ciarlet [26], Theorem 3.2.6).

Lemma 2.36 *Suppose that* $\{\overline{\Omega}^h\}$ *is a family of quasi-uniform Cartesian-product meshes on* $\overline{\Omega} = [0,1] \times [0,1]$, *and let* S^h *be the linear space of continuous piecewise bilinear polynomials defined on the partition of* Ω *induced by* $\overline{\Omega}^h$. *Suppose that* $1 \le q, r \le \infty$. *Then, there exists a positive constant* $C = C(C_\star, q, r)$, *independent of the discretization parameter* h, *such that*

$$\|V\|_{L_q(\Omega)} \le Ch^{\min(0,(2/q)-(2/r))} \|V\|_{L_r(\Omega)} \quad \forall V \in S^h.$$

Proof Consider the rectangle $K_{ij}^- := (x_{i-1}, x_i) \times (y_{j-1}, y_j)$, $1 \le i \le M$, $1 \le j \le N$, and the mapping $(\tilde{x}, \tilde{y}) \mapsto (x, y)$ defined by

$$x = x_{i-1} + \tilde{x} h_i, \qquad y = y_{j-1} + \tilde{y} k_j, \tag{2.122}$$

which maps the unit square $\tilde{K}^+ := (0,1)^2$ onto K_{ij}^-. Let us define

$$\tilde{V}(\tilde{x}, \tilde{y}) := V(x, y),$$

where (x, y) is the image of $(\tilde{x}, \tilde{y})$ under the transformation (2.122). Now

$$\|\tilde{V}\|_{L_r(\tilde{K}^+)} = (h_i k_j)^{-1/r} \|V\|_{L_r(K_{ij}^-)},$$

and

$$\|V\|_{L_q(K_{ij}^-)} = (h_i k_j)^{1/q} \|\tilde{V}\|_{L_q(\tilde{K}^+)}.$$

Let $P(\tilde{K}^+)$ denote the linear space of all bilinear polynomials defined on the square $\tilde{K}^+$:

$$P(\tilde{K}^+) := \{(a + b\tilde{x})(c + d\tilde{y}) : a, b, c, d \in \mathbb{R}, 0 \le \tilde{x}, \tilde{y} \le 1\}.$$

Since $P(\tilde{K}^+)$ is finite-dimensional (in fact, the dimension of $P(\tilde{K}^+)$ is 4), the norms $\|\cdot\|_{L_q(\tilde{K}^+)}$ and $\|\cdot\|_{L_r(\tilde{K}^+)}$ are equivalent on $P(\tilde{K}^+)$. Hence, there is a constant $C_0 = C_0(q, r)$ such that

$$\|\tilde{V}\|_{L_q(\tilde{K}^+)} \le C_0 \|\tilde{V}\|_{L_r(\tilde{K}^+)},$$

for all $\tilde{V}$ in $P(\tilde{K}^+)$. Combining this with the two previous equalities yields

$$\|V\|_{L_q(K_{ij}^-)} \le C_0 (h_i k_j)^{(1/q)-(1/r)} \|V\|_{L_r(K_{ij}^-)},$$

and thus, by defining $C_1 = C_0 C_\star^{\max(0,(2/r)-(2/q))}$, we get

$$\|V\|_{L_q(K_{ij}^-)} \le C_1 h^{(2/q)-(2/r)} \|V\|_{L_r(K_{ij}^-)}. \tag{2.123}$$

Let us suppose that $q = \infty$; then, there exist i_0 and j_0, $1 \le i_0 \le M$, $1 \le j_0 \le N$, such that

$$\|V\|_{L_\infty(\Omega)} = \|V\|_{L_\infty(K_{i_0 j_0})} \le C_1 h^{-2/r} \|V\|_{L_r(K_{i_0 j_0})} \le C_1 h^{-2/r} \|V\|_{L_r(\Omega)},$$

which is the required result in the case of $q = \infty$.

Let us suppose now that $q < \infty$. It follows from (2.123) that

$$\left(\sum_{i,j} \|V\|^q_{L_q(K^-_{ij})}\right)^{1/q} \le C_1 h^{(2/q)-(2/r)} \left(\sum_{i,j} \|V\|^q_{L_r(K^-_{ij})}\right)^{1/q}, \tag{2.124}$$

where the sums are taken over all i and j, $1 \le i \le M$, $1 \le j \le N$.

We shall consider three cases. When $r \le q$, by noting that $s \mapsto (\sum_{i,j} a^s_{ij})^{1/s}$ is monotonic decreasing on $[1, \infty)$ when $0 < a_{ij} \le 1$, we have, with $a_{ij} = \|V\|_{L_r(K^-_{ij})}/\|V\|_{L_r(\Omega)}$, that

$$\left(\sum_{i,j} \|V\|^q_{L_r(K^-_{ij})}\right)^{1/q} \le \left(\sum_{i,j} \|V\|^r_{L_r(K^-_{ij})}\right)^{1/r}.$$

When $q < r < \infty$, Hölder's inequality for finite sums gives

$$\begin{aligned}\left(\sum_{i,j} \|V\|^q_{L_r(K^-_{ij})}\right)^{1/q} &\le (MN)^{(1/q)-(1/r)} \left(\sum_{i,j} \|V\|^r_{L_r(K^-_{ij})}\right)^{1/r} \\ &\le \left(\frac{C_\star}{h}\right)^{(2/q)-(2/r)} \left(\sum_{i,j} \|V\|^r_{L_r(K^-_{ij})}\right)^{1/r}.\end{aligned}$$

Finally, when $r = \infty$, we have that

$$\left(\sum_{i,j} \|V\|^q_{L_\infty(K^-_{ij})}\right)^{1/q} \le \left(\frac{C_\star}{h}\right)^{2/q} \max_{ij} \|V\|_{L_\infty(K^-_{ij})}.$$

It remains to combine (2.124) with one of the three inequalities corresponding to $r \le q$, $q < r < \infty$ and $r = \infty$ respectively to complete the proof. □

Lemma 2.37 *Suppose that $\{\overline{\Omega}^h\}$ is a family of quasi-uniform Cartesian-product meshes, and let S^h_0 be the linear space of continuous piecewise bilinear polynomials defined on the partition of Ω induced by $\overline{\Omega}^h$ that vanish on Γ. Then, there exists a positive constant C, independent of the discretization parameter h, such that,*

$$\|V\|_{L_\infty(\Omega)} \le C|\log h|^{1/2} \|\nabla V\|_{L_2(\Omega)} \quad \forall V \in S^h_0.$$

Proof By Sobolev's embedding theorem on a Lipschitz domain $D \subset \mathbb{R}^n$,

$$\|v\|_{L_p(D)} \leq q\left(\frac{nq}{p}\right)^{1/n+1/p-1} \omega_n^{-1/n} n^{-1/p} \|\nabla v\|_{L_q(D)} \quad \forall v \in \mathring{W}_2^1(D),$$

where $q = np/(n+p)$, and $\omega_n := 2\pi^{n/2}/\Gamma(n/2)$ is the surface area of the unit ball in $\mathbb{R}^n$ (see inequality (2.3.21) in Maz'ya [136]). Specifically, by taking $n = 2$ and $D = \Omega$,

$$\|v\|_{L_p(\Omega)} \leq Cq\left(\frac{2q}{p}\right)^{1/2+1/p-1} 2^{-1/p} \|\nabla v\|_{L_q(\Omega)} \quad \forall v \in \mathring{W}_2^1(\Omega),$$

with $q = 2p/(2+p)$. Also, by the previous lemma,

$$\|V\|_{L_\infty(\Omega)} \leq Ch^{-2/p} \|V\|_{L_p(\Omega)}$$

and, by an analogous argument to that in the proof of the previous lemma,

$$\|\nabla V\|_{L_q(\Omega)} \leq Ch^{\min(0,2/q-1)} \|\nabla V\|_{L_2(\Omega)},$$

for all V in $\mathcal{S}_0^h$. Setting $p = |\log h| (> 1)$, for sufficiently small h, and combining the last three inequalities, we obtain the required result. □

Theorem 2.38 *Suppose that* $\{\overline{\Omega}^h\}$ *is a family of quasi-uniform Cartesian-product meshes, i.e. there exists a positive constant* $C_\star$ *such that*

$$h = \max_{i,j}(h_i, k_j) \leq C_\star \min_{i,j}(h_i, k_j),$$

and let $u \in W_2^{s+1}(\Omega) \cap \mathring{W}_2^1(\Omega)$, $1/2 < s \leq 2$. *Then,*

$$\|u - U\|_{\infty,h} \leq Ch^s |\log h|^{1/2} |u|_{W_2^{s+1}(\Omega)},$$

where $C = C(s)$ *is a positive constant depending on* $C_\star$, *but independent of* u *and the discretization parameter* h.

Proof Let $I^h : \mathring{W}_2^1(\Omega) \cap C(\overline{\Omega}) \to S_0^h$ denote the interpolation projector onto $\mathcal{S}_0^h$ defined by $(I^h u)(x_i, y_j) = u(x_i, y_j)$ for all $(x_i, y_j) \in \overline{\Omega}^h$. Then,

$$\|u - U\|_{\infty,h} = \left\|I^h u - U\right\|_{\infty,h} \leq \left\|I^h u - U\right\|_{L_\infty(\Omega)}.$$

Thanks to Lemma 2.37,

$$\|V\|_{L_\infty(\Omega)} \leq C|\log h|^{1/2} \|V\|_{W_2^1(\Omega)} \quad \forall V \in \mathcal{S}_0^h.$$

Also, the equivalence of the norms $\|\cdot\|_{W_2^1(\Omega)}$ and $\|\cdot\|_{W_2^1(\Omega^h)}$ on $\mathcal{S}_0^h$ implies that

$$\|V\|_{W_2^1(\Omega)} \leq C\|V\|_{W_2^1(\Omega^h)} \quad \forall V \in \mathcal{S}_0^h.$$

Hence,

$$\|u-U\|_{\infty,h} \le C|\log h|^{1/2}\|u-U\|_{W_2^1(\Omega^h)},$$

and therefore Theorem 2.35 yields

$$\|u-U\|_{\infty,h} \le Ch^s|\log h|^{1/2}|u|_{W_2^{s+1}(\Omega)}. \qquad \square$$

In the next section we extend the error analysis developed here to a more general class of schemes.

2.4.2 An Alternative Scheme

Hitherto it was assumed that the trial space $\mathcal{S}^h$ in the finite volume method (which was subsequently rewritten as a finite difference scheme) consisted of continuous piecewise bilinear functions on the rectangular partition of $\overline{\Omega}$ induced by the Cartesian-product mesh $\overline{\Omega}^h$. One can construct an alternative method, based on continuous piecewise linear trial functions on triangles; to this end, we consider a triangulation of $\overline{\Omega}$ obtained from the original rectangular partition by subdividing each rectangle into two triangles by the diagonal of positive slope. Let $\mathcal{S}^h$ denote the set of all continuous piecewise linear functions on this triangulation, and let $\mathcal{S}_0^h$ be the subset of $\mathcal{S}^h$ consisting of all those functions that vanish on Γ.

Similarly to (2.108), we define the finite volume approximation of u as $U \in \mathcal{S}_0^h$ satisfying

$$-\frac{1}{\hbar_i \bar{k}_j}\int_{\partial K_{ij}} \frac{\partial U}{\partial \nu}\,\mathrm{d}s = \frac{1}{\hbar_i \bar{k}_j}(\chi_{ij} * f)(x_i, y_j) \quad \text{for } (x_i, y_j) \in \Omega^h. \tag{2.125}$$

This scheme resembles the finite volume method (2.108). Indeed, a simple calculation reveals that (2.125) can be rewritten as the finite difference scheme

$$-\big(D_x^+D_x^- + D_y^+D_y^-\big)U = T_h^{11} f \quad \text{in } \Omega^h, \tag{2.126}$$

$$U = 0 \quad \text{on } \Gamma^h. \tag{2.127}$$

In fact, both (2.111), (2.112) and (2.126), (2.127) can be embedded in the following one-parameter family of finite difference schemes:

$$-\big(D_x^+D_x^-\mu_y^\theta + D_y^+D_y^-\mu_x^\theta\big)U = T_h^{11} f \quad \text{in } \Omega^h, \tag{2.128}$$

$$U = 0 \quad \text{on } \Gamma^h, \tag{2.129}$$

where $\theta \in [0, 1]$, and

$$\mu_x^\theta U_{ij} := \frac{1}{\hbar_i}\big[\theta h_i U_{i-1,j} + (1-2\theta)\hbar_i U_{ij} + \theta h_{i+1} U_{i+1,j}\big],$$

with μ_y^θ defined analogously. The scheme (2.111), (2.112) (resp. (2.126), (2.127)) is obtained from (2.128), (2.129) with $\theta = 1/8$ (resp. $\theta = 0$). The rest of this section is devoted to the analysis of the one-parameter family of schemes (2.126), (2.127).

By proceeding similarly as in the proofs of Lemmas 2.31, 2.32 and Theorems 2.33 and 2.34 we arrive at the following set of results, whose proofs have been omitted for the sake of brevity.

Lemma 2.39 *Suppose that V is a mesh-function defined on $\overline{\Omega}^h$, and let $\theta \in [0, 1/4)$.*

(a) *If $V = 0$ on Γ_x^h, then*

$$\left(\mu_x^\theta V, V\right]_y \geq (1 - 4\theta)\|V\|^2_{L_2(\Omega_y^h)}.$$

(b) *If $V = 0$ on Γ_y^h, then*

$$\left(\mu_y^\theta V, V\right]_x \geq (1 - 4\theta)\|V\|^2_{L_2(\Omega_x^h)}.$$

Theorem 2.40 *Let $L^h V := -(D_x^+ D_x^- \mu_y^\theta + D_y^+ D_y^- \mu_x^\theta)V$, and suppose that $\theta \in [0, 1/4)$. Then,*

$$\|V\|_{W_2^1(\Omega^h)} \leq \frac{3}{2(1 - 4\theta)}\left\|L^h V\right\|_{W_2^{-1}(\Omega^h)},$$

for any mesh-function V defined on $\overline{\Omega}^h$ and such that $V = 0$ on Γ^h.

Theorem 2.41 *Suppose that $\theta \in [0, 1/4)$. For any $f \in W_2^s(\Omega)$, $s > -1/2$,* (2.128), (2.129) *has a unique solution U. Moreover,*

$$\|U\|_{W_2^1(\Omega^h)} \leq \frac{3}{2(1 - 4\theta)}\left\|T_h^{11} f\right\|_{W_2^{-1}(\Omega^h)}.$$

The central result of this section is the following error bound for the finite difference scheme (2.128), (2.129).

Theorem 2.42 *Suppose that $u \in W_2^3(\Omega) \cap \mathring{W}_2^1(\Omega)$, and let $\theta \in [0, 1/4)$. Then,*

$$\|u - U\|_{W_2^1(\Omega^h)} \leq C h^2 |u|_{W_2^3(\Omega)},$$

where $h = \max_{i,j}(h_i, k_j)$ and $C = C(\theta)$ is a positive constant independent of u and the discretization parameters.

Proof Let us define the global error as $e := u - U$. We then have that

$$-(D_x^+ D_x^- \mu_y^\theta + D_y^+ D_y^- \mu_x^\theta)e = D_x^+ \eta_1^\theta + D_y^+ \eta_2^\theta \quad \text{in } \Omega^h, \tag{2.130}$$

$$e = 0 \quad \text{on } \Gamma^h, \tag{2.131}$$

where

$$\eta_1^\theta := \eta_1 + \left(\frac{1}{8} - \theta\right)\zeta_1, \qquad \eta_2^\theta := \eta_2 + \left(\frac{1}{8} - \theta\right)\zeta_2,$$

and

$$\eta_1 := T_-^{01}\frac{\partial u}{\partial x} - D_x^- \mu_y u, \qquad \eta_2 := T_-^{10}\frac{\partial u}{\partial y} - D_y^- \mu_x u,$$

as in the proof of Theorem 2.35, and

$$(\zeta_1)_{ij} := h_i^2 D_x^- D_y^+ D_y^- u_{ij}, \qquad (\zeta_2)_{ij} := k_j^2 D_y^- D_x^+ D_x^- u_{ij}.$$

By applying Theorem 2.40 to (2.130), (2.131) we then deduce that

$$\|e\|_{W_2^1(\Omega^h)} \le \frac{3}{2(1-4\theta)} \left\| D_x^+ \eta_1^\theta + D_y^+ \eta_2^\theta \right\|_{W_2^{-1}(\Omega^h)}.$$

Consequently,

$$\begin{aligned}\|u - U\|_{W_2^1(\Omega^h)} &\le \frac{3}{2(1-4\theta)}\left(\|\eta_1\|_{L_2(\Omega_x^h)} + \|\eta_2\|_{L_2(\Omega_y^h)}\right)\\ &\quad + \frac{3|1-8\theta|}{16(1-4\theta)}\left(\|\zeta_1\|_{L_2(\Omega_x^h)} + \|\zeta_2\|_{L_2(\Omega_y^h)}\right). \end{aligned} \tag{2.132}$$

The first two terms on the right-hand side have already been bounded in the proof of Theorem 2.35; we showed there that

$$\|\eta_1\|_{L_2(\Omega_x^h)} \le Ch^2\left(\left|\frac{\partial u}{\partial x}\right|_{W_2^{0,2}(\Omega)} + \left|\frac{\partial u}{\partial x}\right|_{W_2^{2,0}(\Omega)}\right) \tag{2.133}$$

and

$$\|\eta_2\|_{L_2(\Omega_y^h)} \le Ch^2\left(\left|\frac{\partial u}{\partial y}\right|_{W_2^{0,2}(\Omega)} + \left|\frac{\partial u}{\partial y}\right|_{W_2^{2,0}(\Omega)}\right). \tag{2.134}$$

It therefore remains to bound the norms of ζ_1 and ζ_2. We observe in passing that for $\theta = 1/8$ the terms involving ζ_1 and ζ_2 are absent from (2.132).

To this end, let $\phi_i(x)$ (resp. $\psi_j(y)$) denote the standard continuous piecewise linear finite element basis function on $\overline{\Omega}_x^h$ (resp. $\overline{\Omega}_y^h$) such that $\phi_i(x_k) = \delta_{ik}$ (resp. $\psi_j(y_k) = \delta_{jk}$); $(\zeta_1)_{ij}$ and $(\zeta_2)_{ij}$ can then be rewritten as

$$(\zeta_1)_{ij} = h_i^2 \frac{1}{h_i \hbar_j}\int_{x_{i-1}}^{x_i}\int_{y_{j-1}}^{y_{j+1}} \psi_j(y)\frac{\partial^3 u}{\partial x \partial y^2}(x, y)\,\mathrm{d}x\,\mathrm{d}y,$$

$$(\zeta_2)_{ij} = k_j^2 \frac{1}{\hbar_i k_j}\int_{x_{i-1}}^{x_{i+1}}\int_{y_{j-1}}^{y_j} \phi_i(x)\frac{\partial^3 u}{\partial x^2 \partial y}(x, y)\,\mathrm{d}x\,\mathrm{d}y.$$

Clearly,

$$\|\zeta_1\|_{L_2(\Omega_x^h)} \le Ch^2 \left|\frac{\partial u}{\partial x}\right|_{W_2^{0,2}(\Omega)} \tag{2.135}$$

and

$$\|\zeta_2\|_{L_2(\Omega_y^h)} \le Ch^2 \left|\frac{\partial u}{\partial y}\right|_{W_2^{2,0}(\Omega)}. \tag{2.136}$$

Inserting (2.133)–(2.136) in (2.132) we obtain the desired error bound. □

On a quasi-uniform mesh the scheme (2.128), (2.129) can be shown to be (almost) optimally accurate in the discrete maximum norm $\|\cdot\|_{\infty,h}$ for any $\theta \in [0, 1/4)$, by proceeding analogously as in the case of $\theta = 1/8$.

Theorem 2.43 *Suppose that* $\{\overline{\Omega}^h\}$ *is a family of quasi-uniform meshes,* $\theta \in [0, 1/4)$, *and let* $u \in W_2^3(\Omega) \cap \mathring{W}_2^1(\Omega)$. *Then,*

$$\|u - U\|_{\infty,h} \le C(\theta)h^2|\log h|^{1/2}|u|_{W_2^3(\Omega)}.$$

The proof of this result is analogous to that of Theorem 2.38.

2.4.3 The Rotated Discrete Laplacian

In the previous section we considered the analysis of a one-parameter family of finite difference schemes, parametrized by θ. For $\theta \in [0, 1/4)$ we showed there that the scheme is stable and we proved optimal-order error bounds in various norms. A natural question is: *what happens when* $\theta = 1/4$? This section is devoted to the analysis of the resulting discretization.

Let us consider the finite difference scheme (2.128), (2.129), with $\theta = 1/4$. For the sake of notational simplicity we define

$$\hat{\mu}_x V_{ij} := \frac{1}{4\hbar_i}(h_i V_{i-1,j} + 2\hbar_i V_{ij} + h_{i+1} V_{i+1,j}),$$

and $\hat{\mu}_y$ is defined analogously. In fact, by introducing

$$\nu_x V_{ij} := \frac{1}{2}(V_{ij} + V_{i-1,j})$$

we can write

$$\hat{\mu}_x V_{ij} = \frac{1}{2\hbar_i}(h_i \nu_x V_{ij} + h_{i+1} \nu_x V_{i+1,j}).$$

Analogously, by letting

$$\nu_y V_{ij} := \frac{1}{2}(V_{ij} + V_{i,j-1})$$

we have that

$$\hat{\mu}_y V_{ij} = \frac{1}{2\hbar_j}(k_j \nu_y V_{ij} + k_{j+1}\nu_y V_{i,j+1}).$$

In terms of this new notation, for $\theta = 1/4$ the finite difference scheme (2.128), (2.129) can be rewritten as follows:

$$-\left(D_x^+ D_x^- \hat{\mu}_y + D_y^+ D_y^- \hat{\mu}_x\right)U = T_h^{11} f \quad \text{in } \Omega^h, \tag{2.137}$$

$$U = 0 \quad \text{on } \Gamma^h. \tag{2.138}$$

In particular, on a uniform mesh of size h, the resulting five-point finite difference operator is given by

$$-\frac{1}{2h^2}(U_{i-1,j-1} + U_{i-1,j+1} + U_{i+1,j-1} + U_{i+1,j+1} - 4U_{ij})$$

and is usually referred to as the *rotated discrete Laplace operator.*

We begin by showing that the scheme (2.137), (2.138) is stable. A preliminary result in this direction stated in the next lemma concerns the averaging operators $\hat{\mu}_x$, ν_x, $\hat{\mu}_y$ and ν_y.

Lemma 2.44 *Suppose that V is a function defined on the mesh $\overline{\Omega}_h$.*

(a) *If $V_{0j} = V_{Mj} = 0$ for $j = 1, \dots, N$, then*

$$(\hat{\mu}_x V, V]_y = \sum_{i=1}^{M}\sum_{j=1}^{N} h_i k_j |\nu_x V_{ij}|^2;$$

(b) *If $V_{i0} = V_{iM} = 0$ for $i = 1, \dots, M$, then*

$$(\hat{\mu}_y V, V]_x = \sum_{i=1}^{M}\sum_{j=1}^{N} h_i k_j |\nu_y V_{ij}|^2.$$

Proof We shall prove (a); the proof of (b) is completely analogous. By noting the definition of $\hat{\mu}_x$ we have that

$$\begin{aligned}(\hat{\mu}_x V, V]_y &= \frac{1}{4}\sum_{i=1}^{M-1}\sum_{j=1}^{N} k_j\left[h_{i+1}V_{i+1,j} + V_{ij}(h_{i+1} + h_i) + h_i V_{i-1,j}\right]V_{ij} \\ &= \frac{1}{4}\sum_{j=1}^{N} k_j\left[\sum_{i=1}^{M-1}(h_{i+1} + h_i)V_{ij}^2 + 2\sum_{i=1}^{M} h_i V_{ij} V_{i-1,j}\right]\end{aligned}$$

$$= \frac{1}{4}\sum_{i=1}^{M}\sum_{j=1}^{N} h_i k_j (V_{ij} + V_{i-1,j})^2,$$

and that completes the proof of (a). □

Lemma 2.45 *Let* $L^h V := -(D_x^+ D_x^- \hat{\mu}_y + D_y^+ D_y^- \hat{\mu}_x)V$. *Then,*

$$\left(L^h V, V\right)_h = \sum_{i=1}^{M}\sum_{j=1}^{N} h_i k_j \left(\left|\nu_y D_x^- V_{ij}\right|^2 + \left|\nu_x D_y^- V_{ij}\right|^2\right)$$

for any mesh-function V *defined on* $\overline{\Omega}^h$ *such that* $V = 0$ *on* Γ^h.

Proof This identity is a straightforward consequence of Lemma 2.44 by observing that

$$\begin{aligned}\left(L^h V, V\right)_h &= \left(D_x^- \hat{\mu}_y V, D_x^- V\right]_x + \left(D_y^- \hat{\mu}_x V, D_y^- V\right]_y \\ &= \left(\hat{\mu}_y D_x^- V, D_x^- V\right]_x + \left(\hat{\mu}_x D_y^- V, D_y^- V\right]_y,\end{aligned}$$

where the first equality follows by summation by parts and the second by noting that D_x^- commutes with $\hat{\mu}_y$ and D_y^- commutes with $\hat{\mu}_x$. □

We deduce from Lemma 2.45 that

$$\sum_{i=1}^{M}\sum_{j=1}^{N} h_i k_j \left(\left|\nu_y D_x^- V_{ij}\right|^2 + \left|\nu_x D_y^- V_{ij}\right|^2\right) = \left(L^h V, V\right)_h,$$

for any function V defined on $\overline{\Omega}^h$ such that $V = 0$ on Γ^h. Therefore, by applying the Cauchy–Schwarz inequality on the right-hand side, noting that

$$V_{ij} = \nu_x V_{ij} + \frac{1}{2} h_i D_x^- V_{ij}, \tag{2.139}$$

and letting

$$W_{ij} := h_i D_x^- V_{ij},$$

we deduce that, for any such mesh-function V,

$$\begin{aligned}&\sum_{i=1}^{M}\sum_{j=1}^{N} h_i k_j \left(\left|\nu_y D_x^- V_{ij}\right|^2 + \left|\nu_x D_y^- V_{ij}\right|^2\right) \\ &\quad \le \left\|L^h V\right\|_{L_2(\Omega^h)} \left(\|\nu_x V\|_{L_2(\Omega^h)} + \frac{1}{2}\|W\|_{L_2(\Omega^h)}\right).\end{aligned} \tag{2.140}$$

Now to complete the stability analysis of the finite difference scheme (2.137), (2.138) it remains to relate the two norms in the brackets on the right-hand side of (2.140) to the expression on the left. To do so, we state and prove two lemmas.

Lemma 2.46 *Suppose that $\{\overline{\Omega}^h\}$ is a family of quasi-uniform meshes, i.e. there exists a positive constant $C_\star$ such that*

$$h = \max_{i,j}(h_i, k_j) \le C_\star \min_{i,j}(h_i, k_j).$$

Let V be a function defined on $\overline{\Omega}^h$ such that $V = 0$ on Γ^h; then,

$$\|\nu_x V\|_{L_2(\Omega^h)} \le \frac{1}{2}(1 + C_\star)^{1/2}\left(\sum_{i=1}^{M}\sum_{j=1}^{N} h_i k_j |\nu_x D_y^- V_{ij}|^2\right)^{1/2}.$$

Proof Let $Z_{ij} = \nu_x V_{ij}$; then, because $Z_{i0} = 0$ for $i = 1, \dots, M$, we have that

$$\begin{aligned}
|Z_{ij}|^2 &= \left(\sum_{n=1}^{j} k_n D_y^- Z_{in}\right)^2 \\
&\le \left(\sum_{n=1}^{j} k_n\right)\left(\sum_{n=1}^{j} k_n |D_y^- Z_{in}|^2\right) \le \sum_{n=1}^{j} k_n |D_y^- Z_{in}|^2
\end{aligned}$$

for $i = 1, \dots, M-1$ and $j = 1, \dots, N$. Hence,

$$\|Z\|^2_{L_2(\Omega^h)} \le \sum_{i=1}^{M-1}\sum_{n=1}^{j} \hbar_i k_n |D_y^- Z_{in}|^2, \quad 1 \le j \le N.$$

Similarly, since $Z_{iN} = 0$ for $i = 1, \dots, M$, we also have that

$$\|Z\|^2_{L_2(\Omega^h)} \le \sum_{i=1}^{M-1}\sum_{n=j+1}^{N} \hbar_i k_n |D_y^- Z_{in}|^2, \quad 0 \le j \le N-1.$$

By adding the last two inequalities we deduce that

$$\|Z\|^2_{L_2(\Omega^h)} \le \frac{1}{2}\sum_{i=1}^{M-1}\sum_{n=1}^{N} \hbar_i k_n |D_y^- Z_{in}|^2.$$

Because ν_x commutes with D_y^- this yields

$$\|\nu_x V\|^2_{L_2(\Omega^h)} \le \frac{1}{2}\sum_{i=1}^{M-1}\sum_{j=1}^{N} \hbar_i k_j |\nu_x D_y^- V_{ij}|^2.$$

Since $\hbar_i \le \frac{1}{2}h_i(1+C_\star)$ it follows that

$$\|\nu_x V\|^2_{L_2(\Omega^h)} \le \frac{1}{4}(1+C_\star)\sum_{i=1}^{M-1}\sum_{j=1}^{N} h_i k_j |\nu_x D_y^- V_{ij}|^2,$$

and hence, by increasing the right-hand side of this inequality further by extending the upper limit of the sum over i from $M-1$ to M, we obtain the desired inequality. □

Our next result is concerned with bounding $W_{ij} := h_i D_x^- V_{ij}$.

Lemma 2.47 *Suppose that $\{\overline{\Omega}^h\}$ is a family of quasi-uniform meshes, i.e. there exists a positive constant $C_\star$ such that*

$$h = \max_{i,j}(h_i, k_j) \le C_\star \min_{i,j}(h_i, k_j).$$

Let V be a function defined on $\overline{\Omega}^h$ such that $V = 0$ on Γ^h and let $W_{ij} = h_i D_x^- V_{ij}$; then,

$$\|W\|_{L_2(\Omega^h)} \le 2C_\star\left(\sum_{i=1}^{M}\sum_{j=1}^{N} h_i k_j |\nu_y D_x^- V_{ij}|^2\right)^{1/2}.$$

Proof By noting that $W_{i0} = 0$ for $i = 1, \dots, M$, we have that

$$W_{ij} = \sum_{n=1}^{j}(-1)^{j-n}(W_{in} + W_{i,n-1})$$

for $i = 1, \dots, M$ and $j = 1, \dots, N$. Therefore,

$$|W_{ij}|^2 \le 4j\sum_{n=1}^{j}|\nu_y W_{in}|^2 \le 4j\sum_{n=1}^{N} h_i^2 |\nu_y D_x^- V_{in}|^2.$$

As $h_i^2 \le h h_i$ and $\hbar_i \le C_\star k_n$ for all $i \in \{1, \dots, M\}$ and all $n \in \{1, \dots, N\}$, and $h\sum_{j=1}^{N} j\hbar_j \le Nh \le C_\star$, we deduce that

$$\|W\|^2_{L_2(\Omega^h)} \le 4C_\star^2 \sum_{i=1}^{M}\sum_{n=1}^{N} h_i k_n |\nu_y D_x^- V_{in}|^2,$$

and hence the desired inequality upon renaming the index n into j. □

By combining (2.139), (2.140) and Lemmas 2.46 and 2.47, we deduce that

$$\sum_{i=1}^{M}\sum_{j=1}^{N} h_i k_j \left(\left|\nu_y D_x^- V_{ij}\right|^2 + \left|\nu_x D_y^- V_{ij}\right|^2\right)$$
$$\leq C\left\|L^h V\right\|_{L_2(\Omega^h)} \left(\sum_{i=1}^{M}\sum_{j=1}^{N} h_i k_j \left(\left|\nu_y D_x^- V_{ij}\right|^2 + \left|\nu_x D_y^- V_{ij}\right|^2\right)\right)^{1/2}.$$

This yields the inequality

$$\left(\sum_{i=1}^{M}\sum_{j=1}^{N} h_i k_j \left(\left|\nu_y D_x^- V_{ij}\right|^2 + \left|\nu_x D_y^- V_{ij}\right|^2\right)\right)^{1/2} \leq C\left\|L^h V\right\|_{L_2(\Omega^h)}, \tag{2.141}$$

and thereby the difference scheme is stable in the discrete W_2^1 norm defined by the left-hand side of this inequality.

Remark 2.11 We note that stability has been proved in a weaker sense here, for $\theta = 1/4$, than in the previous section for $\theta \in [0, 1/4)$. Indeed, for $\theta \in [0, 1/4)$ we deduce from Theorem 2.40 the stronger bound

$$\left[\sum_{i=1}^{M}\sum_{j=1}^{N} h_i k_j \left(\left|D_x^- V_{ij}\right|^2 + \left|D_y^- V_{ij}\right|^2\right)\right]^{1/2} \leq C(\theta)\left\|L^h V\right\|_{W_2^{-1}(\Omega^h)}, \tag{2.142}$$

whose left-hand side is an upper bound on the left-hand side of (2.141).

Worse still, the stability of the scheme (2.137), (2.138) is not robust, in the sense that when the homogeneous Dirichlet boundary condition is replaced by 1-periodic boundary conditions in the two co-ordinate directions, on a uniform mesh with spacing $h = 1/(2M)$, $M > 1$, the resulting difference scheme is ill-posed for any 1-periodic f. To see this, first take $f = 0$ and note that, in addition to the trivial constant solution (which is, incidentally, also a solution to the boundary-value problem), the difference scheme has the oscillatory chequer-board-like solution $U_{ij}^\star = (-1)^{i+j}$. Thus if U is a solution of the difference scheme with $f \neq 0$ subject to 1-periodic boundary conditions in the two co-ordinate directions, then $U + \alpha U^\star$ is also a solution, for any real number α. In other words, the solution is not unique. In fact, the finite difference scheme (2.137), corresponding to the choice of $\theta = 1/4$ in (2.128), with 1-periodic boundary condition, has infinitely many solutions for any f. This is consistent with the fact that, with a 1-periodic boundary condition, the expression appearing on the left-hand side of (2.142) has a nontrivial kernel in the set of mesh-functions defined on a uniform mesh with spacing $h = 1/(2M)$, $M > 1$, and is therefore only a seminorm in that case rather than a norm; and it is also consistent with the fact that, with $\theta \in [0, 1/4)$, the stability constant $C(\theta)$ of the scheme (2.128), (2.129) in the discrete $W_2^1(\Omega^h)$ norm, appearing in (2.142), tends to $+\infty$ as $\theta \to 1/4 - 0$.

2.5 Convergence Analysis in L_p Norms

Hitherto, with the exception of various error bounds in the discrete maximum norm, we have been concerned with the error analysis of finite difference schemes in mesh-dependent analogues of Hilbertian Sobolev norms, i.e. discrete Sobolev norms that are induced by inner products.

In this section we develop a framework for the error analysis of finite difference schemes in mesh-dependent versions of the Sobolev and Bessel-potential norms W_p^s and H_p^s, respectively. For the sake of simplicity, we shall confine ourselves to finite difference approximations of the homogeneous Dirichlet boundary-value problem for Poisson's equation on an open square Ω, assuming that the weak solution of the boundary-value problem belongs to $W_p^s(\Omega)$, $0 \le s \le 4$, $1 < p < \infty$. We shall make extensive use of the theory of discrete Fourier multipliers to investigate the stability of the difference schemes considered, in conjunction with the Bramble–Hilbert lemma in fractional-order Sobolev spaces to derive error bounds of optimal order. The presentation in this section is based on the following sources: the journal papers by Mokin [140] and Süli, Jovanović, Ivanović [173] and the monograph of Samarskiĭ, Lazarov and Makarov [160].

2.5.1 Discrete Fourier Multipliers

In previous sections we relied on the use of energy estimates based on Hilbert space techniques to show the stability of the finite difference schemes considered. In order to extend these stability results to L_p norms, $p \neq 2$, we require a new tool – discrete Fourier multipliers. To this end, we shall state and prove below a discrete counterpart of the Marcinkiewicz multiplier theorem. First, however, we shall introduce the notion of *discrete Fourier transform*.

Suppose that N is a positive integer and $h = \pi/N$. We consider the mesh

$$\mathbb{R}_h^n = h\mathbb{Z}^n := \left\{x \in \mathbb{R}^n : x = hk,\ k \in \mathbb{Z}^n\right\}$$

and the set of all 2π-periodic mesh-functions defined on $\mathbb{R}_h^n$. We let

$$\mathbb{I} := \{-N+1, \dots, -1, 0, \dots, N\}.$$

Then, any 2π-periodic function V defined on $\mathbb{R}_h^n$ is completely determined by its values on the 'basic cell'

$$\omega^h = h\mathbb{I}^n := \left\{hk : k \in \mathbb{I}^n\right\}.$$

With each mesh-function V defined on ω^h we associate its *discrete Fourier transform* $\mathcal{F}V$ given by

$$(\mathcal{F}V)(k) := h^n \sum_{x \in \omega^h} V(x) \mathrm{e}^{-\imath x \cdot k}, \quad k \in \mathbb{I}^n. \tag{2.143}$$

In order to distinguish the discrete Fourier transform from its integral counterpart F defined in Chap. 1 we have used the calligraphic letter $\mathcal{F}$ here instead of F. Clearly, $\mathcal{F}V$ is a $2N$-periodic function of its variables $k_1, \dots, k_n$, and $2N$ is the minimum period; thus it suffices to consider $\mathcal{F}V$ on the basic cell $\mathbb{I}^n$. Hence our choice of $k \in \mathbb{I}^n$ in (2.143).

For $x \in \omega^h$ the following *discrete Fourier inversion formula* holds:

$$V(x) = \frac{1}{(2\pi)^n} \sum_{k \in \mathbb{I}^n} (\mathcal{F}V)(k) \mathrm{e}^{\imath x \cdot k}. \tag{2.144}$$

Indeed, substituting (2.143) into the right-hand side of (2.144), we have that

$$\frac{1}{(2\pi)^n} \sum_{k \in \mathbb{I}^n} \mathrm{e}^{\imath x \cdot k} \sum_{y \in \omega^h} h^n V(y) \mathrm{e}^{-\imath y \cdot k} = \frac{1}{(2\pi)^n} \sum_{y \in \omega^h} h^n V(y) \sum_{k \in \mathbb{I}^n} \mathrm{e}^{\imath (x-y) \cdot k}.$$

However, for any $x, y \in \omega^h$ we have that

$$\sum_{k \in \mathbb{I}^n} \mathrm{e}^{\imath (x-y) \cdot k} = \begin{cases} (2N)^n & \text{if } x = y, \\ 0 & \text{otherwise,} \end{cases}$$

and hence (2.144), by noting that $h^n (2N)^n = (2\pi)^n$.

We can write (2.144) as $V = \mathcal{F}^{-1}\mathcal{F}V$ where, for a sequence $a = \{a(k)\}_{k \in \mathbb{I}^n}$, the *inverse discrete Fourier transform* $\mathcal{F}^{-1}a$ of a is defined by

$$\left(\mathcal{F}^{-1}a\right)(x) := \frac{1}{(2\pi)^n} \sum_{k \in \mathbb{I}^n} a(k) \mathrm{e}^{\imath x \cdot k}, \quad x \in \omega^h.$$

Assuming that V is a function defined on the mesh ω^h, we consider the trigonometric polynomial T_V given by

$$T_V(x) = \frac{1}{(2\pi)^n} \sum_{k \in \mathbb{I}^n} (\mathcal{F}V)(k) \mathrm{e}^{\imath x \cdot k}, \quad x \in (-\pi, \pi]^n. \tag{2.145}$$

According to the discrete Fourier inversion formula,

$$T_V(x) = V(x) \quad \forall x \in \omega^h;$$

in other words, T_V *interpolates* V over the mesh ω^h.

Next we introduce the space $L_p(\omega^h)$, $1 \le p < \infty$, consisting all mesh-functions V defined on ω^h such that, for some constant M, independent of the discretization parameter h,

$$\|V\|_{L_p(\omega^h)} = \left(h^n \sum_{x \in \omega^h} \left| V(x) \right|^p \right)^{1/p} \le M.$$

The following lemma establishes a useful relationship between the L_p norm of a mesh-function V defined on ω^h and the L_p norm of the associated trigonometric interpolant T_V on $\omega = \mathbb{T}^n := (-\pi, \pi)^n$.

Lemma 2.48 *Suppose that $V \in L_p(\omega^h)$, $1 \le p < \infty$, and let $\omega = (-\pi, \pi)^n$. Then,*

$$\|V\|_{L_p(\omega^h)} \le (1+\pi)^n \|T_V\|_{L_p(\omega)}.$$

Proof Let us suppose for simplicity that $n = 1$; for $n > 1$ the proof follows from the case of $n = 1$ by induction over n. We shall first show that there exists a real number ξ_0 in the interval $(-h, 0)$ such that

$$\|T_V\|_{L_p(\omega)} = \left(h \sum_{x\in\omega^h} |T_V(x+\xi_0)|^p\right)^{1/p}, \tag{2.146}$$

where now $\omega = (-\pi, \pi)$ and $\omega^h = h\mathbb{I}$.

Indeed,

$$\begin{aligned}
\|T_V\|^p_{L_p(\omega)} &= \int_{-\pi}^{\pi} |T_V(x)|^p \,\mathrm{d}x = \sum_{k=-N+1}^{N} \int_{x_{k-1}}^{x_k} |T_V(x)|^p \,\mathrm{d}x \\
&= \sum_{k=-N+1}^{N} \int_0^h |T_V(y + x_{k-1})|^p \,\mathrm{d}y \\
&= \int_0^h \sum_{k\in\mathbb{I}} |T_V(x_k + y - h)|^p \,\mathrm{d}y.
\end{aligned}$$

Now the integrand is a continuous function of y on $[0, h]$; therefore, by the integral mean-value theorem, there exists a ξ in $(0, h)$ such that

$$\int_0^h \sum_{k\in\mathbb{I}} |T_V(x_k + y - h)|^p \,\mathrm{d}y = h \sum_{k\in\mathbb{I}} |T_V(x_k + \xi - h)|^p.$$

Letting $\xi_0 := \xi - h$ and noting that $k \in \mathbb{I}$ if, and only if, $x = x_k \in \omega^h = h\mathbb{I}$, we deduce (2.146).

Now consider

$$\mathcal{D} := \left| \left(h \sum_{x\in\omega^h} |V(x)|^p\right)^{1/p} - \left(h \sum_{x\in\omega^h} |T_V(x+\xi_0)|^p\right)^{1/p} \right|.$$

We shall prove that

$$\mathcal{D} \le h \left\| T_V' \right\|_{L_p(\omega)}. \tag{2.147}$$

This follows by noting that $V(x) = T_V(x)$ for x in ω^h, and observing that by the reverse triangle inequality, the Newton–Leibniz formula and Hölder's inequality we have that

$$\begin{aligned}\mathcal{D} &\le \left(\sum_{x\in\omega^h} h\left|T_V(x) - T_V(x+\xi_0)\right|^p\right)^{1/p} \\ &= \left(\sum_{x\in\omega^h} h\left|\int_{x+\xi_0}^{x} T_V'(t)\,\mathrm{d}t\right|^p\right)^{1/p} \\ &\le \left(\sum_{x\in\omega^h} h\left(\int_{x-h}^{x} \left|T_V'(t)\right|\mathrm{d}t\right)^p\right)^{1/p} \\ &\le \left(\sum_{x\in\omega^h} h h^{p-1}\int_{x-h}^{x} \left|T_V'(t)\right|^p\mathrm{d}t\right)^{1/p} = h\left\|T_V'\right\|_{L_p(\omega)}.\end{aligned}$$

Now using (2.146) and (2.147) we deduce that

$$\begin{aligned}\|V\|_{L_p(\omega^h)} &= \|V\|_{L_p(\omega^h)} - \left(h\sum_{x\in\omega^h}\left|T_V(x+\xi_0)\right|^p\right)^{1/p} + \|T_V\|_{L_p(\omega)} \\ &\le \mathcal{D} + \|T_V\|_{L_p(\omega)} \\ &\le h\left\|T_V'\right\|_{L_p(\omega)} + \|T_V\|_{L_p(\omega)}.\end{aligned}$$

We bound the first term on the right-hand side further by applying Bernstein's inequality to the trigonometric polynomial T_V of degree N (see, Nikol'skiĭ [144], p. 115):

$$\left\|T_V'\right\|_{L_p(\omega)} \le N\|T_V\|_{L_p(\omega)},$$

and noting that $hN = \pi$. Hence the required result for $n = 1$. □

After this brief preparation, we are now ready to discuss a discrete counterpart of the Marcinkiewicz multiplier theorem, Theorem 1.75, due to Mokin [140] (see also Samarskiĭ, Lazarov, Makarov [160]), which will be our main tool in the stability analysis of finite difference schemes in discrete L_p norms. In order to state it, we require the notion of *total variation*. For a $2N$-periodic function a defined on $\mathbb{Z}^n$, the total variation of a over $\mathbb{I}^n$ is defined by

$$\mathrm{var}(a) := \sup_{k\in\mathbb{Z}^n}\ \max_{0\neq\alpha\in\{0,1\}^n} \sum_{\nu}^{\alpha}\left|\Delta^\alpha a(\nu)\right|.$$

Here $\Delta^\alpha := \Delta_1^{\alpha_1} \cdots \Delta_n^{\alpha_n}$, as in Theorem 1.75, and, for $\alpha \in \{0, 1\}^n$, we have used the multi-index notation

$$\sum_{\nu}^{\alpha} := \sum_{\nu_1}^{\alpha_1} \cdots \sum_{\nu_n}^{\alpha_n}$$

where now, in contrast with the notational convention in Theorem 1.75,

$$\sum_{\nu_j}^{\alpha_j} := \begin{cases} \max_{\nu_j=\pm 2^{|k_j|-1},\dots,\pm 2^{|k_j|}-1 \text{ such that } \nu_j \in \mathbb{I}} & \text{if } \alpha_j = 0, \\ \sum_{\nu_j=\pm 2^{|k_j|-1},\dots,\pm 2^{|k_j|}-1 \text{ such that } \nu_j \in \mathbb{I}} & \text{if } \alpha_j = 1. \end{cases}$$

In order to distinguish the total variation of a $2N$-periodic function over $\mathbb{I}^n$ defined here from total variation of a function on $\mathbb{Z}^n$ (as in the statement of the Marcinkiewicz multiplier theorem, Theorem 1.75, stated in the previous chapter), we have used the symbol 'var' here instead of our earlier notation 'Var'. The set of $k \in \mathbb{Z}^n$ for which the index set of $\sum_\nu^\alpha$ is nonempty is finite. Therefore, 'sup' in the definition of var(a) can be replaced with 'max'.

Theorem 2.49 (Discrete Marcinkiewicz Multiplier Theorem) *Let a be a $2N$-periodic function defined on $\mathbb{Z}^n$, and suppose that one of the following two conditions holds*:

(a) *a is a bounded function on $\mathbb{I}^n$ with bounded variation; i.e. there exists a constant M_0 such that*

$$\max_{k\in\mathbb{I}^n} \big|a(k)\big| \le M_0, \quad \text{var}(a) \le M_0;$$

(b) *a can be extended to a function, still denoted by a, which is defined and continuous on $[-N+1, N]^n$, with $\partial^\alpha a \in C([-N+1, N]^n \setminus \mathbb{I}^n)$ for every multi-index $\alpha \in \{0, 1\}^n$, and such that $\xi^\alpha \partial^\alpha a(\xi)$ is bounded for every $\alpha \in \{0, 1\}^n$; i.e. there exists a constant M_0 such that*

$$\max_{\alpha\in\{0,1\}^n} \sup_{\xi\in[-N+1,N]^n\setminus\mathbb{I}^n} \big|\xi^\alpha \partial^\alpha a(\xi)\big| \le M_0.$$

Then, a is a discrete Fourier multiplier on $L_p(\omega^h)$, $1 < p < \infty$; that is,

$$\big\|\mathcal{F}^{-1}(a\mathcal{F}V)\big\|_{L_p(\omega^h)} \le C\|V\|_{L_p(\omega^h)},$$

for all V in $L_p(\omega^h)$, where $C = C_p M_0$ and C_p is a positive constant, independent of a, h and V.

A simple sufficient condition for var$(a) \le M_0$ in part (a) of this theorem is that var$_*(a) \le M_0$, where var$_*(a)$ is defined analogously to var(a), except that $\sum_{\nu_j}^{\alpha_j}$ is defined as $\max_{\nu_j\in\mathbb{I}}$ when $\alpha_j = 0$ and as $\sum_{\nu_j\in\mathbb{I}}$ when $\alpha_j = 1$. As there is then no dependence on the diadic sets $\{\pm(2^{|k_j|} - 1), \dots, \pm(2^{|k_j|} - 1)\}$, the symbol $\sup_{k\in\mathbb{Z}^n}$ can be omitted from the definition of var$_*(a)$.

The proof of the theorem relies on the following result.

Lemma 2.50 *Let* $\omega = \mathbb{T}^n := (-\pi, \pi)^n$.

1. *Suppose that* $a(k)$ *satisfies the hypotheses in part* (a) *of Theorem* 2.49. *Then, the sequence* $\{\tilde{a}(k)\}_{k\in\mathbb{Z}^n}$ *defined by*

$$\tilde{a}(k) = \begin{cases} a(k) & \text{for } k \in \mathbb{I}^n, \\ 0 & \text{otherwise,} \end{cases}$$

is a Fourier multiplier on $L_p(\omega)$, $1 < p < \infty$.
2. *Consider the sequence* $\{\tilde{b}(k)\}_{k\in\mathbb{Z}^n}$ *defined by* $\tilde{b}(k) = b(k_1)\cdots b(k_n)$, *with*

$$b(m) = \begin{cases} 1 & \text{if } m = 0, \\ \frac{mh/2}{\sin(mh/2)} & \text{if } m \in \mathbb{I} \setminus \{0\}, \\ \pi/2 & \text{otherwise.} \end{cases}$$

Then, $\{\tilde{b}(k)\}_{k\in\mathbb{Z}^n}$ *is a Fourier multiplier on* $L_p(\omega)$, $1 < p < \infty$.
3. *The sequence* $\{\tilde{a}(k)\tilde{b}(k)\}_{k\in\mathbb{Z}^n}$ *is a Fourier multiplier on* $L_p(\omega)$, $1 < p < \infty$.

Proof The proof of this lemma is straightforward and proceeds as follows.

1. The stated result is obtained by noting that

$$\sup_{k\in\mathbb{Z}^n} |\tilde{a}(k)| = \max_{k\in\mathbb{I}^n} |a(k)| \le M_0,$$

and

$$\mathrm{Var}(\tilde{a}) \le \max\Big\{\max_{k\in\mathbb{I}^n} |a(k)|, \mathrm{var}(a)\Big\} \le M_0 =: M_0(a),$$

and by applying Theorem 1.75 to the sequence $\tilde{a} = \{\tilde{a}(k)\}_{k\in\mathbb{Z}^n}$.

2. The result is proved by noting that

$$\sup_{k\in\mathbb{Z}^n} |\tilde{b}(k)| \le \left(\frac{\pi}{2}\right)^n,$$

and

$$\mathrm{Var}(\tilde{b}) \le \left(\frac{\pi^2}{2}\right)^n =: M_0(b),$$

and applying Theorem 1.75 to the sequence $\tilde{b} = \{\tilde{b}(k)\}_{k\in\mathbb{Z}^n}$.

3. The stated result follows by observing that

$$\sup_{k\in\mathbb{Z}^n} |\tilde{a}(k)\tilde{b}(k)| \le \left(\frac{\pi}{2}\right)^n \max_{k\in\mathbb{I}^n} |a(k)| \le M_0(a)M_0(b),$$

and

$$\mathrm{Var}(\tilde{a}\tilde{b}) \le 2^n M_0(a)M_0(b) = \pi^{2n} M_0(a) =: M_0(ab),$$

and applying Theorem 1.75 to the sequence $\tilde{a}\tilde{b} = \{\tilde{a}(k)\tilde{b}(k)\}_{k\in\mathbb{Z}^n}$. □

We are now ready to prove Theorem 2.49.

Proof of Theorem 2.49 (a) Let us suppose that u is defined on ω^h, and consider its piecewise constant extension w to $\mathbb{R}^n$, defined as follows:

$$w(x) := \begin{cases} u(y), & \text{for } \|x-y\|_\infty < h/2,\ y \in \mathbb{I}^n, \\ 2\pi\text{-periodically extended to } \mathbb{R}^n, & \end{cases}$$

where $\|\cdot\|_\infty$ denotes the norm on $\mathbb{R}^n$ defined by $\|x\|_\infty := \max_{1\le j\le n} |x_j|$. Clearly,

$$\|w\|_{L_p(\omega)} = \|u\|_{L_p(\omega^h)}, \qquad \omega = \mathbb{T}^n := (-\pi,\pi)^n.$$

Furthermore, with the same notational conventions as in Sect. 1.9.5.1, w has the Fourier series expansion

$$w(x) = \frac{1}{(2\pi)^n} \sum_{k\in\mathbb{Z}^n} \hat{w}(k) \mathrm{e}^{\imath x\cdot k}, \qquad x \in \omega,$$

with Fourier coefficients

$$\hat{w}(k) = \int_\omega w(x)\mathrm{e}^{-\imath x\cdot k}\,\mathrm{d}x = \tilde{c}(k) h^n \sum_{x\in\omega^h} u(x)\mathrm{e}^{-\imath x\cdot k},$$

where $\tilde{c}(k) = c(k_1)\cdots c(k_n)$ and

$$c(m) = \begin{cases} 1 & \text{if } m = 0, \\ \frac{\sin(mh/2)}{mh/2} & \text{if } m \in \mathbb{Z}\setminus\{0\}. \end{cases}$$

By noting from Lemma 2.50, part (2), that $c(k) = 1/b(k)$ for $k \in \mathbb{I}$ and therefore $\tilde{c}(k) = 1/\tilde{b}(k)$ for $k \in \mathbb{I}^n$, we have that

$$\hat{w}(k) = \tilde{c}(k)(\mathcal{F}u)(k) = \frac{1}{\tilde{b}(k)}(\mathcal{F}u)(k) \quad \text{for } k \in \mathbb{I}^n.$$

Now, the trigonometric polynomial of degree N defined by

$$T_V : x \in \omega \mapsto \frac{1}{(2\pi)^n} \sum_{k\in\mathbb{I}^n} a(k)(\mathcal{F}u)(k)\mathrm{e}^{\imath x\cdot k}, \quad x \in (-\pi,\pi]^n,$$

is the trigonometric interpolant of the mesh-function $V := \mathcal{F}^{-1}(a\mathcal{F}u)$ defined on ω^h. Therefore, by Lemma 2.48, we have that

$$\begin{aligned} \left\|\mathcal{F}^{-1}(a\mathcal{F}u)\right\|_{L_p(\omega^h)} &\le (1+\pi)^n \left\| \frac{1}{(2\pi)^n} \sum_{k\in\mathbb{I}^n} a(k)(\mathcal{F}u)(k)\mathrm{e}^{\imath x\cdot k} \right\|_{L_p(\omega)} \\ &= (1+\pi)^n \left\| \frac{1}{(2\pi)^n} \sum_{k\in\mathbb{I}^n} a(k)\tilde{b}(k)\hat{w}(k)\mathrm{e}^{\imath x\cdot k} \right\|_{L_p(\omega)} \end{aligned}$$

$$= (1+\pi)^n \left\| \frac{1}{(2\pi)^n} \sum_{k\in\mathbb{Z}^n} \tilde{a}(k)\tilde{b}(k)\hat{w}(k)\mathrm{e}^{\imath x\cdot k} \right\|_{L_p(\omega)}$$

$$= (1+\pi)^n \left\| (\tilde{a}\tilde{b}\hat{w})^\vee \right\|_{L_p(\omega)}.$$

Here $\hat{\cdot}$ and $\cdot^\vee$ denote the Fourier transform of a periodic distribution and its inverse transform, defined in Sect. 1.9.5. Finally, by recalling from Lemma 2.50, part (3), that the sequence $\{\tilde{a}(k)\tilde{b}(k)\}_{k\in\mathbb{Z}^n}$ is a Fourier multiplier on $L_p(\omega)$ it follows that

$$\begin{aligned} \left\|\mathcal{F}^{-1}(a\mathcal{F}u)\right\|_{L_p(\omega^h)} &\le (1+\pi)^n C_p M_0(ab)\|w\|_{L_p(\omega)} \\ &= (1+\pi)^n C_p M_0(ab)\|u\|_{L_p(\omega^h)}, \end{aligned}$$

where C_p is as in Theorem 1.75 and $M_0(ab) = \pi^{2n} M_0(a)$, as in the proof of Lemma 2.50. Thus we have shown that

$$\left\|\mathcal{F}^{-1}(a\mathcal{F}u)\right\|_{L_p(\omega^h)} \le C_1 M_0 \|u\|_{L_p(\omega^h)},$$

where $C_1 = (1+\pi)^n \pi^{2n} C_p$ is a positive constant and $M_0 = M_0(a)$ is the constant from the statement of the theorem.

(b) This is a direct consequence of part (a), using the mean-value theorem in those variables x_j for which $\alpha_j = 1$ for a certain $\alpha \in \{0,1\}^n$. □

We shall now prove the converse of the inequality stated in Lemma 2.48, which will be required in our subsequent considerations.

Lemma 2.51 *Suppose that V is a mesh-function defined on ω^h, and let T_V be its trigonometric interpolant defined by* (2.145). *Then, for $1 < p < \infty$, there exists a positive constant C_p, independent of h and V, such that*

$$\|T_V\|_{L_p(\omega)} \le C_p \|V\|_{L_p(\omega^h)}.$$

Proof We shall prove this result in one dimension ($n = 1$); the case of $n > 1$ is dealt with by induction over n, starting from $n = 1$. In the proof of Lemma 2.48 we showed that there exists a ξ_0 in the interval $(-h_0, 0)$ such that

$$\begin{aligned} \|T_V\|_{L_p(\omega)} &= \left(h \sum_{x\in\omega^h} |T_V(x+\xi_0)|^p \right)^{1/p} = \left\| T_V(\cdot + \xi_0) \right\|_{L_p(\omega^h)} \\ &= \left\| \sum_{k\in\mathbb{I}} (\mathcal{F}V)(k)\mathrm{e}^{\imath xk}\mathrm{e}^{\imath\xi_0 k} \right\|_{L_p(\omega^h)}. \end{aligned} \tag{2.148}$$

Next we shall prove that the sequence $\{\lambda(k)\}_{k\in\mathbb{I}}$, with $\lambda(k) := \mathrm{e}^{\imath\xi_0 k}$, is a discrete Fourier multiplier on $L_p(\omega^h)$. First, note that $|\mathrm{e}^{\imath\xi_0 k}| = 1$; furthermore,

$$\sum_{k=-N+1}^{N} \left|\mathrm{e}^{\imath\xi_0 k} - \mathrm{e}^{\imath\xi_0 (k-1)}\right| = \sum_{k=-N+1}^{N} \left|1 - \mathrm{e}^{-\imath\xi_0}\right|$$
$$\leq 2N|\xi_0| \leq 2Nh = 2\pi.$$

Hence, $\mathrm{var}(\lambda) \leq 2\pi$ and, by Theorem 2.49, $\{\lambda(k)\}_{k\in\mathbb{I}}$ is a discrete Fourier multiplier on $L_p(\omega^h)$. Thanks to (2.148) we then have that

$$\|T_V\|_{L_p(\omega)} = \left\|\mathcal{F}^{-1}(\lambda\mathcal{F}V)\right\|_{L_p(\omega^h)} \leq 2\pi C_p\|V\|_{L_p(\omega^h)},$$

where C_p is a positive constant, and hence the required result (with the constant $2\pi C_p$ relabelled as C_p). □

After this interlude on discrete Fourier multipliers, we are ready to embark on the error analysis of finite difference approximations to our elliptic model problem in discrete L_p spaces.

2.5.2 The Model Problem and Its Approximation

Suppose that $\Omega = (0,\pi)^2$. For $f \in W_2^{-1}(\Omega)$, we consider the homogeneous Dirichlet boundary-value problem

$$-\Delta u = f \quad \text{in } \Omega, \tag{2.149}$$

$$u = 0 \quad \text{on } \Gamma = \partial\Omega. \tag{2.150}$$

Throughout the section we shall suppose that the unique weak solution $u \in \mathring{W}_2^1(\Omega)$ of (2.149), (2.150) belongs to $W_p^s(\Omega)$ for some $s \geq 0$ and $p \in (1,\infty)$ (other than $s = 1$ and $p = 2$, of course).

For a nonnegative integer $N \geq 2$ let $h := \pi/N$, and define the meshes:

$$\Omega^h := \left\{(x_i, y_j) : x_i = ih,\ y_j = jh,\ 1 \leq i, j \leq N-1\right\},$$
$$\overline{\Omega}^h := \left\{(x_i, y_j) : x_i = ih,\ y_j = jh,\ 0 \leq i, j \leq N\right\},$$
$$\Gamma^h := \overline{\Omega}^h \setminus \Omega^h.$$

In addition to these, we shall also require the following meshes:

$$\Gamma_x^h := \Gamma^h \cap \left(\{0,\pi\} \times (0,\pi)\right),$$
$$\Gamma_y^h := \Gamma^h \cap \left((0,\pi) \times \{0,\pi\}\right),$$

$$\Gamma_+^h := \Gamma^h \cap \big(\{\pi\} \times (0,\pi) \cup (0,\pi) \times \{\pi\}\big),$$
$$\Omega_+^h := \Omega^h \cup \Gamma_+^h,$$
$$\Omega_x^h := \Omega^h \cup \big(\Gamma_+^h \cap \Gamma_x^h\big),$$
$$\Omega_y^h := \Omega^h \cup \big(\Gamma_+^h \cap \Gamma_y^h\big).$$

As before, we approximate the Laplace operator $\Delta = \frac{\partial^2}{\partial x^2} + \frac{\partial^2}{\partial y^2}$ by

$$D_x^+ D_x^- + D_y^+ D_y^-.$$

Since f has not been assumed to be a continuous function on Ω, we shall mollify it before sampling it at the mesh-points. To do so, we shall use the mollifier $T^\nu = T_h^\nu$ with $\nu = (\nu_1, \nu_2)$ and $h = \pi/N$, defined in (1.35); for the sake of notational simplicity, we shall write $T_h^{\nu_1\nu_2}$, or simply $T^{\nu_1\nu_2}$, instead of the more cumbersome symbol $T_h^{(\nu_1,\nu_2)}$.

First we shall suppose that the weak solution of the boundary-value problem (2.149), (2.150) belongs to $W_p^s(\Omega)$, $s > 2/p$, $1 < p < \infty$; then, by Sobolev's embedding theorem, u is almost everywhere on $\overline{\Omega}$ equal to a continuous function on $\overline{\Omega}$, and

$$\left(T_h^{20}\frac{\partial^2 u}{\partial x^2}\right)(x,y) = D_x^+ D_x^- u(x,y), \quad (x,y) \in \Omega^h,$$

$$\left(T_h^{02}\frac{\partial^2 u}{\partial y^2}\right)(x,y) = D_y^+ D_y^- u(x,y), \quad (x,y) \in \Omega^h.$$

Therefore,

$$-\big(D_x^+ D_x^- T_h^{02} + D_y^+ D_y^- T_h^{20}\big)u = T_h^{22} f \quad \text{on } \Omega^h, \tag{2.151}$$

$$u = 0 \quad \text{on } \Gamma^h. \tag{2.152}$$

This identity motivates us to consider the difference scheme

$$-\big(D_x^+ D_x^- + D_y^+ D_y^-\big)U = T_h^{22} f \quad \text{on } \Omega^h, \tag{2.153}$$

$$U = 0 \quad \text{on } \Gamma^h. \tag{2.154}$$

The rest of this section is devoted to the error analysis of the finite difference scheme (2.153), (2.154). First we introduce the natural discrete analogues of the L_p spaces on Ω^h.

A function V defined on Ω^h (or on $\overline{\Omega}^h$ and equal to zero on Γ^h) is said to belong to $L_p(\Omega^h)$, $1 < p < \infty$, if there exists a positive constant M, independent of h, such that

$$\|V\|_{L_p(\Omega^h)} := \left(h^2 \sum_{(x,y)\in\Omega^h} |V(x,y)|^p\right)^{1/p} \le M.$$

If V is defined on Ω_+^h (or on $\overline{\Omega}^h$ and equal to zero on $\Gamma^h \setminus \Gamma_+^h$), the norm $\|\cdot\|_{L_p(\Omega^h)}$ is replaced by

$$\|V\|_{L_p(\Omega_+^h)} := \left(h^2 \sum_{(x,y)\in\Omega_+^h} |V(x,y)|^p\right)^{1/p}.$$

For mesh-functions defined on Ω_x^h and Ω_y^h the norms $\|\cdot\|_{L_p(\Omega_x^h)}$ and $\|\cdot\|_{L_p(\Omega_y^h)}$ are defined analogously.

The discrete analogues of the Sobolev norms $W_p^1(\Omega)$ and $W_p^2(\Omega)$ are defined, respectively, by

$$\|V\|_{W_p^1(\Omega^h)} := \left(\|V\|_{L_p(\Omega^h)}^p + |V|_{W_p^1(\Omega^h)}^p\right)^{1/p},$$

where

$$|V|_{W_p^1(\Omega^h)} := \left(\|D_x^- V\|_{L_p(\Omega_x^h)}^p + \|D_y^- V\|_{L_p(\Omega_y^h)}^p\right)^{1/p};$$

and

$$\|V\|_{W_p^2(\Omega^h)} := \left(\|V\|_{W_p^1(\Omega^h)}^p + |V|_{W_p^2(\Omega^h)}^p\right)^{1/p},$$

where

$$|V|_{W_p^2(\Omega^h)} := \left(\|D_x^+ D_x^- V\|_{L_p(\Omega^h)}^p + \|D_x^- D_y^- V\|_{L_p(\Omega_+^h)}^p + \|D_y^+ D_y^- V\|_{L_p(\Omega^h)}^p\right)^{1/p}.$$

Let us recall the notion of discrete Fourier transform from the previous section. However, as we are now working on $(0,\pi)^2$ rather than $(-\pi,\pi)^2$ and the functions we shall be dealing with will satisfy a homogeneous Dirichlet boundary condition rather then a periodic boundary condition, some adjustments have to be made before the techniques developed in the previous section can be applied.

Suppose that V is defined on Ω^h (or on $\overline{\Omega}^h$ and equal to zero on Γ^h). We shall consider the *odd extension* $\tilde{V}$ of the mesh-function V to the mesh

$$\omega^h = h\mathbb{I}^2 = \{(x_i, y_j) : x_i = ih,\, y_j = jh,\; i, j = -N+1, \dots, N\}$$

contained in $(-\pi,\pi]^2$. Thus

$$\tilde{V}(-x,y) = -\tilde{V}(x,y) \quad \text{and} \quad \tilde{V}(x,-y) = -\tilde{V}(x,y) \quad \text{for all } (x,y) \text{ in } \Omega^h.$$

After such an extension, $\tilde{V}$ is further extended 2π-periodically in each co-ordinate direction to the whole of $h\mathbb{Z}^2$. Let us note that

$$\|\tilde{V}\|_{L_p(\omega^h)} = 4^{1/p}\|V\|_{L_p(\Omega^h)}. \tag{2.155}$$

Lemma 2.52 *Let us suppose that V is defined on Ω^h (or on $\overline{\Omega}^h$ and equal to zero on Γ^h), and consider its odd extension $\tilde{V}$. The discrete Fourier transform $\mathcal{F}\tilde{V}$ has the following properties:*

1. *For any $k = (k_1, k_2) \in \mathbb{I}^2$,*

$$\mathcal{F}\tilde{V}(k_1, k_2) = -4h^2 \sum_{i=1}^{N-1}\sum_{j=1}^{N-1} V(x_i, y_j)\sin(k_1 x_i)\sin(k_2 y_j);$$

2. *$\mathcal{F}\tilde{V}$ is an odd function on $\mathbb{I}^2$; that is,*

$$\mathcal{F}\tilde{V}(-k_1, k_2) = -\mathcal{F}\tilde{V}(k_1, k_2) \quad \text{and} \quad \mathcal{F}\tilde{V}(k_1, -k_2) = -\mathcal{F}\tilde{V}(k_1, k_2)$$

for all $k = (k_1, k_2) \in \mathbb{I}^2$. Also, $\mathcal{F}\tilde{V}(0, k_2) = \mathcal{F}\tilde{V}(k_1, 0) = \mathcal{F}\tilde{V}(0, 0) = 0$;

3. *For $1 \le i, j \le N - 1$,*

$$V(x_i, y_j) = -\frac{1}{\pi^2}\sum_{k_1=1}^{N-1}\sum_{k_2=1}^{N-1} \mathcal{F}\tilde{V}(k_1, k_2)\sin(k_1 x_i)\sin(k_2 y_j).$$

The proof of this result is elementary and is left to the reader.

Lemma 2.52 implies that the values of $\mathcal{F}\tilde{V}$ on $\mathbb{I}^2$ are completely determined by the values of V on Ω^h; conversely, V can be completely characterized on Ω^h (and $\tilde{V}$ on ω^h) by the values $\mathcal{F}\tilde{V}(k_1, k_2)$, $k_1, k_2 = 1, \ldots, N - 1$. Consequently, it is meaningful to consider the *discrete Fourier sine-transform* $\mathcal{F}_\sigma V$ of a mesh-function V defined on Ω^h (or on $\overline{\Omega}^h$ and equal to zero on Γ^h). Indeed, we let

$$\mathcal{F}_\sigma V := -\frac{1}{4}\mathcal{F}\tilde{V},$$

and, for a function W defined on the set $\{(i, j) : 1 \le i, j \le N - 1\}$ with odd extension $\tilde{W}$ to $\mathbb{I}^2$, we put

$$\mathcal{F}_\sigma^{-1} W := -4\mathcal{F}^{-1}\tilde{W}.$$

Thus,

$$\mathcal{F}_\sigma V(k_1, k_2) = h^2 \sum_{i=1}^{N-1}\sum_{j=1}^{N-1} V(x_i, y_j)\sin(k_1 x_i)\sin(k_2 y_j)$$

and

$$\mathcal{F}_\sigma^{-1} W(x, y) = \left(\frac{2}{\pi}\right)^2 \sum_{k_1=1}^{N-1}\sum_{k_2=1}^{N-1} W(k_1, k_2)\sin(k_1 x)\sin(k_2 y).$$

In order to derive error bounds for the finite difference scheme under consideration we shall need the following stability result.

Lemma 2.53 *Suppose that η_1 and η_2 are two functions defined on $\overline{\Omega}^h$ that vanish on Γ^h. Further, let e be the solution to the problem*

$$-(D_x^+D_x^- + D_y^+D_y^-)e = D_x^+D_x^-\eta_1 + D_y^+D_y^-\eta_2 \quad \text{in } \Omega^h, \tag{2.156}$$

$$e = 0 \quad \text{on } \Gamma^h. \tag{2.157}$$

Then, for any $p \in (1, \infty)$,

$$\|e\|_{L_p(\Omega^h)} \le C_p\big(\|\eta_1\|_{L_p(\Omega^h)} + \|\eta_2\|_{L_p(\Omega^h)}\big), \tag{2.158}$$

$$|e|_{W_p^1(\Omega^h)} \le C_p\big(\big\|D_x^-\eta_1\big\|_{L_p(\Omega_x^h)} + \big\|D_y^-\eta_2\big\|_{L_p(\Omega_y^h)}\big), \tag{2.159}$$

$$|e|_{W_p^2(\Omega^h)} \le C_p\big(\big\|D_x^+D_x^-\eta_1\big\|_{L_p(\Omega^h)} + \big\|D_y^+D_y^-\eta_2\big\|_{L_p(\Omega^h)}\big), \tag{2.160}$$

where C_p is a positive constant, independent of h, e, η_1 and η_2.

Proof (1) Let us first prove (2.158). As

$$\mathcal{F}_\sigma\big(D_x^+D_x^-e\big) = -\lambda_1^2\mathcal{F}_\sigma e \quad \text{and} \quad \mathcal{F}_\sigma\big(D_y^+D_y^-e\big) = -\lambda_2^2\mathcal{F}_\sigma e,$$

where

$$\lambda_1 = \lambda_1(k_1) := \frac{2}{h}\sin\frac{k_1 h}{2} \quad \text{and} \quad \lambda_2 = \lambda_1(k_2) := \frac{2}{h}\sin\frac{k_2 h}{2},$$

with $k := (k_1, k_2)$, $1 \le k_1, k_2 \le N-1$, it follows that

$$e = \mathcal{F}_\sigma^{-1}(a_1\mathcal{F}_\sigma\eta_1) + \mathcal{F}_\sigma^{-1}(a_2\mathcal{F}_\sigma\eta_2),$$

where

$$a_l(k_1, k_2) := \frac{\lambda_l^2(k_l)}{\lambda_1^2(k_1) + \lambda_2^2(k_2)}, \qquad 1 \le k_1, k_2 \le N-1,\ l = 1, 2.$$

We note that $a_1(k_1, k_2)$ and $a_2(k_1, k_2)$ can be defined for all $k \in \mathbb{I}^2 \setminus \{0\}$ by letting

$$a_l(-k_1, k_2) := a_l(k_1, k_2),$$
$$a_l(k_1, -k_2) := a_l(k_1, k_2),$$
$$a_l(-k_1, -k_2) := a_l(k_1, k_2),$$

for all $k = (k_1, k_2)$, $1 \le k_1, k_2 \le N-1$, $l = 1, 2$.

Let $\tilde{e}$, $\tilde{\eta}_1$ and $\tilde{\eta}_2$ denote the odd extensions of the mesh-functions e, η_1 and η_2, respectively, from $\overline{\Omega}_h$ to ω^h. Then,

$$\tilde{e} = \mathcal{F}^{-1}(a_1\mathcal{F}\tilde{\eta}_1) + \mathcal{F}^{-1}(a_2\mathcal{F}\tilde{\eta}_2).$$

The fact that a_1 and a_2 are not defined at $(0,0)$ is of no significance, since

$$(\mathcal{F}\tilde{\eta}_l)(0,0) = h^2 \sum_{i=-N+1}^{N} \sum_{j=-N+1}^{N} \tilde{\eta}_i(x_i, y_j) = 0, \quad l = 1, 2,$$

which follows from our assumption that η_l, $l = 1, 2$, vanish on Γ^h, by noting that $\tilde{\eta}_l$ is the odd extension of η_l.

Hence, by the triangle inequality,

$$\|\tilde{e}\|_{L_p(\omega^h)} \leq \left\|\mathcal{F}^{-1}(a_1\mathcal{F}\tilde{\eta}_1)\right\|_{L_p(\omega^h)} + \left\|\mathcal{F}^{-1}(a_2\mathcal{F}\tilde{\eta}_2)\right\|_{L_p(\omega^h)}.$$

Next we show that a_1 and a_2 are discrete Fourier multipliers on $L_p(\omega^h)$.

Clearly $0 \leq a_1 \leq 1$ on $\mathbb{I}^2$. Further, as $a_1 + a_2 = 1$,

$$x\frac{\partial a_1}{\partial x} = 2a_1(1-a_1)\frac{xh}{2}\cot\frac{xh}{2}.$$

Thus, noting that $|t \cot t| \leq 1$ for $|t| \leq \pi/2$, we have that

$$\left|x\frac{\partial a_1}{\partial x}(x,y)\right| \leq \frac{1}{2} \quad \text{for } (x,y) \in \mathbb{I}^2.$$

Similarly, noting again that $a_1 + a_2 = 1$,

$$y\frac{\partial a_1}{\partial y} = -2a_1(1-a_1)\frac{yh}{2}\cot\frac{yh}{2}.$$

Therefore,

$$\left|y\frac{\partial a_1}{\partial y}(x,y)\right| \leq \frac{1}{2} \quad \text{for } (x,y) \in \mathbb{I}^2.$$

Finally,

$$xy\frac{\partial^2 a_1}{\partial x \partial y} = 4\left(y\frac{\partial a_1}{\partial y}\right)\frac{xh}{2}\cot\frac{xh}{2},$$

and so,

$$\left|xy\frac{\partial^2 a_1}{\partial x \partial y}(x,y)\right| \leq 2 \quad \text{for } (x,y) \in \mathbb{I}^2.$$

Hence, by Theorem 2.49, a_1 is a discrete Fourier multiplier on $L_p(\omega^h)$. By symmetry, the same is true of a_2.

Therefore,

$$\|\tilde{e}\|_{L_p(\omega^h)} \leq C_p\big(\|\tilde{\eta}_1\|_{L_p(\omega^h)} + \|\tilde{\eta}_2\|_{L_p(\omega^h)}\big),$$

from which (2.158) immediately follows by noting (2.155).

(2) As we have seen in part (1),

$$\mathcal{F}\tilde{e} = a_1\mathcal{F}\tilde{\eta}_1 + a_2\mathcal{F}\tilde{\eta}_2.$$

Multiplying this identity by $(1 - \exp(-\imath k_1 h))/h$ we deduce that

$$D_x^- \tilde{e} = \mathcal{F}^{-1}\big(a_1\mathcal{F}(D_x^- \tilde{\eta}_1)\big) + \mathcal{F}^{-1}\big(b_2\mathcal{F}(D_y^- \tilde{\eta}_2)\big),$$

where

$$b_2(k_1, k_2) := a_2(k_1, k_2)\frac{1 - \mathrm{e}^{-\imath k_1 h}}{1 - \mathrm{e}^{-\imath k_2 h}}.$$

We have already shown in part (1) that a_1 and a_2 are discrete Fourier multipliers on $L_p(\omega^h)$. Similarly, using Theorem 2.49 we deduce that the same is true of $(1 - \mathrm{e}^{-\imath k_1 h})/(1 - \mathrm{e}^{-\imath k_2 h})$, and therefore of b_2. Hence,

$$\|D_x^- \tilde{e}\|_{L_p(\omega^h)} \le C_p\big(\|D_x^- \tilde{\eta}_1\|_{L_p(\omega^h)} + \|D_y^- \tilde{\eta}_2\|_{L_p(\omega^h)}\big),$$

which yields

$$\|D_x^- e\|_{L_p(\Omega_x^h)} \le C_p\big(\|D_x^- \eta_1\|_{L_p(\Omega_x^h)} + \|D_y^- \eta_2\|_{L_p(\Omega_y^h)}\big).$$

An identical bound holds for $\|D_y^- e\|_{L_p(\Omega_y^h)}$, which, when added to the last inequality, yields (2.159).

(3) To prove (2.160), we note, by recalling the definitions of a_1 and a_2 from part (1) of the proof, that

$$-\lambda_1^2\mathcal{F}\tilde{e} = \frac{\lambda_1^2}{\lambda_1^2 + \lambda_2^2}(-\lambda_1^2\mathcal{F}\tilde{\eta}_1) + \frac{\lambda_1^2}{\lambda_1^2 + \lambda_2^2}(-\lambda_2^2\mathcal{F}\tilde{\eta}_2).$$

Thus,

$$\mathcal{F}(D_x^+ D_x^- \tilde{e}) = \frac{\lambda_1^2}{\lambda_1^2 + \lambda_2^2}\mathcal{F}(D_x^+ D_x^- \tilde{\eta}_1) + \frac{\lambda_1^2}{\lambda_1^2 + \lambda_2^2}\mathcal{F}(D_y^+ D_y^- \tilde{\eta}_2).$$

Equivalently,

$$D_x^+ D_x^- \tilde{e} = \mathcal{F}^{-1}\left(\frac{\lambda_1^2}{\lambda_1^2 + \lambda_2^2}\mathcal{F}(D_x^+ D_x^- \tilde{\eta}_1)\right) + \mathcal{F}^{-1}\left(\frac{\lambda_1^2}{\lambda_1^2 + \lambda_2^2}\mathcal{F}(D_y^+ D_y^- \tilde{\eta}_2)\right).$$

As $\lambda_l^2/(\lambda_1^2 + \lambda_2^2)$, $l = 1, 2$, are discrete Fourier multipliers on $L_p(\omega^h)$, it follows that

$$\|D_x^+ D_x^- \tilde{e}\|_{L_p(\omega^h)} \le C_p\big(\|D_x^+ D_x^- \tilde{\eta}_1\|_{L_p(\omega^h)} + \|D_y^+ D_y^- \tilde{\eta}_2\|_{L_p(\omega^h)}\big),$$

which gives

$$\|D_x^+ D_x^- e\|_{L_p(\Omega^h)} \le C_p\big(\|D_x^+ D_x^- \eta_1\|_{L_p(\Omega^h)} + \|D_y^+ D_y^- \eta_2\|_{L_p(\Omega^h)}\big).$$

Identical bounds hold for $\|D_y^+ D_y^- e\|_{L_p(\Omega^h)}$ and $\|D_x^- D_y^- e\|_{L_p(\Omega_+^h)}$, which, when added to the last inequality, yield (2.160). □

It is possible to derive bounds analogous to (2.159) and (2.160), but with e measured in a norm rather than a seminorm. To see this, we need the following preliminary result that relates discrete Sobolev seminorms to the corresponding discrete Sobolev norms.

Lemma 2.54 *Suppose that V is a function defined on the mesh $\overline{\Omega}^h$ such that $V = 0$ on Γ^h. Then, the following bounds hold*:

(a) *Assuming that* $1 < p < \infty$,

$$\|V\|_{L_p(\Omega^h)} \le 2^{-1/p}\pi |V|_{W_p^1(\Omega^h)};$$

(b) *There exists a constant C_p, independent of V and h, such that*

$$|V|_{W_p^1(\Omega^h)} \le C_p |V|_{W_p^2(\Omega^h)}, \quad 1 < p < \infty;$$

(c) *Assuming that* $1 < p < \infty$,

$$\|V\|_{W_p^1(\Omega^h)} \le \left(1 + \frac{1}{2}\pi^p\right)^{1/p} |V|_{W_p^1(\Omega^h)};$$

(d) *With C_p denoting the constant from part (b)*,

$$\|V\|_{W_p^2(\Omega^h)} \le \left(1 + \left(1 + \frac{1}{2}\pi^p\right) C_p^p\right)^{1/p} |V|_{W_p^2(\Omega^h)}, \quad 1 < p < \infty.$$

Proof Part (c) is a direct consequence of (a), while (d) follows by combining (c) and (b). We note that (c) is a discrete Friedrichs inequality, which generalizes Lemma 2.19. It remains to prove (a) and (b).

(a) As $V = 0$ on Γ^h, we can write

$$V_{ij} = \sum_{k=1}^{i} h D_x^- V_{kj}.$$

By Hölder's inequality for finite sums,

$$|V_{ij}|^p \le (ih)^{p/q} \sum_{k=1}^{i} h |D_x^- V_{kj}|^p, \quad \text{where } \frac{1}{p} + \frac{1}{q} = 1.$$

Multiplying by h^2, increasing the upper limit in the sum on the right to N, and summing through $i, j = 1, \dots, N-1$, we get that

$$\|V\|_{L_p(\Omega^h)}^p \le h^p \left(\sum_{i=1}^{N-1} i^{p/q}\right) \|D_x^- V\|_{L_p(\Omega_x^h)}^p.$$

Now

$$\sum_{i=1}^{N-1} i^{p/q} \le 1 + \int_1^{N-1} x^{p/q}\,\mathrm{d}x = 1 + \frac{(N-1)^{(p/q)+1} - 1}{(p/q)+1} \le N^{(p/q)+1} = N^p,$$

and therefore, since $Nh = \pi$, we deduce that

$$\|V\|^p_{L_p(\Omega^h)} \le \pi^p \big\|D_x^- V\big\|^p_{L_p(\Omega^h_x)}.$$

Analogously,

$$\|V\|^p_{L_p(\Omega^h)} \le \pi^p \big\|D_y^- V\big\|^p_{L_p(\Omega^h_y)}.$$

By adding the last two inequalities we deduce (a).

(b) Let $W := -(D_x^+ D_x^- + D_y^+ D_y^-)V$. Using the same technique and the same notation as in the proof of Lemma 2.53, and observing that

$$\left(\frac{1 - \mathrm{e}^{-\imath k_1 x}}{h}\right)\left(\frac{1}{\lambda_1^2(k_1, k_2) + \lambda_2^2(k_1, k_2)}\right)$$

and

$$\left(\frac{1 - \mathrm{e}^{-\imath k_2 y}}{h}\right)\left(\frac{1}{\lambda_1^2(k_1, k_2) + \lambda_2^2(k_1, k_2)}\right)$$

are discrete Fourier multipliers on $L_p(\omega^h)$, we deduce from Theorem 2.49 that

$$\big\|D_x^- V\big\|_{L_p(\Omega^h_x)} \le C_p \|W\|_{L_p(\Omega^h)},$$

and

$$\big\|D_y^- V\big\|_{L_p(\Omega^h_y)} \le C_p \|W\|_{L_p(\Omega^h)}.$$

Hence

$$|V|_{W^1_p(\Omega^h)} \le 2^{1/p} C_p \|W\|_{L_p(\Omega^h)},$$

and therefore, by the triangle inequality,

$$|V|_{W^1_p(\Omega^h)} \le 2^{1/p} C_p \big(\big\|D_x^+ D_x^- V\big\|_{L_p(\Omega^h)} + \big\|D_y^+ D_y^- V\big\|_{L_p(\Omega^h)}\big).$$

Thus, by noting the inequality $a + b \le 2^{1-(1/p)}(a^p + b^p)^{1/p}$ for $a, b \ge 0$,

$$|V|_{W^1_p(\Omega^h)} \le 2C_p |V|_{W^2_p(\Omega^h)}.$$

Renaming the constant $2C_p$ into C_p then yields the stated inequality. □

Combining the last two lemmas, we arrive at the following result.

Lemma 2.55 *Suppose that η_1 and η_2 are two functions defined on $\overline{\Omega}^h$ that vanish on Γ^h. Let further e denote the solution of the problem*

$$-(D_x^+ D_x^- + D_y^+ D_y^-)e = D_x^+ D_x^- \eta_1 + D_y^+ D_y^- \eta_2 \quad \text{in } \Omega^h, \tag{2.161}$$

$$e = 0 \quad \text{on } \Gamma^h. \tag{2.162}$$

Then, there exists a positive constant C_p, independent of h, such that

$$\|e\|_{L_p(\Omega^h)} \le C_p\big(\|\eta_1\|_{L_p(\Omega^h)} + \|\eta_2\|_{L_p(\Omega^h)}\big), \tag{2.163}$$

$$\|e\|_{W_p^1(\Omega^h)} \le C_p\big(\big\|D_x^- \eta_1\big\|_{L_p(\Omega_x^h)} + \big\|D_y^- \eta_2\big\|_{L_p(\Omega_y^h)}\big), \tag{2.164}$$

$$\|e\|_{W_p^2(\Omega^h)} \le C_p\big(\big\|D_x^+ D_x^- \eta_1\big\|_{L_p(\Omega^h)} + \big\|D_y^+ D_y^- \eta_2\big\|_{L_p(\Omega^h)}\big). \tag{2.165}$$

Now we are ready to state the main result of this section.

Theorem 2.56 *Let u be the weak solution of the boundary-value problem* (2.149), (2.150), *let U be the solution of the finite difference scheme* (2.153), (2.154) *and suppose that $m \in \{0, 1, 2\}$. Assuming that u belongs to $W_p^s(\Omega)$, with $m \le s$, $2/p < s \le m+2$, $1 < p < \infty$, the following error bound holds*:

$$\|u - U\|_{W_p^m(\Omega^h)} \le C h^{s-m} |u|_{W_p^s(\Omega)},$$

with a positive constant $C = C(p, m, s)$, independent of h.

Proof (a) Let us first suppose that $m = 2$ and $s \ge 2$. We define the global error e on $\overline{\Omega}^h$ by $e_{ij} := u(x_i, y_j) - U_{ij}$. It follows from (2.151)–(2.154) that e satisfies (2.161), (2.162) with

$$\eta_1 = u - T_h^{02} u \quad \text{and} \quad \eta_2 = u - T_h^{20} u.$$

Now η_1 (resp. η_2) is defined on the mesh $\Omega^h \cup \Gamma_x^h$ (resp. $\Omega^h \cup \Gamma_y^h$) and equal to zero on Γ_x^h (resp. Γ_y^h). According to (2.165), in order to obtain the desired error bound for $m = 2$, it suffices to estimate $\|D_x^+ D_x^- \eta_1\|_{L_p(\Omega^h)}$ and $\|D_y^+ D_y^- \eta_2\|_{L_p(\Omega^h)}$. To do so, we define the squares

$$K_{ij}^0 := (x_{i-1}, x_{i+1}) \times (y_{j-1}, y_{j+1}),$$

$$\tilde{K}^0 := (-1, 1) \times (-1, 1),$$

and consider the affine mapping $(x, y) \in K_{ij}^0 \mapsto (\tilde{x}, \tilde{y}) \in \tilde{K}^0$, where

$$x = x(\tilde{x}) := (i + \tilde{x})h, \qquad y = y(\tilde{y}) := (j + \tilde{y})h.$$

Let $\tilde{u}(\tilde{x}, \tilde{y}) = u(x(\tilde{x}), y(\tilde{y}))$. We then have the following equalities:

$$\begin{aligned}
&\left(D_x^+ D_x^- \eta_1\right)_{ij} \\
&\quad = \frac{u(x_{i+1}, y_j) - 2u(x_i, y_j) + u(x_{i-1}, y_j)}{h^2} \\
&\qquad - \int_{-1}^{1} \theta_2(\tilde{y}) \frac{u(x_{i+1}, y_j + \tilde{y}h) - 2u(x_i, y_j + \tilde{y}h) + u(x_{i-1}, y_j + \tilde{y}h)}{h^2} \, \mathrm{d}\tilde{y} \\
&\quad = \frac{1}{h^2} \Big\{ \tilde{u}(1,0) - 2\tilde{u}(0,0) + \tilde{u}(-1,0) \\
&\qquad - \int_{-1}^{1} \theta_2(\tilde{y}) \big[\tilde{u}(1,\tilde{y}) - 2\tilde{u}(0,\tilde{y}) + \tilde{u}(-1,\tilde{y})\big] \, \mathrm{d}\tilde{y} \Big\},
\end{aligned}$$

where $\theta_2(\tilde{y}) = 1 - |\tilde{y}|$, $\tilde{y} \in (-1, 1)$.

Now $(D_x^+ D_x^- \eta_1)_{ij}$ is a bounded linear functional on $W_p^s(\tilde{K}^0)$, $s > 2/p$, whose kernel contains $\mathcal{P}_3(\tilde{K}^0)$. According to the Bramble–Hilbert lemma,

$$\left|\left(D_x^+ D_x^- \eta_1\right)_{ij}\right| \leq C h^{-2} |\tilde{u}|_{W_p^s(\tilde{K}^0)}$$

for $2/p < s \leq 4$. Thus, by changing from the $(\tilde{x}, \tilde{y})$ to the (x, y) co-ordinate system, we have that

$$\left|\left(D_x^+ D_x^- \eta_1\right)_{ij}\right| \leq C h^{-2} h^{s-2/p} |u|_{W_p^s(K_{ij}^0)}$$

for $2/p < s \leq 4$. Hence,

$$\left\| D_x^+ D_x^- \eta_1 \right\|_{L_p(\Omega^h)} \leq C h^{s-2} |u|_{W_p^s(\Omega)}, \quad 2/p < s \leq 4.$$

Likewise,

$$\left\| D_y^+ D_y^- \eta_2 \right\|_{L_p(\Omega^h)} \leq C h^{s-2} |u|_{W_p^s(\Omega)}, \quad 2/p < s \leq 4,$$

which, after insertion into (2.165), completes the proof for the case $m = 2$.

(b) Let $m = 1$ and $s \geq 1$. By (2.164) it suffices to bound $\|D_x^- \eta_1\|_{L_p(\Omega_x^h)}$ and $\|D_y^- \eta_2\|_{L_p(\Omega_y^h)}$. We proceed in the same way as in part (a) to deduce that

$$\left(D_x^- \eta_1\right)_{ij} = \frac{1}{h} \left\{ \tilde{u}(1,0) - \tilde{u}(0,0) - \int_{-1}^{1} \theta_2(\tilde{y}) \big[\tilde{u}(1,\tilde{y}) - \tilde{u}(0,\tilde{y})\big] \, \mathrm{d}\tilde{y} \right\}$$

is a bounded linear functional on $W_p^s(\tilde{K}^0)$, $s > 2/p$, whose kernel contains $\mathcal{P}_2(\tilde{K}^0)$. Therefore,

$$\left\| D_x^- \eta_1 \right\|_{L_p(\Omega_x^h)} \leq C h^{s-1} |u|_{W_p^s(\Omega)}, \quad 2/p < s \leq 3,$$

and, similarly,

$$\left\| D_y^- \eta_2 \right\|_{L_p(\Omega_y^h)} \leq C h^{s-1} |u|_{W_p^s(\Omega)}, \quad 2/p < s \leq 3.$$

Inserting these into (2.164) we obtain the desired error bound for $m = 1$.

(c) Let $m = 0$ and $s \geq 0$. We need to estimate $\|\eta_1\|_{L_p(\Omega^h)}$ and $\|\eta_2\|_{L_p(\Omega^h)}$. Since

$$(\eta_1)_{ij} = \tilde{u}(0,0) - \int_{-1}^{1} \theta_2(\tilde{y})\tilde{u}(0,\tilde{y})\,\mathrm{d}\tilde{y}$$

is a bounded linear functional on $W_p^s(\tilde{K}^0)$, $s > 2/p$, whose kernel contains $\mathcal{P}_1(\tilde{K}^0)$, it follows that

$$\|\eta_1\|_{L_p(\Omega^h)} \leq Ch^s|u|_{W_p^s(\Omega)}, \quad 2/p < s \leq 2,$$

and, likewise,

$$\|\eta_2\|_{L_p(\Omega^h)} \leq Ch^s|u|_{W_p^s(\Omega)}, \quad 2/p < s \leq 2.$$

Substituting these into (2.163) we obtain the desired error bound for $m = 0$. That completes the proof of the theorem. □

In the remainder of this section we shall discuss the rate of convergence of the finite difference scheme (2.153), (2.154) in the case when $0 \leq s < 1 + 1/p$, which also covers the case $0 \leq s \leq 2/p$. Let us define the function space $\tilde{W}_p^s(\Omega)$, $1 < p < \infty$, by

$$\tilde{W}_p^s(\Omega) = \begin{cases} W_p^s(\Omega), & 0 \leq s \leq 1/p, \\ \{w : w \in W_p^s(\Omega),\ w = 0 \text{ on } \Gamma\}, & 1/p < s < 1 + 1/p. \end{cases}$$

We observe that if u, the weak solution of the boundary-value problem (2.149), (2.150) belongs $W_p^s(\Omega)$ then $u \in \tilde{W}_p^s(\Omega)$. Let $\Omega^* := (-\pi, 2\pi) \times (-\pi, 2\pi)$; the extension of u by 0 is a continuous linear operator from $\tilde{W}_p^s(\Omega)$ into $W_p^s(\Omega^*)$, $0 \leq s < 1 + 1/p$, $s \neq 1/p$, $1 < p < \infty$ (cf. Triebel [182], Sect. 2.10.2, Lemma and Remark 1 on p. 227 and Theorem 1 on p. 228). Hence

$$u \mapsto u^* = \text{odd extension of } u$$

is a continuous mapping from $\tilde{W}_p^s(\Omega)$ into $W_p^s(\Omega^*)$, $0 \leq s < 1 + 1/p$, $s \neq 1/p$, $1 < p < \infty$. Moreover, $(T_h^{11}u^*)(x, y) = 0$ for $(x, y) \in \Gamma^h$.

Theorem 2.57 *Let u be the weak solution of the boundary-value problem* (2.149), (2.150), *let U be the solution of the finite difference scheme* (2.153), (2.154) *and suppose that $m \in \{0, 1\}$. Assuming that u belongs to $W_p^s(\Omega)$ with $m \leq s$, $0 \leq s < 1 + 1/p$, $s \neq 1/p$ and $1 < p < \infty$, the following error bound holds*:

$$\|T_h^{11}u - U\|_{W_p^m(\Omega^h)} \leq Ch^{s-m}|u|_{W_p^s(\Omega)},$$

with a positive constant $C = C(p, m, s)$, independent of h.

Proof The proof is completely analogous to that of Theorem 2.56, except that we now define the global error e on $\overline{\Omega}^h$ by

$$e_{ij} = \big(T_h^{11}u^*\big)(x_i, y_j) - U_{ij}.$$

Clearly $e_{ij} = 0$ for $(x_i, y_j) \in \Gamma^h$, and $e_{ij} = (T_h^{11}u)(x_i, y_j) - U_{ij}$ when $(x_i, y_j) \in \Omega^h$. In addition, it follows from (2.151)–(2.154) that e satisfies (2.156), (2.157) with

$$\eta_1 = T_h^{11}u^* - T_h^{02}u^* \quad \text{and} \quad \eta_2 = T_h^{11}u^* - T_h^{20}u^*.$$

Again, η_1 (resp. η_2) is defined on the mesh $\Omega^h \cup \Gamma_x^h$ (resp. $\Omega^h \cup \Gamma_y^h$) and is equal to zero on Γ_x^h (resp. Γ_y^h). The rest of the proof is the same as in the case of Theorem 2.56, except that now $s \in [0, 1/p) \cup (1/p, 1 + 1/p)$. □

2.5.3 *Convergence in Discrete Bessel-Potential Norms*

This section is devoted to error estimation in discrete Bessel-potential norms. A function v defined on $\Omega^h \subset (0, \pi)^2$ (or on $\overline{\Omega}^h \subset [0, \pi]^2$ and equal to zero on Γ^h) is said to belong to the discrete Bessel-potential space $H_p^s(\Omega^h)$, with $-\infty < s < \infty$, $1 < p < \infty$, if there exists a function $V \in L_p(\Omega^h)$ such that

$$v = I_{s,h}V := \mathcal{F}_\sigma^{-1}\big((1 + |k|^2)^{-s/2}\mathcal{F}_\sigma V\big) = \mathcal{F}^{-1}\big((1 + |k|^2)^{-s/2}\mathcal{F}\tilde{V}\big),$$

where $\tilde{V}$ is the odd extension of V from Ω^h to $\omega^h = h\mathbb{I}^2$, defined to be zero on Γ^h, and further extended 2π-periodically to the whole of $h\mathbb{Z}^2$. We then define (compare with the definition in Sect. 1.9.5.3)

$$\|v\|_{H_p^s(\Omega^h)} := \|V\|_{L_p(\Omega^h)} = 4^{-1/p}\|\tilde{V}\|_{L_p(\omega^h)},$$

where the last equality is a consequence of (2.155).

First we shall prove equivalence of the discrete Sobolev norm $\|\cdot\|_{W_p^m(\Omega^h)}$ and the norm $\|\cdot\|_{H_p^m(\Omega^h)}$ for integer m; then, the error bounds in discrete Bessel-potential norms of integer order will follow from the error bounds derived in Theorems 2.56 and 2.57. Error bounds in fractional-order discrete Bessel-potential norms will be derived from these by function space interpolation. We need the following preliminary result in the univariate case.

Lemma 2.58 *Let W be a mesh-function defined on $\omega^h = h\mathbb{I}$, where $\mathbb{I} = \{-N + 1, \ldots, N\}$, and let T_W be the trigonometric interpolant of W on $(-\pi, \pi]$ given by* (2.145), *with $n = 1$. Then, there exists a constant C_p, independent of h and W, such that the following inequalities hold, with $\omega = (-\pi, \pi)$:*

(a) $\|D_x^- W\|_{L_p(\omega^h)} \le \|T_W'\|_{L_p(\omega)} \le C_p\|D_x^- W\|_{L_p(\omega^h)}$;

(b) $\|D_x^+ D_x^- W\|_{L_p(\omega^h)} \leq \|T_W''\|_{L_p(\omega)} \leq C_p \|D_x^+ D_x^- W\|_{L_p(\omega^h)}$.

Proof (a) Since W and T_W coincide at the mesh-points,

$$D_x^- W(x_i) = D_x^- T_W(x_i) = \frac{1}{h}\int_{x_{i-1}}^{x_i} T_W'(x)\,\mathrm{d}x.$$

Thus,

$$h\big|D_x^- W(x_i)\big|^p \leq \int_{x_{i-1}}^{x_i} \big|T_W'(x)\big|^p\,\mathrm{d}x.$$

Summing over all x_i in ω^h, we deduce that

$$\big\|D_x^- W\big\|_{L_p(\omega^h)} \leq \big\|T_W'\big\|_{L_p(\omega)}.$$

To deduce the second inequality, let us note that

$$T_W'(x) = \frac{1}{2\pi}\sum_{k\in\mathbb{I}} (\imath k)\mathcal{F}W(k)\mathrm{e}^{\imath xk},$$

and

$$\mathcal{F}\big(D_x^- W\big)(k) = \frac{1-\mathrm{e}^{-\imath kh}}{h}\mathcal{F}W(k).$$

Therefore,

$$T_W'(x) = \frac{1}{2\pi}\sum_{k\in\mathbb{I}} \frac{\imath kh}{1-\mathrm{e}^{-\imath kh}}\mathcal{F}\big(D_x^- W\big)(k)\mathrm{e}^{\imath xk}.$$

Since T_W' is a trigonometric polynomial of degree N, it follows from (2.146) that there is a ξ_0 in $(-h, 0)$ such that

$$\big\|T_W'\big\|_{L_p(\omega)} = \big\|T_W'(\cdot + \xi_0)\big\|_{L_p(\omega^h)}.$$

Letting

$$\lambda(kh) := \frac{\imath kh}{1-\mathrm{e}^{-\imath kh}}$$

and

$$\mu(k) := \lambda(kh)\mathrm{e}^{\imath k\xi_0},$$

the last equality can be rewritten as follows:

$$\big\|T_W'\big\|_{L_p(\omega)} = \big\|\mathcal{F}^{-1}\big(\mu\mathcal{F}\big(D_x^- W\big)\big)\big\|_{L_p(\omega^h)}.$$

A simple calculation shows that both λ and $\mathrm{var}(\lambda)$ are bounded by a constant, independent of h. It remains to apply part (a) of Theorem 2.49 to deduce that λ is a discrete Fourier multiplier on $L_p(\omega^h)$, and therefore the same is true of μ. Hence the upper bound in part (a).

(b) Let us define $Z = D_x^+ W$. Then, $D_x^+ D_x^- W = D_x^- Z$ and by part (a) of this lemma we have that

$$\|D_x^+ D_x^- W\|_{L_p(\omega^h)} = \|D_x^- Z\|_{L_p(\omega^h)} \leq \|T_Z'\|_{L_p(\omega)}.$$

Since

$$\mathcal{F}Z(k) = \mathcal{F}(D_x^+ W)(k) = \frac{e^{\imath kh} - 1}{h} \mathcal{F}W(k),$$

it follows that

$$\begin{aligned} T_Z(x) &= \frac{1}{2\pi} \sum_{k \in \mathbb{I}} \mathcal{F}Z(k) e^{\imath xk} \\ &= \frac{1}{2\pi} \sum_{k \in \mathbb{I}} \frac{e^{\imath kh} - 1}{h} \mathcal{F}W(k) = D_x^+ T_W(x). \end{aligned}$$

By noting that $T_W(x)$ is a 2π-periodic function of x we deduce that

$$\begin{aligned} \|T_Z'\|_{L_p(\omega)}^p &= \|D_x^+ T_W'\|_{L_p(\omega)}^p = h^{-p} \int_{-\pi}^{\pi} |T_W'(x+h) - T_W'(x)|^p \, dx \\ &= h^{-p} \int_{-\pi}^{\pi} \left| \int_x^{x+h} T_W''(\xi) \, d\xi \right|^p dx \leq \frac{1}{h} \int_{-\pi}^{\pi} \int_x^{x+h} |T_W''(t)|^p \, dt \, dx \\ &= \frac{1}{h} \int_{-\pi}^{\pi} |T_W''(t)|^p \left(\int_{t-h}^{t} dx \right) dt = \int_{-\pi}^{\pi} |T_W''(t)|^p \, dt = \|T_W''\|_{L_p(\omega)}^p. \end{aligned}$$

Hence we obtain the first inequality in (b). The second inequality is proved in the same way as in part (a), by observing that

$$T_W''(x) = \frac{1}{2\pi} \sum_{k \in \mathbb{I}} \frac{(\imath kh)^2}{(e^{\imath kh} - 1)(1 - e^{-\imath kh})} \mathcal{F}(D_x^+ D_x^- W)(k) e^{\imath xk}.$$

Thus, by noting that with $\xi_0 \in (-h, 0]$ as in part (a) the function μ_1 defined on $\mathbb{I}$ by

$$\mu_1(k) := \frac{(\imath kh)^2}{(e^{\imath kh} - 1)(1 - e^{-\imath kh})} e^{\imath k \xi_0} = \left(\frac{\frac{kh}{2}}{\sin \frac{kh}{2}} \right)^2 e^{\imath k \xi_0}$$

is bounded by $\pi/2$ and $\mathrm{var}(\mu_1)$ is bounded by a constant, independent of h, it follows from part (a) of Theorem 2.49 that μ_1 is a discrete Fourier multiplier on $L_p(\omega^h)$, and hence the upper bound stated in part (b). □

Lemma 2.58 has the following extension to two space dimensions.

Lemma 2.59 *Let W be a mesh-function defined on $\omega^h = h\mathbb{I}^2$, where $\mathbb{I} = \{-N+1, \dots, N\}$, and let T_W be the trigonometric interpolant of W on $(-\pi, \pi]^2$ given by (2.145), with $n = 2$. There is a constant $C_p > 0$, independent of h and W, such that the following inequalities hold, with $\omega = (-\pi, \pi)^2$:*

(a)

$$\frac{1}{1+\pi}\|D_x^- W\|_{L_p(\omega^h)} \le \left\|\frac{\partial}{\partial x}T_W\right\|_{L_p(\omega)} \le C_p\|D_x^- W\|_{L_p(\omega^h)}$$

and

$$\frac{1}{1+\pi}\|D_y^- W\|_{L_p(\omega^h)} \le \left\|\frac{\partial}{\partial y}T_W\right\|_{L_p(\omega)} \le C_p\|D_y^- W\|_{L_p(\omega^h)};$$

(b)

$$\frac{1}{1+\pi}\|D_x^+ D_x^- W\|_{L_p(\omega^h)} \le \left\|\frac{\partial^2}{\partial x^2}T_W\right\|_{L_p(\omega)} \le C_p\|D_x^+ D_x^- W\|_{L_p(\omega^h)},$$

$$\|D_x^- D_y^- W\|_{L_p(\omega^h)} \le \left\|\frac{\partial^2}{\partial x \partial y}T_W\right\|_{L_p(\omega)} \le C_p\|D_x^- D_y^- W\|_{L_p(\omega^h)}$$

and

$$\frac{1}{1+\pi}\|D_y^+ D_y^- W\|_{L_p(\omega^h)} \le \left\|\frac{\partial^2}{\partial y^2}T_W\right\|_{L_p(\omega)} \le C_p\|D_y^+ D_y^- W\|_{L_p(\omega^h)}.$$

Proof The proof of this result is a straightforward consequence of Lemma 2.58, and Lemmas 2.48 and 2.51 with $n = 1$; Lemma 2.58 is applied in the co-ordinate direction in which differentiation has taken place, and Lemmas 2.48 and 2.51 in the other direction. □

Lemma 2.60 *The norms $\|\cdot\|_{W_p^m(\Omega^h)}$ and $\|\cdot\|_{H_p^m(\Omega^h)}$ are equivalent, uniformly in h, for $m = 0, 1, 2$ and $1 < p < \infty$; i.e. there exist two constants C_1 and C_2, independent of h, such that for all functions V defined on Ω^h (or on $\overline{\Omega}^h$ and equal to zero of Γ^h),*

$$C_1\|V\|_{W_p^m(\Omega^h)} \le \|V\|_{H_p^m(\Omega^h)} \le C_2\|V\|_{W_p^m(\Omega^h)}.$$

Proof The statement is obviously true for $m = 0$ with $C_1 = C_2 = 1$. Now for $m = 1, 2$ we shall proceed as follows. Let $\tilde{V}$ denote the odd extension of V to $\omega^h = h\mathbb{I}^2$, where $\mathbb{I} = \{-N+1, \dots, N\}$. Further, let $T_{\tilde{V}}$ denote the trigonometric interpolant of $\tilde{V}$ defined by (2.145) with $n = 2$. By applying Lemma 2.59 with $W = \tilde{V}$, we deduce the existence of two positive constants C_1 and C_2, independent of V and h, (with $C_2 = C_2(p)$ and C_1 independent of p), such that

$$C_1\|V\|_{W_p^1(\Omega^h)} \le \|T_{\tilde{V}}\|_{W_p^1(\omega)} \le C_2\|V\|_{W_p^1(\Omega^h)}$$

and

$$C_1\|V\|_{W_p^2(\Omega^h)} \le \|T_{\tilde{V}}\|_{W_p^2(\omega)} \le C_2\|V\|_{W_p^2(\Omega^h)}.$$

For $p \in (1,\infty)$ and a nonnegative integer m the Sobolev norm $\|\cdot\|_{W_p^m(\omega)}$ on $\omega = \mathbb{T}^2$, is equivalent to the periodic Bessel-potential norm $\|\cdot\|_{H_p^m(\omega)}$ defined by

$$\|v\|_{H_p^m(\omega)} := \left\|\left(\left(1+|k|^2\right)^{m/2}\hat{v}\right)^{\vee}\right\|_{L_p(\omega)}$$

(see, Schmeiser and Triebel [162]), where $\hat{\cdot}$ and $\cdot^{\vee}$ denote the Fourier transform of a periodic distribution and its inverse, defined in Sect. 1.9.5; therefore,

$$C_1\|V\|_{W_p^m(\Omega^h)} \le \left\|\left(\left(1+|k|^2\right)^{m/2}\widehat{T_{\tilde{V}}}\right)^{\vee}\right\|_{L_p(\omega)} \le C_2\|V\|_{W_p^m(\Omega^h)}, \quad m=1,2.$$

Finally, since $((1+|k|^2)^{m/2}\widehat{T_{\tilde{V}}})^{\vee} = T_{\mathcal{F}^{-1}((1+|k|^2)^{m/2}\mathcal{F}\tilde{V})}$ on ω, we have by Lemmas 2.48 and 2.51 that

$$C_1\|V\|_{W_p^m(\Omega^h)} \le \left\|\mathcal{F}^{-1}\left(\left(1+|k|^2\right)^{m/2}\mathcal{F}\tilde{V}\right)\right\|_{L_p(\omega^h)} \le C_2\|V\|_{W_p^m(\Omega^h)}, \quad m=1,2,$$

from which the result follows by noting that

$$\begin{aligned}\left\|\mathcal{F}^{-1}\left(\left(1+|k|^2\right)^{m/2}\mathcal{F}\tilde{V}\right)\right\|_{L_p(\omega^h)} &= 4^{1/p}\left\|\mathcal{F}_\sigma^{-1}\left(\left(1+|k|^2\right)^{m/2}\mathcal{F}_\sigma V\right)\right\|_{L_p(\Omega^h)} \\ &= 4^{1/p}\|V\|_{H_p^m(\Omega^h)}, \quad m=1,2. \end{aligned}$$ □

We shall now use function space interpolation to obtain scales of error bounds in fractional-order discrete Bessel-potential norms. We start with a generalization of an interpolation inequality of Mokin (cf. Theorem 5 in [141]).

Lemma 2.61 *Let α and β be two nonnegative real numbers such that $\alpha < \beta$ and suppose that $1 < p < \infty$. There exists a positive constant C, independent of h, such that for any real number r, $\alpha \le r \le \beta$,*

$$\|V\|_{H_p^r(\Omega^h)} \le C\|V\|_{H_p^\alpha(\Omega^h)}^{1-\mu}\|V\|_{H_p^\beta(\Omega^h)}^{\mu} \quad \forall V \in H_p^\beta(\Omega^h),$$

where $\mu = (r-\alpha)/(\beta-\alpha)$.

Proof Let us first prove the result for $\alpha = 0$. We define $W := I_{-r,h}V$; then

$$\begin{aligned}\|V\|_{H_p^r(\Omega^h)} &= \|W\|_{L_p(\Omega^h)} = 4^{-1/p}\|\tilde{W}\|_{L_p(\omega^h)} \\ &\le 4^{-1/p}(1+\pi)^2\|T_{\tilde{W}}\|_{L_p(\omega)} = 4^{-1/p}(1+\pi)^2\|T_{\tilde{V}}\|_{H_p^r(\omega)}.\end{aligned}$$

Also,

$$\|T_{\tilde{V}}\|_{H_p^r(\omega)} \le C\|T_{\tilde{V}}\|_{L_p(\omega)}^{1-(r/\beta)}\|T_{\tilde{V}}\|_{H_p^\beta(\omega)}^{r/\beta},$$

(see, Nikol'skiĭ [144], p. 310) where $C = C(p,r,s)$ is a positive constant, and by Lemma 2.51 we have that

$$\|T_{\tilde{V}}\|_{L_p(\omega)} \le 4^{1/p} C \|V\|_{L_p(\Omega^h)} \quad \text{and} \quad \|T_{\tilde{V}}\|_{H_p^\beta(\omega)} \le 4^{1/p} C \|V\|_{H_p^\beta(\Omega^h)}.$$

Combining the last four inequalities, we deduce the statement of the lemma in the case of $\alpha = 0$.

For $\alpha > 0$, let us define $W := I_{-\alpha,h} V$. Then,

$$\|W\|_{H_p^{\beta-\alpha}(\Omega^h)} = \|V\|_{H_p^\beta(\Omega^h)} \quad \text{and} \quad \|W\|_{H_p^{r-\alpha}(\Omega^h)} = \|V\|_{H_p^r(\Omega^h)};$$

moreover, as $0 \le r - \alpha \le \beta - \alpha$, it follows from the case of $\alpha = 0$ above that

$$\|W\|_{H_p^{r-\alpha}(\Omega^h)} \le C \|W\|^{1-\mu}_{L_p(\Omega^h)} \|W\|^{\mu}_{H_p^{\beta-\alpha}(\Omega^h)},$$

and hence the desired inequality. □

Lemma 2.61 will play a key role in the proof of the next theorem, concerned with optimal error bounds in fractional-order discrete Bessel-potential norms.

Theorem 2.62 *Let u be the weak solution of the boundary-value problem* (2.149), (2.150), *let U be the solution of the finite difference scheme* (2.153), (2.154). *If u belongs to $W_p^s(\Omega)$, $2/p < s \le 2$ and $0 \le r \le 2$, or $2/p < s \le 3$ and $1 \le r \le 2$, with $1 < p < \infty$ and $r \le s$, then we have that*

$$\|u - U\|_{H_p^r(\Omega^h)} \le C h^{s-r} |u|_{W_p^s(\Omega)},$$

with a positive constant C, dependent on p, r and s, but independent of h.

Proof Let us suppose that u belongs to $W_p^s(\Omega)$, $2/p < s \le 2$, $1 < p < \infty$ and $0 \le r \le 2$. We apply Lemma 2.61 with $\alpha = 0$, $\beta = 2$ and Theorem 2.56 to obtain the error bound.

Similarly, if u belongs to $W_p^s(\Omega)$, $2/p < s \le 3$, $1 < p < \infty$ and $1 \le r \le 2$, then we take $\alpha = 1$ and $\beta = 2$ in Lemma 2.61 in combination with Theorem 2.56 to deduce the error bound. □

By invoking Lemma 2.61 with $\alpha = 0$ and $\beta = 1$, we obtain from Theorem 2.57, using function space interpolation, the following scale of error bounds in fractional-order discrete Bessel-potential norms.

Theorem 2.63 *Let u be the weak solution of the boundary-value problem* (2.149), (2.150), *let U be the solution of the finite difference scheme* (2.151), (2.152). *If u belongs to $W_p^s(\Omega)$, $0 \le s < 1 + 1/p$, $s \ne 1/p$, $1 < p < \infty$, $0 \le r \le 1$ and $r \le s$, then*

$$\left\|T_h^{11} u - U\right\|_{H_p^r(\Omega^h)} \le C h^{s-r} |u|_{W_p^s(\Omega)},$$

with a positive constant C, dependent on p, r and s, but independent of h.

The error bounds stated in Theorems 2.56, 2.57, 2.62, and 2.63 cover the range of possible Sobolev indices, $s \in [0,4]$, for which the solution U of the difference scheme (2.151), (2.152) converges to the weak solution u (or its mollification $T_h^{11}u$) of the boundary-value problem (2.149), (2.150), provided that $u \in W_p^s(\Omega)$. To conclude, we note that to derive these results it is not essential that u is *weak* solution: indeed, if we assume that $u \in W_p^s(\Omega)$ with $s > 1/p$ is a solution of the boundary-value problem in the sense of distributions and that it satisfies a homogeneous Dirichlet boundary condition in the sense of the trace theorem, the error bounds obtained above still hold.

2.6 Approximation of Second-Order Elliptic Equations with Variable Coefficients

Hitherto we have been concerned with the construction and error analysis of finite difference schemes for second-order linear elliptic equations of the form $-\Delta u + c(x,y)u = f(x,y)$. In particular, we derived optimal-order error bounds under minimal smoothness requirements on the solution. Here we shall extend these results to elliptic equations with variable coefficients in the principal part of the differential operator, under minimal regularity hypotheses on the solution and the coefficients.

In Sect. 2.6.1 we consider the Dirichlet problem for a second-order elliptic equation with variable coefficients in the principal part of the operator. The finite difference approximation of this problem is shown to be convergent, with optimal order, in the discrete W_2^1 norm. In Sects. 2.6.2 and 2.6.3 similar results are proved in the discrete W_2^2 norm and in the discrete L_2 norm; then, using function space interpolation, these bounds are extended to fractional-order discrete W_2^r norms, with $r \in [0,2]$, in Sect. 2.6.4. In Sect. 2.6.5 we focus on elliptic equations with separated variables and derive optimal bounds in the discrete L_2 norm, which are compatible with our hypotheses on the smoothness of the data.

2.6.1 Convergence in the Discrete W_2^1 Norm

As a model problem, we shall consider the following homogeneous Dirichlet boundary-value problem for a second-order linear elliptic equation with variable coefficients on the open unit square $\Omega = (0,1)^2$:

$$\begin{aligned} \mathcal{L}u := -\sum_{i,j=1}^{2} \partial_i(a_{ij}\partial_j u) + au = f \quad \text{in } \Omega, \\ u = 0 \quad \text{on } \Gamma = \partial\Omega. \end{aligned} \tag{2.166}$$

For the sake of notational simplicity, we have denoted the two independent variables here by x_1 and x_2 instead of x and y.

We shall suppose that (2.166) has a solution in $W_2^s(\Omega)$, which satisfies the partial differential equation in the sense of distributions and the boundary condition in the sense of the trace theorem, with the right-hand side f being an element of $W_2^{s-2}(\Omega)$. In order for the solution of this problem to have a well-defined trace on $\partial\Omega$ it is necessary to assume that $s > 1/2$. It is then natural to require that the coefficients a_{ij} and a belong to appropriate spaces of multipliers; that is,

$$a_{ij} \in M\big(W_2^{s-1}(\Omega)\big), \quad a \in M\big(W_2^s(\Omega) \to W_2^{s-2}(\Omega)\big).$$

According to the results in Sect. 1.8 the following conditions are sufficient in order to ensure that this is the case:

(a) if $|s-1| > 1$, then

$$a_{ij} \in W_2^{|s-1|}(\Omega), \quad a \in W_2^{|s-1|-1}(\Omega);$$

(b) if $0 \leq |s-1| \leq 1$, then

$$a_{ij} \in W_p^{|s-1|+\delta}(\Omega), \quad a = a_0 + \sum_{i=1}^{2} \partial_i a_i,$$

$$a_0 \in L_{2+\varepsilon}(\Omega), \quad a_i \in W_p^{|s-1|+\delta}(\Omega),$$

where $\varepsilon > 0$; and $\delta > 0$, $p \geq 2/|s-1|$ for $0 < |s-1| < 1$; $\delta = 0$, $p > 2$ for $s = 0$; $\delta = 0$, $p = \infty$ when $s = 1$.

In addition to these assumptions on the smoothness of the data, we shall adopt the following structural hypotheses on the coefficients a_{ij} and a:

- there exists a $c_0 > 0$ such that

$$\sum_{i,j=1}^{2} a_{ij}(x)\xi_i\xi_j \geq c_0 \sum_{i=1}^{2} \xi_i^2 \quad \forall x \in \overline{\Omega}, \ \forall \xi = (\xi_1, \xi_2) \in \mathbb{R}^2;$$

- the matrix $(a_{ij}) \in \mathbb{R}^{2\times 2}$ is symmetric, i.e.

$$a_{ij} = a_{ji}, \quad i, j = 1, 2;$$

- the coefficient a is nonnegative in the sense of distributions; i.e.

$$\langle a\varphi, \varphi\rangle_{\mathcal{D}'\times\mathcal{D}} \geq 0 \quad \forall \varphi \in \mathcal{D}(\Omega).$$

We shall construct a finite difference approximation of this boundary-value problem on the uniform mesh $\overline{\Omega}^h := \Omega^h \cup \Gamma^h$ of mesh-size $h := 1/N$, with $N \geq 2$, defined in Sect. 2.2.4.

When $s \le 3$, our hypotheses on the smoothness of the data do not guarantee that the forcing function f and the coefficient a are continuous on Ω: it is therefore necessary to mollify them so as to ensure that they have well-defined values at the mesh-points.

These observations lead us to consider the following finite difference approximation of the boundary-value problem:

$$\begin{aligned} \mathcal{L}_h U &= T_h^{22} f \quad \text{on } \Omega^h, \\ U &= 0 \quad \text{on } \Gamma^h, \end{aligned} \tag{2.167}$$

with

$$\mathcal{L}_h U := -\frac{1}{2} \sum_{i,j=1}^{2} \left[D_{x_i}^{+}(a_{ij} D_{x_j}^{-} U) + D_{x_i}^{-}(a_{ij} D_{x_j}^{+} U) \right] + (T_h^{22} a) U,$$

where $D_{x_i}^{\pm} V$, $i = 1, 2$, are the divided difference operators in the x_i co-ordinate direction defined in Sect. 2.2.4, and T_h^{22} is the mollifier with mesh-size h defined in Sect. 1.9.2.

It is helpful to note that the two-dimensional mollifier T_h^{22} can be expressed in terms of the one-dimensional mollifiers $T_1 = T_{1,h}$ and $T_2 = T_{2,h}$, acting in the x_1 and x_2 co-ordinate direction, respectively, as

$$T_h^{22} = T_1^2 T_2^2.$$

For a locally integrable function w defined on Ω,

$$T_1 w(x_1, x_2) := \frac{1}{h} \int_{x_1 - h/2}^{x_1 + h/2} w(\xi_1, x_2) \, \mathrm{d}\xi_1,$$

$$T_1^2 w(x_1, x_2) := \frac{1}{h} \int_{x_1 - h}^{x_1 + h} \left(1 - \left| \frac{x_1 - \xi_1}{h} \right| \right) w(\xi_1, x_2) \, \mathrm{d}\xi_1;$$

$T_2 w$ and $T_2^2 w$ can be represented analogously. When w is a distribution, T_i and T_i^2 are defined as convolutions of w with the scaled univariate B-splines θ_h^1 and θ_h^2, respectively, as explained in Sect. 1.9.

We note that (2.167) is the standard symmetric seven-point difference scheme with mollified right-hand side and mollified coefficient a.

With the notations from Sect. 2.2.4, we consider the discrete L_2 inner product $(V, W)_h$ (see (2.48)) in the linear space S_0^h of real-valued mesh-functions defined on $\overline{\Omega}^h$ that vanish on Γ^h, the associated discrete L_2 norm $\|V\|_{L_2(\Omega^h)}$, and the discrete Sobolev norms $\|V\|_{W_2^1(\Omega^h)}$ and $\|V\|_{W_2^2(\Omega^h)}$.

The error bounds stated in the next theorem are compatible with the smoothness hypotheses (a) and (b) formulated above, for the coefficients appearing in the partial differential equation.

Theorem 2.64 *The difference scheme* (2.167) *satisfies the following error bounds in the* $W_2^1(\Omega^h)$ *norm*:

$$\|u - U\|_{W_2^1(\Omega^h)} \le Ch^{s-1}\Big(\max_{i,j} \|a_{ij}\|_{W_2^{s-1}(\Omega)} + \|a\|_{W_2^{s-2}(\Omega)}\Big)\|u\|_{W_2^s(\Omega)},$$

$$\text{for } 2 < s \le 3, \tag{2.168}$$

and

$$\|u - U\|_{W_2^1(\Omega^h)} \le Ch^{s-1}\Big(\max_{i,j} \|a_{ij}\|_{W_p^{s-1+\delta}(\Omega)} + \max_{i} \|a_i\|_{W_p^{s-1+\delta}(\Omega)}$$

$$+ \|a_0\|_{L_{2+\varepsilon}(\Omega)}\Big)\|u\|_{W_2^s(\Omega)}, \quad \text{for } 1 < s \le 2, \tag{2.169}$$

where p, δ *and* ε *are as in condition* (b) *above, and* C *is a positive constant, independent of* h.

Before embarking on the proofs of these error bounds we shall make some preliminary observations. Let u denote the solution of the boundary-value problem (2.166) and let U be the solution of the finite difference scheme (2.167). When $s > 1$, as in Theorem 2.64, the function u is continuous on $\overline{\Omega}$ and therefore the *global error* $e := u - U$ is correctly defined on the uniform mesh $\overline{\Omega}^h$. In addition, it is easily seen that

$$\mathcal{L}_h e = \sum_{i,j=1}^{2} D_{x_i}^{-}\eta_{ij} + \eta \quad \text{on } \Omega^h,$$

$$e = 0 \quad \text{on } \Gamma^h, \tag{2.170}$$

where

$$\eta_{ij} := T_i^{+}T_{3-i}^{2}(a_{ij}\partial_j u) - \frac{1}{2}\big(a_{ij}D_{x_j}^{+}u + a_{ij}^{+i}D_{x_j}^{-}u^{+i}\big), \quad i = 1, 2,$$

and

$$\eta := \big(T_1^2T_2^2 a\big)u - T_1^2T_2^2(au).$$

Here, for a locally integrable function w defined on Ω, we have used the asymmetric mollifiers $T_i^{\pm}w$, defined at $x = (x_1, x_2)$ by

$$\big(T_i^{\pm}w\big)(x) := (T_i w)\Big(x \pm \frac{1}{2}h\mathrm{e}_i\Big), \quad \text{with } \mathrm{e}_i := (\delta_{i1}, \delta_{i2}),\ i = 1, 2.$$

By taking the $(\cdot,\cdot)_h$ inner product of $L_h e$ with e and performing summations by parts in the leading terms on the left- and right-hand sides, in exactly the same manner as in the argument that led to the estimate (2.83) stated in Lemma 2.24, we arrive at the following result.

Lemma 2.65 *The difference scheme* (2.170) *is stable, in the sense that*

$$\|e\|_{W_2^1(\Omega^h)} \leq C\left(\sum_{i,j=1}^{2} \|\eta_{ij}\|_{L_2(\Omega_i^h)} + \|\eta\|_{L_2(\Omega^h)}\right), \tag{2.171}$$

where C is a positive constant, independent of h.

The error analysis of the finite difference scheme (2.167) is thereby reduced to estimating the right-hand side in the inequality (2.171). To this end, we decompose η_{ij} as follows:

$$\eta_{ij} = \eta_{ij1} + \eta_{ij2} + \eta_{ij3} + \eta_{ij4},$$

where

$$\eta_{ij1} := T_i^{+} T_{3-i}^{2}(a_{ij}\partial_j u) - (T_i^{+} T_{3-i}^{2} a_{ij})(T_i^{+} T_{3-i}^{2} \partial_j u),$$

$$\eta_{ij2} := \left[T_i^{+} T_{3-i}^{2} a_{ij} - \frac{1}{2}(a_{ij} + a_{ij}^{+i})\right](T_i^{+} T_{3-i}^{2} \partial_j u),$$

$$\eta_{ij3} := \frac{1}{2}(a_{ij} + a_{ij}^{+i})\left[T_i^{+} T_{3-i}^{2} \partial_j u - \frac{1}{2}(D_{x_j}^{+} u + D_{x_j}^{-} u^{+i})\right]$$

and

$$\eta_{ij4} := -\frac{1}{4}(a_{ij} - a_{ij}^{+i})(D_{x_j}^{+} u - D_{x_j}^{-} u^{+i}).$$

We shall also perform a decomposition of η, but the form of this decomposition will depend on whether $1 < s \leq 2$ or $2 < s \leq 3$.

When $1 < s \leq 2$, we shall write

$$\eta = \eta_0 + \eta_1 + \eta_2,$$

where

$$\eta_0 := (T_1^2 T_2^2 a_0)u - T_1^2 T_2^2 (a_0 u)$$

and

$$\eta_i := (T_1^2 T_2^2 \partial_i a_i)u - T_1^2 T_2^2 (u \partial_i a_i), \quad i = 1, 2.$$

Whereas if $2 < s \leq 3$, we shall use the decomposition

$$\eta = \eta_3 + \eta_4,$$

where

$$\eta_3 := (T_1^2 T_2^2 a)(u - T_1^2 T_2^2 u)$$

and

$$\eta_4 := \bigl(T_1^2 T_2^2 a\bigr)\bigl(T_1^2 T_2^2 u\bigr) - T_1^2 T_2^2 (au).$$

Proof of Theorem 2.64 We introduce the 'elementary rectangles'

$$K^0 = K^0(x) := \bigl\{y = (y_1, y_2) : |y_j - x_j| < h,\ j = 1, 2\bigr\}$$

and

$$K^i = K^i(x) := \bigl\{y : x_i < y_i < x_i + h,\ |y_{3-i} - x_{3-i}| < h\bigr\}, \quad i = 1, 2.$$

The linear transformation $y = x + h\tilde{x}$ defines a bijective mapping of the 'canonical rectangles'

$$\tilde{K}^0 := \bigl\{\tilde{x} = (\tilde{x}_1, \tilde{x}_2) : |\tilde{x}_j| < 1,\ j = 1, 2\bigr\}$$

and

$$\tilde{K}^i := \{\tilde{x} : 0 < \tilde{x}_i < 1,\ |\tilde{x}_{3-i}| < 1\}, \quad i = 1, 2,$$

onto K^0 and K^i, respectively. Further, we define

$$\tilde{a}_{ij}(\tilde{x}) := a_{ij}(x + h\tilde{x}), \qquad \tilde{u}(\tilde{x}) := u(x + h\tilde{x}),$$

and so on. The value of η_{ij1} at a mesh-point $x \in \Omega_i^h$ can be expressed as

$$\eta_{ij1}(x) = \frac{1}{h}\biggl[\int_{\tilde{K}^i} \bigl(1 - |\tilde{x}_{3-i}|\bigr)\tilde{a}_{ij}(\tilde{x})\frac{\partial \tilde{u}}{\partial \tilde{x}_j}\,\mathrm{d}\tilde{x} - \int_{\tilde{K}^i} \bigl(1 - |\tilde{x}_{3-i}|\bigr)\tilde{a}_{ij}(\tilde{x})\,\mathrm{d}\tilde{x} \times \int_{\tilde{K}^i} \bigl(1 - |\tilde{x}_{3-i}|\bigr)\frac{\partial \tilde{u}}{\partial \tilde{x}_j}\,\mathrm{d}\tilde{x}\biggr].$$

Hence we deduce that $\eta_{ij1}(x)$ is a bounded bilinear functional of the argument

$$(\tilde{a}_{ij}, \tilde{u}) \in W_q^\lambda\bigl(\tilde{K}^i\bigr) \times W_{2q/(q-2)}^\mu\bigl(\tilde{K}^i\bigr),$$

where $\lambda \geq 0$, $\mu \geq 1$ and $q > 2$. Furthermore, $\eta_{ij1} = 0$ whenever $\tilde{a}_{ij}$ is a constant function or $\tilde{u}$ is a polynomial of degree 1. By applying the bilinear version of the Bramble–Hilbert lemma (cf. Lemma 2.30 with $m = 2$), we deduce that

$$\bigl|\eta_{ij1}(x)\bigr| \leq \frac{C}{h}|\tilde{a}_{ij}|_{W_q^\lambda(\tilde{K}^i)}|\tilde{u}|_{W_{2q/(q-2)}^\mu(\tilde{K}^i)}, \quad 0 \leq \lambda \leq 1,\ 1 \leq \mu \leq 2.$$

Returning from the canonical variables $(\tilde{x}_1, \tilde{x}_2)$ to the original variables (x_1, x_2) we obtain

$$|\tilde{a}_{ij}|_{W_q^\lambda(\tilde{K}^i)} = h^{\lambda - 2/q}|a_{ij}|_{W_q^\lambda(K^i)}$$

and

$$|\tilde{u}|_{W^{\mu}_{2q/(q-2)}(\tilde{K}^i)} = h^{\mu-(q-2)/q}|u|_{W^{\mu}_{2q/(q-2)}(K^i)}.$$

Therefore,

$$\left|\eta_{ij1}(x)\right| \leq Ch^{\lambda+\mu-2}|a_{ij}|_{W^{\lambda}_{q}(K^i)}|u|_{W^{\mu}_{2q/(q-2)}(K^i)}, \quad 0 \leq \lambda \leq 1, 1 \leq \mu \leq 2.$$

By summing through the mesh-points in Ω_i^h and applying Hölder's inequality we then deduce, for $0 \leq \lambda \leq 1$ and $1 \leq \mu \leq 2$, the bound

$$\|\eta_{ij1}\|_{L_2(\Omega_i^h)} \leq Ch^{\lambda+\mu-1}|a_{ij}|_{W^{\lambda}_{q}(\Omega)}|u|_{W^{\mu}_{2q/(q-2)}(\Omega)}. \tag{2.172}$$

Let us choose $\lambda = s-1$, $\mu = 1$ and $q = p$. Thanks to the Sobolev embedding theorem (cf. Theorem 1.34),

$$W_p^{s-1+\delta}(\Omega) \hookrightarrow W_p^{s-1}(\Omega) \quad \text{and} \quad W_2^s(\Omega) \hookrightarrow W^1_{2p/(p-2)}(\Omega), \quad 1 < s \leq 2.$$

Thus, (2.172) yields

$$\|\eta_{ij1}\|_{L_2(\Omega_i^h)} \leq Ch^{s-1}\|a_{ij}\|_{W_p^{s-1+\delta}(\Omega)}\|u\|_{W_2^s(\Omega)}, \quad 1 < s \leq 2. \tag{2.173}$$

Analogous bounds hold for η_{ij2}, η_{ij4}, η_1 and η_2. Now suppose that $q > 2$; then, the following Sobolev embeddings hold:

$$W_2^{\lambda+\mu-1}(\Omega) \hookrightarrow W_q^{\lambda}(\Omega) \quad \text{for } \mu > 2 - 2/q$$

and

$$W_2^{\lambda+\mu}(\Omega) \hookrightarrow W^{\mu}_{2q/(q-2)}(\Omega) \quad \text{for } \lambda > 2/q.$$

Setting $\lambda + \mu = s$ in (2.172) yields

$$\|\eta_{ij1}\|_{L_2(\Omega_i^h)} \leq Ch^{s-1}\|a_{ij}\|_{W_2^{s-1}(\Omega)}\|u\|_{W_2^s(\Omega)}, \quad 2 < s \leq 3. \tag{2.174}$$

The functional η_{ij4} is bounded in a similar fashion.

For $s > 2$, η_{ij2} is a bilinear functional of the argument

$$(a_{ij}, u) \in W_2^{s-1}(K^i) \times W^1_{\infty}(K^i)$$

and $\eta_{ij2} = 0$ whenever a_{ij} is a polynomial of degree 1 or if u is a constant function. By applying Lemma 2.65 and the embedding $W_2^s(\Omega) \hookrightarrow W^1_{\infty}(\Omega)$ we obtain a bound on η_{ij2}, which is of the form (2.174).

By a similar argument, $\eta_{ij3}(x)$ is a bounded bilinear functional of the argument

$$(a_{ij}, u) \in C(\overline{K}^i) \times W_2^s(K^i)$$

for $s > 1$ and it vanishes whenever u is a polynomial of degree 2. By noting the embeddings

$$W_p^{s-1+\delta}(\Omega) \hookrightarrow C(\overline{\Omega}) \quad \text{for } 1 < s \leq 2$$

and

$$W_2^{s-1}(\Omega) \hookrightarrow C(\overline{\Omega}) \quad \text{for } s > 2,$$

we obtain bounds of the form (2.173) and (2.174) for η_{ij3}.

Let $2 < q < 2/(3-s)$. When $2 < s \leq 3$, $\eta_3(x)$ is a bounded bilinear functional of the argument

$$(a, u) \in L_q\big(K^0\big) \times W_{2q/(q-2)}^{s-1}\big(K^0\big).$$

Moreover, $\eta_3 = 0$ when u is a polynomial of degree 1. By noting the Bramble–Hilbert lemma and the Sobolev embeddings

$$W_2^{s-2}(\Omega) \hookrightarrow L_q(\Omega) \quad \text{and} \quad W_2^s(\Omega) \hookrightarrow W_{2q/(q-2)}^{s-1}(\Omega)$$

we obtain

$$\|\eta_3\|_{L_2(\Omega^h)} \leq C h^{s-1} \|a\|_{W_2^{s-2}(\Omega)} \|u\|_{W_2^s(\Omega)}, \quad 2 < s \leq 3. \tag{2.175}$$

When $2 < s \leq 3$, η_4 is a bounded bilinear functional of

$$(a, u) \in W_2^{s-2}\big(K^0\big) \times W_\infty^1\big(K^0\big)$$

and $\eta_4 = 0$ whenever a or u is a constant function. Using the same technique as before, together with the embedding

$$W_2^s(\Omega) \hookrightarrow W_\infty^1(\Omega),$$

we obtain a bound of the form (2.175) for η_4.

Finally, let $2 < q < \min\{2+\varepsilon, 2/(2-s)\}$. Then, for $1 < s \leq 2$, $\eta_0(x)$ is a bounded bilinear functional of the argument

$$(a_0, u) \in L_q\big(K^0\big) \times W_{2q/(q-2)}^{s-1}\big(K^0\big)$$

and it vanishes when u is a constant function. By noting the embeddings

$$L_{2+\varepsilon}(\Omega) \hookrightarrow L_q(\Omega) \quad \text{and} \quad W_2^s(\Omega) \hookrightarrow W_{2q/(q-2)}^{s-1}(\Omega)$$

we obtain

$$\|\eta_0\|_{L_2(\Omega^h)} \leq C h^{s-1} \|a_0\|_{L_{2+\varepsilon}(\Omega)} \|u\|_{W_2^s(\Omega)}, \quad 1 < s \leq 2. \tag{2.176}$$

Finally, by combining (2.171) with (2.172)–(2.176) we deduce the desired bounds on the global error. □

2.6.2 Convergence in the Discrete W_2^2 Norm

In this section we consider the error analysis of the scheme (2.167) in the discrete W_2^2 norm (2.50).

From the error bound (2.168) in the $W_2^1(\Omega^h)$ norm derived in the previous section for the difference scheme (2.167) and the *inverse inequality*

$$|V|_{W_2^2(\Omega^h)} \leq \frac{\sqrt{6}}{h}|V|_{W_2^1(\Omega^h)} \quad \forall V \in S_0^h, \tag{2.177}$$

we immediately deduce, with $V = e$, the following error bound in the $W_2^2(\Omega^h)$ norm

$$\|u - U\|_{W_2^2(\Omega^h)} \leq Ch^{s-2}\Big(\max_{i,j} \|a_{ij}\|_{W_2^{s-1}(\Omega)} + \|a\|_{W_2^{s-2}(\Omega)}\Big)\|u\|_{W_2^s(\Omega)}$$
$$\text{for } 2 < s \leq 3. \tag{2.178}$$

In order to derive an analogous error bound when $3 < s \leq 4$ it is necessary to establish the discrete counterpart of the elliptic regularity result

$$\|v\|_{W_2^2(\Omega)} \leq C\|\mathcal{L}v\|_{L_2(\Omega)} \quad \forall v \in W_2^2(\Omega) \cap \mathring{W}_2^1(\Omega),$$

called the *second fundamental inequality*, following the terminology of Ladyzhenskaya and Ural'tseva [118]. A result of this kind was proved for the finite difference operator $\mathcal{L}_h$ by D'yakonov [39]; it states that

$$|V|_{W_2^2(\Omega^h)} \leq C\|\mathcal{L}_h V\|_{L_2(\Omega^h)} \quad \forall V \in S_0^h, \tag{2.179}$$

where

$$C := C(a_{11}, a_{12}, a_{22}, a) = C_0\big(1 + \big\|T_h^{22}a\big\|_{L_q(\Omega^h)}\big)\Big(1 + \max_{i,j} \|a_{ij}\|_{W_q^1(\Omega^h)}^{q/(q-2)}\Big),$$

with $2 < q \leq \infty$; here $\|\cdot\|_{L_q(\Omega^h)}$ and $\|\cdot\|_{W_q^1(\Omega^h)}$ are mesh-dependent norms defined, for $q < \infty$, by

$$\|V\|_{L_q(\Omega^h)} := \Big(h^2 \sum_{x\in\Omega^h} |V(x)|^q\Big)^{1/q},$$

$$\|V\|_{W_q^1(\Omega^h)} := \left(\|V\|_{L_q(\Omega^h)}^q + \sum_{i=1}^{2}\|D_{x_i}^+ V\|_{L_q(\Omega_i^h)}^q\right)^{1/q},$$

where $\|\cdot\|_{L_q(\Omega_i^h)}$ is defined in the same way as $\|\cdot\|_{L_q(\Omega^h)}$, except that the sum is taken over mesh-points in Ω_i^h instead of Ω^h. When $q = \infty$,

$$\|V\|_{L_\infty(\Omega^h)} = \|V\|_{\infty,h} := \max_{x\in\Omega^h} \big|V(x)\big|,$$

with an analogous definition of $\|V\|_{W^1_\infty(\Omega^h)}$.

By applying the Bramble–Hilbert lemma it is easily shown that

$$\|a_{ij}\|_{W^1_q(\Omega^h)} \le C_1 \|a_{ij}\|_{W^1_q(\Omega)} \quad \text{and} \quad \left\|T_1^2 T_2^2 a\right\|_{L_q(\Omega^h)} \le C_2 \|a\|_{L_q(\Omega)},$$

where C_1 and C_2 are independent of h. Thus we can assume in (2.179) that

$$\begin{aligned} C &= C(a_{11}, a_{12}, a_{22}, a) \\ &:= C_3(1 + \|a\|_{L_q(\Omega)})\Big(1 + \max_{i,j} \|a_{ij}\|^{q/(q-2)}_{W^1_q(\Omega)}\Big), \quad 2 < q \le \infty. \end{aligned}$$

Following the terminology of Ladyzhenskaya and Ural'tseva again, we note that the discrete version of the *first fundamental inequality* is

$$c_0 |V|^2_{W^1_2(\Omega^h)} \le (\mathcal{L}_h V, V)_h \quad \forall V \in S^h_0. \tag{2.180}$$

For the difference operator $\mathcal{L}_h$ appearing in (2.167) the first fundamental inequality is easily shown using summation by parts, in the same way as in the case of the result stated in Lemma 2.65.

Now we are ready to consider the error analysis of the difference scheme (2.167) in the norm $W^2_2(\Omega^h)$ for $u \in W^s_2(\Omega)$ when $3 < s \le 4$.

It follows from (2.170) and (2.177) that

$$\|e\|_{W^2_2(\Omega^h)} \le C\left(\sum_{i,j=1}^{2} \|D^-_{x_i} \eta_{ij}\|_{L_2(\Omega^h_i)} + \|\eta\|_{L_2(\Omega^h)}\right), \tag{2.181}$$

where C is a positive constant, independent of h. By bounding $D^-_{x_i}\eta_{ij}$ and η analogously as in the previous section, we obtain the error bound (2.178) for $3 < s \le 4$. Thus we deduce that (2.178) holds for $2 < s \le 4$ (see also Berikelashvili [10]).

2.6.3 Convergence in the Discrete L_2 Norm

The derivation of an optimal error bound in the $L_2(\Omega^h)$ norm is based on a technique that is usually referred to as a *duality argument*: it uses the adjoint of the difference operator $\mathcal{L}_h$ and the second fundamental inequality for the adjoint of the difference operator $\mathcal{L}_h$. Since in our case the difference operator $\mathcal{L}_h$ is symmetric and, more specifically, selfadjoint on the finite-dimensional space S^h_0 of real-valued mesh-functions defined on $\overline{\Omega}^h$ that vanish on Γ^h, equipped with the inner product of $L_2(\Omega^h)$, the second fundamental inequality for the adjoint of $\mathcal{L}_h$ is, in fact, identical to the second fundamental inequality for $\mathcal{L}_h$, stated in (2.179).

For the sake of simplicity, we shall restrict ourselves to the case when $a(x) \equiv 0$; the boundary-value problem (2.166) then becomes

$$-\sum_{i,j=1}^{2} \partial_i (a_{ij}\partial_j u) = f \quad \text{in } \Omega, \qquad u = 0 \quad \text{on } \Gamma = \partial\Omega, \tag{2.182}$$

and the corresponding finite difference scheme is

$$\mathcal{L}_h U := -\frac{1}{2}\sum_{i,j=1}^{2} \left[D_{x_i}^{+}(a_{ij}D_{x_j}^{-}U) + D_{x_i}^{-}(a_{ij}D_{x_j}^{+}U)\right] = T_h^{22} f \quad \text{in } \Omega^h,$$
$$U = 0 \quad \text{on } \Gamma^h. \tag{2.183}$$

The error analysis of this scheme in the $L_2(\Omega^h)$ norm is based on the observation that the global error $e := u - U$ is the solution of the difference scheme

$$\mathcal{L}_h e = \sum_{i,j=1}^{2} D_{x_i}^{-}\eta_{ij} \quad \text{in } \Omega^h, \qquad e = 0 \quad \text{on } \Gamma^h, \tag{2.184}$$

where the η_{ij} are the same as in (2.170). The right-hand side can be rewritten as follows:

$$\sum_{i,j=1}^{2} D_{x_i}^{-}\eta_{ij} = \sum_{i=1}^{2}\left(\mathcal{L}_{ii}\xi_{ii} + \mathcal{K}_i\chi_i + \sum_{j=1}^{2} D_{x_i}^{-}\upsilon_{ij}\right), \tag{2.185}$$

where

$$\mathcal{L}_{ii}V := -D_{x_i}^{-}\left[(T_i^{+}T_{3-i}^{2}a_{ii})D_{x_i}^{+}V\right], \qquad \mathcal{K}_i V := D_{x_i}^{-}\left[(T_i^{+}T_{3-i}^{2}a_{i,3-i})D_{x_{3-i}}^{+}V\right],$$

and

$$\xi_{ij} := u - \frac{1}{2}(T_{3-i}^{-}T_{3-j}^{+}u + T_{3-i}^{+}T_{3-j}^{-}u),$$

$$\chi_i := \varrho_i - \frac{1}{2}(\xi_{i,3-i} + \xi_{i,3-i}^{+i,-(3-i)}),$$

$$\varrho_i := \frac{1}{4}\left[(T_{3-i}^{-}T_i^{+}u - T_{3-i}^{+}T_i^{-}u) - (T_{3-i}^{-}T_i^{+}u - T_{3-i}^{+}T_i^{-}u)^{+i,-(3-i)}\right],$$

$$\upsilon_{ij} := T_i^{+}T_{3-i}^{2}(a_{ij}\partial_j u) - (T_i^{+}T_{3-i}^{2}a_{ij})(T_i^{+}T_{3-i}^{2}\partial_j u)$$
$$+ \frac{1}{2}\left[(T_i^{+}T_{3-i}^{2}a_{ij})(D_{x_j}^{+}u + D_{x_j}^{-}u^{+i}) - a_{ij}D_{x_j}^{+}u - a_{ij}^{+i}D_{x_j}^{-}u^{+i}\right].$$

Here we have assumed that the solution $u \in W_2^s(\Omega) \cap \mathring{W}_2^1(\Omega)$, $0 \le s \le 2$, has been extended, preserving its Sobolev class, to the square $(-h_0, 1+h_0)^2$ where h_0 is a fixed positive constant such that $h < h_0$.

Lemma 2.66 *Suppose that $a_{ij} \in W_q^1(\Omega)$, $q > 2$. The solution of the finite difference scheme* (2.184) *then satisfies the bound*

$$\|e\|_{L_2(\Omega^h)} \le C \sum_{i=1}^{2} \Bigg(\|\xi_{ii}\|_{L_2(\Omega^h)} + \|\xi_{i,3-i}\|_{L_2(\overline{\Omega}^h)} + \|\varrho_i\|_{L_2(\Omega^h_{i-1,2-i})} + \sum_{j=1}^{2} \|\upsilon_{ij}\|_{L_2(\Omega^h_i)} \Bigg), \qquad (2.186)$$

where C is a positive constant, independent of h.

Proof The proof is based on a duality argument. Let us consider the auxiliary function W, defined as the solution of the finite difference scheme

$$\mathcal{L}_h W = e \quad \text{in } \Omega^h, \qquad W = 0 \quad \text{on } \Gamma^h.$$

We note in passing that in general one would have written $(\mathcal{L}_h)^*$, the adjoint of $\mathcal{L}_h$, on the left-hand side instead of $\mathcal{L}_h$; however, in our case $\mathcal{L}_h$ is selfadjoint. Thus, crucially, $(e, \mathcal{L}_h W)_h = (\mathcal{L}_h e, W)_h$. It then follows from (2.184) and (2.185) that

$$\begin{aligned}
\|e\|^2_{L_2(\Omega^h)} &= (e, \mathcal{L}_h W)_h = (\mathcal{L}_h e, W)_h \\
&= \sum_{i=1}^{2} \left[(\mathcal{L}_{ii}\xi_{ii}, W)_h + (\mathcal{K}_i \chi_i, W)_h + \sum_{j=1}^{2} \big(D^-_{x_i} \upsilon_{ij}, W\big)_h \right] \\
&= \sum_{i=1}^{2} \left[(\xi_{ii}, \mathcal{L}_{ii} W)_h + \big(\chi_i, \mathcal{K}_i^* W\big)_{i-1,2-i,h} - \sum_{j=1}^{2} \big(\upsilon_{ij}, D^+_{x_i} W\big)_{i,h} \right] \\
&\le \sum_{i=1}^{2} \Bigg(\|\xi_{ii}\|_{L_2(\Omega^h)} \|\mathcal{L}_{ii} W\|_{L_2(\Omega^h)} + \|\chi_i\|_{L_2(\Omega^h_{i-1,2-i})} \big\|\mathcal{K}_i^* W\big\|_{L_2(\Omega^h_{i-1,2-i})} \\
&\qquad + \sum_{j=1}^{2} \|\upsilon_{ij}\|_{L_2(\Omega^h_i)} \big\|D^+_{x_i} W\big\|_{L_2(\Omega^h_i)} \Bigg),
\end{aligned}$$

where

$$\mathcal{K}_i^* W = D^-_{x_{3-i}} \big[\big(T_i^+ T_{3-i}^2 a_{i,3-i}\big) D^+_{x_i} W \big].$$

The second fundamental inequality (2.179) implies that

$$\|\mathcal{L}_{ii} W\|_{L_2(\Omega^h)}, \qquad \big\|\mathcal{K}_i^* W\big\|_{L_2(\Omega^h_{i-1,2-i})}, \qquad \big\|D^+_{x_i} W\big\|_{L_2(\Omega^h_i)}$$

are all bounded by

$$C\|\mathcal{L}_h W\|_{L_2(\Omega^h)} = C\|e\|_{L_2(\Omega^h)},$$

and hence, after substitution of the defining expression for χ_i, we deduce the inequality (2.186). □

We observe that for the second fundamental inequality to hold it is necessary that $a_{ij} \in W_q^1(\Omega)$, $q > 2$; thus we can only expect a sharp error bound when $s = 2$. Let us assume that this is indeed the case, and we proceed to estimate the terms that appear on the right-hand side of the inequality (2.186).

We begin by noting that ξ_{ij} and ϱ_i are bounded linear functionals on $W_2^2(\Omega)$ that vanish on all polynomials of degree 1. By the Bramble–Hilbert lemma,

$$\|\xi_{ii}\|_{L_2(\Omega^h)}, \quad \|\xi_{i,3-i}\|_{L_2(\overline{\Omega}^h)}, \quad \|\varrho_i\|_{L_2(\Omega^h_{i-1,2-i})} \leq Ch^2 \|u\|_{W_2^2(\Omega)}. \tag{2.187}$$

Arguing in the same way as in the previous section, υ_{ij} is decomposed into three terms that are bounded by means of the Bramble–Hilbert lemma to obtain:

$$\|\upsilon_{ij}\|_{L_2(\Omega_i^h)} \leq Ch^2 \big(\|a_{ij}\|_{W^1_\infty(\Omega)} \|u\|_{W_2^2(\Omega)} + \|a_{ij}\|_{W^2_\infty(\Omega)} \|u\|_{W_2^1(\Omega)}\big). \tag{2.188}$$

From (2.186)–(2.188) we deduce the following error bound for the difference scheme (2.183):

$$\|u - U\|_{L_2(\Omega^h)} \leq Ch^2 \max_{i,j} \|a_{ij}\|_{W^2_\infty(\Omega)} \|u\|_{W_2^2(\Omega)}. \tag{2.189}$$

While the power of h in the error bound (2.189) is optimal in the sense that it is compatible with the smoothness of u, the bound is not entirely satisfactory as the coefficients a_{ij} are required to belong to $W^2_\infty(\Omega)$, which, in the light of the hypotheses (a) and (b) from the beginning of Sect. 2.6.1, can be seen as an excessively strong assumption on the regularity of the coefficients a_{ij}. The requirement for the additional smoothness of the coefficients a_{ij} can be attributed to our crude bound on $D^-_{x_i}\upsilon_{ij}$ in (2.186).

An improved estimate can be obtained by considering an alternative scheme where the coefficients a_{ij} have been mollified:

$$\begin{aligned} \hat{\mathcal{L}}_h U &:= \sum_{i,j=1}^{2} \mathcal{L}_{ij} U = T_h^{22} f \quad \text{in } \Omega^h, \\ U &= 0 \quad \text{on } \Gamma^h, \end{aligned} \tag{2.190}$$

where

$$\mathcal{L}_{ij} U := -\frac{1}{2} D^-_{x_i}\big[\big(T_i^+ T_{3-i}^2 a_{ij}\big) D^+_{x_j}\big(U + U^{+i,-j}\big)\big].$$

For this scheme the global error $e := u - U$ satisfies

$$\hat{\mathcal{L}}_h e = \sum_{i=1}^{2} \left(\mathcal{L}_{ii}\xi_{ii} + \mathcal{K}_i \chi_i + \sum_{j=1}^{2} D^-_{x_i} \eta_{ij1} \right) \quad \text{in } \Omega^h, \qquad z = 0 \quad \text{on } \Gamma^h,$$

where ξ_{ii}, χ_i and η_{ij1} are as before. Assuming that $a_{ij} \in W_q^1(\Omega)$, $q > 2$, and proceeding in the same manner as in the case of our previous scheme where the coefficients a_{ij} were not mollified, we obtain the bound

$$\|e\|_{L_2(\Omega^h)} \leq C \sum_{i=1}^{2} \Bigg(\|\xi_{ii}\|_{L_2(\Omega^h)} + \|\xi_{i,3-i}\|_{L_2(\overline{\Omega}^h)} + \|\varrho_i\|_{L_2(\Omega^h_{i-1,2-i})} + \sum_{j=1}^{2} \|\eta_{ij1}\|_{L_2(\Omega_i^h)} \Bigg).$$

Using the estimates (2.187) and (2.172) derived earlier and slightly strengthening the smoothness requirements on the a_{ij} by demanding that $a_{ij} \in W_\infty^1(\Omega)$, we arrive at the error bound

$$\|u - U\|_{L_2(\Omega^h)} \leq Ch^2 \max_{i,j} \|a_{ij}\|_{W_\infty^1(\Omega)} \|u\|_{W_2^2(\Omega)}, \tag{2.191}$$

which is now *almost compatible* with the smoothness of the data in the sense that we assumed $a_{ij} \in W_\infty^1(\Omega)$ instead of the minimal smoothness requirement $a_{ij} \in W_q^1(\Omega)$, $q > 2$.

Let us now discuss the case when u belongs to the fractional-order Sobolev space $W_2^s(\Omega)$, $1 < s \leq 2$. Allowing some incompatibility between the smoothness of the coefficients and the corresponding solution by assuming instead of our initial hypothesis

$$u \in W_2^s(\Omega), \qquad a_{ij} \in W_p^{|s-1|+\delta}(\Omega), \quad 1 < s \leq 2,$$

that

$$u \in W_2^s(\Omega), \quad 1 < s \leq 2; \qquad a_{ij} \in W_\infty^1(\Omega)$$

and arguing as above, instead of (2.191) we arrive at the error bound

$$\|u - U\|_{L_2(\Omega^h)} \leq Ch^s \max_{i,j} \|a_{ij}\|_{W_\infty^1(\Omega)} \|u\|_{W_2^s(\Omega)}, \; 1 < s \leq 2.$$

This error bound is again incompatible with the smoothness of the data, except in the case of $s = 2$ when it coincides with (2.191).

2.6.4 Convergence in Discrete Fractional-Order Norms

By noting our error bounds in integer-order discrete Sobolev norms and the interpolation inequalities (2.54) we can obtain new bounds in fractional-order discrete

Sobolev norms. Thus, for example, for the scheme (2.167) from (2.168) and (2.178), we have that

$$\|u - U\|_{W_2^r(\Omega^h)} \le Ch^{s-r}\Big(\max_{i,j} \|a_{ij}\|_{W_2^{s-1}(\Omega)} + \|a\|_{W_2^{s-2}(\Omega)}\Big)\|u\|_{W_2^s(\Omega)},$$

$$\text{for } 1 \le r \le 2 < s \le 3.$$

From (2.169), (2.177) and (2.54) we deduce that

$$\|u - U\|_{W_2^r(\Omega^h)} \le Ch^{s-r}\Big(\max_{i,j} \|a_{ij}\|_{W_p^{s-1+\delta}(\Omega)} + \max_{i} \|a_i\|_{W_p^{s-1+\delta}(\Omega)}$$

$$+ \|a_0\|_{L_{2+\varepsilon}(\Omega)}\Big)\|u\|_{W_2^s(\Omega)}, \quad \text{for } 1 \le r < s \le 2.$$

Similarly, from (2.191), (2.54), the *inverse inequality*

$$|V|_{W_2^1(\Omega^h)} \le \frac{2\sqrt{2}}{h}\|V\|_{L_2(\Omega^h)} \quad \forall V \in S_0^h$$

with $V = e$ and (2.177) we obtain the following error bound for the difference scheme (2.190):

$$\|u - U\|_{W_2^r(\Omega^h)} \le Ch^{2-r} \max_{i,j} \|a_{ij}\|_{W_\infty^1(\Omega)}\|u\|_{W_2^2(\Omega)}, \quad 0 \le r \le 2.$$

In the next section we shall further sharpen these error bounds in the special case of an equation where the off-diagonal entries in the coefficient matrix (a_{ij}) are identically zero.

2.6.5 Convergence in the Discrete L_2 Norm: Separated Variables

In Sect. 2.6.3 we saw that the derivation of optimal error bounds in the $L_2(\Omega^h)$ norm under minimal smoothness requirements on the coefficients a_{ij} is associated with technical difficulties. The error bounds that we obtained are satisfactory in this respect only when $s = 2$, while for $s < 2$ they are incompatible with the natural minimal regularity requirements on the coefficients. These results can be improved in the case of a differential equation that separates the two variables; that is, when

$$-\sum_{i=1}^{2} \partial_i(a_i \partial_i u) = f \quad \text{in } \Omega, \qquad u = 0 \quad \text{on } \Gamma = \partial\Omega, \tag{2.192}$$

where

$$a_i = a_i(x_i), \quad i = 1, 2,$$

are such that there exist positive constants c_0 and c_1 with

$$0 < c_0 \le a_i(x_i) \le c_1 \quad \text{for all } x_i \in (0,1),\ i = 1,2.$$

In order to ensure that the a_i belong to the function space of multipliers $M(W_2^{s-1}(\Omega))$, we shall suppose that

$$a_i \in W_p^{|s-1|+\delta}(0,1),$$

where the real numbers s, p and δ are assumed to satisfy the following conditions:

$$\begin{aligned} &p = 2, \quad \delta = 0 \quad \text{when } |s-1| > 1/2,\\ &p > 2, \quad \delta > 0 \quad \text{when } s = 1/2 \text{ or } s = 3/2,\\ &p \ge 1/|s-1|, \quad \delta > 0 \quad \text{when } 0 < |s-1| < 1/2,\\ &p = \infty, \quad \delta = 0 \quad \text{when } s = 1. \end{aligned} \tag{2.193}$$

Let us introduce the following univariate mollifiers:

$$(S_i f)(x) := \frac{1}{h}\int_{x_i-h}^{x_i+h} \kappa_i(t) f\big(x + (t - x_i)\mathrm{e}_i\big)\,\mathrm{d}t, \quad i = 1,2,$$

where

$$\kappa_i(t) := \begin{cases} \int_{x_i-h}^{t} \frac{\mathrm{d}\tau}{a_i(\tau)} \Big/ \int_{x_i-h}^{x_i} \frac{\mathrm{d}\tau}{a_i(\tau)}, & t \in (x_i - h, x_i),\\ \int_{t}^{x_i+h} \frac{\mathrm{d}\tau}{a_i(\tau)} \Big/ \int_{x_i}^{x_i+h} \frac{\mathrm{d}\tau}{a_i(\tau)}, & t \in (x_i, x_i + h). \end{cases}$$

These operators satisfy the identity

$$S_i\big(\partial_i(a_i\partial_i u)\big) = D_{x_i}^-\big(\hat{a}_i D_{x_i}^+ u\big),$$

where $\hat{a}_i$ is the harmonic average of a_i, defined by

$$\hat{a}_i(x_i) := \left(\frac{1}{h}\int_{x_i}^{x_i+h} \frac{\mathrm{d}\tau}{a_i(\tau)}\right)^{-1}, \quad i = 1,2.$$

In particular when $a_i(x_i) \equiv 1$, we have that

$$S_i = T_i^2 = T_i^+ T_i^-.$$

We approximate the boundary-value problem (2.192) by the following finite difference scheme:

$$-\sum_{i=1}^{2} b_{3-i} D_{x_i}^-\big(\hat{a}_i D_{x_i}^+ U\big) = S_1 S_2 f \quad \text{in } \Omega^h, \tag{2.194}$$

$$U = 0 \quad \text{on } \Gamma^h, \tag{2.195}$$

where $b_i := S_i(1)$, $i = 1, 2$. We define the global error by

$$e := \overline{u} - U, \quad \text{where } \overline{u} := \begin{cases} T_h^{11}u, & \text{if } 0 < s \leq 1, \\ u, & \text{if } 1 < s \leq 2. \end{cases}$$

Then, e is easily seen to be a solution of the following finite difference scheme on the mesh $\overline{\Omega}^h$:

$$-\sum_{i=1}^{2} b_{3-i} D_{x_i}^-(\hat{a}_i D_{x_i}^+ e) = \sum_{i=1}^{2} D_{x_i}^-(\hat{a}_i D_{x_i}^+ \psi_i) \quad \text{in } \Omega^h,$$
$$e = 0 \quad \text{on } \Gamma^h,$$

where $\psi_i = S_{3-i}(u) - b_{3-i}\overline{u}$, $i = 1, 2$. It is easy to show by a duality argument (cf. the proof of Lemma 2.66) that

$$\|e\|_{L_2(\Omega^h)} \leq C\big(\|\psi_1\|_{L_2(\Omega^h)} + \|\psi_2\|_{L_2(\Omega^h)}\big). \tag{2.196}$$

The task of deriving an error bound for the difference scheme (2.194) has thus been reduced to estimating the expression on the right-hand side of (2.196). We shall discuss the cases $1/2 < s \leq 1$ and $1 < s \leq 2$ separately.

First suppose that $1/2 < s \leq 1$. Clearly, the value of ψ_i at a node $x \in \Omega^h$ is a bounded linear functional of $u \in W_2^s(K^0)$, $s > 1/2$, where

$$K^0 = K^0(x) = \big\{y = (y_1, y_2) : |y_j - x_j| < h,\ j = 1, 2\big\}.$$

Moreover, $\psi_i = 0$ when u is a constant function. By applying the Bramble–Hilbert lemma we deduce that

$$|\psi_i| \leq Ch^{s-1}|u|_{W_2^s(K^0)}, \quad 1/2 < s \leq 1.$$

Summing over the nodes of the mesh Ω^h we obtain, for $i = 1, 2$, that

$$\|\psi_i\|_{L_2(\Omega^h)} \leq Ch^s|u|_{W_2^s(\Omega)}, \quad 1/2 < s \leq 1. \tag{2.197}$$

Now let us consider the case $1 < s \leq 2$. The key difficulty in obtaining an error bound is that ψ_{3-i} represents a nonlinear functional of a_i, $i = 1, 2$; nevertheless ψ_{3-i}, $i = 1, 2$, may be conveniently decomposed and, thereby, the nonlinear terms can be directly estimated. Let us write

$$\psi_{3-i} = \psi_{3-i,1} + \psi_{3-i,2} + \psi_{3-i,3},$$

where

$$\psi_{3-i,1} := \int_0^1 \big[u(x + h\tau e_i) - 2u(x) + u(x - h\tau e_i)\big]\left(\int_{x_i-h}^{x_i-h\tau} \frac{d\sigma}{a_i(\sigma)}\right)$$

$$\times \left(\int_{x_i-h}^{x_i} \frac{d\sigma}{a_i(\sigma)} \right)^{-1} d\tau,$$

$$\psi_{3-i,2} := \int_0^1 \big[u(x+h\tau e_i) - u(x)\big] \left(\int_{x_i}^{x_i+h} \frac{d\sigma}{a_i(\sigma)} \int_{x_i-h}^{x_i} \frac{d\sigma}{a_i(\sigma)} \right)^{-1}$$
$$\times \left(\int_{x_i+h\tau}^{x_i+h} \frac{d\sigma}{a_i(\sigma)} \right) h^{-1} \int_{x_i-h}^{x_i} \int_{x_i}^{x_i+h} \frac{a_i(t) - a_i(t')}{a_i(t) a_i(t')} \, dt \, dt' \, d\tau,$$

$$\psi_{3-i,3} := \int_0^1 \big[u(x+h\tau e_i) - u(x)\big] \left(\int_{x_i-h}^{x_i} \frac{d\sigma}{a_i(\sigma)} \right)^{-1}$$
$$\times h^{-1}(1-\tau)^{-1} \int_{x_i-h}^{x_i-h\tau} \int_{x_i+h\tau}^{x_i+h} \frac{a_i(t) - a_i(t')}{a_i(t) a_i(t')} \, dt \, dt' \, d\tau.$$

The value of $\psi_{3-i,1}$ at $x \in \Omega^h$ is a bounded linear functional of $u \in W_2^s(K^0)$, $s > 1$, which vanishes whenever u is a polynomial of degree 1. Using the Bramble–Hilbert lemma we obtain

$$\|\psi_{3-i,1}\|_{L_2(\Omega^h)} \le C h^s |u|_{W_2^s(\Omega)}, \quad 1 < s \le 2. \tag{2.198}$$

For $3/2 < s \le 2$, $\psi_{3-i,2}$ is a bounded linear functional of $u \in W_2^s(K^0)$:

$$|\psi_{3-i,2}| \le C h^{\lambda - 1/2} \big(h^{-1} \|u\|_{L_2(K^0)} + |u|_{W_2^1(K^0)}$$
$$+ h^{s-1} |u|_{W_2^s(K^0)} \big) |a_i|_{W_2^\lambda(I^0)}, \ \lambda > 0,$$

where $I^0 = I^0(x_i) := (x_i - h, x_i + h)$. Moreover, $\psi_{3-i,2} = 0$ when u is a constant function, and therefore the term $h^{-1}\|u\|_{L_2(K^0)}$ on the right-hand side can be eliminated by applying the Bramble–Hilbert lemma. Summing over the nodes in the mesh Ω^h yields

$$\|\psi_{3-i,2}\|_{L_2(\Omega^h)} \le C h^{\lambda+1/2} \Big(\max_{x_i} |u|_{W_2^1(\Omega_{h,i})} + h^{s-1} |u|_{W_2^s(\Omega)} \Big) |a_i|_{W_2^\lambda(0,1)},$$

where

$$\Omega_{h,i} = \Omega_{h,i}(x) := \big\{ y \in \mathbb{R}^2 : \ x_i - h < y_i < x_i + h, \ 0 < y_{3-i} < 1 \big\}.$$

Choosing $\lambda = s - 1$ and invoking the boundary-layer estimate (see Oganesyan and Rukhovets [148], Chap. I, §8)

$$\|v\|_{L_2(0,\varepsilon)} \le C F(\varepsilon) \|v\|_{W_2^s(0,1)}, \quad 0 < \varepsilon < 1, \ 0 \le s \le 1, \tag{2.199}$$

where

$$F(\varepsilon) := \begin{cases} \varepsilon^s & 0 \le s < 1/2, \\ \varepsilon^{1/2} |\log \varepsilon| & s = 1/2, \\ \varepsilon^{1/2} & 1/2 < s \le 1, \end{cases}$$

which implies that

$$|u|_{W_2^1(\Omega_{h,i})} \le Ch^{1/2}\|u\|_{W_2^s(\Omega)}, \quad s > 3/2,$$

we thus obtain the bound

$$\|\psi_{3-i,2}\|_{L_2(\Omega^h)} \le Ch^s\|a_i\|_{W_2^{s-1}(0,1)}\|u\|_{W_2^s(\Omega)}, \quad 3/2 < s \le 2. \qquad (2.200)$$

Similarly,

$$\|\psi_{3-i,2}\|_{L_2(\Omega^h)} \le Ch^s\|a_i\|_{W_p^{s-1+\delta}(0,1)}\|u\|_{W_2^s(\Omega)}, \quad 1 < s \le 3/2, \qquad (2.201)$$

with p as in (2.193). An analogous bound holds for $\psi_{3-i,3}$. Combining (2.196) with (2.197), (2.198), (2.200) and (2.201) we thus obtain the following result.

Theorem 2.67 *Suppose that $u \in W_2^s(\Omega)$ and $a_i \in W_p^{|s-1|}(\Omega)$, $i = 1, 2$, with $1/2 < s \le 2$ and p as in* (2.193). *Then, the finite difference scheme* (2.194) *satisfies the error bound*

$$\|u - U\|_{L_2(\Omega^h)} \le Ch^s \max_i \|a_i\|_{W_p^{|s-1|+\delta}(0,1)}\|u\|_{W_2^s(\Omega)}, \qquad (2.202)$$

where C is a positive constant, independent of h.

Unlike our earlier optimal error bounds in the $L_2(\Omega^h)$ norm, (2.202) is now also compatible with the smoothness of the coefficients.

We note that for $0 < s \le 1/2$ the function S_1S_2f, with $f \in W_2^{s-2}(\Omega)$, is not necessarily continuous on $\overline{\Omega}$; in this case the right-hand side of the difference scheme (2.194) is not defined at the mesh-points. A more fundamental difficulty is that $u \in W_2^s(\Omega)$ does not have a trace on $\Gamma = \partial\Omega$ when $s \le 1/2$, and it makes no sense, therefore, to demand that it satisfies a homogeneous Dirichlet boundary condition on Γ.

2.7 Fourth-Order Elliptic Equations

This section is devoted to boundary-value problems for fourth-order elliptic equations with variable coefficients of the form

$$\mathcal{L}u := \partial_1^2 M_1(u) + 2\partial_1\partial_2 M_3(u) + \partial_2^2 M_2(u) = f(x), \quad x \in \Omega, \qquad (2.203)$$

where $\Omega = (0, 1)^2$ and

$$M_1(u) := a_1\partial_1^2 u + a_0\partial_2^2 u,$$

$$M_2(u) := a_0\partial_1^2 u + a_2\partial_2^2 u,$$

$$M_3(u) := a_3 \partial_1 \partial_2 u.$$

We shall assume that

$$\begin{aligned} &a_i \geq c_0 > 0, \quad i = 1,2,3, \qquad a_1 a_2 - a_0^2 \geq c_1 > 0, \quad x \in \Omega, \\ &u \in W_2^s(\Omega), \qquad f \in W_2^{s-4}(\Omega), \quad 2 < s \leq 4. \end{aligned} \tag{2.204}$$

In order for (2.204) to hold it is necessary that the coefficients a_i belong to the multiplier space $M(W_2^{s-2}(\Omega))$. According to the results in Sect. 1.8, the following conditions are sufficient for that to be the case:

$$a_i \in W_p^{s-2+\varepsilon}(\Omega), \quad i = 0,1,2,3, \tag{2.205}$$

where

$$\begin{aligned} &p = 2, \qquad \varepsilon = 0 \quad \text{when } 3 < s \leq 4, \\ &p > 2, \qquad \varepsilon = 0 \quad \text{when } s = 3, \\ &p \geq 2/(s-2), \qquad \varepsilon > 0 \quad \text{when } 2 < s < 3. \end{aligned}$$

We begin by considering the partial differential equation (2.203) subject to the boundary conditions

$$\begin{aligned} u &= 0 \quad \text{on } \Gamma = \partial\Omega; \\ \partial_i^2 u &= 0 \quad \text{on } \Gamma_{i0} \cup \Gamma_{i1}, i = 1,2, \end{aligned} \tag{2.206}$$

where

$$\Gamma_{ik} := \{x \in \Gamma : x_i = k, 0 < x_{3-i} < 1\}, \quad i,k = 0,1.$$

By adopting the same notation as in Sects. 2.2.4 and 2.7 we approximate the boundary-value problem (2.203), (2.206) by the finite difference scheme

$$\mathcal{L}_h U = T_h^{22} f, \quad \text{on } \Omega^h, \tag{2.207}$$

$$\begin{aligned} U &= 0, \quad \text{on } \Gamma^h, \\ D_{x_i}^+ D_{x_i}^- U &= 0, \quad \text{on } \Gamma_{i0}^h \cup \Gamma_{i1}^h, \; i = 1,2, \end{aligned} \tag{2.208}$$

where $\Gamma_{ik}^h = \Gamma_{ik} \cap \Gamma^h$,

$$\mathcal{L}_h U := D_{x_1}^+ D_{x_1}^- m_1(U) + 2 D_{x_1}^- D_{x_2}^- m_3(U) + D_{x_2}^+ D_{x_2}^- m_2(U),$$

and

$$\begin{aligned} m_1(U) &:= a_1 D_{x_1}^+ D_{x_1}^- U + a_0 D_{x_2}^+ D_{x_2}^- U, \\ m_2(U) &:= a_0 D_{x_1}^+ D_{x_1}^- U + a_2 D_{x_2}^+ D_{x_2}^- U, \end{aligned}$$

$$m_3(U) := \hat{a}_3 D^+_{x_1} D^+_{x_2} U,$$

with

$$\hat{a}_3(x) := a_3\left(x_1 + \frac{1}{2}h, x_2 + \frac{1}{2}h\right).$$

Let us note that the difference scheme also involves mesh-points in $h\mathbb{Z}^2$ that are contained in $[-h, 1+h]^2$. Thus we shall suppose that the solution u and the coefficients a_i have been extended onto the larger square $(-h_0, 1+h_0)^2$ preserving their Sobolev class; here h_0 is a positive constant, $h_0 > h$.

Next we develop the error analysis of this finite difference scheme. The global error $e := u - U$ is easily seen to satisfy the following difference scheme:

$$\mathcal{L}_h e = D^+_{x_1} D^-_{x_1}\varphi_1 + 2D^-_{x_1} D^-_{x_2}\varphi_3 + D^+_{x_2} D^-_{x_2}\varphi_2, \quad x \in \Omega^h, \tag{2.209}$$

$$\begin{aligned} e &= 0, \quad x \in \Gamma^h, \\ D^+_{x_i} D^-_{x_i} e &= D^+_{x_i} D^-_{x_i} u, \quad x \in \Gamma^h_{i0} \cup \Gamma^h_{i1},\ i = 1, 2, \end{aligned} \tag{2.210}$$

where

$$\varphi_i := m_i(u) - T^2_{3-i} M_i(u), \quad i = 1, 2; \qquad \varphi_3 := m_3(u) - T^+_1 T^+_2 M_3(u).$$

Thus (2.206), (2.208) and (2.210) imply that

$$m_i(e) = \varphi_i, \quad x \in \Gamma^h_{i0} \cup \Gamma^h_{i1},\ i = 1, 2.$$

By taking the inner product of (2.209) with e, performing summations by parts and applying the Cauchy–Schwarz inequality we get

$$\|e\|^2_{W^2_2(\Omega^h)} \le C\left(\|\varphi_1\|^2_{L_2(\Omega^h)} + \|\varphi_2\|^2_{L_2(\Omega^h)} + \|\varphi_3\|^2_{L_2(\Omega^h_{00})}\right). \tag{2.211}$$

Theorem 2.68 *Assuming that the data and the corresponding solution of the boundary-value problem* (2.203), (2.206) *obey the conditions* (2.204) *and* (2.205), *the difference scheme* (2.207), (2.208) *satisfies the error bound*

$$\|u - U\|_{W^2_2(\Omega^h)} \le Ch^{s-2} \max_i \|a_i\|_{W^{s-2+\varepsilon}_p(\Omega)} \|u\|_{W^s_2(\Omega)}, \quad 5/2 < s \le 4. \tag{2.212}$$

Proof In order to prove the error bound (2.212) it suffices to bound the terms on the right-hand side of the inequality (2.211). Let us begin by representing φ_1 as the sum

$$\varphi_1 = \sum_{j=1}^{8} \varphi_{1,j},$$

where

$$\begin{aligned}
\varphi_{1,k} &:= a_{2-k}\big(D^+_{x_k}D^-_{x_k}u - T_1^2T_2^2\partial_k^2 u\big),\\
\varphi_{1,k+2} &:= \big(a_{2-k} - T_1^2T_2^2 a_{2-k}\big)\big(T_1^2T_2^2\partial_k^2 u\big),\\
\varphi_{1,k+4} &:= \big(T_1^2T_2^2 a_{2-k}\big)\big(T_1^2T_2^2\partial_k^2 u\big) - T_1^2T_2^2\big(a_{2-k}\partial_k^2 u\big),\\
\varphi_{1,k+6} &:= T_1^2T_2^2\big(a_{2-k}\partial_k^2 u\big) - T_2^2\big(a_{2-k}\partial_k^2 u\big), \quad k = 1, 2,
\end{aligned}$$

with an analogous representation of φ_2. Further, let

$$\varphi_3 = \varphi_{3,1} + \varphi_{3,2},$$

where

$$\begin{aligned}
\varphi_{3,1} &:= \big(\hat{a}_3 - T_1^+T_2^+ a_3\big)D^+_{x_1}D^+_{x_2}u,\\
\varphi_{3,2} &:= \big(T_1^+T_2^+ a_3\big)D^+_{x_1}D^+_{x_2}u - T_1^+T_2^+(a_3\partial_1\partial_2 u).
\end{aligned}$$

When $s \geq 2$, the value of $\varphi_{1,1}$ at a mesh-point $x \in \Omega^h$ is a bounded linear functional of $u \in W_2^s(K^0)$:

$$|\varphi_{1,1}| \leq C(h)\|a_1\|_{C(\overline{\Omega})}\|u\|_{W_2^s(K^0)}.$$

Moreover, $\varphi_{1,1} = 0$ when u is a polynomial of degree 3. By the Bramble–Hilbert lemma,

$$|\varphi_{1,1}| \leq Ch^{s-3}\|a_1\|_{C(\overline{\Omega})}|u|_{W_2^s(K^0)}, \quad 2 \leq s \leq 4.$$

By noting the Sobolev embedding $W_p^{s-2+\varepsilon}(K^0) \hookrightarrow C(\overline{K^0})$, $s > 2$, and summing over the mesh-points in Ω^h we thus obtain

$$\|\varphi_{1,1}\|_{L_2(\Omega^h)} \leq Ch^{s-2}\|a_1\|_{W_p^{s-2+\varepsilon}(\Omega)}|u|_{W_2^s(\Omega)}, \quad 2 \leq s \leq 4. \tag{2.213}$$

The term $\varphi_{1,2}$ is bounded in the same way. Next $\varphi_{1,3}(x)$, $x \in \Omega^h$, is a bounded bilinear functional of $(a_1, u) \in W_p^\lambda(K^0) \times W_q^2(K^0)$, with $\lambda p > 2$; $q = \infty$ when $p = 2$; and $q = 2p/(p-2)$ when $p > 2$. Moreover, $\varphi_{1,3} = 0$ when either a_1 or u is a polynomial of degree 1. From the bilinear version of the Bramble–Hilbert lemma (cf. Lemma 2.30 with $m = 2$) we deduce that

$$|\varphi_{1,3}| \leq Ch^{\lambda-1}\|a_1\|_{W_p^\lambda(K^0)}|u|_{W_q^2(K^0)}, \quad 2/p < \lambda \leq 2,$$

and thereby

$$\|\varphi_{1,3}\|_{L_2(\Omega^h)} \leq Ch^\lambda\|a_1\|_{W_p^\lambda(\Omega)}\|u\|_{W_q^2(\Omega)}.$$

By choosing $\lambda = s - 2 + \varepsilon$ and noting the Sobolev embeddings

$$W_2^s(\Omega) \hookrightarrow W_\infty^2(\Omega), \quad s > 3,$$

and

$$W_2^s(\Omega) \hookrightarrow W_{2p/(p-2)}^2(\Omega), \quad 2 < s \le 3,$$

we obtain

$$\|\varphi_{1,3}\|_{L_2(\Omega^h)} \le Ch^{s-2} \|a_1\|_{W_p^{s-2+\varepsilon}(\Omega)} \|u\|_{W_2^s(\Omega)}, \quad 2 < s \le 4. \tag{2.214}$$

The terms $\varphi_{1,4}$ and $\varphi_{3,1}$ are bounded in the same way.

For $\lambda \ge 0$, $\mu \ge 2$ and $q > 2$ the value of $\varphi_{1,5}(x)$ at $x \in \Omega^h$ is a bounded bilinear functional of $(a_1, u) \in W_q^\lambda(K^0) \times W_{2q/(q-2)}^\mu(K^0)$. Furthermore, $\varphi_{1,5} = 0$ when a_1 is a constant function or when u is a polynomial of degree 2. By the bilinear version of the Bramble–Hilbert lemma,

$$\|\varphi_{1,5}\|_{L_2(\Omega^h)} \le Ch^{\lambda+\mu-2} \|a_1\|_{W_q^\lambda(\Omega)} \|u\|_{W_{2q/(q-2)}^\mu(\Omega)},$$

where $0 \le \lambda \le 1$ and $2 \le \mu \le 3$. Now let $\lambda + \mu = s$. When $\lambda + \mu > 3$, there exists a $q = q(\lambda, \mu)$ such that $\lambda \ge 2/q \ge 3 - \mu$; then,

$$W_p^{s-2+\varepsilon}(\Omega) = W_2^{\lambda+\mu-2+\varepsilon}(\Omega) \hookrightarrow W_q^\lambda(\Omega)$$

and

$$W_2^s(\Omega) = W_2^{\lambda+\mu}(\Omega) \hookrightarrow W_{2q/(q-2)}^\mu(\Omega).$$

Analogously, when $2 < \lambda + \mu \le 3$, there exists a real number q such that $\lambda \ge 2/q \ge 2/p - (\mu - 2)$. In this case,

$$W_p^{s-2+\varepsilon}(\Omega) = W_p^{\lambda+\mu-2+\varepsilon}(\Omega) \hookrightarrow W_q^\lambda(\Omega)$$

and

$$W_2^s(\Omega) = W_2^{\lambda+\mu}(\Omega) \hookrightarrow W_{2q/(q-2)}^\mu(\Omega).$$

It follows from these embeddings that

$$\|\varphi_{1,5}\|_{L_2(\Omega^h)} \le Ch^{s-2} \|a_1\|_{W_p^{s-2+\varepsilon}(\Omega)} \|u\|_{W_2^s(\Omega)}, \quad 2 < s \le 4. \tag{2.215}$$

The terms $\varphi_{1,6}$ and $\varphi_{3,2}$ are bounded in the same way.

When $\lambda > 1/2$, the value of $\varphi_{1,7}(x)$ at $x \in \Omega^h$ is a bounded linear functional of $a_1 \partial_1^2 u \in W_2^\lambda(K^0)$, which vanishes on all polynomials of degree 1. By the Bramble–Hilbert lemma, we have that

$$\|\varphi_{1,7}\|_{L_2(\Omega^h)} \le Ch^\lambda |a_1 \partial_1^2 u|_{W_2^\lambda(\Omega)}, \quad 1/2 < \lambda \le 2.$$

By choosing $\lambda = s - 2$, the inequality

$$|a_1 \partial_1^2 u|_{W_2^\lambda(\Omega)} \le C \|a_1\|_{W_p^{\lambda+\varepsilon}(\Omega)} \|\partial_1^2 u\|_{W_2^\lambda(\Omega)}$$

implies that

$$\|\varphi_{1,7}\|_{L_2(\Omega^h)} \le C h^{s-2} \|a_1\|_{W_p^{s-2+\varepsilon}(\Omega)} \|u\|_{W_2^s(\Omega)}, \quad 5/2 < s \le 4. \tag{2.216}$$

The term $\varphi_{1,8}$ is bounded in the same way. Finally (2.213)–(2.216) and (2.211) yield the desired error bound (2.212). □

We note that for $2 < s \le 5/2$ the function $T_h^{22} f$ is not necessarily continuous on $\overline{\Omega}$ and therefore the right-hand side in the difference equation (2.207) is not defined for this range of values of the Sobolev index s. In fact, for $s \le 5/2$, the second-normal derivative of $u \in W_2^s(\Omega)$ does not have a trace on $\Gamma_{i0} \cup \Gamma_{i1}$ and therefore the boundary-value problem (2.203)–(2.206) is not meaningful as stated for this range of s.

Now let us consider the partial differential equation (2.203) subject to the homogeneous Dirichlet boundary conditions

$$\begin{aligned} u &= 0 \quad \text{on } \Gamma, \\ \partial_i u &= 0 \quad \text{on } \Gamma_{i0} \cup \Gamma_{i1},\ i = 1, 2. \end{aligned} \tag{2.217}$$

With the notational conventions from Sects. 2.2.4 and 2.7 equation (2.203) is again approximated by (2.207), and the boundary conditions (2.217) are discretized as follows:

$$\begin{aligned} U &= 0 \quad \text{on } \Gamma^h, \\ D_{x_i}^0 U &= 0 \quad \text{on } \Gamma_{i0}^h \cup \Gamma_{i1}^h,\ i = 1, 2. \end{aligned} \tag{2.218}$$

The error $e := u - U$ satisfies (2.209) and the boundary conditions

$$\begin{aligned} e &= 0 \quad \text{on } \Gamma^h, \\ D_{x_i}^0 e &= D_{x_i}^0 u \quad \text{on } \Gamma_{i0}^h \cup \Gamma_{i1}^h,\ i = 1, 2. \end{aligned} \tag{2.219}$$

Defining $\zeta_i = \zeta_i(x)$ by

$$\zeta_i := \left(D_{x_i}^0 u - \partial_i u\right)/h, \quad i = 1, 2,$$

the derivative boundary condition in (2.219) can be rewritten as

$$D_{x_i}^0 e = h\zeta_i, \quad x \in \Gamma_{i0}^h \cup \Gamma_{i1}^h,\ i = 1, 2.$$

Theorem 2.69 *The following bound holds on the global error $e := u - U$ between the analytical solution u and its finite difference approximation U:*

$$\|u-U\|_{W_2^2(\Omega^h)} \le Ch^{\min\{s-2,3/2\}}|\log h|^{1-|\mathrm{sgn}(s-7/2)|}$$
$$\times \max_i \|a_i\|_{W_p^{s-2+\varepsilon}(\Omega)}\|u\|_{W_2^s(\Omega)}, \quad 5/2<s<4. \qquad (2.220)$$

Proof We begin by noting that

$$\|e\|^2_{W_2^2(\Omega^h)} \le C\Bigg(\|\varphi_1\|^2_{L_2(\Omega_1^h\cup\Gamma_{11}^h)} + \|\varphi_2\|^2_{L_2(\Omega_2^h\cup\Gamma_{21}^h)} + \|\varphi_3\|^2_{L_2(\Omega_{00}^h)}$$
$$+\sum_{i=1}^{2} h^2 \sum_{x\in\Gamma_{i0}^h\cup\Gamma_{i1}^h} \zeta_i^2(x)\Bigg). \qquad (2.221)$$

The first three terms on the right-hand side of (2.221) are bounded in the same way as in the case of the boundary-value problem (2.203), (2.204) considered earlier. The only new ingredient in the analysis is the estimation of the last term in (2.221), which we discuss below.

When $s>2$, ζ_i represents a bounded linear functional of $u\in W_2^s(K^0)$, which vanishes on all polynomials of degree 2. By applying the Bramble–Hilbert lemma we obtain

$$\Bigg(h^2\sum_{x\in\Gamma_{i0}^h}\zeta_i^2(x)\Bigg)^{1/2} \le Ch^{s-2}|u|_{W_2^s(\Omega_{i0})}, \quad 2<s\le 3, \qquad (2.222)$$

where

$$\Omega_{i0}=\Omega_{hi}(0):=\{x: -h<x_i<h,\ 0<x_{3-i}<1\}.$$

By noting the boundary-layer estimate (2.199), we deduce from (2.222) that

$$\Bigg(h^2\sum_{x\in\Gamma_{i0}^h}\zeta_i^2\Bigg)^{1/2} \le Ch^{\min\{s-2,3/2\}}|\log h|^{1-|\mathrm{sgn}(s-7/2)|}\|u\|_{W_2^s(\Omega)}, \quad 2<s\le 4. \qquad (2.223)$$

For $x\in\Gamma_{i1}^h$ the terms ζ_i, $i=1,2$, are bounded analogously. From (2.221), (2.223) and our earlier bounds on φ_1,φ_2 and φ_3 we obtain the desired error bound (2.220) for the difference scheme (2.207), (2.218).

For $s<7/2$ the solution of (2.203), (2.217) has an even extension (i.e. an extension as an even function) across Γ that preserves the Sobolev class W_2^s. With such an even extension of u, $\zeta_i=0$ on $\Gamma_{i0}^h\cup\Gamma_{i1}^h$, and (2.220) is then a direct consequence of (2.207)–(2.216) and (2.221). □

Finally, we consider the partial differential equation (2.203) subject to the *natural boundary conditions*

$$
\begin{aligned}
&M_i(u)=0 \quad \text{and} \quad \partial_i M_i(u)+2\partial_{3-i}M_3(u)=0 \quad \text{on } \Gamma_{i0}\cup\Gamma_{i1},\ i=1,2;\\
&M_3(u)=0 \quad \text{on } \Gamma_*=\big\{(0,0),(0,1),(1,0),(1,1)\big\}.
\end{aligned}
\tag{2.224}
$$

The solution of problem (2.203), (2.224) is unique, up to the addition of a polynomial of degree 1. In order to ensure that we have a unique solution, we shall assume that, in addition to (2.224), the values of u at three vertices of Ω have been fixed; that is,

$$
u(0,0)=c_{00}, \qquad u(0,1)=c_{01}, \qquad u(1,0)=c_{10}. \tag{2.225}
$$

With the notational conventions from Sects. 2.2.4 and 2.7, the conditions (2.224), (2.225) are approximated by

$$
\begin{aligned}
m_i(U)=0,\quad &D^0_{x_i}m_i(U)+D^-_{x_{3-i}}\big[m_3(U)+m_3(U)^{-i}\big]=0,\\
&\text{on } \overline{\Gamma}^h_{i0}\cup\overline{\Gamma}^h_{i1},\ i=1,2;
\end{aligned}
\tag{2.226}
$$

$$
\begin{aligned}
&m_3(U)+m_3(U)^{-1}+m_3(U)^{-2}+m_3(U)^{-1,-2}=0 \quad \text{on}\,\Gamma_*;\\
&U(0,0)=c_{00}, \qquad U(0,1)=c_{01}, \qquad U(1,0)=c_{10},
\end{aligned}
\tag{2.227}
$$

where $\overline{\Gamma}^h_{ik}:=\overline{\Gamma}_{ik}\cap\Gamma^h$. Let us observe that the difference scheme also involves points exterior to $\overline{\Omega}$ that are at a distance $\le 2h$ from Γ; therefore (2.203), (2.226), (2.227) has fewer equations than unknowns. In order to account for the missing equations, we also discretize the partial differential equation at the boundary mesh-points. Let us introduce the asymmetric mollifiers

$$
T_i^{2\pm}f:=2\int_0^1(1-t)f(x\pm the_i)\,dt, \quad i=1,2,
$$

and the additional equations

$$
\mathcal{L}_hU=\begin{cases}
T_i^{2+}T_{3-i}^2f & \text{for } x\in\Gamma^h_{i0}\\
T_i^{2-}T_{3-i}^2f & \text{for } x\in\Gamma^h_{i1},\\
T_1^{2+}T_2^{2+}f & \text{for } x=(0,0),\\
\text{and analogously} & \text{for } x=(0,1),(1,0),(1,1).
\end{cases}
\tag{2.228}
$$

Theorem 2.70 *The difference scheme* (2.203), (2.226), (2.228) *satisfies the error bound*

$$
\begin{aligned}
[\![u-U]\!]_{W_2^2(\Omega^h)} &\le Ch^{\min\{s-2,3/2\}}|\log h|^{1-|\mathrm{sgn}(s-7/2)|}\\
&\qquad\times\max_i\|a_i\|_{W_p^{s-2+\varepsilon}(\Omega)}\|u\|_{W_2^s(\Omega)}, \quad 3<s\le 4,
\end{aligned}
$$

where C is a positive constant, independent of h.

Proof The global error $e := u - U$ satisfies the inequality

$$
\begin{aligned}
[\![e]\!]^2_{W_2^2(\Omega^h)} \le C\big([\![\varphi_1]\!]^2_{L_2(\Omega^h)} + [\![\varphi_2]\!]^2_{L_2(\Omega^h)} + \|\varphi_3\|^2_{L_2(\Omega^h_{00})} \\
+ [\![\phi_1]\!]^2_{L_2(\Omega^h)} + [\![\phi_2]\!]^2_{L_2(\Omega^h)}\big),
\end{aligned} \tag{2.229}
$$

where, for $i = 1, 2$,

$$
\phi_i := \begin{cases} T^2_{3-i} M_i(u) - T^{2+}_{3-i} M_i(u) & \text{on } \overline{\Gamma}^h_{i0}, \\ T^2_{3-i} M_i(u) - T^{2-}_{3-i} M_i(u) & \text{on } \overline{\Gamma}^h_{i1}, \\ 0 \quad \text{at the remaining mesh-points.} \end{cases}
$$

The terms φ_1, φ_2 and φ_3 are estimated in the same way as before. Finally, ϕ_i is a bounded linear functional of $M_i(u) \in W_2^\lambda(\Omega)$, $\lambda > 1/2$, which vanishes on all constant functions. Using the Bramble–Hilbert lemma and the boundary-layer estimate (2.199) we obtain

$$
\begin{aligned}
[\![\phi_i]\!]_{L_2(\Omega^h)} \le C h^{\min\{s-2,3/2\}} |\log h|^{1-|\mathrm{sgn}(s-7/2)|} \\
\times \max_j \|a_j\|_{W_p^{s-2+\varepsilon}(\Omega)} \|u\|_{W_2^s(\Omega)}, \quad 3 < s \le 4.
\end{aligned} \tag{2.230}
$$

The desired error bound follows from (2.229), (2.230) and our earlier bounds on the terms φ_1, φ_2 and φ_3. □

2.8 An Elliptic Interface Problem

The technique of convergence analysis introduced in earlier sections of this chapter can be extended to finite difference schemes for more general boundary-value problems. As an example, we consider here a model partial differential equation with a singular coefficient. Problems of the kind discussed here are usually referred to as *interface problems* or *transmission problems*. For further details we refer the reader to Jovanović and Vulkov [101].

Let $\Omega = (0, 1)^2$ and $\Gamma = \partial\Omega$. A typical point in $\overline{\Omega}$ will be denoted by $x = (x_1, x_2)$. Let further Σ be the intersection of the line segment $x_2 = \xi$, $0 < \xi < 1$, with $\overline{\Omega}$. We consider the Dirichlet boundary-value problem

$$
\mathcal{L}u + k(x)\delta_\Sigma(x)u = f(x) \quad \text{in } \Omega, \qquad u = 0 \quad \text{on } \Gamma, \tag{2.231}
$$

where $\delta_\Sigma(x) = \delta(x_2 - \xi)$ is the Dirac distribution concentrated on Σ, $k(x) = k(x_1)$ and $\mathcal{L}$ is the symmetric elliptic operator introduced in (2.166); i.e.

$$
\mathcal{L}u := -\sum_{i,j=1}^{2} \partial_i(a_{ij}\partial_j u) + au.
$$

The Dirac distribution δ_Σ belongs to the Sobolev space $W_2^{-\lambda}(\Omega)$, with $\lambda > 1/2$. Equation (2.231) must be therefore understood in a weak sense: we seek $u \in \mathring{W}_2^1(\Omega)$

such that

$$\langle \mathcal{L}u, v\rangle + (k\delta_\Sigma)(uv) = \langle f, v\rangle \quad \forall v \in \mathring{W}_2^1(\Omega), \tag{2.232}$$

where $\langle f, v\rangle$ denotes the duality pairing between the spaces $W_2^{-1}(\Omega)$ and $\mathring{W}_2^1(\Omega)$, and

$$(k\delta_\Sigma)(w) := \int_\Sigma kw\Big|_\Sigma \, \mathrm{d}\Sigma, \quad w \in W_1^1(\Omega),$$

where $w|_\Sigma \in L_1(\Sigma)$ denotes the trace of $w \in W_1^1(\Omega)$ on Σ, and $k \in L_\infty(\Sigma)$.

Alternatively, problem (2.232) can be restated as follows: *find* $u \in \mathring{W}_2^1(\Omega)$ *such that*

$$a(u, v) = \langle f, v\rangle \quad \forall v \in \mathring{W}_2^1(\Omega), \tag{2.233}$$

where

$$a(u, v) = \int_\Omega \left(\sum_{i,j=1}^{2} a_{ij}\partial_j u \partial_i v + auv \right) \mathrm{d}x + \int_\Sigma k(uv)\Big|_\Sigma \, \mathrm{d}\Sigma. \tag{2.234}$$

Thus, (2.233) can be seen as the weak formulation of the boundary-value problem (2.231). A relevant point in this respect is that for the domain $\Omega = (0, 1)^2 \subset \mathbb{R}^2$ the product uv of $u, v \in \mathring{W}_2^1(\Omega)$ belongs to $\mathring{W}_p^1(\Omega)$ for all $p \in [1, 2)$ and thus by Theorem 1.42 (see also Theorem 1.5.1.3 on p. 38 of Grisvard [62] for $p \in (1, 2)$ and Theorem 2.10 on p. 37 of Giusti [54] for $p = 1$), the boundary integral term in (2.234) is meaningful. The following assertion concerning the existence of a unique weak solution is an immediate consequence of the Lax–Milgram theorem and the trace theorem for $W_2^1(\Omega)$.

Lemma 2.71 *Suppose that*

$$f \in W_2^{-1}(\Omega), \quad a_{ij}, a \in L_\infty(\Omega), \quad k \in L_\infty(\Sigma), \quad a_{ij} = a_{ji}, \quad a \geq 0, \quad k \geq 0,$$

$$\exists c_0 > 0 \ \forall \xi = (\xi_1, \xi_2) \in \mathbb{R}^2 \ \forall x \in \Omega : \quad \sum_{i,j=1}^{2} a_{ij}(x)\xi_i\xi_j \geq c_0 \sum_{i=1}^{2} \xi_i^2.$$

Then, there exists a unique weak solution $u \in \mathring{W}_2^1(\Omega)$ *to the boundary-value problem* (2.233), (2.234), *and*

$$\|u\|_{W_2^1(\Omega)} \leq C\|f\|_{W_2^{-1}(\Omega)}.$$

Let us now assume that the coefficients a_{ij}, $i, j = 1, 2$, and a of the differential operator $\mathcal{L}$ belong to the Hölder space $C^{0,\lambda}(\overline{\Omega})$, with $\lambda > |\theta|$ and $|\theta| < 1/2$. The bilinear functional $a(\cdot, \cdot)$ can then be continuously extended to $\mathring{W}_2^{1-\theta}(\Omega) \times \mathring{W}_2^{1+\theta}(\Omega)$. The following assertion can be proved by applying Theorem 3.3 in Nečas [143].

Lemma 2.72 *Suppose that*

$$f \in W_2^{\theta-1}(\Omega), \quad |\theta| < 1/2, \quad a_{ij}, a \in C^{0,\lambda}(\overline{\Omega}), \quad \lambda > |\theta|, \quad k \in L_\infty(\Sigma),$$
$$a_{ij} = a_{ji}, \quad a \geq 0, \quad k \geq 0,$$

$$\exists c_0 > 0\ \forall \xi = (\xi_1, \xi_2) \in \mathbb{R}^2\ \forall x \in \Omega : \quad \sum_{i,j=1}^{2} a_{ij}(x)\xi_i\xi_j \geq c_0 \sum_{i=1}^{2} \xi_i^2.$$

Then, there exists a unique solution $u \in \mathring{W}_2^{1+\theta}(\Omega)$ *to the boundary-value problem* (2.233), (2.234).

In the case when f does not contain a concentrated singularity on Σ, such as δ_Σ, problem (2.233), (2.234) can be shown to be the weak formulation of the following boundary-value problem with transmission (conjugation) conditions on the interface Σ:

$$\begin{aligned} &\mathcal{L}u = f \quad \text{in } \Omega^- \cup \Omega^+, \qquad u = 0 \quad \text{on } \Gamma, \\ &[u]_\Sigma = 0, \quad \left[\sum_{j=1}^{2} a_{2j}\partial_j u\right]_\Sigma = ku|_\Sigma, \end{aligned} \tag{2.235}$$

where $\Omega^- := (0,1) \times (0,\xi)$, $\Omega^+ := (0,1) \times (\xi,1)$, and

$$[u]_\Sigma := u(x_1, \xi + 0) - u(x_1, \xi - 0).$$

In this sense, the boundary-value problems (2.231) and (2.235) are equivalent.

Higher regularity of the solution can be proved under additional assumptions on the data. For $s \geq 2$ we define the subspace $\widehat{W}_2^s(\Omega)$ of $\mathring{W}_2^1(\Omega)$, consisting of all $u \in \mathring{W}_2^1(\Omega)$ such that

$$\begin{aligned} \partial_1^i u &\in L_2(\Omega), \quad i = 0, 1, \ldots, s, \\ \partial_1^{i-1}\partial_2 u &\in L_2(\Omega), \quad i = 1, 2, \ldots, s, \\ \partial_1^{i-j}\partial_2^j u &\in L_2(\Omega^-) \cap L_2(\Omega^+), \quad i = j, j+1, \ldots, s, \ j = 2, 3, \ldots, s, \end{aligned}$$

with the norm $\|\cdot\|_{\widehat{W}_2^s(\Omega)}$ defined by

$$\begin{aligned} \|u\|^2_{\widehat{W}_2^s(\Omega)} &:= \sum_{i=0}^{s} \|\partial_1^i u\|^2_{L_2(\Omega)} + \sum_{i=1}^{s} \|\partial_1^{i-1}\partial_2 u\|^2_{L_2(\Omega)} \\ &+ \sum_{j=2}^{s}\sum_{i=j}^{s} \left(\|\partial_1^{i-j}\partial_2^j u\|^2_{L_2(\Omega^-)} + \|\partial_1^{i-j}\partial_2^j u\|^2_{L_2(\Omega^+)}\right). \end{aligned}$$

Obviously,

$$\widehat{W}_2^s(\Omega) \subset \widetilde{W}_2^s(\Omega) := \mathring{W}_2^1(\Omega) \cap W_2^s(\Omega^-) \cap W_2^s(\Omega^+).$$

Lemma 2.73 *Suppose that in addition to the assumptions of Lemma* 2.71 *we have that*

$$f \in L_2(\Omega), \qquad a_{ij} \in W_\infty^1(\Omega), \qquad k \in W_\infty^1(\Sigma);$$

then, $u \in \widehat{W}_2^2(\Omega)$. *If, in addition,*

$$\partial_1 f \in L_2(\Omega), \qquad \partial_2 f \in L_2(\Omega^\pm), \qquad a_{ij} \in W_\infty^2(\Omega),$$
$$a \in W_\infty^1(\Omega), \qquad k \in W_\infty^2(\Sigma)$$

and

$$f = a_{12} = \partial_1 a_{11} = 0 \quad \textit{for } x_1 = 0 \textit{ and } x_1 = 1,$$

then $u \in \widehat{W}_2^3(\Omega)$.

Proof For $x \in \Omega^- \cup \Omega^+$ (2.235) can be written as

$$a_{11}\partial_1^2 u + 2a_{12}\partial_1\partial_2 u + a_{22}\partial_2^2 u = -\sum_{i,j=1}^{2} \partial_i a_{ij}\partial_j u + au - f. \tag{2.236}$$

Multiplying (2.236) by $\partial_1^2 u$, integrating over Ω and performing partial integration we obtain

$$\int_\Omega \left[a_{11}\left(\partial_1^2 u\right)^2 + 2a_{12}\partial_1^2 u\,\partial_1\partial_2 u + a_{22}(\partial_1\partial_2 u)^2\right] \mathrm{d}x + \int_\Sigma k(\partial_1 u)^2\Big|_\Sigma \,\mathrm{d}\Sigma$$
$$= I_1 + I_2 + I_3,$$

where

$$I_1 := -\int_\Omega \left(\sum_{i,j=1}^{2} \partial_i a_{ij}\partial_j u - au + f\right)\partial_1^2 u \,\mathrm{d}x,$$

$$I_2 := \int_\Omega \left(\partial_2 a_{22}\partial_2 u\,\partial_1^2 u - \partial_1 a_{22}\partial_2 u\,\partial_1\partial_2 u\right) \mathrm{d}x,$$

$$I_3 := -\int_\Sigma k' u\,\partial_1 u \,\mathrm{d}\Sigma.$$

Further,

$$\int_\Omega \left[a_{11}\left(\partial_1^2 u\right)^2 + 2a_{12}\partial_1^2 u\,\partial_1\partial_2 u + a_{22}(\partial_1\partial_2 u)^2\right] \mathrm{d}x + \int_\Sigma k(\partial_1 u)^2 \,\mathrm{d}\Sigma$$
$$\ge c_0\left(\left\|\partial_1^2 u\right\|_{L_2(\Omega)}^2 + \|\partial_1\partial_2 u\|_{L_2(\Omega)}^2\right).$$

The integrals I_1, I_2 and I_3 can be bounded by applying the Cauchy–Schwarz inequality with $\varepsilon \in (0,1)$ as follows:

$$|I_1| \le \varepsilon \|\partial_1^2 u\|^2_{L_2(\Omega)} + \frac{C}{\varepsilon}\big(\|u\|^2_{W_2^1(\Omega)} + \|f\|^2_{L_2(\Omega)}\big).$$

Similarly,

$$|I_2| \le \varepsilon\big(\|\partial_1^2 u\|^2_{L_2(\Omega)} + \|\partial_1\partial_2 u\|^2_{L_2(\Omega)}\big) + \frac{C}{\varepsilon}\|\partial_2 u\|^2_{L_2(\Omega)}$$

and

$$\begin{aligned} |I_3| &\le \varepsilon\|\partial_1 u\|^2_{L_2(\Sigma)} + \frac{C}{\varepsilon}\|u\|^2_{L_2(\Sigma)} \\ &\le C_1\varepsilon\big(\|\partial_1^2 u\|^2_{L_2(\Omega)} + \|\partial_1\partial_2 u\|^2_{L_2(\Omega)}\big) + \frac{C}{\varepsilon}\|u\|^2_{W_2^1(\Omega)}. \end{aligned}$$

Hence, by selecting a sufficiently small $\varepsilon > 0$, we obtain the bound

$$\|\partial_1^2 u\|^2_{L_2(\Omega)} + \|\partial_1\partial_2 u\|^2_{L_2(\Omega)} \le C\|f\|^2_{L_2(\Omega)}.$$

From (2.236) we immediately have that

$$\|\partial_2^2 u\|_{L_2(\Omega^\pm)} \le C\big(\|\partial_1^2 u\|_{L_2(\Omega)} + \|\partial_1\partial_2 u\|_{L_2(\Omega)} + \|u\|_{W_2^1(\Omega)} + \|f\|_{L_2(\Omega)}\big),$$

which proves the first part of the lemma.

When the assumptions of the second part of the lemma are satisfied, we deduce from (2.231) that

$$\partial_1^2 u = 0 \quad \text{on } \Gamma.$$

By differentiating (2.231) one obtains

$$\mathcal{L}\partial_1^2 u + k(x)\delta_\Sigma(x)\partial_1^2 u = f_1(x), \quad x \in \Omega,$$

where

$$\begin{aligned} f_1 := \partial_1^2 f + \sum_{i,j=1}^{2} \partial_i\big(2\partial_1 a_{ij}\partial_1\partial_j u + \partial_1^2 a_{ij}\partial_j u\big) \\ - 2\partial_1 a\partial_1 u - \partial_1^2 a u - 2k'\delta_\Sigma\partial_1 u - k''\delta_\Sigma u \in W_2^{-1}(\Omega). \end{aligned}$$

By applying Lemma 2.71 we then deduce the regularity result stated in the second part of the lemma. □

For further details regarding the analysis of elliptic boundary-value problems in domains with corners we refer to Grisvard [62] and Dauge [28].

2.8.1 Finite Difference Approximation

In the sequel we shall assume that the weak solution of the boundary-value problem (2.231) belongs to $\widetilde{W}_2^s(\Omega)$, $s > 2$, and that the coefficients of the equation satisfy the following regularity hypotheses:

$$a_{ij} \in W_2^{s-1}(\Omega^-) \cap W_2^{s-1}(\Omega^+) \cap C(\overline{\Omega}), \qquad a \in W_2^{s-2}(\Omega^-) \cap W_2^{s-2}(\Omega^+)$$

and

$$k \in W_2^{s-1}(\Sigma).$$

We define

$$\|u\|_{\widetilde{W}_2^s(\Omega)} := \left(\|u\|^2_{W_2^1(\Omega)} + \|u\|^2_{W_2^s(\Omega^-)} + \|u\|^2_{W_2^s(\Omega^+)}\right)^{1/2}.$$

In particular, for $s = 0$ we set

$$\|u\|_{\widetilde{W}_2^0(\Omega)} = \|u\|_{\widetilde{L}_2(\Omega)} := \left(\|u\|^2_{L_2(\Omega)} + \|u\|^2_{L_2(\Sigma)}\right)^{1/2}.$$

For the sake of simplicity we shall also assume that ξ is a rational number. Let $\overline{\Omega}^h$ be a uniform square mesh on $\overline{\Omega}$ with mesh-size $h := 1/N$, where N is an integer such that ξN is also an integer. We shall use the notations from Sect. 2.2 and define

$$\Sigma^h := \Omega^h \cap \Sigma \quad \text{and} \quad \Sigma^h_- := \Sigma^h \cup \{(0, \xi)\}.$$

Let us approximate the boundary-value problem (2.231) on the mesh $\overline{\Omega}^h$ by the following finite difference scheme with mollified right-hand side:

$$\mathcal{L}_h U + k\delta_{\Sigma^h} U = T_1^2 T_2^2 f \quad \text{in } \Omega^h, \qquad U = 0 \quad \text{on } \Gamma^h, \tag{2.237}$$

where

$$\mathcal{L}_h U := -\frac{1}{2} \sum_{i,j=1}^{2} \left[D_{x_i}^+ \left(a_{ij} D_{x_j}^- U\right) + D_{x_i}^- \left(a_{ij} D_{x_j}^+ U\right)\right] + \left(T_1^2 T_2^2 a\right) U$$

and

$$\delta_{\Sigma^h}(x) = \delta_h(x_2 - \xi) := \begin{cases} 0 & \text{for } x \in \Omega^h \setminus \Sigma^h, \\ 1/h & \text{for } x \in \Sigma^h \end{cases}$$

is the discrete Dirac delta-function.

Further, we define the asymmetric mollifiers T_2^{2-} and T_2^{2+} by

$$T_2^{2-} f(x_1, x_2) := \frac{2}{h} \int_{x_2-h}^{x_2} \left(1 + \frac{x_2' - x_2}{h}\right) f(x_1, x_2')\, dx_2',$$

$$T_2^{2+} f(x_1, x_2) := \frac{2}{h} \int_{x_2}^{x_2+h} \left(1 - \frac{x_2' - x_2}{h}\right) f(x_1, x_2')\, \mathrm{d}x_2'.$$

In addition to the discrete inner products and norms defined in Sect. 2.6.1 we introduce

$$(U, V)_{\Sigma^h} := h \sum_{x \in \Sigma^h} U(x)V(x), \qquad \|U\|_{L_2(\Sigma^h)} := (U, U)_{\Sigma^h}^{1/2},$$

$$|U|_{W_2^{1/2}(\Sigma^h)} := \left(h^2 \sum_{x \in \Sigma_-^h} \sum_{x' \in \Sigma_-^h, x' \neq x} \frac{|U(x) - U(x')|^2}{|x_1 - x_1'|^2}\right)^{1/2}.$$

The following lemma holds.

Lemma 2.74 *Let $U \in S_0^h$ and let V be a mesh-function defined on Σ_-^h. Then,*

$$\left|\left(D_{x_1}^- V, U\right)_{\Sigma^h}\right| \leq C \|U\|_{W_2^1(\Omega^h)} |V|_{W_2^{1/2}(\Sigma^h)}.$$

Proof Similarly as in the proof of Lemma 2 in Jovanović and Popović [92], we expand U and V in the following Fourier sums:

$$U(x_1, x_2) = \sum_{k=1}^{N-1} \sum_{l=1}^{N-1} b_{kl} \sin k\pi x_1 \sin l\pi x_2 = \sum_{k=1}^{N-1} B_k(x_2) \sin k\pi x_1, \tag{2.238}$$

$$V(x_1) = \sum_{k=1}^{N-1} a_k \cos k\pi \left(x_1 + \frac{h}{2}\right). \tag{2.239}$$

Hence we have that

$$D_{x_1}^- V(x_1) = -\sum_{k=1}^{N-1} \sqrt{\lambda_k} a_k \sin k\pi x_1, \quad \text{where } \lambda_k := \frac{4}{h^2} \sin^2 \frac{k\pi h}{2}.$$

Using the orthogonality of sine functions, we deduce that

$$\begin{aligned}\left(D_{x_1}^- V, U\right)_{\Sigma^h} &= -\frac{1}{2} \sum_{k=1}^{N-1} \sqrt{\lambda_k} a_k B_k(x_2) \\ &\leq \left(\frac{1}{2} \sum_{k=1}^{N-1} \sqrt{\lambda_k} a_k^2\right)^{1/2} \left(\frac{1}{2} \sum_{k=1}^{N-1} \sqrt{\lambda_k} B_k^2(x_2)\right)^{1/2}. \end{aligned} \tag{2.240}$$

Let us consider the following sum (over mesh-points):

$$N^2(V) := h^2 \sum_{x_1, t_1 = -1, t_1 \neq 0}^{1-h} \left(\frac{V(x_1) - V(x_1 - t_1)}{t}\right)^2, \tag{2.241}$$

where the mesh-function V has been extended outside Σ_-^h by (2.239). Using the periodicity and orthogonality of cosine functions, we then deduce that

$$N^2(V) = h^2 \sum_{x_1=-1}^{1-h} \sum_{0\neq t_1=-1}^{1-h} \frac{-V(x_1+t_1)+2V(x_1)-V(x_1-t_1)}{t_1^2} V(x_1)$$
$$= 4\sum_{k=1}^{N-1} \sqrt{\lambda_k} a_k^2 I_k,$$

where

$$I_k := \frac{\frac{k\pi h}{2}}{\sin\frac{k\pi h}{2}} J_k, \quad J_k := \frac{k\pi h}{2} \sum_{t_1=h}^{1-h} \left(\frac{\sin\frac{k\pi t_1}{2}}{\frac{k\pi t_1}{2}}\right)^2.$$

We note that

$$1 \leq \frac{\frac{k\pi h}{2}}{\sin\frac{k\pi h}{2}} \leq \frac{\pi}{2}$$

and that J_k is a Riemann sum for $\int_0^{k\pi/2} \left(\frac{\sin\tau}{\tau}\right)^2 \mathrm{d}\tau$, which therefore satisfies the following two-sided bound:

$$\frac{1}{\pi} \leq J_k \leq \frac{\pi}{2} + \frac{2}{\pi}.$$

Hence,

$$\frac{4}{\pi} \sum_{k=1}^{N-1} \sqrt{\lambda_k} a_k^2 \leq N^2(V) \leq (\pi^2+4) \sum_{k=1}^{N-1} \sqrt{\lambda_k} a_k^2.$$

From (2.241), using the periodicity of the cosine function, we also have that

$$N^2(V) = h^2 \sum_{x_1, x_1'=-1, x_1\neq x_1'}^{1-h} \left(\frac{V(x_1)-V(x_1')}{x_1-x_1'}\right)^2 \leq 4|V|^2_{W_2^{1/2}(\Sigma^h)},$$

whereby

$$\sum_{k=1}^{N-1} \sqrt{\lambda_k} a_k^2 \leq \pi |V|^2_{W_2^{1/2}(\Sigma^h)}. \tag{2.242}$$

On the other hand, since $B_k(0)=0$, we obtain

$$B_k^2(x_2) = h\sum_{x_2'=0}^{x_2-h} D_{x_2}^+\left(B_k^2(x_2')\right) = h\sum_{x_2'=0}^{x_2-h} \left(D_{x_2}^+ B_k(x_2')\right)\left(B_k(x_2'+h)+B_k(x_2')\right)$$

$$\leq \varepsilon_k h \sum_{x_2'=h}^{1-h} B_k^2(x_2') + \frac{1}{\varepsilon_k} h \sum_{x_2'=0}^{1-h} \left(D_{x_2}^+ B_k(x_2')\right)^2,$$

with $\varepsilon_k > 0$, $k = 1, \ldots, N-1$, to be chosen.

Selecting $\varepsilon_k = \sqrt{\lambda_k}$ for $k = 1, \ldots, N-1$, and using the discrete Parseval identities (2.21) and (2.22), we have that

$$\frac{1}{2} \sum_{k=1}^{N-1} \sqrt{\lambda_k} B_k^2(x_2) \leq \left\| D_{x_1}^+ U \right\|_{L_2(\Omega_1^h)}^2 + \left\| D_{x_2}^+ U \right\|_{L_2(\Omega_2^h)}^2 \leq \| U \|_{W_2^1(\Omega^h)}^2. \quad (2.243)$$

Finally, the assertion follows from the inequalities (2.240), (2.242) and (2.243) with $C = \sqrt{\pi/2}$. That completes the proof. □

2.8.2 Convergence in the Discrete W_2^1 Norm

Let u be the solution of the boundary-value problem (2.231) and let U denote the solution of the finite difference scheme (2.237). The global error $e := u - U$ then satisfies the finite difference scheme

$$\mathcal{L}_h e + k\delta_{\Sigma^h} e = \varphi \text{ in } \Omega^h, \quad e = 0 \text{ on } \Gamma^h, \quad (2.244)$$

where

$$\varphi := \sum_{i,j=1}^{2} D_{x_i}^- \eta_{ij} + \eta + \delta_{\Sigma^h} \mu,$$

$$\eta_{ij} := T_i^+ T_{3-i}^2 (a_{ij} \partial_j u) - \frac{1}{2} \left(a_{ij} D_{x_j}^+ u + a_{ij}^{+i} D_{x_j}^- u^{+i}\right),$$

$$\eta := (T_1^2 T_2^2 a) u - T_1^2 T_2^2 (au),$$

$$\mu := ku - T_1^2 (ku).$$

Let us decompose η_{1j} and η as follows:

$$\eta_{1j} = \tilde{\eta}_{1j} + \delta_{\Sigma^h} \hat{\eta}_{1j} \quad \text{and} \quad \eta = \tilde{\eta} + \delta_{\Sigma^h} \hat{\eta},$$

where

$$\hat{\eta}_{11} := \frac{h^2}{6} T_1^+ \left([a_{11} \partial_1 \partial_2 u + \partial_2 a_{11} \partial_1 u]_\Sigma\right),$$

$$\hat{\eta}_{12} := \frac{h^2}{6} T_1^+ \left(\left[a_{12} \partial_2^2 u + \partial_2 a_{12} \partial_2 u\right]_\Sigma\right) - \frac{h^2}{4} T_1^+ \left(\left[\partial_1 (a_{12} \partial_2 u)\right]_\Sigma\right),$$

$$\hat{\eta} := -\frac{h^2}{6}\big[(T_1^2 a)(T_1^2 \partial_2 u)\big]_\Sigma.$$

By performing summations by parts and applying Lemma 2.74 we deduce the following bound:

$$\|z\|_{W_2^1(\Omega^h)} \le C\Bigg[\sum_{j=1}^{2}\Big(\|\eta_{2j}\|_{L_2(\Omega_2^h)} + \|\tilde{\eta}_{1j}\|_{L_2(\Omega_1^h)} + |\hat{\eta}_{1j}|_{W_2^{1/2}(\Sigma^h)}\Big)$$
$$+ \|\tilde{\eta}\|_{L_2(\Omega^h)} + \|\hat{\eta}\|_{L_2(\Sigma^h)} + \|\mu\|_{L_2(\Sigma^h)}\Bigg]. \tag{2.245}$$

Hence, in order to estimate the convergence rate of the finite difference scheme (2.244), it suffices to bound the terms on the right-hand side of (2.245).

The terms η_{2j}, $j = 1, 2$, have been bounded in Sect. 2.6.1. After summation over the mesh Ω_2^h we obtain

$$\|\eta_{2j}\|_{L_2(\Omega_2^h)} \le Ch^{s-1}\Big(\|a_{2j}\|_{W_2^{s-1}(\Omega^-)}\|u\|_{W_2^s(\Omega^-)}$$
$$+ \|a_{2j}\|_{W_2^{s-1}(\Omega^+)}\|u\|_{W_2^s(\Omega^+)}\Big), \quad 2 < s \le 3. \tag{2.246}$$

The terms $\tilde{\eta}_{1j}$ for $x \in \Omega_1^h \setminus \Sigma_-^h$ can be bounded in the same way. For $x \in \Sigma_-^h$ we set

$$\tilde{\eta}_{11} := \sum_{k=1}^{3}\big(\eta_{11,k}^- + \eta_{11,k}^+\big),$$

$$\tilde{\eta}_{12} := \sum_{k=1}^{4}\big(\eta_{12,k}^- + \eta_{12,k}^+\big),$$

where

$$\eta_{11,1}^\pm := \frac{1}{2}T_1^+T_2^{2\pm}(a_{11}\partial_1 u) - \frac{1}{2}\big(T_1^+T_2^{2\pm}a_{11}\big)\big(T_1^+T_2^{2\pm}\partial_1 u\big)$$
$$\pm\frac{h}{6}\big(T_1^+\partial_2 a_{11}\big)\big[\big(T_1^+T_2^{2\pm}\partial_1 u\big) - \big(T_1^+\partial_1 u\big)\big]$$
$$\pm\frac{h}{6}\bigg[\frac{a_{11}+a_{11}^{+1}}{2}\big(T_1^+\partial_1\partial_2 u\big) - T_1^+(a_{11}\partial_1\partial_2 u)\bigg]$$
$$\pm\frac{h}{6}\big[\big(T_1^+\partial_2 a_{11}\big)\big(T_1^+\partial_1 u\big) - T_1^+(\partial_2 a_{11}\partial_1 u)\big]\big|_{x_2=\xi\pm 0},$$

$$\eta_{11,2}^\pm := \frac{1}{2}\bigg[\big(T_1^+T_2^{2\pm}a_{11}\big) - \frac{a_{11}+a_{11}^{+1}}{2} \mp \frac{h}{3}\big(T_1^+\partial_2 a_{11}\big)\bigg]\big(T_1^+T_2^{2\pm}\partial_1 u\big)\big|_{x_2=\xi\pm 0},$$

$$\eta^{\pm}_{11,3} := \frac{a_{11}+a^{+1}_{11}}{4}\Big[\big(T_1^+T_2^{2\pm}\partial_1 u\big) - u_{x_1} \mp \frac{h}{3}\big(T_1^+\partial_1\partial_2 u\big)\Big]\Big|_{x_2=\xi\pm 0},$$

$$\eta^{\pm}_{12,1} := \frac{1}{2}T_1^+T_2^{2\pm}(a_{12}\partial_2 u) - \frac{1}{2}\big(T_1^+T_2^{2\pm}a_{12}\big)\big(T_1^+T_2^{2\pm}\partial_2 u\big)$$
$$\pm\frac{h}{6}\big(T_1^+\partial_2 a_{12}\big)\big[\big(T_1^+T_2^{2\pm}\partial_2 u\big) - \big(T_1^+\partial_2 u\big)\big]$$
$$\pm\frac{h}{6}\Big[\frac{a_{12}+a^{+1}_{12}}{2}\big(T_1^+\partial_2^2 u\big) - T_1^+\big(a_{12}\partial_2^2 u\big)\Big]$$
$$\pm\frac{h}{6}\big[\big(T_1^+\partial_2 a_{12}\big)\big(T_1^+\partial_2 u\big) - T_1^+(\partial_2 a_{12}\partial_2 u)\big]$$
$$\pm\frac{h}{4}T_1^+\big(\partial_1 a_{12}\big(T_2^{\pm}\partial_2 u - \partial_2 u\big)\big)\big|_{x_2=\xi\pm 0},$$

$$\eta^{\pm}_{12,2} := \frac{1}{2}\Big[\big(T_1^+T_2^{2\pm}a_{12}\big) - \frac{a_{12}+a^{+1}_{12}}{2} \mp \frac{h}{3}\big(T_1^+\partial_2 a_{12}\big)\Big]\big(T_1^+T_2^{2\pm}\partial_2 u\big)\big|_{x_2=\xi\pm 0},$$

$$\eta^{+}_{12,3} := \frac{a_{12}+a^{+1}_{12}}{4}\Big[\big(T_1^+T_2^{2+}\partial_2 u\big) - \frac{D^+_{x_2}u + D^+_{x_2}u^{+1}}{2} - \frac{h}{3}\big(T_1^+\partial_2^2 u\big)\Big]$$
$$+\frac{h}{4}T_1^+\big(a_{12}\big(T_2^+\partial_1\partial_2 u - \partial_1\partial_2 u\big)\big)\big|_{x_2=\xi+0},$$

$$\eta^{-}_{12,3} := \frac{a_{12}+a^{+1}_{12}}{4}\Big[\big(T_1^+T_2^{2-}\partial_2 u\big) - \frac{D^+_{x_2}u + D^+_{x_2}u^{+1}}{2} - \frac{h}{3}\big(T_1^+\partial_2^2 u\big)\Big]$$
$$+\frac{h}{4}T_1^+\big(a_{12}\big(T_2^-\partial_1\partial_2 u - \partial_1\partial_2 u\big)\big)\big|_{x_2=\xi-0},$$

$$\eta^{+}_{12,4} := -\frac{1}{8}\big(a^{+1}_{12} - a_{12}\big)\big(D^+_{x_2}u^{+1} - u_{x_2}\big)\big|_{x_2=\xi+0},$$

$$\eta^{-}_{12,4} := -\frac{1}{8}\big(a^{+1}_{12} - a_{12}\big)\big(D^-_{x_2}u^{+1} - u_{\bar{x}_2}\big)\big|_{x_2=\xi-0}.$$

The terms $\eta^{\pm}_{1j,k}$ can be bounded analogously to the corresponding terms $\eta_{1j,k}$ considered in Sect. 2.6.1. Thus we obtain:

$$\|\tilde{\eta}_{1j}\|_{L_2(\Omega_1^h)} \le Ch^{s-1}\big(\|a_{1j}\|_{W_2^{s-1}(\Omega^-)}\|u\|_{W_2^s(\Omega^-)}$$
$$+\|a_{1j}\|_{W_2^{s-1}(\Omega^+)}\|u\|_{W_2^s(\Omega^+)}\big), \quad 2.5 < s \le 3. \tag{2.247}$$

For $x \in \Omega^h \setminus \Sigma^h$, the term $\tilde{\eta}$ can be bounded in the same way as the corresponding term η in Sect. 2.6.1. For $x \in \Sigma^h$ we use the following decomposition:

$$\tilde{\eta} = \eta^+_{(1)} + \eta^-_{(1)} + \eta^+_{(2)} + \eta^-_{(2)},$$

where

$$\eta_{(1)}^{\pm} := \frac{1}{2}(T_1^2 T_2^{2\pm} a)\Big[u - (T_1^2 T_2^{2\pm} u) \pm \frac{h}{3}(T_1^2 \partial_2 u)\Big]\Big|_{x_2=\xi\pm 0},$$

$$\eta_{(2)}^{\pm} := \frac{1}{2}\Big[(T_1^2 T_2^{2\pm} a)(T_1^2 T_2^{2\pm} u) - T_1^2 T_2^{2\pm}(au)$$

$$\pm \frac{h}{6}\big((T_1^2 a) - (T_1^2 T_2^{2\pm} a)\big)(T_1^2 \partial_2 u)\Big]\Big|_{x_2=\xi\pm 0}.$$

These terms can be bounded analogously to the terms η_3 and η_4 discussed in Sect. 2.6.1. Hence we deduce that

$$\|\tilde{\eta}\|_{L_2(\Omega^h)} \le C h^{s-1}\big(\|a\|_{W_2^{s-2}(\Omega^-)}\|u\|_{W_2^s(\Omega^-)}$$

$$+ \|a\|_{W_2^{s-2}(\Omega^+)}\|u\|_{W_2^s(\Omega^+)}\big), \quad 2 < s \le 3. \tag{2.248}$$

The value of μ at the node $(x_1, \xi) \in \Sigma^h$ is a bounded linear functional of $ku \in W_2^{s-1}(\iota)$, $\iota = (x_1 - h, x_1 + h) \times \{\xi\}$, $s > 3/2$, which vanishes on all linear polynomials. Using the Bramble–Hilbert lemma one then obtains that

$$\|\mu\|_{L_2(\Sigma^h)} \le C h^{s-1}\|ku\|_{W_2^{s-1}(\Sigma)}$$

$$\le C h^{s-1}\|k\|_{W_2^{s-1}(\Sigma)}\big(\|u\|_{W_2^s(\Omega^-)} + \|u\|_{W_2^s(\Omega^+)}\big), \quad 1.5 < s \le 3. \tag{2.249}$$

The term $\hat{\eta}$ can be bounded directly:

$$\|\hat{\eta}\|_{L_2(\Sigma^h)} \le C h^2\big(\|a\|_{L_2(\Sigma^+)}\|\partial_2 u\|_{C(\overline{\Omega}^+)} + \|a\|_{L_2(\Sigma^-)}\|\partial_2 u\|_{C(\overline{\Omega}^-)}\big)$$

$$\le C h^2\big(\|a\|_{W_2^{s-2}(\Omega^+)}\|u\|_{W_2^s(\Omega^+)} + \|a\|_{W_2^{s-2}(\Omega^-)}\|u\|_{W_2^s(\Omega^-)}\big), \quad s > 2.5, \tag{2.250}$$

where we have used the following notation:

$$\|a\|_{L_2(\Sigma^\pm)} := \big\|a(\cdot, \xi \pm 0)\big\|_{L_2(0,1)}.$$

For a function $\varphi \in W_2^\lambda(\Sigma)$, $0 < \lambda \le 1/2$, the seminorm $|T_1^+ \varphi|_{W_2^{1/2}(\Sigma^h)}$ can be estimated directly:

$$\big|T_1^+ \varphi\big|_{W_2^{1/2}(\Sigma^h)} \le 2^{\lambda+1/2} h^{\lambda-1/2}|\varphi|_{W_2^\lambda(\Sigma)} \le C h^{\lambda-1/2}\|\varphi\|_{W_2^{\lambda+1/2}(\Omega^\pm)}.$$

We thus deduce that

$$
\begin{aligned}
|\hat{\eta}_{11}|_{W_2^{1/2}(\Sigma^h)} &\le Ch^{s-1}\big(\|a_{11}\partial_1\partial_2 u\|_{W_2^{s-2}(\Omega^+)} + \|a_{11}\partial_1\partial_2 u\|_{W_2^{s-2}(\Omega^-)} \\
&\quad + \|\partial_2 a_{11}\partial_1 u\|_{W_2^{s-2}(\Omega^+)} + \|\partial_2 a_{11}\partial_1 u\|_{W_2^{s-2}(\Omega^-)}\big) \\
&\le Ch^{s-1}\big(\|a_{11}\|_{W_2^{s-1}(\Omega^+)}\|u\|_{W_2^s(\Omega^+)} + \|a_{11}\|_{W_2^{s-1}(\Omega^-)}\|u\|_{W_2^s(\Omega^-)}\big),
\end{aligned} \tag{2.251}
$$

for $2.5 < s \le 3$, and analogously

$$
\begin{aligned}
|\hat{\eta}_{12}|_{W_2^{1/2}(\Sigma^h)} &\le Ch^{s-1}\big(\|a_{12}\|_{W_2^{s-1}(\Omega^+)}\|u\|_{W_2^s(\Omega^+)} \\
&\quad + \|a_{12}\|_{W_2^{s-1}(\Omega^-)}\|u\|_{W_2^s(\Omega^-)}\big), \quad \text{for } 2.5 < s \le 3.
\end{aligned} \tag{2.252}
$$

Hence, from (2.245)–(2.252) we obtain the main result of this section.

Theorem 2.75 *Suppose that the solution of the boundary-value problem* (2.231) *belongs to the function space* $\widetilde{W}_2^s(\Omega)$, *and that the coefficients of the equation* (2.231) *satisfy the following regularity hypotheses*:

$$
a_{ij} \in W_2^{s-1}(\Omega^+) \cap W_2^{s-1}(\Omega^-) \cap C(\overline{\Omega}),
$$
$$
a \in W_2^{s-2}(\Omega^+) \cap W_2^{s-2}(\Omega^-), \quad k \in W_2^{s-1}(\Sigma).
$$

Then, the finite difference scheme (2.244) *converges and the following error bound holds*:

$$
\begin{aligned}
&\|u - U\|_{W_2^1(\Omega^h)} \\
&\le Ch^{s-1}\Big(\max_{i,j}\|a_{ij}\|_{W_2^{s-1}(\Omega^+)} + \max_{i,j}\|a_{ij}\|_{W_2^{s-1}(\Omega^-)} \\
&\quad + \|a\|_{W_2^{s-2}(\Omega^+)} + \|a\|_{W_2^{s-2}(\Omega^-)} + \|k\|_{W_2^{s-1}(\Sigma)}\Big)\|u\|_{\widetilde{W}_2^s(\Omega)}, \quad 2.5 < s \le 3,
\end{aligned}
$$

where $C = C(s)$ is a positive constant, independent of h.

2.9 Bibliographical Notes

The principal purpose of this chapter has been to develop a technique for the derivation of error bounds, which are compatible with the smoothness of the data, for finite difference approximations of boundary-value problems for second- and fourth-order linear elliptic partial differential equations. The technique is based on the Bramble–Hilbert lemma and its generalizations (see Bramble and Hilbert [20, 21], Dupont and Scott [37], Dražić [32], Jovanović [79]).

According to the definition of Lazarov, Makarov and Samarskiĭ [125], an error bound of the form

$$
\|u - U\|_{W_2^r(\Omega^h)} \le Ch^{s-r}\|u\|_{W_2^s(\Omega)}, \quad s > r, \tag{2.253}
$$

is said to be *compatible with the smoothness of the solution* to the boundary–value problem. Similar error bounds, in 'continuous' norms, of the form

$$\|u - u^h\|_{W_2^r(\Omega)} \le Ch^{s-r}\|u\|_{W_2^s(\Omega)}, \quad 0 \le r \le 1 < s \le p+1,$$

are typical for finite elements methods (see e.g. Strang and Fix [169], Ciarlet [26], Brenner and Scott [23]) and are usually referred to as *optimal*; here u^h denotes the finite element approximation of the analytical solution u using continuous piecewise polynomials of degree p.

In the case of equations with variable coefficients the constant C in the error bound (2.253) depends on norms of the coefficients. One of our main objectives in this chapter has therefore been to understand this dependence in the case of various second-order and fourth-order linear elliptic model problems with variable coefficients. Specifically, we proved error bounds that are of the typical form

$$\|u - U\|_{W_2^r(\Omega^h)} \le Ch^{s-r}\Big(\max_{i,j}\|a_{ij}\|_{W_p^{s-1}(\Omega)} + \|a\|_{W_p^{s-2}(\Omega)}\Big)\|u\|_{W_2^s(\Omega)}.$$

To the best of our knowledge, error bounds of the form (2.253) were first derived by Weinelt [195], for $r = 1$ and $s = 2, 3$, in case of Poisson's equation. Subsequently, bounds of the form (2.253) were obtained by Lazarov, Makarov, Samarskiĭ, Weinelt, Jovanović, Ivanović, Süli, Gavrilyuk, Voĭtsekhovskiĭ, Berikelashvili and others, by systematic use of the Bramble–Hilbert lemma.

For example, families of finite difference schemes for Poisson's equation and the generalized Poisson equation with mollified right-hand sides were introduced by Jovanović [111] and Ivanović, Jovanović and Süli [75, 106], and scales of error bounds of the form (2.253) were established in the case of both integer and fractional values of s.

A procedure for determining the constant in the Bramble–Hilbert lemma, using the mapping of elementary rectangles on a canonical rectangle, was proposed by Lazarov [119]; see also [37] and [38] for related issues.

In the papers of Lazarov [119], Lazarov and Makarov [123] and Makarov and Ryzhenko [130, 131], the convergence of various difference schemes was examined for Poisson's equation in cylindrical, polar and spherical coordinates, and error bounds of the form (2.253) were derived under the assumption that the analytical solutions to these problems belong to appropriate weighted Sobolev spaces. Finite difference approximations of Poisson's equation by special classes of finite volume and finite difference schemes on nonuniform meshes were studied by Süli [171] and Jovanović and Matus [73]. In particular, the results in Sects. 2.4 and 2.4.2 are based on the paper [171]. The analysis presented in Sect. 2.4.3 was stimulated by discussions with Professor Rupert Klein, Free University Berlin. For related work, we refer to the paper of Oevermann and Klein [147].

A finite difference scheme with enhanced accuracy for second-order elliptic equations with constant coefficients was derived by Jovanović, Süli and Ivanović [108], and similar results were obtained later by Voĭtsekhovskiĭ and Novichenko [188].

Difference schemes for the biharmonic equation with a nonsmooth source term were considered by Lazarov [120], Gavrilyuk, Lazarov, Makarov and Pirnazarov [50], Ivanović, Jovanović and Süli [76], and for systems of partial differential equations in linear elasticity theory by Kalinin and Makarov [114, 129] and Voĭtsekhovskiĭ and Kalinin [187].

The convergence of the so-called *exact difference schemes* was investigated by Lazarov, Makarov and Samarskiĭ [125].

The error analysis of finite difference schemes for linear partial differential equations with variable coefficients was developed later. The first attempts in this direction were focused on finite difference schemes for the generalized Poisson equation with a variable coefficient in the lowest-order term (Lazarov, Makarov and Weinelt [126, 196], Voĭtsekhovskiĭ, Makarov and Shabliĭ [189]); subsequently, problems with variable coefficients in the principal part of the partial differential operator were considered (Godev and Lazarov [58], Jovanović, Ivanović and Süli [110], Jovanović [83]). Partial differential equations where the coefficient of the lowest-order term belongs to a negative Sobolev space were considered by Voĭtsekhovskiĭ, Makarov and Rybak [192], and Jovanović [83]. Zlotnik [203, 205] obtained different error estimates for discretizations of elliptic problems with variable coefficients.

Fourth-order equations with variable coefficients were studied by Gavrilyuk, Prikazchikov and Khimich [51], and Jovanović [84]. Quasilinear equations in arbitrary domains, solved by a combination of finite difference and fictitious domain methods, were studied by Voĭtsekhovskiĭ and Gavrilyuk [186], Voĭtsekhovskiĭ, Gavrilyuk and Makarov [191] and Jovanović [80, 81].

The technique described above was also used for the solution of eigenvalue problems (Prikazchikov and Khimich [151]), variational inequalities (Voĭtsekhovskiĭ, Gavrilyuk and Sazhenyuk [190], Gavrilyuk and Sazhenyuk [49]) and in the analysis of supraconvergence on nonuniform meshes (Marletta [134]). Berikeshvili systematically used the same technique for the numerical approximation of a general class of elliptic problems, including equations of higher order, systems of elliptic equations, problems with nonlocal boundary conditions, etc.; for further details, we refer to the survey paper [11], which also contains an extensive bibliography. Berikeshvili, Gupta and Mirianashvili [12] investigated the convergence of fourth-order compact difference schemes for three-dimensional convection-diffusion equations. Jovanović and Vulkov [101] studied the finite difference approximation of elliptic interface problems with variable coefficients.

Recently, a group of mathematicians (Barbeiro, Ferreira, Emmrich, Grigorieff et al.) exploited the techniques discussed in this chapter for the analysis of super- and supraconvergence effects in finite-difference and finite-element schemes (see Barbeiro [5], Barbeiro, Ferreira and Grigorieff [6], Emmrich [44], Emmrich and Grigorieff [45] and Ferreira and Grigorieff [47]).

There has also been work on the convergence analysis of finite difference schemes in discrete W^k_p norms, for $p \neq 2$; see, for example, Lazarov and Mokin [124], Lazarov [121], Godev and Lazarov [57], Drenska [33, 34], Süli, Jovanović and Ivanović [173, 174]. In this case, the derivation of a priori estimates is technically more complex—the theory of discrete Fourier multipliers, developed by

Mokin [140], is used instead of standard discrete energy estimates. Error bounds for the difference schemes under consideration are then obtained by combining these a priori estimates with the use of the Bramble–Hilbert lemma, as we have described in this chapter.

An alternative technique for the derivation of error bounds of the form (2.253) in fractional-order norms is based on function space interpolation, and was used by Jovanović [89].

Our goal in the rest of the book is to extend the methodology developed in the present chapter to time-dependent problems. In Chap. 3 we shall be concerned with parabolic partial differential equations, while in Chap. 4 we focus on hyperbolic equations.

Chapter 3
Finite Difference Approximation of Parabolic Problems

In Chap. 2 we considered finite difference methods for the approximate solution of elliptic equations. The present chapter is devoted to the analysis of finite difference schemes for parabolic equations.

In Sect. 3.1 we discuss the question of well-posedness of initial-boundary-value problems for second-order parabolic equations. In Sect. 3.2 we review some classical results concerning standard finite difference schemes for the heat equation. Section 3.3 is devoted to the convergence analysis of difference schemes for the heat equation with nonsmooth data. In Sect. 3.4 we extend these ideas to a linear second-order parabolic equation with variable coefficients and derive error bounds in the mesh-dependent anisotropic Sobolev norm $W_2^{1,1/2}$ that are compatible with the smoothness of the data. In Sects. 3.5 and 3.6 we shall be concerned with the finite difference approximation of interface and transmission problems for second-order linear parabolic equations. We conclude with some comments on the literature.

3.1 Parabolic Equations

We begin with a brief account of the theory of existence and uniqueness of solutions to evolution equations of the general form

$$\frac{\partial u}{\partial t} + A(t)u = f(x,t), \quad (x,t) \in \Omega \times (0,T],$$

subject to an *initial condition*

$$u(x,0) = u_0(x), \quad x \in \Omega,$$

with f and u_0 specified, and suitable boundary conditions for u on $\partial\Omega$, where u is a function of $x \in \Omega$ and $t \in [0,T]$, with $T > 0$, and Ω is a Lipschitz domain in $\mathbb{R}^n$; $A(t)$ will denote a linear elliptic partial differential operator. We shall suppose that $[0,T]$ is a bounded interval, that is $T < \infty$. An alternative viewpoint to considering

B.S. Jovanović, E. Süli, *Analysis of Finite Difference Schemes*, Springer Series in Computational Mathematics 46,
DOI 10.1007/978-1-4471-5460-0_3,

u as a function of x and t is to consider the mapping $t \in [0, T] \mapsto u(\cdot, t)$ with values in a Banach space or, more specifically, a Hilbert space, which is typically a Sobolev space of functions defined on Ω. The partial differential equation can then be viewed as an ordinary differential equation in a Banach or Hilbert space. The technical details of this alternative viewpoint are discussed in the next section.

3.1.1 Abstract Parabolic Initial-Value Problems

Suppose that $\mathcal{H}$ is a separable real Hilbert space with inner product $(\cdot, \cdot)$ and associated norm $\|\cdot\| = \|\cdot\|_{\mathcal{H}}$, and let $\mathcal{V}$ be another separable real Hilbert space with inner product $(\cdot, \cdot)_{\mathcal{V}}$ and norm $\|\cdot\|_{\mathcal{V}}$, which is continuously and densely embedded in $\mathcal{H}$. By the Riesz representation theorem $\mathcal{H}$ can be identified with its dual space $\mathcal{H}'$. The dual space of $\mathcal{V}$ is denoted by $\mathcal{V}'$. Thus we have

$$\mathcal{V} \hookrightarrow \mathcal{H} \equiv \mathcal{H}' \hookrightarrow \mathcal{V}'$$

with continuous and dense embeddings. Such a triple of spaces $\mathcal{V}, \mathcal{H}, \mathcal{V}'$, regardless of whether the spaces are separable or not, is called a *Gelfand triple* (or rigged Hilbert space). The duality pairing between $\mathcal{V}'$ and $\mathcal{V}$ will be denoted by $\langle \cdot, \cdot \rangle$. For $t \in [0, T]$ we consider the bilinear functional $a(t; \cdot, \cdot) : \mathcal{V} \times \mathcal{V} \to \mathbb{R}$, such that the following hypotheses hold:

(a) The function $t \mapsto a(t; w, v)$ is measurable on $[0, T]$, for any fixed $w, v \in \mathcal{V}$;
(b) There exists a real number $c_1 > 0$ such that

$$\left|a(t; w, v)\right| \le c_1 \|w\|_{\mathcal{V}} \|v\|_{\mathcal{V}} \quad \text{for all } t \in [0, T] \text{ and } w, v \in \mathcal{V};$$

(c) There exist real numbers $\lambda \ge 0$, $c_0 > 0$ such that

$$a(t; v, v) + \lambda \|v\|_{\mathcal{H}}^2 \ge c_0 \|v\|_{\mathcal{V}}^2 \quad \text{for all } t \in [0, T] \text{ and } v \in \mathcal{V}.$$

As in the case of elliptic problems considered in the previous chapter, condition (c) is referred to as *Gårding's inequality.*

Thanks to condition (b), for any $t \in [0, T]$ and $w \in \mathcal{V}$ fixed, the mapping $\ell_{t,w} : v \in \mathcal{V} \mapsto a(t; w, v) \in \mathbb{R}$ is a bounded linear functional on $\mathcal{V}$. Thus, $\ell_{t,w} \in \mathcal{V}'$. As $a(t; \cdot, v)$ is a bounded linear functional on $\mathcal{V}$, it follows that for each $t \in [0, T]$ the mapping $A(t) : w \in \mathcal{V} \mapsto \ell_{t,w} = A(t)w \in \mathcal{V}'$ is a bounded linear operator on $\mathcal{V}$ with $a(t; w, v) = \langle A(t)w, v\rangle$, $t \in [0, T]$, $w, v \in \mathcal{V}$. However, under our assumptions a more precise statement can be made.

Lemma 3.1 *Suppose that hypotheses (a) and (b) hold; then, the operator $A(t)$ associated with the bilinear functional $a(t; \cdot, \cdot)$ is bounded and linear as a map*

$$A : L_2\big((0, T); \mathcal{V}\big) \to L_2\big((0, T); \mathcal{V}'\big).$$

Here, for $g \in L_2((0, T); \mathcal{V})$, $A(g)$ denotes the function $t \mapsto A(t)(g(t)) \in \mathcal{V}'$.

Proof It follows from (a) and (b) that the functions $t \mapsto A(t)(g(t))$ and $t \mapsto A(t)$ are measurable on $[0, T]$. The proof of this statement is based on results from measure theory and is beyond the scope of the discussion here; the interested reader is referred to Wloka [199], Lemma 26.1 on p. 395. As $A(t)g(t) \in \mathcal{V}'$ for all $t \in [0, T]$, we have that

$$\|A(t)g(t)\|_{\mathcal{V}'} = \sup_{0 \neq v \in \mathcal{V}} \frac{\langle A(t)g(t), v\rangle}{\|v\|_{\mathcal{V}}} \leq c_1 \|g(t)\|_{\mathcal{V}}$$

for all $t \in [0, T]$, where the definition of $A(t)$ and hypothesis (b) have been used. Thus

$$\int_0^T \|A(t)g(t)\|_{\mathcal{V}'}^2 \, dt \leq c_1^2 \int_0^T \|g(t)\|_{\mathcal{V}}^2 \, dt,$$

which implies that the linear operator $A : L_2((0, T); \mathcal{V}) \to L_2((0, T); \mathcal{V}')$ is bounded and therefore also continuous. □

Let us define the space

$$\mathcal{W}(0, T) := \left\{ v : v \in L_2\big((0, T); \mathcal{V}\big), \frac{dv}{dt} \in L_2\big((0, T); \mathcal{V}'\big) \right\},$$

equipped with the inner product

$$(v, w)_{\mathcal{W}} := \int_0^T \big(v(t), w(t)\big)_{\mathcal{V}} \, dt + \int_0^T \left(\frac{dv(t)}{dt}, \frac{dw(t)}{dt} \right)_{\mathcal{V}'} dt,$$

where $(\cdot, \cdot)_{\mathcal{V}'}$ is the inner product of $\mathcal{V}'$. It is a straightforward matter to show that $\mathcal{W}$ is complete under the norm induced by this inner product, and therefore $\mathcal{W}$ is a Hilbert space. Moreover, $\mathcal{W}(0, T) \hookrightarrow C([0, T]; \mathcal{H})$, and for any $u \in \mathcal{W}(0, T)$ the following equality holds in the sense of distributions on the interval $(0, T)$:

$$\frac{d}{dt}\big(\|u\|_{\mathcal{H}}^2\big) = 2\left\langle \frac{du}{dt}, u \right\rangle \tag{3.1}$$

(see Theorems 25.4 and 25.5, pp. 392–395 in Wloka [199]).

Let us now consider the following problem (P): *Let* $f \in L_2((0, T); \mathcal{V}')$ *and* $u_0 \in \mathcal{H}$; *find*

$$u \in \mathcal{W}(0, T)$$

such that $u(0) = u_0$, *and*

$$\frac{du}{dt} + A(\cdot)u = f(\cdot) \quad \text{in } L_2\big((0, T); \mathcal{V}'\big),$$

that is,

$$\left\langle \frac{du}{dt}, v \right\rangle + \big\langle A(\cdot)u, v \big\rangle = \big\langle f(\cdot), v \big\rangle \quad \forall v \in \mathcal{V}.$$

as an equality in $L_2(0, T)$.

Thanks to the embedding $\mathcal{W}(0, T) \hookrightarrow C([0, T]; \mathcal{H})$, the initial condition $u(0) = u_0$ with $u_0 \in \mathcal{H}$ is meaningful.

We state the following existence and uniqueness result for problem (P) (see Wloka [199], Theorem 26.1 on p. 397).

Theorem 3.2 *Suppose that hypotheses* (a), (b) *and* (c) *hold, and assume that* $T < \infty$. *Then, problem* (P) *has a unique solution* $u \in \mathcal{W}(0, T)$ *satisfying the initial condition* $u(0) = u_0$, *and* u *depends continuously on the data* f *and* u_0; *that is, the map*

$$(f, u_0) \mapsto u, \quad \text{where } u \text{ is the solution of (P)},$$

is continuous from $L_2((0, T); \mathcal{V}') \times \mathcal{H}$ *into* $\mathcal{W}(0, T)$.

3.1.2 Some a Priori Estimates

In the previous section we treated $A = A(t)$ as a *bounded* linear operator from the Hilbert space $L_2((0, T); \mathcal{V})$ into its dual space $L_2(0, T, \mathcal{V}')$. Here we shall discuss an alternative perspective, in the special case when $A = A(t)$ is independent of t. Instead of viewing A as a bounded linear operator from $\mathcal{V}$ into $\mathcal{V}'$, we shall consider A as an *unbounded* densely defined linear operator on a real Hilbert space $\mathcal{H}$ with inner product $(\cdot, \cdot)_{\mathcal{H}} = (\cdot, \cdot)$ and norm $\|\cdot\|_{\mathcal{H}} = \|\cdot\|$ (i.e. the domain $D(A)$ of A is assumed to be dense in $\mathcal{H}$). We shall confine ourselves to the special case when A is a *selfadjoint* and *positive definite* operator; the latter means that $\inf_{v \in D(A)\setminus\{0\}} (Av, v)/\|v\|^2 > 0$.

The bilinear functional $(w, v) \in D(A) \times D(A) \mapsto (w, v)_A := (Aw, v)$, $w, v \in D(A)$, satisfies the axioms of inner product. The completion of the inner product space $D(A)$ in the induced norm $\|v\|_A := (v, v)_A^{1/2}$ is a real Hilbert space referred to as the *energy space* of A, denoted by $\mathcal{H}_A$, which is continuously and densely embedded in $\mathcal{H}$. By spectral decomposition of the selfadjoint, positive definite, densely defined linear operator A one can then define the power A^α of A for any real number α. In particular, $D(A^{1/2}) = \mathcal{H}_A$ and $(w, v)_A = (A^{1/2}w, A^{1/2}v)$ for all $w, v \in D(A)$.

Similarly, by completion in the norm $\|v\|_{A^{-1}} = (v, v)_{A^{-1}}^{1/2}$ induced by the inner product $(w, v)_{A^{-1}} = (A^{-1}w, v)$, $w, v \in D(A^{-1})$, we obtain the energy space $\mathcal{H}_{A^{-1}}$. Then, $\mathcal{H} \equiv \mathcal{H}'$ is continuously and densely embedded in $(\mathcal{H}_A)' = \mathcal{H}_{A^{-1}}$. The spaces $\mathcal{H}_A$, $\mathcal{H}$ and $\mathcal{H}_{A^{-1}}$ form a *Gelfand triple*: $\mathcal{H}_A \hookrightarrow \mathcal{H} \hookrightarrow \mathcal{H}_{A^{-1}}$. The inner product $(\cdot, \cdot)$ can be continuously extended to a duality pairing $\langle \cdot, \cdot \rangle$ on $\mathcal{H}_{A^{-1}} \times \mathcal{H}_A$ and the operator $A : D(A) \subset \mathcal{H}_A \to \mathcal{H}$ can be extended to a bounded linear operator (still denoted by A) from $\mathcal{H}_A$ into $\mathcal{H}_{A^{-1}}$.

We consider the bilinear functional $a(t; w, v) = a(w, v) := (A^{1/2}w, A^{1/2}v)$ for $w, v \in \mathcal{V} := D(A^{1/2})$. The assumptions (a), (b) and (c) from Sect. 3.1.1 are then satisfied with $c_0 = c_1 = 1$ and $\lambda = 0$. Problem (P) amounts to finding $u \in \mathcal{W}(0,T)$ such that

$$\frac{du}{dt} + Au = f(t), \quad u(0) = u_0, \tag{3.2}$$

with $u_0 \in \mathcal{H}$ and $f \in L_2((0,T); \mathcal{H}_{A^{-1}})$. The unique solution $u \in \mathcal{W}(0,T)$ satisfies the following *Hadamard inequality* (cf. Wloka [199], p. 403):

$$\int_0^T \left(\|u(t)\|_A^2 + \left\| \frac{du(t)}{dt} \right\|_{A^{-1}}^2 \right) dt \leq \|u_0\|^2 + \int_0^T \|f(t)\|_{A^{-1}}^2 \, dt. \tag{3.3}$$

Indeed, it follows from (3.2) that

$$\|f\|_{A^{-1}}^2 = \left\| \frac{du}{dt} + Au \right\|_{A^{-1}}^2 = \left\| \frac{du}{dt} \right\|_{A^{-1}}^2 + \|u\|_A^2 + 2\left\langle \frac{du}{dt}, u \right\rangle.$$

By integrating this equality with respect to t from 0 to T and using the relation (3.1) we obtain (3.3).

In the rest of this section we shall develop, in a nonrigorous manner, a collection of energy inequalities satisfied by the solution of problem (P). The purpose of the discussion that follows is merely to motivate our subsequent derivation of discrete counterparts of these estimates, on *finite-dimensional* Hilbert spaces, which can be proved, in a completely rigorous fashion, by mimicking the nonrigorous arguments here. Those discrete energy inequalities will then form the basis of our error analysis of finite difference approximations to parabolic problems.

By applying to (3.2) the operator $A^{1/2}$, resp. $A^{-1/2}$, and noting the inequality (3.3), we (formally) obtain the following *a priori* estimates:

$$\int_0^T \left(\|Au(t)\|^2 + \left\| \frac{du(t)}{dt} \right\|^2 \right) dt \leq \|u_0\|_A^2 + \int_0^T \|f(t)\|^2 \, dt \tag{3.4}$$

and

$$\int_0^T \left(\|u(t)\|^2 + \left\| A^{-1} \frac{du(t)}{dt} \right\|^2 \right) dt \leq \|u_0\|_{A^{-1}}^2 + \int_0^T \|A^{-1} f(t)\|^2 \, dt, \tag{3.5}$$

assuming that the right-hand sides of these inequalities are finite.

Let us now turn our attention to initial-value problems of the form

$$B\frac{du}{dt} + Au = f(t), \quad u(0) = u_0, \tag{3.6}$$

where B and $A = A_0 + A_1$, A_0, A_1 are unbounded, densely defined linear operators on a real separable Hilbert space $\mathcal{H}$. Let us further suppose that A_0 and B are selfadjoint and assume that there exist positive constants $m_i > 0$, $i = 1, 2, 3$, such that

$$(Bu, u) \geq m_1 \|u\|^2 \quad \forall u \in D(B), \qquad (A_0 u, u) \geq m_2 (Bu, u) \quad \forall u \in D(A_0) \cap D(B),$$

and that, for all $u \in D(A_0) \cap D(A_1)$ and all $v \in D(A_0)$,

$$(A_1 u, v)^2 \leq m_3 \|u\| \|u\|_{A_0} \|v\| \|v\|_{A_0}. \tag{3.7}$$

By applying the operator $B^{-1/2}$ to (3.6) we have that

$$\frac{\mathrm{d}\tilde{u}}{\mathrm{d}t} + \tilde{A}\tilde{u} = \tilde{f}(t), \quad \tilde{u}(0) = \tilde{u}_0,$$

where we have used the notations

$$\tilde{u} := B^{1/2} u, \qquad \tilde{u}_0 := B^{1/2} u_0, \qquad \tilde{A} := B^{-1/2} A B^{-1/2}, \qquad \tilde{f} := B^{-1/2} f.$$

Let us further define $\tilde{A}_i := B^{-1/2} A_i B^{-1/2}$, $i = 0, 1$. We observe that the linear operator $\tilde{A}_0$ is selfadjoint and positive definite on $\mathcal{H}$, i.e.

$$\begin{aligned}(\tilde{A}_0 v, v) &= \left(B^{-1/2} A_0 B^{-1/2} v, v\right) = \left(A_0 B^{-1/2} v, B^{-1/2} v\right) \\ &\geq m_2 \left(B B^{-1/2} v, B^{-1/2} v\right) = m_2 \|v\|^2 \quad \forall v \in D(\tilde{A}_0).\end{aligned}$$

Further, we have that

$$\begin{aligned}(\tilde{A}_1 v, w)^2 &= \left(A_1 B^{-1/2} v, B^{-1/2} w\right)^2 \\ &\leq m_3 \left\|B^{-1/2} v\right\| \left\|B^{-1/2} v\right\|_{A_0} \left\|B^{-1/2} w\right\| \left\|B^{-1/2} w\right\|_{A_0} \\ &\leq \frac{m_3}{m_1 m_2} \left\|B^{-1/2} v\right\|_{A_0}^2 \left\|B^{-1/2} w\right\|_{A_0}^2 = \frac{m_3}{m_1 m_2} \|v\|_{\tilde{A}_0}^2 \|w\|_{\tilde{A}_0}^2\end{aligned} \tag{3.8}$$

for all $v \in D(\tilde{A}_0) \cap D(\tilde{A}_1)$ and $w \in D(\tilde{A}_0)$; and, for all $v \in D(\tilde{A}_0) \cap D(\tilde{A}_1)$,

$$\begin{aligned}\left|(\tilde{A}_1 v, v)\right| &\leq \sqrt{m_3} \left\|B^{-1/2} v\right\| \left\|B^{-1/2} v\right\|_{A_0} \leq \sqrt{\frac{m_3}{m_1}} \|v\| \left\|B^{-1/2} v\right\|_{A_0} \\ &= \sqrt{\frac{m_3}{m_1}} \|v\| \|v\|_{\tilde{A}_0} \leq \frac{1}{2} \|v\|_{\tilde{A}_0}^2 + \frac{m_3}{2m_1} \|v\|^2.\end{aligned} \tag{3.9}$$

Thus, by taking $\mathcal{V} := \mathcal{H}_{\tilde{A}_0}$ and $a(t; v, w) := (\tilde{A} v, w)$ we deduce that the conditions (b) and (c) above hold with $c_1 = 1 + \sqrt{\frac{m_3}{m_1 m_2}}$, $c_0 = 1/2$ and $\lambda = \frac{m_3}{2m_1}$, while condition (a) holds trivially. Returning to the original notation, we thus deduce that if $u_0 \in \mathcal{H}_B$ and $f \in L_2((0,T); \mathcal{H}_{A_0^{-1}})$ then problem (3.6) has a unique solution $u \in L_2((0,T); \mathcal{H}_{A_0})$ with $B \frac{\mathrm{d}u}{\mathrm{d}t} \in L_2((0,T); \mathcal{H}_{A_0^{-1}})$.

We now turn to the (formal) derivation of a priori bounds on the solution of problem (3.6). Denoting by $\langle \cdot, \cdot \rangle$ the duality pairing between $(\mathcal{H}_{A_0})' = \mathcal{H}_{A_0^{-1}}$ and $\mathcal{H}_{A_0}$, we have from (3.6) that

$$\frac{\mathrm{d}}{\mathrm{d}t}\|u\|_B^2 + 2\|u\|_{A_0}^2 = 2\langle f, u\rangle - 2(A_1 u, u)$$

$$\leq 2\|u\|_{A_0}\|f\|_{A_0^{-1}} + 2\sqrt{m_3}\|u\|_{A_0}\frac{1}{\sqrt{m_1}}\|u\|_B$$

$$\leq \|u\|_{A_0}^2 + 2\|f\|_{A_0^{-1}}^2 + 2C_1\|u\|_B^2, \qquad C_1 = \frac{m_3}{m_1}. \tag{3.10}$$

We multiply this inequality by the nonnegative function $t \mapsto 2\mathrm{e}^{-2C_1 t}$ to deduce that

$$\frac{\mathrm{d}}{\mathrm{d}t}(\mathrm{e}^{-2C_1 t}\|u\|_B^2) + \mathrm{e}^{-2C_1 t}\|u\|_{A_0}^2 \leq 2\mathrm{e}^{-2C_1 t}\|f\|_{A_0^{-1}}^2,$$

which, after integration over $t \in (0, T)$ and noting that $\mathrm{e}^{-2C_1 T} \leq \mathrm{e}^{-2C_1 t} \leq 1$ for all $t \in [0, T]$, yields

$$\int_0^T \|u(t)\|_{A_0}^2\,\mathrm{d}t \leq 2\mathrm{e}^{2C_1 T}\left(\|u_0\|_B^2 + \int_0^T \|f(t)\|_{A_0^{-1}}^2\,\mathrm{d}t\right). \tag{3.11}$$

Now, (3.6) directly implies that

$$\left\|B\frac{\mathrm{d}u}{\mathrm{d}t}\right\|_{A_0^{-1}} = \|-A_0 u - A_1 u + f\|_{A_0^{-1}} \leq \|u\|_{A_0} + \|A_1 u\|_{A_0^{-1}} + \|f\|_{A_0^{-1}}.$$

Using (3.7) we deduce that

$$\|A_1 u\|_{A_0^{-1}}^4 = \left(A_0^{-1}A_1 u, A_1 u\right)^2 \leq m_3\|u\|\|u\|_{A_0}\left\|A_0^{-1}A_1 u\right\|\left\|A_0^{-1}A_1 u\right\|_{A_0}$$

$$\leq \frac{m_3}{\sqrt{m_1 m_2}}\|u\|_{A_0}^2\frac{1}{\sqrt{m_1 m_2}}\left\|A_0^{-1}A_1 u\right\|_{A_0}^2$$

$$= \frac{m_3}{m_1 m_2}\|u\|_{A_0}^2\|A_1 u\|_{A_0^{-1}}^2, \tag{3.12}$$

which implies that

$$\|A_1 u\|_{A_0^{-1}} \leq \sqrt{\frac{m_3}{m_1 m_2}}\|u\|_{A_0}.$$

From (3.11) and the subsequent inequalities we deduce the following analogue of the a priori estimate (3.3):

$$\int_0^T\left(\|u(t)\|_{A_0}^2 + \left\|B\frac{\mathrm{d}u(t)}{\mathrm{d}t}\right\|_{A_0^{-1}}^2\right)\mathrm{d}t \leq C\left(\|u_0\|_B^2 + \int_0^T \|f(t)\|_{A_0^{-1}}^2\,\mathrm{d}t\right), \tag{3.13}$$

where C is a computable constant, which depends on T:

$$C \leq 2\mathrm{e}^{2C_1 T} + \left[1 + \sqrt{2}\left(1 + \sqrt{\frac{m_3}{m_1 m_2}}\right)\mathrm{e}^{C_1 T}\right]^2 \leq C_2\mathrm{e}^{2C_1 T}.$$

We shall now derive a similar bound on the Sobolev seminorm of order $1/2$ with respect to the variable t. To this end, we shall use the Fourier series expansions

$$u(t) = \frac{a_0}{2} + \sum_{j=1}^{\infty} a_j \cos\frac{j\pi t}{T}; \qquad u(t) = \sum_{j=1}^{\infty} b_j \sin\frac{j\pi t}{T}, \qquad t \in (0, T),$$

of $u \in L_2((0, T); \mathcal{H})$, where

$$a_j = a_j[u] = \frac{2}{T}\int_0^T u(t') \cos\frac{j\pi t'}{T}\,\mathrm{d}t' \in \mathcal{H},$$
$$b_j = b_j[u] = \frac{2}{T}\int_0^T u(t') \sin\frac{j\pi t'}{T}\,\mathrm{d}t' \in \mathcal{H},$$

and the integrals are to be understood in the sense of Bochner (see, Wloka [199], p. 384). Direct calculations show that

$$\int_0^T \|u(t)\|^2\,\mathrm{d}t = \frac{T}{2}\left(\frac{\|a_0[u]\|^2}{2} + \sum_{j=1}^{\infty}\|a_j[u]\|^2\right) = \frac{T}{2}\sum_{j=1}^{\infty}\|b_j[u]\|^2. \tag{3.14}$$

An analogous result holds if the $\mathcal{H}$ norm $\|\cdot\|$ is replaced by an energy norm.

Let us multiply (3.6) by $\sin k\pi t/T$ and (formally) integrate the resulting equality over $t \in [0, T]$. Using the expansion

$$\frac{\mathrm{d}u(t)}{\mathrm{d}t} = -\sum_{j=1}^{\infty} a_j[u]\frac{j\pi}{T}\sin\frac{j\pi t}{T}$$

and the orthogonality of the sine functions in the above expansion over the interval $(0, T)$, we deduce that

$$\frac{k\pi}{T}Ba_k[u] = A_0 b_k[u] - b_k[f] + A_1 b_k[u].$$

By taking the inner product in $\mathcal{H}$ of the resulting equality with $a_k[u]$ and summing over k we get

$$\begin{aligned}
\frac{\pi}{T}\sum_{k=1}^{\infty} k\|a_k[u]\|_B^2 &= \sum_{k=1}^{\infty}\big[\big(A_0 b_k[u], a_k[u]\big) - \big(b_k[f], a_k[u]\big) + \big(A_1 b_k[u], a_k[u]\big)\big] \\
&\le \sum_{k=1}^{\infty}\|b_k[u]\|_{A_0}\|a_k[u]\|_{A_0} + \|a_k[u]\|_{A_0}\|b_k[f]\|_{A_0^{-1}}^2 \\
&\quad + \sqrt{\frac{m_3}{m_1 m_2}}\|b_k[u]\|_{A_0}\|a_k[u]\|_{A_0} \\
&\le \frac{1}{2}\sum_{k=1}^{\infty}\bigg[\bigg(1 + \sqrt{\frac{m_3}{m_1 m_2}}\bigg)\big(\|b_k[u]\|_{A_0}^2 + \|a_k[u]\|_{A_0}^2\big)
\end{aligned}$$

$$+ \left\| a_k[u] \right\|_{A_0}^2 + \left\| b_k[f] \right\|_{A_0^{-1}}^2 \Big].$$

Using (3.14), we deduce that

$$\sum_{k=1}^{\infty} k \left\| a_k[u] \right\|_B^2 \le \frac{1}{\pi} \int_0^T \left[\left(3 + 2\sqrt{\frac{m_3}{m_1 m_2}} \right) \left\| u(t) \right\|_{A_0}^2 + \left\| f(t) \right\|_{A_0^{-1}}^2 \right] \mathrm{d}t. \quad (3.15)$$

Let us consider the expression

$$J_1 := \int_{-T}^{T} \int_{-T}^{T} \frac{\| u(t) - u(t-s) \|_B^2}{|s|^2} \, \mathrm{d}s \, \mathrm{d}t,$$

assuming that $t \mapsto u(t)$ has been extended as an even function outside the interval $[0, T]$:

$$u(t) = \begin{cases} u(t) & \text{for } t \in [0, T], \\ u(-t) & \text{for } t \in [-T, 0], \\ u(2T - t) & \text{for } t \in [T, 2T], \\ \text{and so on.} & \end{cases}$$

Using the periodicity of $t \mapsto u(t)$ and an expansion in cosines, we deduce that

$$J_1 = \int_{-T}^{T} \left[\int_{-T}^{T} \big(u(t), -u(t+s) + 2u(t) - u(t-s) \big)_B \, \mathrm{d}t \right] \frac{\mathrm{d}s}{s^2}$$

$$= \int_{-T}^{T} \int_{-T}^{T} \Bigg(\frac{a_0[u]}{2} + \sum_{j=1}^{\infty} a_j[u] \cos \frac{j\pi t}{T}, \sum_{k=1}^{\infty} a_k[u] \Bigg(-\cos \frac{k\pi(t+s)}{T}$$

$$+ 2\cos \frac{k\pi t}{T} - \cos \frac{k\pi(t-s)}{T} \Bigg) \Bigg)_B \mathrm{d}t \frac{\mathrm{d}s}{s^2}$$

$$= \int_{-T}^{T} \int_{-T}^{T} \left(\frac{a_0[u]}{2} + \sum_{j=1}^{\infty} a_j[u] \cos \frac{j\pi t}{T}, \sum_{k=1}^{\infty} 4 a_k[u] \sin^2 \frac{k\pi s}{2T} \cos \frac{k\pi t}{T} \right)_B \mathrm{d}t \frac{\mathrm{d}s}{s^2}$$

$$= 4T \sum_{k=1}^{\infty} \left\| a_k[u] \right\|_B^2 \int_{-T}^{T} \sin^2 \frac{k\pi s}{2T} \frac{\mathrm{d}s}{s^2}.$$

Furthermore,

$$\int_{-T}^{T} \sin^2 \frac{k\pi s}{2T} \frac{\mathrm{d}s}{s^2} = 2 \int_0^T \sin^2 \frac{k\pi s}{2T} \frac{\mathrm{d}s}{s^2} = \frac{k\pi}{T} \int_0^{k\pi/2} \frac{\sin^2 \theta}{\theta^2} \, \mathrm{d}\theta$$

$$\le \frac{k\pi}{T} \int_0^{\infty} \frac{\sin^2 \theta}{\theta^2} \, \mathrm{d}\theta = \frac{k\pi^2}{2T},$$

which implies that

$$J_1 \le 2\pi^2 \sum_{k=1}^{\infty} k \big\| a_k[u] \big\|_B^2. \tag{3.16}$$

By noting (3.15), (3.16) and the obvious inequality

$$J_2 := \int_0^T \int_0^T \frac{\|u(t) - u(t')\|_B^2}{|t - t'|^2} \, dt \, dt' \le \frac{1}{2} J_1,$$

we get that

$$J_2 \le \pi \int_0^T \bigg[\bigg(3 + m_3 + \frac{1}{m_1 m_2} \bigg) \big\| u(t) \big\|_{A_0}^2 + \big\| f(t) \big\|_{A_0^{-1}}^2 \bigg] dt.$$

Hence, by (3.11), we obtain

$$\int_0^T \int_0^T \frac{\|u(t) - u(t')\|_B^2}{|t - t'|^2} \, dt \, dt' \le C \bigg(\|u_0\|_B^2 + \int_0^T \big\| f(t) \big\|_{A_0^{-1}}^2 \, dt \bigg). \tag{3.17}$$

Now, (3.11) and (3.17) imply that

$$\begin{aligned} &\int_0^T \big\| u(t) \big\|_{A_0}^2 \, dt + \int_0^T \int_0^T \frac{\|u(t) - u(t')\|_B^2}{|t - t'|^2} \, dt \, dt' \\ &\quad \le C \bigg(\|u_0\|_B^2 + \int_0^T \big\| f(t) \big\|_{A_0^{-1}}^2 \, dt \bigg). \end{aligned} \tag{3.18}$$

When $f(t) = dg(t)/dt$, we obtain, in a similar way,

$$\begin{aligned} &\int_0^T \big\| u(t) \big\|_{A_0}^2 \, dt + \int_0^T \int_0^T \frac{\|u(t) - u(t')\|_B^2}{|t - t'|^2} \, dt \, dt' \\ &\quad \le C \bigg[\|u_0\|_B^2 + \int_0^T \int_0^T \frac{\|g(t) - g(t')\|_{B^{-1}}^2}{|t - t'|^2} \, dt \, dt' \\ &\qquad + \int_0^T \bigg(\frac{1}{t} + \frac{1}{T - t} \bigg) \big\| g(t) \big\|_{B^{-1}}^2 \, dt \bigg]. \end{aligned} \tag{3.19}$$

If, instead of (3.7), the following condition is assumed to hold:

$$\exists m_3 > 0 \; \forall u \in D(A_0) \cap D(A_1) \cap D\big(A_1^*\big) \quad \max\big(\|A_1 u\|^2, \|A_1^* u\|^2\big) \le m_3 (A_0 u, u),$$

where A_1^* is the adjoint of the linear operator A_1, then one can derive similar bounds in both stronger and weaker norms. For example,

$$\int_0^T \bigg(\big\| A_0 u(t) \big\|_{B^{-1}}^2 + \bigg\| \frac{du(t)}{dt} \bigg\|_B^2 \bigg) dt \le C \bigg(\|u_0\|_{A_0}^2 + \int_0^T \big\| f(t) \big\|_{B^{-1}}^2 \, dt \bigg), \tag{3.20}$$

$$\int_0^T \left(\|u(t)\|_B^2 + \left\| A_0^{-1} B \frac{\mathrm{d}u(t)}{\mathrm{d}t} \right\|_B^2 \right) \mathrm{d}t$$

$$\leq C \left(\|Bu_0\|_{A_0^{-1}}^2 + \int_0^T \|A_0^{-1} f(t)\|_B^2 \, \mathrm{d}t \right) \tag{3.21}$$

and

$$\int_0^T \|u(t)\|_B^2 \, \mathrm{d}t \leq C \left(\|Bu_0 - g(0)\|_{A_0^{-1}}^2 + \int_0^T \|g(t)\|_{B^{-1}}^2 \, \mathrm{d}t \right). \tag{3.22}$$

In all of these bounds C is a computable constant such that $C \leq C_3 e^{2C_1 T}$.

In particular, when the operator A is positive definite, we can take $A_0 = A$, $A_1 = 0$ and $m_3 = C_1 = 0$, and the constants C in the inequalities (3.11)–(3.13), and (3.17)–(3.22) are then independent of T, the length of the time interval $[0, T]$.

3.1.3 Application to Parabolic Partial Differential Equations

As an application of the abstract results discussed in the previous sections of this chapter, we consider the existence and uniqueness of solutions to an initial-boundary-value problem for the partial differential equation

$$\frac{\partial u}{\partial t} + P(x, t, \partial)u = f(x, t) \quad \text{in } \Omega \times (0, T], \tag{3.23}$$

with

$$P(x, t, \partial)u := \sum_{0 \leq |\alpha|, |\beta| \leq k} (-1)^{|\alpha|} \partial^\alpha \big(a_{\alpha\beta}(x, t) \partial^\beta u \big), \tag{3.24}$$

subject to the boundary conditions

$$\partial_\nu^m u = 0 \quad \text{on } \partial\Omega \times (0, T], \text{ for } 0 \leq m \leq k - 1, \tag{3.25}$$

and the initial condition

$$u(x, 0) = u_0(x), \quad x \in \Omega, \tag{3.26}$$

where Ω is a Lipschitz domain in $\mathbb{R}^n$. We shall suppose that the partial differential operator $P(x, t, \partial)$ satisfies the uniform ellipticity condition (2.2) with a positive constant $\tilde{c} > 0$, for all $x \in \overline{\Omega}$ and $t \in [0, T]$. Under these conditions, the partial differential operator

$$\frac{\partial}{\partial t} + P(x, t, \partial)$$

is called *uniformly parabolic*. Suppose further that

$$a_{\alpha\beta}(\cdot, t) \in C(\Omega) \quad \text{for } |\alpha| = |\beta| = k$$

for each $t \in [0, T]$, and

$$a_{\alpha\beta} \in L_\infty\big(\Omega \times (0, T)\big) \quad \text{for } |\alpha|, |\beta| \leq k.$$

Then, by Theorem 2.4, there exist constants $c_0 > 0$ and $\lambda \geq 0$ such that

$$a(t; v, v) + \lambda \|v\|^2_{L_2(\Omega)} \geq c_0 \|v\|^2_{W_2^k(\Omega)} \quad \forall v \in \mathring{W}_2^k(\Omega),\ t \in [0, T]. \tag{3.27}$$

Furthermore, a straightforward application of the Cauchy–Schwarz inequality implies the existence of a constant $c_1 > 0$ such that

$$a(t; w, v) \leq c_1 \|w\|_{W_2^k(\Omega)} \|v\|_{W_2^k(\Omega)} \quad \forall w, v \in \mathring{W}_2^k(\Omega),\ t \in [0, T].$$

Thus, by Theorem 3.2 with $\mathcal{V} = \mathring{W}_2^k(\Omega)$ and $\mathcal{H} = L_2(\Omega)$, the parabolic initial-boundary-value problem (3.23), (3.25), (3.26) has a unique solution

$$u \in L_2\big((0, T); \mathring{W}_2^k(\Omega)\big), \quad \frac{\mathrm{d}u}{\mathrm{d}t} \in L_2\big((0, T); W_2^{-k}(\Omega)\big),$$

provided that $f \in L_2((0, T); W_2^{-k}(\Omega))$ and $u_0 \in L_2(\Omega)$.

For a second-order uniformly parabolic differential operator of the form

$$u \mapsto \frac{\partial u}{\partial t} - \sum_{i,j=1}^{n} \frac{\partial}{\partial x_j}\left(a_{ij}(x, t)\frac{\partial u}{\partial x_i}\right) + \sum_{i=1}^{n}\left[-\frac{\partial}{\partial x_i}\big(a_i(x, t)u\big) + b_i(x, t)\frac{\partial u}{\partial x_i}\right] + c(x, t)u$$

the bilinear functional $a(t; \cdot, \cdot)$ is given, for $t \in [0, T]$, by

$$a(t; w, v) := \sum_{i,j=1}^{n} \int_\Omega a_{ij}(x, t)\frac{\partial w}{\partial x_i}\frac{\partial v}{\partial x_j}\,\mathrm{d}x + \sum_{i=1}^{n} a_i(x, t) w \frac{\partial v}{\partial x_i}\,\mathrm{d}x + \int_\Omega b_i(x, t)\frac{\partial w}{\partial x_i} v\,\mathrm{d}x + \int_\Omega c(x, t) w v\,\mathrm{d}x, \quad w, v \in \mathring{W}_2^1(\Omega).$$

In this case, Gårding's inequality can be proved under relaxed smoothness hypotheses on the coefficients in the principal part: the a_{ij} need not be continuous functions of x on Ω; it suffices to assume that $a_{ij} \in L_\infty(\Omega \times (0, T))$ for $i, j = 1, \ldots, n$. To be more precise, suppose that $\Omega \subset \mathbb{R}^n$ is a Lipschitz domain, and let $P(x, t, \partial)$ be the second-order linear partial differential operator defined by (2.3) where a_{ij}, a_i, $b_j \in L_\infty(\Omega \times (0, T))$, $i, j = 1, \ldots, n$, and $c \in L_\infty(\Omega \times (0, T))$ are such that, for some $\tilde{c} > 0$, the uniform ellipticity condition (2.4) holds. Then, according to Theorem 2.5, there exist constants $c_0 > 0$ and $\lambda \geq 0$ such that

$$a(t; v, v) + \lambda \|v\|^2_{L_2(\Omega)} \geq c_0 \|v\|^2_{W_2^1(\Omega)} \quad \text{for a.e. } t \in [0, T] \text{ and all } v \in \mathring{W}_2^1(\Omega).$$

Thus we deduce the existence of a unique solution u to the corresponding second-order parabolic initial-boundary-value problem, with

$$u \in L_2\big((0,T);\, \mathring{W}_2^1(\Omega)\big), \quad \frac{\mathrm{d}u}{\mathrm{d}t} \in L_2\big((0,T);\, W_2^{-1}(\Omega)\big),$$

whenever $f \in L_2((0,T);\, W_2^{-1}(\Omega))$ and $u_0 \in L_2(\Omega)$.

In fact, by Remark 2.2 this statement holds under even weaker hypotheses on a_{ij}, a_i and b_i. In particular, it suffices to assume that

$$\begin{aligned} a_{ij} &\in L_\infty\big((0,T);\, M\big(L_2(\Omega) \to L_2(\Omega)\big)\big), \quad i,j = 1,\dots,n, \\ a_i, b_i &\in L_\infty\big((0,T);\, M\big(W_2^1(\Omega) \to L_2(\Omega)\big)\big), \quad i = 1,\dots,n, \\ c &\in L_\infty\big((0,T);\, M\big(W_2^1(\Omega) \to L_p(\Omega)\big)\big), \end{aligned}$$

where $p = 2n/(n+2)$ if $n > 2$; $p > 1$ (but arbitrarily close to 1) if $n = 2$; and $p = 1$ if $n = 1$.

We conclude this section with a brief comment on the physical implications of Gårding's inequality (3.27). Assuming that (3.27) holds for some constants $c_0 > 0$ and $\lambda \geq 0$, it follows that the solution u to problem (P) satisfies the inequality

$$\frac{1}{2}\frac{\mathrm{d}}{\mathrm{d}t}\|u(\cdot,t)\|^2_{L_2(\Omega)} + c_0\|u(\cdot,t)\|^2_{W_2^k(\Omega)} \leq \langle f(\cdot,t), u(\cdot,t)\rangle + \lambda\|u(\cdot,t)\|^2_{L_2(\Omega)}. \tag{3.28}$$

Bounding the first term on the right-hand side by

$$|\langle f(\cdot,t), u(\cdot,t)\rangle| \leq \frac{1}{2c_0}\|f(\cdot,t)\|^2_{W_2^{-k}(\Omega)} + \frac{c_0}{2}\|u(\cdot,t)\|^2_{W_2^k(\Omega)},$$

it follows that

$$\frac{\mathrm{d}}{\mathrm{d}t}\|u(\cdot,t)\|^2_{L_2(\Omega)} + c_0\|u(\cdot,t)\|^2_{W_2^k(\Omega)} \leq \frac{1}{c_0}\|f(\cdot,t)\|^2_{W_2^{-k}(\Omega)} + 2\lambda\|u(\cdot,t)\|^2_{L_2(\Omega)}.$$

Now we multiply both sides of this inequality by $\mathrm{e}^{-2\lambda t}$; thus,

$$\frac{\mathrm{d}}{\mathrm{d}t}\big(\mathrm{e}^{-2\lambda t}\|u(\cdot,t)\|^2_{L_2(\Omega)}\big) + c_0\mathrm{e}^{-2\lambda t}\|u(\cdot,t)\|^2_{W_2^k(\Omega)} \leq \frac{1}{c_0}\mathrm{e}^{-2\lambda t}\|f(\cdot,t)\|^2_{W_2^{-k}(\Omega)},$$

and thereby,

$$\begin{aligned} &\|u(\cdot,t)\|^2_{L_2(\Omega)} + c_0\int_0^t \mathrm{e}^{2\lambda(t-s)}\|u(\cdot,s)\|^2_{W_2^k(\Omega)}\,\mathrm{d}s \\ &\quad \leq \mathrm{e}^{2\lambda t}\|u_0\|^2_{L_2(\Omega)} + \frac{1}{c_0}\int_0^t \mathrm{e}^{2\lambda(t-s)}\|f(\cdot,s)\|^2_{W_2^{-k}(\Omega)}\,\mathrm{d}s \end{aligned} \tag{3.29}$$

for all $t \in (0,T]$, which expresses the continuous dependence of the solution u on the data $u_0 \in L_2(\Omega)$ and $f \in L_2((0,T);\, W_2^{-k}(\Omega))$. Now suppose that, instead of

$c_0 > 0$, the slightly stronger hypothesis $c_0 > 2\lambda \geq 0$ is assumed to hold for the constants c_0 and λ in Gårding's inequality. Then, bounding from below in (3.28) the $W_2^k(\Omega)$ norm of u by its $L_2(\Omega)$ norm and following the same route as above, we deduce that

$$\|u(\cdot,t)\|^2_{L_2(\Omega)} \leq \mathrm{e}^{-2Kt}\|u_0\|^2_{L_2(\Omega)} + \frac{1}{c_0}\int_0^t \mathrm{e}^{-2K(t-s)}\|f(\cdot,s)\|^2_{W_2^{-k}(\Omega)}\,\mathrm{d}s,$$

where $K := \frac{1}{2}c_0 - \lambda$. In particular when $f = 0$, which physically corresponds to considering the evolution of the solution from the initial state u_0 in the absence of external forces (or heat sources, in the case of a second-order parabolic equation modelling the diffusion of heat in Ω), we have that

$$\|u(\cdot,t)\|^2_{L_2(\Omega)} \leq \mathrm{e}^{-2Kt}\|u_0\|^2_{L_2(\Omega)}, \quad t \geq 0. \tag{3.30}$$

In other words, the "energy" $\frac{1}{2}\|u(\cdot,t)\|^2_{L_2(\Omega)}$ is dissipated exponentially fast, and the rate of dissipation depends on the (positive) difference $\frac{1}{2}c_0 - \lambda$.

In the next section we consider a class of two-level operator-difference schemes for the numerical solution of parabolic equations.

3.1.4 Abstract Two-Level Operator-Difference Schemes

Let $\mathcal{H}^h$ be a finite-dimensional Hilbert space over the field of real numbers, equipped with the inner product $(\cdot,\cdot)_h$ and induced norm $\|\cdot\|_h := \|\cdot\|_{\mathcal{H}^h}$. Suppose further that $[0,T]$ is a bounded nonempty closed interval of the real line, and let $\overline{\Omega}^\tau := \{t_m := m\tau : m = 0,1,\ldots,M\}$ be a uniform mesh on the interval $[0,T]$ with mesh-size $\tau := T/M$, where M is a positive integer. Let us define $\Omega^\tau := \overline{\Omega}^\tau \cap (0,T)$, $\Omega^\tau_- := \Omega^\tau \cup \{0\}$ and $\Omega^\tau_+ := \Omega^\tau \cup \{T\}$. The forward and backward divided differences of a function $U : \overline{\Omega}^\tau \to \mathcal{H}^h$ are defined by

$$D_t^+ U := (\hat{U} - U)/\tau, \qquad D_t^- U := (U - \check{U})/\tau,$$

respectively, where we have used the notation $U := U(t)$, $\hat{U} := U(t+\tau)$, $\check{U} := U(t-\tau)$. We shall also write: $U^m := U(t_m) = U(m\tau)$, $m = 0,1,\ldots,M$.

In this section we shall consider the following family of two-level operator-difference schemes:

$$B_h\big(D_t^+ U\big) + A_h U = F, \quad t \in \Omega^\tau_-; \qquad U(0) = U^0. \tag{3.31}$$

Here, $F : \Omega^\tau_- \to \mathcal{H}^h$ and $U_0 \in \mathcal{H}^h$ are given functions, $U : \overline{\Omega}^\tau \to \mathcal{H}^h$ is the unknown function, and A_h and B_h are selfadjoint linear operators that are *positive definite, uniformly with respect to* h, on $\mathcal{H}^h$; i.e. there exist positive constants c_a and c_b, independent of h, such that $(A_h V, V)_h \geq c_a\|V\|_h^2$ for all $V \in \mathcal{H}^h$ and $(B_h V, V)_h \geq c_b\|V\|_h^2$ for all $V \in \mathcal{H}^h$.

Since $\mathcal{H}^h$ is a *finite-dimensional* Hilbert space, it is understood that all linear operators under consideration are defined on the entire space $\mathcal{H}^h$.

The purpose of this section is to investigate the stability of this class of schemes by mimicking the formal arguments developed in the previous section for the partial differential equation. To this end, we take the inner product of (3.31) with $2\tau D_t^+ U = 2(\hat{U} - U)$ and use the identity

$$U = \frac{1}{2}(\hat{U} + U) - \frac{1}{2}\tau D_t^+ U \tag{3.32}$$

to obtain

$$\begin{aligned} 2\tau\left(\left(B_h - \frac{1}{2}\tau A_h\right) D_t^+ U, D_t^+ U\right)_h &+ (A_h \hat{U}, \hat{U})_h \\ - (A_h U, U)_h &= 2\tau \big(F, D_t^+ U\big)_h. \end{aligned} \tag{3.33}$$

Similarly, by taking the inner product of (3.31) with $2\tau\hat{U}$ and using the identity (3.32) and that

$$\hat{U} = \frac{1}{2}(\hat{U} + U) + \frac{1}{2}\tau D_t^+ U,$$

we obtain

$$\begin{aligned} (B_h \hat{U}, \hat{U})_h - (B_h U, U)_h &+ \frac{1}{2}\tau \big(A_h(\hat{U} + U), \hat{U} + U\big)_h \\ + \tau^2\left(\left(B_h - \frac{1}{2}\tau A_h\right) D_t^+ U, D_t^+ U\right)_h &= 2\tau (F, \hat{U})_h. \end{aligned} \tag{3.34}$$

Thus, in particular, when $F = 0$ and

$$B_h - \frac{1}{2}\tau A_h \geq 0, \tag{3.35}$$

we obtain from (3.33) and (3.34) that

$$\|\hat{U}\|_{A_h} \leq \|U\|_{A_h}, \qquad \|\hat{U}\|_{B_h} \leq \|U\|_{B_h},$$

where $\|U\|_{A_h}$ and $\|U\|_{B_h}$ denote the energy norms $\|U\|_{A_h} := (A_h U, U)_h^{1/2}$ and $\|U\|_{B_h} := (B_h U, U)_h^{1/2}$, respectively. Hence we deduce by induction that

$$\big\|U^m\big\|_{A_h} \leq \big\|U^0\big\|_{A_h}, \qquad \big\|U^m\big\|_{B_h} \leq \|U^0\|_{B_h}. \tag{3.36}$$

The inequalities (3.36) imply the stability of the homogeneous operator-difference scheme

$$B_h\big(D_t^+ U\big) + A_h U = 0, \quad t \in \Omega_-^\tau; \qquad U(0) = U^0, \tag{3.37}$$

with respect to perturbations of the initial condition, in the energy norms $\|\cdot\|_{A_h}$ and $\|\cdot\|_{B_h}$, the condition (3.35) being both necessary and sufficient for the stability of the operator-difference scheme (3.37) (see, Samarskiĭ [159], Sect. 6.2). More precisely, the following statement holds (see Samarskiĭ [159], p. 404).

Lemma 3.3 *Suppose, as above, that A_h and B_h are linear selfadjoint positive definite operators, uniformly with respect to h, on the real Hilbert space $\mathcal{H}^h$. Suppose further that A_h and B_h commute, i.e. $A_h B_h = B_h A_h$. Then, the condition* (3.35) *is both necessary and sufficient for the stability of the operator-difference scheme* (3.31) *in the norm $\|\cdot\|_{D_h}$, where D_h is an arbitrary linear selfadjoint positive definite operator, uniformly with respect to h, on $\mathcal{H}^h$ that commutes with both A_h and B_h.*

The scheme (3.31) can be interpreted as a numerical approximation of the abstract parabolic problem (3.6). Let us express (3.31) in the form

$$\bar{B}_h(D_t^- U) + A_h U = \check{F}, \quad t \in \Omega_+^\tau; \qquad U(0) = U^0, \tag{3.38}$$

where $\bar{B}_h := B_h - \tau A_h$. Assuming that the operator $\bar{B}_h$ is positive definite on $\mathcal{H}^h$, uniformly with respect to h, we have that

$$\begin{aligned}\|\check{F}\|^2_{A_h^{-1}} &= \|\bar{B}_h(D_t^- U) + A_h U\|^2_{A_h^{-1}}\\ &= \|\bar{B}_h(D_t^- U)\|^2_{A_h^{-1}} + \|U\|^2_{A_h} + 2(\bar{B}_h(D_t^- U), U)_h.\end{aligned}$$

As

$$2(\bar{B}_h(D_t^- U), U)_h = D_t^-(\|U\|^2_{\bar{B}_h}) + \tau\|D_t^- U\|^2_{\bar{B}_h},$$

we arrive at the following discrete analogue of Hadamard's inequality (3.3):

$$\tau \sum_{t\in\Omega_+^\tau} \left(\|U(t)\|^2_{A_h} + \|\bar{B}_h(D_t^- U(t))\|^2_{A_h^{-1}}\right) \le \|U^0\|^2_{\bar{B}_h} + \tau \sum_{t\in\Omega_-^\tau} \|F(t)\|^2_{A_h^{-1}}. \tag{3.39}$$

From (3.39), proceeding in the same way as in the 'continuous' case in the previous section, we deduce that

$$\tau \sum_{t\in\Omega_+^\tau} \left(\|A_h U(t)\|^2_{\bar{B}_h^{-1}} + \|D_t^- U(t)\|^2_{\bar{B}_h}\right) \le \|U^0\|^2_{A_h} + \tau \sum_{t\in\Omega_-^\tau} \|F(t)\|^2_{\bar{B}_h^{-1}} \tag{3.40}$$

and

$$\begin{aligned}&\tau \sum_{t\in\Omega_+^\tau} \left(\|U(t)\|^2_{\bar{B}_h} + \|A_h^{-1}\bar{B}_h(D_t^- U(t))\|^2_{\bar{B}_h}\right)\\ &\le \|\bar{B}_h U^0\|^2_{A_h^{-1}} + \tau \sum_{t\in\Omega_-^\tau} \|A_h^{-1} F(t)\|^2_{\bar{B}_h}.\end{aligned} \tag{3.41}$$

The requirement that A_h is selfadjoint and positive definite, uniformly with respect to h, can be weakened in the same way as in the 'continuous' case. Suppose that in (3.38) $\bar{B}_h$ is a selfadjoint linear operator that is positive definite, uniformly with respect to h, on $\mathcal{H}^h$; suppose further that $A_h = A_{0h} + A_{1h}$, where A_{0h} is a selfadjoint positive definite linear operator, uniformly with respect to h, on $\mathcal{H}^h$, and that

$$\begin{aligned} &(\bar{B}_h U, U)_h \geq m_1 \|U\|_h^2, \qquad (A_{0h} U, U)_h \geq m_2 (\bar{B}_h U, U)_h, \\ &(A_{1h} U, V)_h^2 \leq m_3 \|U\|_h \|U\|_{A_{0h}} \|V\|_h \|V\|_{A_{0h}}, \quad m_i > 0, i = 1, 2, 3. \end{aligned} \tag{3.42}$$

By taking the inner product of equation (3.38) with $2U$ we have that

$$\begin{aligned} &D_t^- \left(\|U\|_{\bar{B}_h}^2\right) + \tau \left\|D_t^- U\right\|_{\bar{B}_h}^2 + 2\|U\|_{A_{0h}}^2 \\ &\quad = 2(U, \check{F})_h - 2(A_{1h} U, U)_h \\ &\quad \leq 2\|U\|_{A_{0h}} \|\check{F}\|_{A_{0h}^{-1}} + 2\sqrt{m_3} \|U\|_{A_{0h}} \frac{1}{\sqrt{m_1}} \|U\|_{\bar{B}_h} \\ &\quad \leq \|U\|_{A_{0h}}^2 + 2\|\check{F}\|_{A_{0h}^{-1}}^2 + 2C_1 \|U\|_{\bar{B}_h}^2, \quad C_1 = m_3/m_1. \end{aligned}$$

Let us suppose that $\tau < 1/(2C_1)$, and multiply the inequality above by $\varphi(t-\tau)$, where

$$\varphi(t) := (1 - 2C_1 \tau)^{t/\tau}.$$

After simple rearrangements we thus obtain for $t \geq \tau$ that

$$D_t^- \left(\varphi(t) \left\|U(t)\right\|_{\bar{B}_h}^2\right) + \varphi(t-\tau) \left\|U(t)\right\|_{A_{0h}}^2 \leq 2\varphi(t-\tau) \left\|F(t-\tau)\right\|_{A_{0h}^{-1}}^2.$$

Summing over the points of the mesh Ω_+^τ and noting that

$$(1 - 2C_1 \tau)^{T/\tau} = \varphi(T) \leq \varphi(t) \leq \varphi(0) = 1 \quad \text{for } 0 \leq t \leq T,$$

we obtain

$$\tau \sum_{t \in \Omega_+^\tau} \left\|U(t)\right\|_{A_{0h}}^2 \leq (1 - 2C_1 \tau)^{-T/\tau} \left(\left\|U^0\right\|_{\bar{B}_h}^2 + 2\tau \sum_{t \in \Omega_-^\tau} \left\|F(t)\right\|_{A_{0h}^{-1}}^2 \right). \tag{3.43}$$

Now, (3.38) and (3.42) imply that

$$\begin{aligned} \left\|\bar{B}_h \left(D_t^- U\right)\right\|_{A_{0h}^{-1}} &= \left\|-A_{0h} U - A_{1h} U + \check{F}\right\|_{A_{0h}^{-1}} \\ &\leq \left(1 + \sqrt{\frac{m_3}{m_1 m_2}}\right) \|U\|_{A_{0h}} + \|\check{F}\|_{A_{0h}^{-1}}, \end{aligned}$$

which, on noting (3.43), implies that

$$\tau \sum_{t\in\Omega_+^\tau} \left(\|U(t)\|^2_{A_{0h}} + \|\bar{B}_h(D_t^- U)\|_{A_{0h}^{-1}}\right)$$
$$\leq C\left(\|U^0\|^2_{\bar{B}_h} + \tau \sum_{t\in\Omega_-^\tau} \|F(t)\|^2_{A_{0h}^{-1}}\right), \tag{3.44}$$

where C is a computable constant that depends on T and is defined by $C := C_2(1 - 2C_1\tau)^{-T/\tau} < C_2 e^{C_3 T}$.

By using Fourier series, as in Sect. 3.1.2, we obtain the following analogue of the inequality (3.17):

$$\tau^2 \sum_{t\in\overline{\Omega}^\tau} \sum_{t'\in\overline{\Omega}^\tau, t'\neq t} \frac{\|U(t)-U(t')\|^2_{\bar{B}_h}}{|t-t'|^2}$$
$$\leq C\left(\|U^0\|^2_{\bar{B}_h} + \tau\|U^0\|^2_{A_{0h}} + \tau \sum_{t\in\Omega_-^\tau} \|F(t)\|^2_{A_{0h}^{-1}}\right).$$

Combining this with (3.44), we deduce that

$$\tau \sum_{t\in\Omega_+^\tau} \|U(t)\|^2_{A_{0h}} + \tau^2 \sum_{t\in\overline{\Omega}^\tau} \sum_{t'\in\overline{\Omega}^\tau, t'\neq t} \frac{\|U(t)-U(t')\|^2_{\bar{B}_h}}{|t-t'|^2}$$
$$\leq C\left(\|U^0\|^2_{\bar{B}_h} + \tau\|U^0\|^2_{A_{0h}} + \tau \sum_{t\in\Omega_-^\tau} \|F(t)\|^2_{A_{0h}^{-1}}\right). \tag{3.45}$$

Similarly, in the case of $F(t) = D_t^+ G$ we obtain

$$\tau \sum_{t\in\Omega_+^\tau} \|U(t)\|^2_{A_{0h}} + \tau^2 \sum_{t\in\overline{\Omega}^\tau} \sum_{t'\in\overline{\Omega}^\tau, t'\neq t} \frac{\|U(t)-U(t')\|^2_{\bar{B}_h}}{|t-t'|^2}$$
$$\leq C\Bigg[\|U^0\|^2_{\bar{B}_h} + \tau\|U^0\|^2_{A_{0h}} + \tau^2 \sum_{t\in\overline{\Omega}^\tau} \sum_{t'\in\overline{\Omega}^\tau, t'\neq t} \frac{\|G(t)-G(t')\|^2_{\bar{B}_h^{-1}}}{|t-t'|^2}$$
$$+ \tau \sum_{t\in\Omega^\tau} \left(\frac{1}{t} + \frac{1}{T-t}\right)\|G(t)\|^2_{\bar{B}_h^{-1}}\Bigg]. \tag{3.46}$$

Under our assumptions on the linear operator $A_h = A_{0h} + A_{1h}$, the a priori estimates (3.45) and (3.46) also hold for the difference scheme (3.31), provided that we take $\bar{B}_h := B_h - \tau A_{0h}$. If the original linear operator B_h is selfadjoint and positive definite on $\mathcal{H}^h$, uniformly with respect to h, i.e. there exists an $m_4 > 0$, independent of h, such that $(B_hU,U)_h \geq m_4\|U\|_h^2$ for all $U \in \mathcal{H}^h$, then the same is true of $\bar{B}_h$, provided that $\tau < m_4/\|A_{0h}\|$, where $\|A_{0h}\| := \sup_{U\in\mathcal{H}^h} \|A_{0h}U\|_h/\|U\|_h$ denotes the usual operator norm of A_{0h}.

If, instead of (3.42), the following stronger condition is assumed:

$$\exists m_3 > 0\ \forall U \in \mathcal{H}^h \quad \max\bigl(\|A_{1h}U\|_h, \|A_{1h}^*U\|_h\bigr) \leq m_3(A_{0h}U,U)_h,$$

then, similarly as in the 'continuous' case, the following a priori estimates can be shown to hold:

$$\tau \sum_{t\in\Omega_+^\tau} \bigl(\|A_{0h}U(t)\|_{\bar{B}_h^{-1}}^2 + \|D_t^-U(t)\|_{\bar{B}_h}^2\bigr) \leq C\Bigl(\|U^0\|_{A_{0h}}^2 + \tau \sum_{t\in\Omega_-^\tau} \|F(t)\|_{\bar{B}_h^{-1}}^2\Bigr), \tag{3.47}$$

$$\tau \sum_{t\in\Omega_+^\tau} \bigl(\|U(t)\|_{\bar{B}_h}^2 + \|A_{0h}^{-1}\bar{B}_h(D_t^-U(t))\|_{\bar{B}_h}^2\bigr) \leq C\Bigl(\|\bar{B}_hU^0\|_{A_{0h}^{-1}}^2 + \tau \sum_{t\in\Omega_-^\tau} \|A_{0h}^{-1}F(t)\|_{\bar{B}_h}^2\Bigr) \tag{3.48}$$

and

$$\tau \sum_{t\in\Omega_+^\tau} \|U(t)\|_{\bar{B}_h}^2 \leq C\Bigl(\|\bar{B}_hU^0 - G(0)\|_{A_{0h}^{-1}}^2 + \tau \sum_{t\in\Omega_+^\tau} \|G(t)\|_{\bar{B}_h^{-1}}^2\Bigr). \tag{3.49}$$

In the a priori estimates (3.45)–(3.49), C signifies a generic computable positive constant of the form $C_4(1-2C_1\tau)^{-T/\tau} < C_4 e^{C_3 T}$.

In the next section we shall consider some simple finite difference schemes for the numerical solution of parabolic initial-boundary-value problems. In order to simplify the presentation we shall begin by discussing the simplest parabolic equation,

$$\frac{\partial u}{\partial t} - \frac{\partial^2 u}{\partial x^2} = f(x,t),$$

the heat equation in one space dimension. We shall then consider multidimensional parabolic equations with nonsmooth coefficients.

3.2 Classical Difference Schemes for the Heat Equation

This section surveys some classical results concerning standard finite difference approximations of the heat equation in one space dimension. We shall assume for the time being that the solution to the initial-boundary-value problem under consideration possesses a sufficient number of continuous and bounded partial derivatives on the space-time domain $(0, 1) \times (0, T]$. In Sect. 3.3 we shall then relax the excessive regularity requirements on the solution, and in Sect. 3.4 we shall extend these results to second-order parabolic equations with variable coefficients.

3.2.1 Explicit and Implicit Schemes

Our first model problem concerns the heat equation in one space dimension. Let $Q = \Omega \times (0, T]$, where $\Omega = (0, 1)$, $T > 0$: find $u = u(x, t)$ such that

$$\begin{aligned} &\frac{\partial u}{\partial t} = \frac{\partial^2 u}{\partial x^2} + f(x, t), \quad x \in (0, 1), \ t \in (0, T], \\ &u(0, t) = 0, \qquad u(1, t) = 0, \quad t \in (0, T], \\ &u(x, 0) = u_0(x), \quad x \in [0, 1]. \end{aligned} \tag{3.50}$$

Physically $u(x, t)$ represents the temperature of a rod of unit length at the point x at time t, which has temperature $u_0(x)$ at time 0; it is kept at zero temperate at its endpoints, $x = 0$ and $x = 1$, and is subject to external heat sources whose distribution in space and time is described by the function f.

We shall discuss two simple schemes for the numerical solution of (3.50). Both schemes involve the same discretization of $\partial^2 u/\partial x^2$; however, while the first scheme (called the explicit scheme) includes a forward difference in t to approximate $\partial u/\partial t$, the second (called the implicit scheme) uses a backward difference in t. It will be assumed that u_0 is compatible with the homogeneous Dirichlet boundary conditions at $x = 0$ and $x = 1$, i.e. $u_0(0) = 0$ and $u_0(1) = 0$.

3.2.1.1 The Explicit Scheme

We begin by constructing a mesh on the rectangle $\overline{Q} = [0, 1] \times [0, T]$. Let $h := 1/N$ be the mesh-size in the x-direction and $\tau := T/M$ the mesh-size in the t-direction; here N and M are two integers, $N \geq 2$, $M \geq 1$. We define the uniform 'space-time mesh' $\overline{Q}_h^\tau$ on $\overline{Q}$ by

$$\overline{Q}_h^\tau := \overline{\Omega}^h \times \overline{\Omega}^\tau = \big\{(x_j, t_m) : \ 0 \leq j \leq N; \ 0 \leq m \leq M\big\},$$

where the 'temporal mesh'

$$\overline{\Omega}^\tau := \{t_m := m\tau : 0 \leq m \leq M\} = \Omega^\tau \cup \{0, T\}$$

has been introduced in Sect. 3.1.4, and the 'spatial mesh'

$$\overline{\Omega}^h := \{x_j := jh : 0 \leq j \leq N\} = \Omega^h \cup \{0, 1\}$$

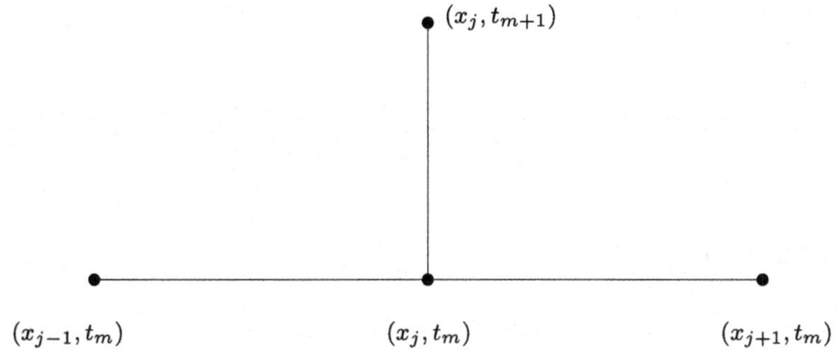

Fig. 3.1 Four-point stencil of the explicit scheme

has been defined in Sect. 2.2.1. On $\overline{Q}_h^\tau$ we approximate (3.50) by the following finite difference scheme: find U_j^m, $j = 0, \ldots, N$, $m = 0, \ldots, M$, such that

$$\begin{aligned} &D_t^+ U_j^m = D_x^+ D_x^- U_j^m + f(x_j, t_m), \\ &\quad j = 1, \ldots, N-1,\ m = 0, \ldots, M-1, \\ &U_0^m = 0, \qquad U_N^m = 0, \quad m = 1, \ldots, M, \\ &U_j^0 = u_0(x_j), \quad j = 0, \ldots, N, \end{aligned} \tag{3.51}$$

where U_j^m represents the numerical approximation of $u(x_j, t_m)$, the value of the analytical solution u at the mesh-point (x_j, t_m), $D_t^+ U_j^m$ is the forward divided difference in the t-direction and $D_x^+ D_x^- U_j^m$ is the second divided (central) difference in the x-direction. Clearly, (3.51) is a four-point difference scheme involving the values of U at the mesh-points

$$(x_{j-1}, t_m), \quad (x_j, t_m), \quad (x_{j+1}, t_m), \quad (x_j, t_{m+1}),$$

shown in Fig. 3.1. The scheme (3.51) is applied as follows. First we set $m = 0$. Since U_{j-1}^0, U_j^0, U_{j+1}^0 are specified by the initial condition $U_j^0 = u_0(x_j)$, $j = 0, \ldots, N$, the values U_j^1, $j = 0, \ldots, N$, can be computed from (3.51):

$$\begin{aligned} &U_j^1 = U_j^0 + \frac{\tau}{h^2}(U_{j+1}^0 - 2U_j^0 + U_{j-1}^0) + \tau f(x_j, t_0), \quad j = 1, \ldots, N-1, \\ &U_0^1 = 0, \qquad U_N^1 = 0. \end{aligned}$$

Suppose that we have already calculated U_j^m, $j = 0, \ldots, N$, the values of U at the time level $t_m = m\tau$. The values of U on the next time level $t_{m+1} = (m+1)\tau$ can be obtained from (3.51) by rewriting it as

$$\begin{aligned} &U_j^{m+1} = U_j^m + \frac{\tau}{h^2}(U_{j+1}^m - 2U_j^m + U_{j-1}^m) + \tau f(x_j, t_m), \\ &\quad j = 1, \ldots, N-1, \end{aligned}$$

$$U_0^{m+1} = 0, \qquad U_N^{m+1} = 0,$$

for any m, $0 \le m \le M-1$.

Clearly, the values of U at $t = t_{m+1}$ can be calculated explicitly from those of $U^m := U|_{t=t_m}$ and the data; hence the name *explicit scheme*.

3.2.1.2 The Implicit Scheme

Alternatively, one can approximate the time derivative by a backward difference, which gives rise to the following *implicit scheme*: find U_j^m, $j = 0, \dots, N$, $m = 0, \dots, M$, such that

$$\begin{aligned}
&D_t^- U_j^{m+1} = D_x^+ D_x^- U_j^{m+1} + f(x_j, t_{m+1}),\\
&\quad j = 1, \dots, N-1,\ m = 0, \dots, M-1,\\
&U_0^{m+1} = 0, \qquad U_N^{m+1} = 0, \quad m = 0, \dots, M-1,\\
&U_j^0 = u_0(x_j), \quad j = 0, \dots, N,
\end{aligned} \tag{3.52}$$

where U_j^m represents the approximation of $u(x_j, t_m)$, the value of u at the mesh-point (x_j, t_m). Unlike the explicit scheme in which the data and the values of the approximate solution U at the previous time level provide an explicit expression for the values of U at the next time level, the implicit scheme necessitates the solution of a system of linear equations on each time level to determine the values of U at the mesh-points on that time level. More precisely, (3.52) can be rewritten as follows:

$$\begin{aligned}
&-\frac{\tau}{h^2} U_{j+1}^{m+1} + \left(\frac{2\tau}{h^2} + 1\right) U_j^{m+1} - \frac{\tau}{h^2} U_{j-1}^{m+1} = U_j^m + \tau f(x_j, t_{m+1}),\\
&\quad j = 1, \dots, N-1,\\
&U_0^{m+1} = 0, \qquad U_N^{m+1} = 0,
\end{aligned} \tag{3.53}$$

for $m = 0, \dots, M-1$.

This is, again, a four-point finite difference scheme, but it now involves the values of U at the mesh-points

$$(x_{j-1}, t_{m+1}), \quad (x_j, t_{m+1}), \quad (x_{j+1}, t_{m+1}), \quad (x_j, t_m),$$

shown in Fig. 3.2. The implicit scheme (3.53) is implemented as follows. First we set $m = 0$; then, (3.53) is a system of linear equations with a tridiagonal matrix, and the right-hand side of the linear system can be computed from the initial datum $U_j^0 = u_0(x_j)$ and the source term $f(x_j, t_1)$. Suppose that we have already computed U_j^m, $j = 1, \dots, N-1$, the values of U on the mth time level, $0 \le m < M$. The values U_j^{m+1}, $j = 0, \dots, N$, of U on the next, $(m+1)$st, time level are then obtained by

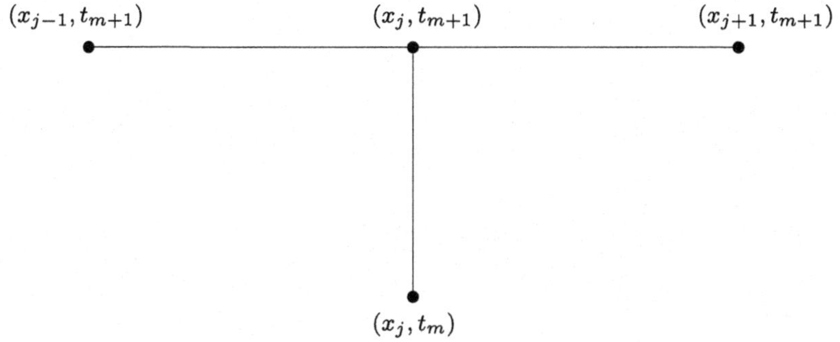

Fig. 3.2 Four-point stencil of the implicit scheme

solving the system of linear equations (3.53), which can be accomplished efficiently, in $\mathcal{O}(N)$ operations, using a simplified form of Gaussian elimination, called the Thomas algorithm (see, for example, Süli and Mayers, Sect. 3.3).

3.2.2 Stability of Explicit and Implicit Schemes

We shall explore the stability of the schemes (3.51) and (3.52) simultaneously, by embedding them into the following one-parameter family of finite difference schemes, called the *θ-scheme*: find U_j^m, $j = 0, \ldots, N$, $m = 0, \ldots, M$, such that

$$\begin{aligned} &D_t^+ U_j^m = D_x^+ D_x^- \left[\theta U_j^{m+1} + (1-\theta) U_j^m\right] + f(x_j, t_{m+\theta}), \\ &\quad j = 1, \ldots, N-1, \; m = 0, \ldots, M-1, \\ &U_0^m = 0, \qquad U_N^m = 0, \quad m = 0, \ldots, M, \\ &U_j^0 = u_0(x_j), \quad j = 0, \ldots, N, \end{aligned} \tag{3.54}$$

where $0 \le \theta \le 1$ and $t_{m+\theta} = t_m + \theta\tau = (m+\theta)\tau$. The most relevant special cases of this are $\theta = 0$ (the explicit scheme, also called the *explicit Euler scheme*), $\theta = 1$ (the implicit scheme, also called the *implicit Euler scheme*), and $\theta = 1/2$, referred to as the *Crank–Nicolson scheme*.

Let us consider the inner product

$$(V, W)_h := \sum_{j=1}^{N-1} h V_j W_j$$

and the associated norm

$$\|V\|_h := (V, V)_h^{1/2}.$$

By taking the inner product of (3.54) with

$$U^{m+\theta} := \theta U^{m+1} + (1 - \theta)U^m,$$

we get

$$\left(\frac{U^{m+1} - U^m}{\tau}, U^{m+\theta}\right)_h - \left(D_x^+ D_x^- U^{m+\theta}, U^{m+\theta}\right)_h = \left(f^{m+\theta}, U^{m+\theta}\right)_h,$$

where $f_j^{m+\theta} = f^{m+\theta}(x_j) = f(x_j, t_{m+\theta})$. Let, as in Sect. 2.2.2.1,

$$\llbracket V \rrbracket_h := \left(\sum_{j=0}^{N-1} h|V_j|^2\right)^{1/2}.$$

By noting that $U_0^{m+\theta} = 0$, $U_N^{m+\theta} = 0$, it is easily seen using summation by parts that

$$-\left(D_x^+ D_x^- U^{m+\theta}, U^{m+\theta}\right)_h = \llbracket D_x^+ U^{m+\theta} \rrbracket_h^2.$$

Thus,

$$\left(\frac{U^{m+1} - U^m}{\tau}, U^{m+\theta}\right)_h + \llbracket D_x^+ U^{m+\theta} \rrbracket_h^2 = \left(f^{m+\theta}, U^{m+\theta}\right)_h.$$

Since

$$U^{m+\theta} = \tau\left(\theta - \frac{1}{2}\right)\frac{U^{m+1} - U^m}{\tau} + \frac{U^{m+1} + U^m}{2},$$

it follows that

$$\tau\left(\theta - \frac{1}{2}\right)\left\|\frac{U^{m+1} - U^m}{\tau}\right\|_h^2 + \frac{\|U^{m+1}\|_h^2 - \|U^m\|_h^2}{2\tau}$$

$$+ \llbracket D_x^+ U^{m+\theta} \rrbracket_h^2 = \left(f^{m+\theta}, U^{m+\theta}\right)_h. \tag{3.55}$$

Suppose that $\theta \in [1/2, 1]$; then

$$\theta - \frac{1}{2} \geq 0, \tag{3.56}$$

and therefore

$$\frac{\|U^{m+1}\|_h^2 - \|U^m\|_h^2}{2\tau} + \llbracket D_x^+ U^{m+\theta} \rrbracket_h^2 \leq \left(f^{m+\theta}, U^{m+\theta}\right)_h$$

$$\leq \left\|f^{m+\theta}\right\|_h \left\|U^{m+\theta}\right\|_h.$$

According to the discrete Friedrichs inequality (2.24),

$$\|U^{m+\theta}\|_h^2 \le \frac{1}{8}\|[D_x^+ U^{m+\theta}\|_h^2,$$

and therefore

$$\frac{\|U^{m+1}\|_h^2 - \|U^m\|_h^2}{2\tau} + 8\|U^{m+\theta}\|_h^2 \le \frac{1}{32}\|f^{m+\theta}\|_h^2 + 8\|U^{m+\theta}\|_h^2,$$

so that

$$\|U^{m+1}\|_h^2 \le \|U^m\|_h^2 + \frac{\tau}{16}\|f^{m+\theta}\|_h^2.$$

Summing the last inequality through $m = 0, \ldots, k-1$ yields

$$\|U^k\|_h^2 \le \|U^0\|_h^2 + \frac{\tau}{16}\sum_{m=0}^{k-1}\|f^{m+\theta}\|_h^2, \tag{3.57}$$

for all k, $1 \le k \le M$.

The inequality (3.57) can be seen as the discrete version of (3.29), when $\lambda = 0$. If follows from (3.57) that

$$\max_{1\le k\le M}\|U^k\|_h^2 \le \|U^0\|_h^2 + \frac{\tau}{16}\sum_{m=0}^{M-1}\|f^{m+\theta}\|_h^2,$$

and therefore

$$\max_{1\le k\le M}\|U^k\|_h \le \left(\|U^0\|_h^2 + \frac{\tau}{16}\sum_{m=0}^{M-1}\|f^{m+\theta}\|_h^2\right)^{1/2}, \tag{3.58}$$

which expresses the stability of the finite difference scheme (3.54): the continuous dependence of the solution of the scheme on the initial datum and the right-hand side, uniformly in the discretization parameters h and τ.

Thus we have proved that for $\theta \in [1/2, 1]$ the scheme (3.54) is stable without any limitations on the time step τ in terms of h. In other words, the scheme (3.54) is *unconditionally stable* for $\theta \in [1/2, 1]$.

Let us now consider the case $\theta \in [0, 1/2)$. First suppose that $f = 0$. Then, according to (3.55),

$$\frac{\|U^{m+1}\|_h^2 - \|U^m\|_h^2}{2\tau} + \|[D_x^+ U^{m+\theta}\|_h^2 = \tau\left(\frac{1}{2} - \theta\right)\left\|\frac{U^{m+1} - U^m}{\tau}\right\|_h^2. \tag{3.59}$$

By (3.54) and our assumption that $f = 0$, it follows that

$$\frac{U^{m+1} - U^m}{\tau} = D_x^+ D_x^- U^{m+\theta}.$$

Moreover, a simple calculation based on the inequality $(a-b)^2 \leq 2a^2 + 2b^2$ shows that

$$\| D_x^+ D_x^- U^{m+\theta} \|_h^2 \leq \frac{4}{h^2} \| D_x^+ U^{m+\theta}]|_h^2. \tag{3.60}$$

Thus, (3.59) implies that

$$\frac{\|U^{m+1}\|_h^2 - \|U^m\|_h^2}{2\tau} + \| D_x^+ U^{m+\theta}]|_h^2 \leq \frac{4\tau}{h^2}\left(\frac{1}{2} - \theta\right) \| D_x^+ U^{m+\theta}]|_h^2,$$

and therefore

$$\frac{\|U^{m+1}\|_h^2 - \|U^m\|_h^2}{2\tau} + \left[1 - \frac{2\tau(1-2\theta)}{h^2}\right] \| D_x^+ U^{m+\theta}]|_h^2 \leq 0.$$

Let us assume that

$$\tau \leq \frac{h^2}{2(1-2\theta)}, \qquad \theta \in [0, 1/2); \tag{3.61}$$

then

$$\|U^{m+1}\|_h^2 \leq \|U^m\|_h^2, \quad m = 0, \ldots, M-1,$$

and hence,

$$\max_{1 \leq k \leq M} \|U^k\|_h \leq \|U^0\|_h.$$

Thus we have shown that for $\theta \in [0, 1/2)$ the scheme (3.54) is stable, provided that (3.61) holds; in other words, for $\theta \in [0, 1/2)$ the scheme is *conditionally stable*, the condition being (3.61) (when $f = 0$).

We shall suppose again that $\theta \in [0, 1/2)$, but will now consider the case when f is not identically zero. We shall prove that the finite difference scheme (3.54) is still only conditionally stable, and, in particular, that the explicit scheme, corresponding to $\theta = 0$, is conditionally stable.

According to (3.55), we have that

$$\frac{\|U^{m+1}\|_h^2 - \|U^m\|_h^2}{2\tau} + \| D_x^+ U^{m+\theta}]|_h^2 \leq \|f^{m+\theta}\|_h \|U^{m+\theta}\|_h + \tau\left(\frac{1}{2} - \theta\right) \left\| \frac{U^{m+1} - U^m}{\tau} \right\|_h^2. \tag{3.62}$$

By (3.54), for any $\varepsilon \in (0, 1)$,

$$\left\| \frac{U^{m+1} - U^m}{\tau} \right\|_h^2 = \| D_x^+ D_x^- U^{m+\theta} + f^{m+\theta} \|_h^2$$

$$\begin{aligned}
&\leq \left(\|D_x^+ D_x^- U^{m+\theta}\|_h + \|f^{m+\theta}\|_h\right)^2 \\
&\leq (1+\varepsilon)\|D_x^+ D_x^- U^{m+\theta}\|_h^2 + (1+\varepsilon^{-1})\|f^{m+\theta}\|_h^2 \\
&\leq (1+\varepsilon)\frac{4}{h^2}\|[D_x^+ U^{m+\theta}\|_h^2 + (1+\varepsilon^{-1})\|f^{m+\theta}\|_h^2,
\end{aligned}$$

where (3.60) was used in the last line. Substituting this into (3.62), we deduce that

$$\begin{aligned}
&\frac{\|U^{m+1}\|_h^2 - \|U^m\|_h^2}{2\tau} + \left[1 - \tau\left(\frac{1}{2}-\theta\right)\frac{4(1+\varepsilon)}{h^2}\right]\|[D_x^+ U^{m+\theta}\|_h^2 \\
&\leq \|f^{m+\theta}\|_h \|U^{m+\theta}\|_h + \tau\left(\frac{1}{2}-\theta\right)(1+\varepsilon^{-1})\|f^{m+\theta}\|_h^2. \qquad (3.63)
\end{aligned}$$

By applying the discrete Friedrichs inequality (2.24) according to which

$$\|U^{m+\theta}\|_h^2 \leq \frac{1}{8}\|[D_x^+ U^{m+\theta}\|_h^2,$$

we have that

$$\begin{aligned}
\|f^{m+\theta}\|_h \|U^{m+\theta}\|_h &\leq \frac{1}{32\varepsilon^2}\|f^{m+\theta}\|_h^2 + 8\varepsilon^2\|U^{m+\theta}\|_h^2 \\
&\leq \frac{1}{32\varepsilon^2}\|f^{m+\theta}\|_h^2 + \varepsilon^2\|[D_x^+ U^{m+\theta}\|_h^2. \qquad (3.64)
\end{aligned}$$

By inserting (3.64) into (3.63) we then deduce that

$$\begin{aligned}
&\frac{\|U^{m+1}\|_h^2 - \|U^m\|_h^2}{2\tau} + \left[1 - \tau\frac{2(1-2\theta)(1+\varepsilon)}{h^2} - \varepsilon^2\right]\|[D_x^+ U^{m+\theta}\|_h^2 \\
&\leq \frac{1}{32\varepsilon^2}\|f^{m+\theta}\|_h^2 + \tau\left(\frac{1}{2}-\theta\right)(1+\varepsilon^{-1})\|f^{m+\theta}\|_h^2.
\end{aligned}$$

Let us suppose that

$$\tau \leq \frac{h^2}{2(1-2\theta)}(1-\varepsilon), \quad \theta \in [0, 1/2),\ \varepsilon \in (0,1), \qquad (3.65)$$

where ε is a fixed real number. Then,

$$1 - \tau\frac{2(1-2\theta)(1+\varepsilon)}{h^2} - \varepsilon^2 \geq 0,$$

and therefore

$$\|U^{m+1}\|_h^2 \leq \|U^m\|_h^2 + \frac{\tau}{16\varepsilon^2}\|f^{m+\theta}\|_h^2 + \tau^2(1-2\theta)(1+\varepsilon^{-1})\|f^{m+\theta}\|_h^2.$$

By letting $c_\varepsilon := 1/(16\varepsilon^2) + T(1-2\theta)(1+\varepsilon^{-1})$, after summation through m this implies that

$$\max_{1\le k\le M} \|U^k\|_h^2 \le \|U^0\|_h^2 + c_\varepsilon \sum_{m=0}^{M-1} \tau \|f^{m+\theta}\|_h^2.$$

We take the square root of both sides to deduce that for $\theta \in [0, 1/2)$ the scheme (3.54) is *conditionally stable* in the sense that

$$\max_{1\le k\le M} \|U^k\|_h \le \left(\|U^0\|_h^2 + c_\varepsilon \sum_{m=0}^{M-1} \tau \|f^{m+\theta}\|_h^2 \right)^{1/2}, \tag{3.66}$$

provided that the condition (3.65) is satisfied.

To summarize, when $\theta \in [1/2, 1]$, the difference scheme (3.54) is unconditionally stable. In particular the implicit Euler scheme corresponding to $\theta = 1$ and the Crank–Nicolson scheme corresponding to $\theta = 1/2$ are both unconditionally stable, and (3.58) holds. When $\theta \in [0, 1/2)$, the scheme (3.54) is conditionally stable, subject to the time step limitation (3.65). In particular the explicit Euler scheme corresponding to $\theta = 0$ is only conditionally stable.

We close this section with a brief discussion about the connection between the "abstract" stability condition (3.35), the condition (3.56) (which in the case of the pure initial-value problem $\frac{du}{dt} = \lambda u$, $u(0) = 1$, $\lambda < 0$, guarantees the A-stability of the θ-scheme; cf. Süli and Mayers [172], for example,) and the requirement (3.61), which guarantees the conditional stability of the θ-scheme for the one-dimensional heat equation in the case of $f = 0$. The scheme (3.54) can be rewritten in the canonical form (3.31), where $\mathcal{H}^h = \mathcal{S}_0^h$ is the set of all mesh-functions defined on $\overline{\Omega}^h$ and equal to 0 on $\overline{\Omega}^h \setminus \Omega^h$,

$$A_h V := \Lambda V = \begin{cases} -D_x^+ D_x^- V & \text{for } x \in \Omega^h, \\ 0 & \text{for } x \in \overline{\Omega}^h \setminus \Omega^h, \end{cases}$$

and

$$B_h := I_h + \left(\theta - \frac{1}{2}\right)\tau \Lambda,$$

where I_h is the identity operator on $\mathcal{S}_0^h$. The operator Λ is linear, selfadjoint and positive definite on $\mathcal{S}_0^h$, uniformly with respect to h, and (cf. (2.18), (2.22)) we have that

$$8\|V\|_h^2 \le (\Lambda V, V)_h < \frac{4}{h^2}\|V\|_h^2, \quad \text{i.e.} \quad 8I_h \le \Lambda < \frac{4}{h^2} I_h.$$

Thus, when $\theta \ge 1/2$ the condition (3.35) is trivially satisfied. When $\theta < 1/2$, we have that

$$I_h + \left(\theta - \frac{1}{2}\right)\tau \Lambda > \left[1 - \left(\frac{1}{2} - \theta\right)\frac{4\tau}{h^2}\right] I_h \ge 0,$$

provided that the condition (3.61) holds.

3.2.3 Error Analysis of Difference Schemes for the Heat Equation

In this section we investigate the accuracy of the finite difference scheme (3.54) for the numerical solution of the initial-boundary-value problem (3.50).

We define the truncation error of the scheme (3.54) by

$$\varphi_j^{m+\theta} := D_t^+ u(x_j, t_m) - D_x^+ D_x^- \left[\theta u(x_j, t_{m+1}) + (1-\theta) u(x_j, t_m)\right] - f(x_j, t_{m+\theta}),$$

for $j = 1, \ldots, N-1$, $m = 0, \ldots, M-1$, and the global error of the scheme by

$$e_j^m := u(x_j, t_m) - U_j^m,$$

for $j = 0, \ldots, N$, $m = 0, \ldots, M$. It is easily seen that e_j^m satisfies the following finite difference scheme:

$$D_t^+ e_j^m - D_x^+ D_x^- \left[\theta e_j^{m+1} + (1-\theta) e_j^m\right] = \varphi_j^{m+\theta}, \quad \begin{cases} j = 1, \ldots, N-1, \\ m = 0, \ldots, M-1. \end{cases}$$

$$e_0^m = 0, \qquad e_N^m = 0, \quad m = 0, \ldots, M,$$

$$e_j^0 = 0, \quad j = 0, \ldots, N.$$

Thanks to the stability results proved in the previous section,

$$\max_{1 \le m \le M} \left\| u^m - U^m \right\|_h \le \left(\frac{\tau}{16} \sum_{k=0}^{M-1} \left\| \varphi^{k+\theta} \right\|_h^2 \right)^{1/2}, \quad \theta \in [1/2, 1], \tag{3.67}$$

by (3.58). Also, by (3.66),

$$\max_{1 \le m \le M} \left\| u^m - U^m \right\|_h \le \left(c_\varepsilon \tau \sum_{k=0}^{M-1} \left\| \varphi^{k+\theta} \right\|_h^2 \right)^{1/2}, \quad \theta \in [0, 1/2), \tag{3.68}$$

provided that (3.65) holds. In either case we have to estimate $\|\varphi^{m+\theta}\|_h$ in order to complete the error analysis. By recalling the differential equation $\partial u/\partial t = \partial^2 u/\partial x^2 + f$ satisfied by u, we deduce that $\varphi_j^{m+\theta}$ can be written as

$$\varphi_j^{m+\theta} = \left[\frac{u(x_j, t_{m+1}) - u(x_j, t_m)}{\tau} - \frac{\partial u}{\partial t}(x_j, t_{m+\theta}) \right] + \left[\frac{\partial^2 u}{\partial x^2}(x_j, t_{m+\theta}) - D_x^+ D_x^- \left(\theta u(x_j, t_{m+1}) + (1-\theta) u(x_j, t_m)\right) \right]. \tag{3.69}$$

In order to estimate the size of the truncation error, $\varphi_j^{m+\theta}$, we expand it in a Taylor series about the point $(x_j, t_{m+1/2})$, assuming that u is sufficiently smooth. To this end, we begin by noting that

$$u_j^{m+1} = \left[u + \frac{1}{2}\tau\frac{\partial u}{\partial t} + \frac{1}{2}\left(\frac{\tau}{2}\right)^2\frac{\partial^2 u}{\partial t^2} + \frac{1}{6}\left(\frac{\tau}{2}\right)^3\frac{\partial^3 u}{\partial t^3} + \cdots\right]_j^{m+1/2},$$

$$u_j^{m} = \left[u - \frac{1}{2}\tau\frac{\partial u}{\partial t} + \frac{1}{2}\left(\frac{\tau}{2}\right)^2\frac{\partial^2 u}{\partial t^2} - \frac{1}{6}\left(\frac{\tau}{2}\right)^3\frac{\partial^3 u}{\partial t^3} + \cdots\right]_j^{m+1/2}.$$

By subtracting the second of these expansions from the first we obtain

$$\frac{u(x_j, t_{m+1}) - u(x_j, t_m)}{\tau} = \left[\frac{\partial u}{\partial t} + \frac{1}{24}\tau^2\frac{\partial^3 u}{\partial t^3} + \cdots\right]_j^{m+1/2}. \tag{3.70}$$

Also, since

$$D_x^+ D_x^- u(x_j, t_{m+1}) = \left[\frac{\partial^2 u}{\partial x^2} + \frac{1}{12}h^2\frac{\partial^4 u}{\partial x^4} + \frac{2}{6!}h^4\frac{\partial^6 u}{\partial x^6} + \cdots\right]_j^{m+1},$$

by expanding the right-hand side about the point $(x_j, t_{m+1/2})$ yields that

$$\begin{aligned} D_x^+ D_x^- u(x_j, t_{m+1}) &= \left[\frac{\partial^2 u}{\partial x^2} + \frac{1}{12}h^2\frac{\partial^4 u}{\partial x^4} + \frac{2}{6!}h^4\frac{\partial^6 u}{\partial x^6} + \cdots\right]_j^{m+1/2} \\ &\quad + \frac{\tau}{2}\left[\frac{\partial^3 u}{\partial x^2 \partial t} + \frac{1}{12}h^2\frac{\partial^5 u}{\partial x^4 \partial t} + \cdots\right]_j^{m+1/2} \\ &\quad + \frac{1}{2}\left(\frac{\tau}{2}\right)^2\left[\frac{\partial^4 u}{\partial x^2 \partial t^2} + \cdots\right]_j^{m+1/2}. \end{aligned}$$

There is a similar expansion for $D_x^+ D_x^- u(x_j, t_m)$. Combining these gives

$$\begin{aligned} &D_x^+ D_x^- \left[\theta u(x_j, t_{m+1}) + (1-\theta) u(x_j, t_m)\right] \\ &= \left[\frac{\partial^2 u}{\partial x^2} + \frac{1}{12}h^2\frac{\partial^4 u}{\partial x^4} + \frac{2}{6!}h^4\frac{\partial^6 u}{\partial x^6} + \cdots\right]_j^{m+1/2} \\ &\quad + \left(\theta - \frac{1}{2}\right)\tau\left[\frac{\partial^3 u}{\partial x^2 \partial t} + \frac{1}{12}h^2\frac{\partial^5 u}{\partial x^4 \partial t} + \cdots\right]_j^{m+1/2} \\ &\quad + \frac{1}{8}\tau^2\left[\frac{\partial^4 u}{\partial x^2 \partial t^2} + \cdots\right]_j^{m+1/2}. \end{aligned} \tag{3.71}$$

We also require the following Taylor expansions:

$$\frac{\partial u}{\partial t}(x_j, t_{m+\theta}) = \left[\frac{\partial u}{\partial t} + \left(\theta - \frac{1}{2}\right)\tau\frac{\partial^2 u}{\partial t^2} + \left(\theta - \frac{1}{2}\right)^2 \frac{\tau^2}{2}\frac{\partial^3 u}{\partial t^3} + \cdots\right]_j^{m+1/2}, \tag{3.72}$$

$$\frac{\partial^2 u}{\partial x^2}(x_j, t_{m+\theta}) = \left[\frac{\partial^2 u}{\partial x^2} + \left(\theta - \frac{1}{2}\right)\tau\frac{\partial^3 u}{\partial t\partial x^2} + \left(\theta - \frac{1}{2}\right)^2 \frac{\tau^2}{2}\frac{\partial^4 u}{\partial t^2\partial x^2} + \cdots\right]_j^{m+1/2}. \tag{3.73}$$

Substituting (3.70)–(3.73) into (3.69) yields that

$$\begin{aligned}\varphi_j^{m+\theta} &= \left[\left(\frac{1}{2} - \theta\right)\tau\frac{\partial^2 u}{\partial t^2} - \frac{h^2}{12}\frac{\partial^4 u}{\partial x^4}\right]_j^{m+1/2} \\ &+ \frac{\tau^2}{2}\left[\left(-\theta^2 + \theta - \frac{1}{6}\right)\frac{\partial^3 u}{\partial t^3} + \left(\theta^2 - \theta\right)\frac{\partial^4 u}{\partial t^2\partial x^2}\right]_j^{m+1/2} \\ &+ \frac{\tau h^2}{12}\left(\frac{1}{2} - \theta\right)\frac{\partial^5 u}{\partial t\partial x^4}\bigg|_j^{m+1/2} - \frac{h^4}{360}\frac{\partial^6 u}{\partial x^6}\bigg|_j^{m+1/2} + \cdots .\end{aligned}$$

Hence,

$$\left|\varphi_j^{m+\theta}\right| \le \frac{h^2}{12}M_{4x} + \frac{\tau^2}{24}(M_{3t} + 3M_{2x2t}) + \text{H.O.T.}, \quad \theta = \frac{1}{2}, \tag{3.74}$$

$$\left|\varphi_j^{m+\theta}\right| \le \frac{h^2}{12}M_{4x} + \left|\frac{1}{2} - \theta\right|\tau M_{2t} + \text{H.O.T.}, \quad \theta \ne \frac{1}{2}, \tag{3.75}$$

with

$$M_{kxlt} := \max_{(x,t)\in\overline{Q}}\left|\frac{\partial^{k+l}}{\partial x^k\partial t^l}u(x,t)\right|,$$

$M_{kx} := M_{kx0t}$, $M_{lt} := M_{0xlt}$, and we assume that M_{4x}, M_{3t}, M_{2x2t} and M_{2t} are finite. H.O.T. signifies terms of higher order than h^2 and τ^2, and h^2 and τ, respectively. Substituting (3.74) into (3.67) and (3.75) into (3.67) or (3.68), and absorbing terms of higher order into lower order terms and altering the constants if necessary, we obtain the following error bounds:

$$\max_{1\le m\le M}\left\|u^m - U^m\right\|_h \le C_1\left(h^2 + \tau^2\right), \quad \theta = \frac{1}{2}, \tag{3.76}$$

where C_1 is a positive constant, independent of h and τ;

$$\max_{1\le m\le M} \|u^m - U^m\|_h \le C_2(h^2+\tau), \quad \theta \in (1/2, 1], \tag{3.77}$$

where C_2 is a positive constant, independent of h and τ. Moreover,

$$\max_{1\le m\le M} \|u^m - U^m\|_h \le C_3(h^2+\tau), \quad \theta \in [0, 1/2), \tag{3.78}$$

where $C_3 = (c_\varepsilon)^{1/2} C_2$, provided that the condition (3.65) is fulfilled. Thus we deduce that the Crank–Nicolson scheme ($\theta = 1/2$) converges in the norm $\|\cdot\|_h$ unconditionally, with a global error of size $\mathcal{O}(h^2+\tau^2)$. For $\theta \in (1/2, 1]$ the scheme converges unconditionally, with a global error of size $\mathcal{O}(h^2+\tau)$. For $\theta \in [0, 1/2)$ the difference scheme converges with a global error of size $\mathcal{O}(h^2+\tau)$, but only conditionally. These error bounds have been derived under quite restrictive requirements on the smoothness of the solution. In the next section we shall be concerned with the error analysis of the difference schemes described above when the solution is less regular.

3.3 The Heat Equation with Nonsmooth Data

It is frequently the case in physical applications that the initial datum $u_0 = u_0(x)$ and the source term $f = f(x,t)$ in the heat equation are nonsmooth functions. In such instances the error analysis described in the previous section no longer applies, as the solution $u(x,t)$ may not have sufficiently many derivatives bounded and continuous on $[0,1] \times [0,T]$. In this section, we address this issue in the case of the implicit Euler scheme corresponding to $\theta = 1$ in the one-parameter family of schemes considered in Sect. 3.2.

3.3.1 The Initial-Boundary-Value Problem and Its Discretization

We consider the initial-boundary-value problem (3.50) in the space-time domain $Q := (0,1) \times (0,T]$. The mesh $\overline{Q}_h^\tau$ is defined in the same way as in Sect. 3.2. We shall also retain the other notations introduced in Sect. 3.2.

We begin with a general discussion concerning the construction of finite difference approximations to our model problem. As we shall be concerned with nonsmooth data and, more specifically, with $f \in L_2((0,T); L_2(\Omega))$ and $u_0 \in L_2(\Omega)$, we mollify these functions so that the resulting mollified functions are continuous and have, therefore, well-defined values at the mesh-points. For this purpose we consider, for any function v that is defined and sufficiently smooth on Q,

$$T_x^2 v(x,t) := \frac{1}{h}\int_{x-h}^{x+h}\left(1 - \left|\frac{x-x'}{h}\right|\right) v(x',t)\,\mathrm{d}x',$$

and

$$T_t^- v(x,t) := \frac{1}{\tau}\int_{t-\tau}^{t} v(x,t')\,\mathrm{d}t' = T_t^+ v(x,t-\tau).$$

If v is a distribution on Q, T_x^2 and $T_t^{\mp}$ should be interpreted as convolutions (cf. Sect. 1.9.4). The mollifiers T_x^2 and T_t^- have the following properties:

$$T_x^2\left(\frac{\partial^2 v}{\partial x^2}\right) = D_x^+ D_x^- v \quad \text{and} \quad T_t^-\left(\frac{\partial v}{\partial t}\right) = D_t^- v.$$

Suppose that u, the weak solution of the initial-boundary-value problem (3.50), belongs to the anisotropic Sobolev space $W_2^{s,s/2}(Q)$, $s > 1$. Then, $T_x^2 u$ and $T_t^- u$ are continuous functions on $\overline{Q}$, and by applying $T = T_x^2 T_t^- = T_t^- T_x^2$ to the heat equation in (3.50) we thus obtain

$$D_t^-(T_x^2 u)_j^m = D_x^+ D_x^-(T_t^- u)_j^m + (T_x^2 T_t^- f)_j^m.$$

This identity motivates our definition of the finite difference approximation of problem (3.50): find a real-valued function U defined on the mesh $\overline{Q}_h^{\tau}$ such that

$$\begin{aligned} &D_t^- U_j^m = D_x^+ D_x^- U_j^m + (T_x^2 T_t^- f)_j^m, \quad j = 1,\dots,N-1,\ m = 1,\dots,M, \\ &U_0^m = 0, \qquad U_N^m = 0, \quad m = 1,\dots,M, \end{aligned} \tag{3.79}$$

and subject to one of the following initial conditions:

$$U_j^0 = u_0(x_j), \quad j = 1,\dots,N-1, \tag{3.80}$$

or

$$U_j^0 = T_x^2 u_0(x_j), \quad j = 1,\dots,N-1, \tag{3.81}$$

the choice being dependent on the smoothness of the initial datum. It will be clear from the error bounds that will be derived below which of the two initial conditions is appropriate in each particular instance.

3.3.2 Error Analysis

Let us define the global error of the scheme in the usual way:

$$e_j^m := u(x_j, t_m) - U_j^m.$$

It is easily seen that in the case of the scheme (3.79), (3.80) the global error thus defined satisfies

$$\begin{aligned} &D_t^- e_j^m - D_x^+ D_x^- e_j^m = D_t^- \psi_j^m - D_x^+ D_x^- \eta_j^m, \\ &\quad j = 1, \ldots, N-1, \; m = 1, \ldots, M, \\ &e_0^m = 0, \qquad e_N^m = 0, \quad m = 1, \ldots, M, \\ &e_j^0 = 0, \quad j = 1, \ldots, N-1, \end{aligned} \tag{3.82}$$

where we have used the notations

$$\psi := u - T_x^2 u, \qquad \eta := u - T_t^- u.$$

Similarly, in the case of the scheme (3.79), (3.81) the global error satisfies

$$\begin{aligned} &D_t^- e_j^m - D_x^+ D_x^- e_j^m = D_t^- \psi_j^m - D_x^+ D_x^- \eta_j^m, \\ &\quad j = 1, \ldots, N-1, \; m = 1, \ldots, M, \\ &e_0^m = 0, \qquad e_N^m = 0, \quad m = 1, \ldots, M, \\ &e_j^0 = \psi_j^0, \quad j = 1, \ldots, N-1. \end{aligned} \tag{3.83}$$

Let us define the following mesh-dependent anisotropic Sobolev norms:

$$\|V\|^2_{L_2(Q_h^\tau)} := \tau \sum_{t \in \Omega_+^\tau} \|V(\cdot,t)\|_h^2,$$

$$\begin{aligned} \|V\|^2_{W_2^{1,1/2}(Q_h^\tau)} &:= \tau \sum_{t \in \Omega_+^\tau} \left(\|V(\cdot,t)\|_h^2 + [\![D_x^+ V(\cdot,t)]\!]_h^2 \right) \\ &\quad + \tau^2 \sum_{t \in \overline{\Omega}^\tau} \sum_{t' \in \overline{\Omega}^\tau,\, t' \neq t} \frac{\|V(\cdot,t) - V(\cdot,t')\|_h^2}{|t-t'|^2}, \end{aligned}$$

$$\begin{aligned} \|V\|^2_{W_2^{2,1}(Q_h^\tau)} &:= \tau \sum_{t \in \Omega_+^\tau} \left(\|V(\cdot,t)\|_h^2 + [\![D_x^+ V(\cdot,t)]\!]_h^2 \right. \\ &\quad \left. + \|D_x^+ D_x^- V(\cdot,t)\|_h^2 + \|D_t^- V(\cdot,t)\|_h^2 \right). \end{aligned}$$

The scheme (3.82) can be rewritten as an operator-difference scheme (3.31), where $\mathcal{H}^h = S_0^h$, $A_h = \Lambda$ and $F = D_t^- \psi - D_x^+ D_x^- \eta$. Thus we deduce from (3.40) that

$$\tau \sum_{t \in \Omega_+^\tau} \left(\|D_x^+ D_x^- e(\cdot,t)\|_h^2 + \|D_t^- e(\cdot,t)\|_h^2 \right)$$

$$\leq \tau \sum_{t\in\Omega_+^\tau} \left\| D_t^- \psi(\cdot,t) - D_x^+ D_x^- \eta(\cdot,t)\right\|_h^2.$$

Hence, by applying the discrete Friedrichs inequality (2.24) we have that

$$\|e\|^2_{W_2^{2,1}(Q_h^\tau)} \leq C\tau \sum_{t\in\Omega_+^\tau} \left(\left\|D_x^+ D_x^- \eta(\cdot,t)\right\|_h^2 + \left\|D_t^- \psi(\cdot,t)\right\|_h^2\right), \tag{3.84}$$

where $C = 2(1 + 1/8 + 1/64) = 73/32$. Similarly, (3.31), (3.44) and (3.46) imply that

$$\tau \sum_{t\in\Omega_+^\tau} \left\|e(\cdot,t)\right\|_A^2 + \tau^2 \sum_{t\in\overline{\Omega}^\tau} \sum_{t'\in\overline{\Omega}^\tau,\ t'\neq t} \frac{\|e(\cdot,t) - e(\cdot,t')\|_h^2}{|t-t'|^2}$$
$$\leq C\Bigg[\tau \sum_{t\in\Omega_+^\tau} \left\|\eta(\cdot,t)\right\|_A^2 + \tau^2 \sum_{t\in\overline{\Omega}^\tau} \sum_{t'\in\overline{\Omega}^\tau,\ t'\neq t} \frac{\|\psi(\cdot,t) - \psi(\cdot,t')\|_h^2}{|t-t'|^2}$$
$$+ \tau \sum_{t\in\Omega^\tau} \left(\frac{1}{t} + \frac{1}{T-t}\right) \left\|\psi(\cdot,t)\right\|_h^2\Bigg].$$

Thus, by noting (2.22) and (2.24), we deduce that

$$\|e\|^2_{W_2^{1,1/2}(Q_h^\tau)} \leq C\Bigg[\tau \sum_{t\in\Omega_+^\tau} \left\|D_x^+ \eta(\cdot,t)\right\|_h^2 + \tau^2 \sum_{t\in\overline{\Omega}^\tau} \sum_{t'\in\overline{\Omega}^\tau,\ t'\neq t} \frac{\|\psi(\cdot,t) - \psi(\cdot,t')\|_h^2}{|t-t'|^2}$$
$$+ \tau \sum_{t\in\Omega^\tau} \left(\frac{1}{t} + \frac{1}{T-t}\right) \left\|\psi(\cdot,t)\right\|_h^2\Bigg]. \tag{3.85}$$

Similarly, in the case of (3.83), using (3.41) and (3.49) we have that

$$\|e\|^2_{L_2(Q_h^\tau)} \leq C\tau \sum_{t\in\Omega_+^\tau} \left(\left\|\eta(\cdot,t)\right\|_h^2 + \left\|\psi(\cdot,t)\right\|_h^2\right). \tag{3.86}$$

Now, to derive error bounds for the finite difference scheme (3.79), (3.80) in the mesh-dependent $W_2^{2,1}$ and $W_2^{1,1/2}$ norms, and for the finite difference scheme (3.79), (3.81) in the mesh-dependent L_2 norm, it suffices to bound the norms of η and ψ appearing in the expressions on the right-hand sides of the inequalities (3.84), (3.85) and (3.86), respectively.

Suppose that (x_j, t_m) is an arbitrary node of the mesh Q_h^τ and consider the associated 'elementary rectangle' $G_j^m = (x_{j-1}, x_{j+1}) \times (t_{m-1}, t_m)$. By using the linear transformation

$$x = x_j + h\tilde{x}, \qquad t = t_m + \tau\tilde{t}, \quad -1 < \tilde{x} < 1,\ -1 < \tilde{t} < 0,$$

G_j^m is bijectively mapped onto the canonical rectangle $\tilde{G} := (-1,1) \times (-1,0)$. By defining $\tilde{u}(\tilde{x},\tilde{t}) := u(x_j + h\tilde{x}, t_m + \tau\tilde{t})$ we have that

$$D_x^+ D_x^- \eta_j^m = \frac{1}{h^2}\Big\{\tilde{u}(-1,0) - 2\tilde{u}(0,0) + \tilde{u}(1,0) - \int_{-1}^{0} \big[\tilde{u}(-1,\tilde{t}) - 2\tilde{u}(0,\tilde{t}) + \tilde{u}(1,\tilde{t})\big]\,\mathrm{d}\tilde{t}\Big\}.$$

Using the Sobolev embedding theorem, we deduce that

$$|D_x^+ D_x^- \eta_j^m| \le \frac{8}{h^2}\|\tilde{u}\|_{C(\bar{\tilde{G}})} \le \frac{C}{h^2}\|\tilde{u}\|_{W_2^{s,s/2}(\tilde{G})}, \quad s > 3/2.$$

Thus we have shown that $D_x^+ D_x^- \eta_j^m$ is a bounded linear functional of $\tilde{u} \in W_2^{s,s/2}(\tilde{G})$ for $s > 3/2$. It can be directly verified that $D_x^+ D_x^- \eta_j^m$ vanishes on all monomials of the form $\tilde{u} = \tilde{x}^\alpha \tilde{t}^\beta$, where α and β are nonnegative integers such that $\alpha + 2\beta < 4$. The Bramble–Hilbert lemma thus yields that

$$|D_x^+ D_x^- \eta_j^m| \le \frac{C}{h^2}|\tilde{u}|_{W_2^{s,s/2}(\tilde{G})}, \quad 3/2 < s \le 4. \tag{3.87}$$

Returning to the original variables x and t we deduce that

$$|D_x^+ D_x^- \eta_j^m| \le \frac{C}{h^2\sqrt{h\tau}}(h^2+\tau)^{s/2}|u|_{W_2^{s,s/2}(G_j^m)}, \quad 3/2 < s \le 4.$$

One can show in the same way that

$$|D_t^- \psi_j^m| \le \frac{C}{\tau\sqrt{h\tau}}(h^2+\tau)^{s/2}|u|_{W_2^{s,s/2}(G_j^m)}, \quad 3/2 < s \le 4.$$

Summing over the nodes of the mesh Q_h^τ gives

$$\tau \sum_{t\in\Omega_+^\tau} \big(\|D_x^+ D_x^- \eta(\cdot,t)\|_h^2 + \|D_t^- \psi(\cdot,t)\|_h^2\big) \le C(h^2+\tau)^s\Big(\frac{1}{h^4} + \frac{1}{\tau^2}\Big)|u|^2_{W_2^{s,s/2}(Q)}.$$

If the mesh-sizes h and τ satisfy the condition $\tau \asymp h^2$, i.e.

$$c_1 h^2 \le \tau \le c_2 h^2, \quad c_1, c_2 = \text{Const.} > 0, \tag{3.88}$$

we further have that

$$\tau \sum_{t\in\Omega_+^\tau} \big(\|D_x^+ D_x^- \eta(\cdot,t)\|_h^2 + \|D_t^- \psi(\cdot,t)\|_h^2\big) \le C h^{2s-4}|u|^2_{W_2^{s,s/2}(Q)}.$$

From this inequality and (3.84), limiting ourselves to the values $s \in (2,4]$, we obtain the following bound on the global error of the finite difference scheme (3.79), (3.80):

$$\|u - U\|_{W_2^{2,1}(Q_h^\tau)} \le C h^{s-2}\|u\|_{W_2^{s,s/2}(Q)}, \quad 2 < s \le 4. \tag{3.89}$$

Similarly, $D_x^+\eta_j^m$ is a bounded linear functional of $u \in W_2^{s,s/2}(G_{j+}^m)$, with $G_{j+}^m := (x_j, x_{j+1}) \times (t_{m-1}, t_m)$, for $s > 3/2$, which vanishes on all monomials of the form $u = x^\alpha t^\beta$, $\alpha + 2\beta < 3$. Assuming that the condition (3.88) holds, using the Bramble–Hilbert lemma we have that

$$\left|D_x^+\eta_j^m\right| \leq Ch^{s-5/2}|u|_{W_2^{s,s/2}(G_{j+}^m)}, \quad 3/2 < s \leq 3,$$

whence, by summing over the nodes of the mesh, we deduce that

$$\tau \sum_{t\in\Omega_+^\tau} \|D_x^+\eta(\cdot,t)\|_h^2 \leq Ch^{2s-2}|u|_{W_2^{s,s/2}(Q)}^2, \quad 3/2 < s \leq 3. \tag{3.90}$$

Further,

$$\tau^2 \sum_{t\in\overline{\Omega}^\tau} \sum_{t'\in\overline{\Omega}^\tau,\ t'\neq t} \frac{\|\psi(\cdot,t) - \psi(\cdot,t')\|_h^2}{|t-t'|^2}$$
$$= 2h\tau^2 \sum_{m=1}^{M}\sum_{k=0}^{m-1}\sum_{j=1}^{N-1} \frac{|\psi_j^m - \psi_j^k|^2}{|t_m - t_k|^2}$$
$$\leq 6h\tau^2 \sum_{m=1}^{M}\sum_{k=0}^{m-1}\sum_{j=1}^{N-1} \frac{|\psi_j^m - T_t^-\psi_j^m|^2}{|t_m - t_k|^2} + 6h\tau^2 \sum_{m=1}^{M}\sum_{k=0}^{m-1}\sum_{j=1}^{N-1} \frac{|T_t^-\psi_j^m - T_t^+\psi_j^k|^2}{|t_m - t_k|^2}$$
$$+ 6h\tau^2 \sum_{m=1}^{M}\sum_{k=0}^{m-1}\sum_{j=1}^{N-1} \frac{|T_t^+\psi_j^k - \psi_j^k|^2}{|t_m - t_k|^2} =: J_1 + J_2 + J_3.$$

Let us estimate separately each of J_1, J_2 and J_3. Clearly,

$$J_1 = 6h \sum_{m=1}^{M}\sum_{k=0}^{m-1}\sum_{j=1}^{N-1} \frac{|\psi_j^m - T_t^-\psi_j^m|^2}{|m-k|^2} \leq \frac{\pi^2}{\tau} h\tau \sum_{m=1}^{M}\sum_{j=1}^{N-1} |\psi_j^m - T_t^-\psi_j^m|^2.$$

It is easily seen that the expression $\psi_j^m - T_t^-\psi_j^m$ is a bounded linear functional of $u \in W_2^{s,s/2}(G_j^m)$, $s > 3/2$, which vanishes on all monomials of the form $u = x^\alpha t^\beta$, $\alpha + 2\beta < 4$. Similarly as in the previous cases we deduce that

$$\left|\psi_j^m - T_t^-\psi_j^m\right| \leq Ch^{s-3/2}|u|_{W_2^{s,s/2}(G_j^m)}, \quad 3/2 < s \leq 4,$$

and then, by summing over the nodes of the mesh, we arrive at the bound

$$J_1 \leq Ch^{2s-2}|u|_{W_2^{s,s/2}(Q)}^2, \quad 3/2 < s \leq 4. \tag{3.91}$$

An identical bound holds for J_3. Let us now bound J_2:

$$J_2 = \sum_{m=1}^{M}\sum_{k=0}^{m-1}\sum_{j=1}^{N-1} \frac{6h\tau^2}{|t_m - t_k|^2}\left\{\frac{1}{\tau^2}\int_{t_{m-1}}^{t_m}\int_{t_k}^{t_{k+1}} [\psi(x_j,t) - \psi(x_j,t')]\,\mathrm{d}t'\,\mathrm{d}t\right\}^2$$

$$\leq \sum_{m=1}^{M}\sum_{k=0}^{m-1}\sum_{j=1}^{N-1} \frac{6h\tau^{-2}}{|t_m - t_k|^2}\left\{\int_{t_{m-1}}^{t_m}\int_{t_k}^{t_{k+1}} \frac{[\psi(x_j,t) - \psi(x_j,t')]^2}{|t-t'|^{1+2\lambda}}\,\mathrm{d}t'\,\mathrm{d}t\right\}$$

$$\times\left\{\int_{t_{m-1}}^{t_m}\int_{t_k}^{t_{k+1}} |t-t'|^{1+2\lambda}\,\mathrm{d}t'\,\mathrm{d}t\right\}, \quad \lambda > 0.$$

By choosing $\lambda \in (0, 1/2]$ we then have that

$$J_2 \leq 6h\tau^{2\lambda-1}\sum_{m=1}^{M}\sum_{k=0}^{m-1}\sum_{j=1}^{N-1}\int_{t_{m-1}}^{t_m}\int_{t_k}^{t_{k+1}} \frac{[\psi(x_j,t) - \psi(x_j,t')]^2}{|t-t'|^{1+2\lambda}}\,\mathrm{d}t'\,\mathrm{d}t$$

$$\leq 6h\tau^{2\lambda-1}\sum_{j=1}^{N-1}\int_0^T\int_0^T \frac{[\psi(x_j,t) - \psi(x_j,t')]^2}{|t-t'|^{1+2\lambda}}\,\mathrm{d}t'\,\mathrm{d}t.$$

Hence, by noting the integral representation

$$\psi(x_j,t) = \frac{1}{h}\int_{x_{j-1}}^{x_{j+1}}\int_x^{x_j}\int_{x_j}^{x'}\left(1 - \frac{|x-x_j|}{h}\right)\frac{\partial^2 u}{\partial x^2}(x'',t)\,\mathrm{d}x''\,\mathrm{d}x'\,\mathrm{d}x$$

and the condition (3.88), we arrive at the bound

$$J_2 \leq Ch^{2+4\lambda}|u|^2_{W_2^{2+2\lambda,1+\lambda}(Q)}, \quad 0 < \lambda \leq 1/2. \tag{3.92}$$

Similarly, by using the integral representation

$$\psi(x_j,t) = \frac{1}{h}\int_{x_{j-1}}^{x_{j+1}}\int_x^{x_j}\left(1 - \frac{|x-x_j|}{h}\right)\frac{\partial u}{\partial x}(x',t)\,\mathrm{d}x'\,\mathrm{d}x,$$

we deduce that

$$J_2 \leq Ch^{4\lambda}|u|^2_{W_2^{1+2\lambda,1/2+\lambda}(Q)}, \quad 0 < \lambda \leq 1/2. \tag{3.93}$$

From (3.92) and (3.93), taking in the first case $s = 2 + 2\lambda$ and in the second case $s = 1 + 2\lambda$, we have that

$$J_2 \leq Ch^{2s-2}|u|^2_{W_2^{s,s/2}(Q)}, \quad 1 < s \leq 3. \tag{3.94}$$

Suppose that $t = t_m \in \Omega_+^\tau$ is fixed, and let us consider $u(\cdot, t_m)$ as a function of the variable x. The expression $\psi_j^m = u(x_j, t_m) - T_x^2 u(x_j, t_m)$ is a bounded linear functional of $u(\cdot, t_m) \in W_2^r(x_{j-1}, x_{j+1})$, $r > 1/2$, which vanishes on all polynomials (in x) of degree ≤ 2. Thanks to the Bramble–Hilbert lemma we thus have that

$$|\psi_j^m| \leq C h^{r-1/2} |u(\cdot, t_m)|_{W_2^r(x_{j-1}, x_{j+1})}, \quad 1/2 < r \leq 2.$$

Summing over the nodes of the mesh Ω^h and applying the trace theorem, we obtain the following bound:

$$\|\psi^m\|_h \leq C h^r |u(\cdot, t_m)|_{W_2^r(0,1)} \leq C h^r \|u\|_{W_2^{r+1,(r+1)/2}(Q)}, \quad 1/2 < r \leq 2.$$

Finally, by writing $r+1 = s$ and summing over the mesh Ω^τ we obtain that

$$\tau \sum_{t \in \Omega^\tau} \left(\frac{1}{t} + \frac{1}{T-t} \right) \|\psi(\cdot, t)\|_h^2 \leq C h^{2s-2} \log \frac{1}{\tau} \|u\|^2_{W_2^{s,s/2}(Q)}, \tag{3.95}$$

where $3/2 < s \leq 3$.

From (3.85), (3.88), (3.90), (3.91), (3.94) and (3.95) we have the following bound on the global error of the finite difference scheme (3.79), (3.80):

$$\|u - U\|_{W_2^{1,1/2}(Q_h^\tau)} \leq C h^{s-1} \sqrt{\log(1/h)} \|u\|_{W_2^{s,s/2}(Q)}, \quad 3/2 < s \leq 3. \tag{3.96}$$

The bound (3.96) is 'almost' compatible with the smoothness of the solution, the slight shortfall from full compatibility being due to the presence of the term $\sqrt{\log(1/h)}$, $h \in (0, 1/2]$, which increases, albeit very slowly, as $h \to 0_+$.

Let us finally bound the global error of the finite difference scheme (3.79), (3.81) in the mesh-dependent L_2 norm. The expressions ψ_j^m and η_j^m are bounded linear functionals of $u \in W_2^{s,s/2}(G_j^m)$, $s > 3/2$, which vanish on monomials of the form $u = x^\alpha t^\beta$, where α and β are nonnegative integers and $\alpha + 2\beta < 2$. By applying the Bramble–Hilbert lemma, under the condition (3.88), we get that

$$|\psi_j^m|, |\eta_j^m| \leq C h^{s-3/2} |u|_{W_2^{s,s/2}(G_j^m)}, \quad 3/2 < s \leq 2.$$

Hence, by summing over the nodes of the mesh, we deduce that

$$\tau \sum_{t \in \Omega_+^\tau} \left(\|\psi(\cdot, t)\|_h^2 + \|\eta(\cdot, t)\|_h^2 \right) \leq C h^{2s} |u|^2_{W_2^{s,s/2}(Q)}, \quad 3/2 < s \leq 2.$$

By noting the inequality (3.86), we obtain the following bound on the global error of the finite difference scheme (3.79), (3.81):

$$\|u - U\|_{L_2(Q_h^\tau)} \leq C h^s \|u\|_{W_2^{s,s/2}(Q)}, \quad 3/2 < s \leq 2. \tag{3.97}$$

Here, and in each of the previous error bounds $C = C(s)$ is a positive constant, independent of h and τ.

The error bounds (3.96) and (3.97) have been derived under the regularity hypothesis that the solution u to the initial-boundary-value problem (3.50) belongs to the function space $W_2^{s,s/2}(Q)$ for $s > 3/2$. This restriction is natural since for $s \le 3/2$ the solution need not be a continuous function, and therefore the definition of the global error as $e_j^m := u(x_j, t_m) - U_j^m$ is then not meaningful. In that case, one can instead define the global error of the scheme as (for example) the difference of the mollified analytical solution and its finite difference approximation:

$$\tilde{e}_j^m := T_x^2 T_t^- u(x_j, t_m) - U_j^m .$$

In fact, for small values of the regularity index s one may need to use even stronger mollification of the analytical solution u in the definition of the global error as well as of the initial datum u_0 and the source term f.

3.3.3 The Case of Independent Mesh-Sizes

The error bounds (3.89), (3.96) and (3.97) above have been derived under the assumption (3.88), which links the temporal mesh-size τ to the spatial mesh-size h in our error analysis, despite the fact that stability of the scheme is unconditional, and therefore from the point of view of stability at least there should be no limitation on the choice of τ in terms of h. We shall show here that in certain cases by careful study of the functionals η and ψ one can avoid linking the mesh-sizes τ and h. Let us suppose, for example, that $s = 4$. From (3.87), by expanding the seminorm $|\tilde{u}|_{W_2^{4,2}(\tilde{G})}$ and returning to the original variables x and t, we have that

$$|D_x^+ D_x^- \eta_j^m| \le \frac{C}{h^2\sqrt{h\tau}} \left(h^8 \left\| \frac{\partial^4 u}{\partial x^4} \right\|^2_{L_2(G_j^m)} + h^4\tau^2 \left\| \frac{\partial^3 u}{\partial x^2 \partial t} \right\|^2_{L_2(G_j^m)} + \tau^4 \left\| \frac{\partial^2 u}{\partial t^2} \right\|^2_{L_2(G_j^m)} \right)^{1/2} . \tag{3.98}$$

Similarly, we obtain that

$$|D_t^- \psi_j^m| \le \frac{C}{\tau\sqrt{h\tau}} \left(h^8 \left\| \frac{\partial^4 u}{\partial x^4} \right\|^2_{L_2(G_j^m)} + h^4\tau^2 \left\| \frac{\partial^3 u}{\partial x^2 \partial t} \right\|^2_{L_2(G_j^m)} + \tau^4 \left\| \frac{\partial^2 u}{\partial t^2} \right\|^2_{L_2(G_j^m)} \right)^{1/2} . \tag{3.99}$$

Thus we observe that the need to link the mesh-sizes h and τ arises because of the presence of the norm of $\partial^2 u/\partial t^2$ on the right-hand side of (3.98) and the norm of $\partial^4 u/\partial x^4$ on the right-hand side of (3.99).

On the other hand, it is easily seen that in the case of $u \in W_2^{4,2}(Q)$ the following integral representations hold:

$$D_x^+ D_x^- \eta_j^m = \frac{1}{h\tau} \int_{x_{j-1}}^{x_{j+1}} \int_{t_{m-1}}^{t_m} \int_{t'}^{t_m} \left(1 - \frac{|x' - x_j|}{h}\right) \frac{\partial^3 u}{\partial x^2 \partial t}(x', t'')\, \mathrm{d}t''\, \mathrm{d}t'\, \mathrm{d}x',$$

$$D_t^- \psi_j^m = \frac{1}{h\tau} \int_{t_{m-1}}^{t_m} \int_{x_{j-1}}^{x_{j+1}} \int_{x'}^{x_j} \int_{x_j}^{x''} \left(1 - \frac{|x' - x_j|}{h}\right) \frac{\partial^3 u}{\partial x^2 \partial t}(x''', t')\, \mathrm{d}x'''\, \mathrm{d}x''\, \mathrm{d}x'\, \mathrm{d}t',$$

and we then directly deduce that

$$\tau \sum_{t \in \Omega_+^\tau} \left(\|D_x^+ D_x^- \eta(\cdot, t)\|_h^2 + \|D_t^- \psi(\cdot, t)\|_h^2\right) \le C(h^4 + \tau^2) \left\| \frac{\partial^3 u}{\partial x^2 \partial t} \right\|_{L_2(Q)}^2.$$

By applying the inequality (3.84) we arrive at the error bound

$$\|u - U\|_{W_2^{2,1}(Q_h^\tau)} \le C(h^2 + \tau) \left\| \frac{\partial^3 u}{\partial x^2 \partial t} \right\|_{L_2(Q)} \le C(h^2 + \tau) \|u\|_{W_2^{4,2}(Q)},$$

without having had to link τ to h. We thus see that the need to link the mesh-sizes h and τ in our original argument based on the use of the Bramble–Hilbert lemma arises because the Bramble–Hilbert lemma invokes a larger number of partial derivatives than is necessary.

3.4 Parabolic Problems with Variable Coefficients

3.4.1 Formulation of the Problem

As our model problem we now consider, in $Q := \Omega \times (0, T] = (0, 1)^2 \times (0, T]$, the following initial-boundary-value problem for a symmetric second-order parabolic equation with variable coefficients:

$$\begin{aligned} \frac{\partial u}{\partial t} + \mathcal{L}u = f, &\quad (x, t) = (x_1, x_2, t) \in Q, \\ u = 0, &\quad (x, t) \in \Gamma \times (0, T] = \partial\Omega \times (0, T], \\ u(x, 0) = u_0(x), &\quad x \in \Omega, \end{aligned} \tag{3.100}$$

where

$$\mathcal{L}u := -\sum_{i,j=1}^{2} \partial_i(a_{ij} \partial_j u) + au, \quad \partial_i := \frac{\partial}{\partial x_i}.$$

We shall suppose that the solution of problem (3.100) belongs to the anisotropic Sobolev space $W_2^{s,\,s/2}(Q)$, $1 < s \le 3$, that the source term f lies in $W_2^{s-2,\,s/2-1}(Q)$ and that the coefficients $a_{ij} = a_{ij}(x)$ and $a = a(x)$ satisfy the same conditions as in the elliptic case considered in Chap. 2; i.e.

(a) if $s > 2$, then

$$a_{ij} \in W_2^{s-1}(\Omega), \quad a \in W_2^{s-2}(\Omega),$$

(b) if $1 < s \le 2$, then

$$a_{ij} \in W_p^{s-1+\delta}(\Omega), \qquad a = a_0 + \sum_{i=1}^{2} \partial_i a_i,$$

$$a_0 \in L_{2+\varepsilon}(\Omega), \qquad a_i \in W_p^{s-1+\delta}(\Omega),$$

where $\varepsilon > 0$,

$$\delta = 0, \qquad p > 2, \quad \text{for } s = 2, \quad \text{and}$$

$$\delta > 0, \qquad p \ge 2/(s-1), \quad \text{for } 1 < s < 2.$$

These conditions ensure that the coefficients a_{ij} and a belong to appropriate spaces of multipliers:

$$a_{ij} \in M\big(W_2^{s-1,\,(s-1)/2}(Q)\big),$$
$$a \in M\big(W_2^{s,\,s/2}(Q) \to W_2^{s-2,\,(s-2)/2}(Q)\big).$$

We shall also assume that

$$a_{ij} = a_{ji} \quad \text{for } i, j = 1, 2,$$

$$\exists c_0 > 0\, \forall x \in \Omega\ \forall \xi \in \mathbb{R}^2 \quad \sum_{i,j=1}^{2} a_{ij}(x)\xi_i\xi_j \ge c_0 \sum_{i=1}^{2} \xi_i^2,$$

$$a(x) \ge 0 \quad \text{in the sense of distributions on } \Omega,$$

$$\text{i.e.} \quad \langle a\varphi, \varphi\rangle_{\mathcal{D}'\times\mathcal{D}} \ge 0 \quad \forall \varphi \in \mathcal{D}(\Omega),$$

as well as appropriate compatibility conditions between the initial and the boundary conditions, and that $u \in W_2^{s,\,s/2}(Q)$.

3.4.2 The Finite Difference Scheme

Let $N, M \in \mathbb{N}$, $N \ge 2$, $M \ge 1$, $h := 1/N$ and $\tau := T/M$. We consider the uniform spatial mesh Ω^h with mesh-size h on Ω and the uniform temporal mesh Ω^τ with mesh-size τ on $(0, T)$. Using the notational conventions introduced in Sects. 2.6 and 3.1.4 we define the space-time meshes

$$Q_h^\tau := \Omega^h \times \Omega^\tau, \qquad Q_h^{\tau\pm} := \Omega^h \times \Omega_\pm^\tau \quad \text{and} \quad \overline{Q}_h^\tau := \overline{\Omega}^h \times \overline{\Omega}^\tau.$$

It will be assumed that the mesh-sizes h and τ satisfy the condition (3.88). For a function V defined on $\overline{Q}_h^\tau$, we shall define the divided differences $D_{x_i}^\pm V$ as in Sect. 2.6 and $D_t^\pm V$ as in Sect. 3.1.4. Finally, we consider the Steklov mollifiers T_i, T_i^+ and T_i^- in the x_i-direction, $i = 1, 2$, (see Sect. 2.6) and the mollifiers T_t^+, T_t^- in the t-direction (see Sect. 3.3).

The initial-boundary-value problem (3.100) will be approximated on $\overline{Q}_h^\tau$ by the finite difference scheme

$$\begin{aligned} D_t^- U + \mathcal{L}_h U = T_1^2 T_2^2 T_t^- f, \quad &\text{on } Q_h^{\tau+}, \\ U = 0 \quad &\text{on } \Gamma^h \times \overline{\Omega}^\tau, \\ U = P u_0 \quad &\text{on } \Omega^h \times \{0\}, \end{aligned} \tag{3.101}$$

where

$$\mathcal{L}_h U := -\frac{1}{2} \sum_{i,j=1}^{2} \left[D_{x_i}^+ \left(a_{ij} D_{x_j}^- U\right) + D_{x_i}^- \left(a_{ij} D_{x_j}^+ U\right) \right] + \left(T_1^2 T_2^2 a\right) U,$$

and

$$Pu := \begin{cases} u & \text{when } 2 < s \leq 3, \\ T_1^2 T_2^2 u & \text{when } 1 < s \leq 2. \end{cases}$$

The scheme (3.101) is a standard symmetric implicit finite difference scheme (see Samarskiĭ [159]) with a mollified right-hand side and lowest coefficient. When $u \in W^{s,s/2}(Q)$ with $s \leq 4$, a scheme of this kind cannot be used without mollification of the source term $f = f(x, t)$, because f is not necessarily continuous and it then makes no sense to sample it at the mesh-points. Similarly, the coefficient $a = a(x)$ need not be continuous when $s \leq 3$. As we are interested in approximating solutions u with low regularity, i.e. ones with Sobolev index $s \in (1, 3]$, we have mollified both a and f in our definition of the finite difference scheme (3.101).

3.4.3 Error Analysis

Let u be the solution of the initial-boundary-value problem (3.100) and let U denote the solution of the difference scheme (3.101). For $1 < s \leq 2$ the solution $u \in W^{s,s/2}(Q)$ is not necessarily a continuous function, although it still possesses an integrable trace on $\Omega \times \{t\}$ for each fixed $t \in [0, T]$. In what follows we shall assume that, for $1 < s \le 2$, the solution $u \in W_2^{s,s/2}(Q)$, with $u|_{\partial\Omega\times(0,T]} = 0$, has been extended outside Q as an odd function in x_1 and x_2. (For the values of s under consideration such an extension preserves the class $W_2^{s,s/2}$; cf. (4.49)–(4.54) in Haroske and Triebel [69] for $s = 2$). Let us define the global error by

$$e := Pu - U,$$

with P as in Sect. 3.4.2. The global error, thus defined, satisfies:

$$\begin{aligned} D_t^- e + \mathcal{L}_h e &= \sum_{i,j=1}^{2} D_{x_i}^- \eta_{ij} + \eta + D_t^- \psi \quad \text{in } Q_h^{\tau+}, \\ e &= 0 \quad \text{on } \Gamma^h \times \Omega_+^\tau, \\ e &= 0 \quad \text{on } \overline{\Omega}^h \times \{0\}, \end{aligned} \tag{3.102}$$

where

$$\eta_{ij} := T_i^+ T_{3-i}^2 T_t^- (a_{ij}\partial_j u) - \frac{1}{2}\big[a_{ij} D_{x_j}^+(Pu) + a_{ij}^{+i}\big(D_{x_j}^-(Pu)\big)^{+i}\big], \quad i, j = 1, 2,$$

$$\eta := \big(T_1^2 T_2^2 a\big)(Pu) - T_1^2 T_2^2 T_t^-(au), \quad \text{and}$$

$$\psi := Pu - T_1^2 T_2^2 u.$$

Analogously as in Sect. 3.3 we introduce the mesh-dependent anisotropic Sobolev norms

$$\|V\|_{L_2(Q_h^\tau)}^2 = \|V\|_{h\tau}^2 := \tau \sum_{t\in\Omega_+^\tau} \|V(\cdot,t)\|_h^2, \qquad \|V\|_{i,h\tau}^2 := \tau \sum_{t\in\Omega_+^\tau} \|V(\cdot,t)\|_{i,h}^2,$$

$$\|V\|_{W_2^{1,1/2}(Q_h^\tau)}^2 := \tau \sum_{t\in\Omega_+^\tau} \big(\|V(\cdot,t)\|_h^2 + \|D_{x_1}^+ V(\cdot,t)\|_{1,h}^2 + \|D_{x_2}^+ V(\cdot,t)\|_{2,h}^2\big)$$

$$+ \tau^2 \sum_{t\in\overline{\Omega}^\tau} \sum_{t'\in\overline{\Omega}^\tau, t'\neq t} \frac{\|V(\cdot,t) - V(\cdot,t')\|_h^2}{|t-t'|^2},$$

with $\|\cdot\|_h$ and $\|\cdot\|_{i,h}$, $i = 1, 2$, denoting the norms defined in Sect. 2.6.

Defining $\mathcal{L}_h V = 0$ on Γ^h the finite difference scheme (3.102) can be rewritten as an operator-difference scheme (3.31), where $\mathcal{H}^h = \mathcal{S}_0^h$ is the set of mesh-functions defined on $\overline{\Omega}^h$ that vanish on Γ^h, equipped with the inner product

$$(V, W)_h := h^2 \sum_{x\in\Omega^h} V(x)W(x).$$

From (3.44) and (3.46), using the relations

$$\|V\|_{\mathcal{L}_h}^2 := (\mathcal{L}_h V, V)_h \ge c_0\big(\|D_{x_1}^+ V\|_{1,h}^2 + \|D_{x_2}^+ V\|_{2,h}^2\big),$$

$$\|D_{x_i}^- V\|_{\mathcal{L}_h^{-1}} := \sup_{W\in\mathcal{S}_0^h} \frac{|(D_{x_i}^- V, W)_h|}{\|W\|_{\mathcal{L}_h}}$$

$$\leq \sup_{W\in\mathcal{S}_0^h} \frac{\|V\|_{i,h}\|D_{x_i}^+ W\|_{i,h}}{\sqrt{c_0(\|D_{x_1}^+ W\|_{1,h}^2 + \|D_{x_2}^+ W\|_{2,h}^2)}} \leq c_0^{-1/2}\|V\|_{i,h}$$

and the discrete Friedrichs inequality (2.53), we obtain the following a priori estimate for the finite difference scheme (3.102):

$$\|e\|^2_{W_2^{1,1/2}(Q_h^\tau)} \leq C\Bigg[\tau^2 \sum_{t\in\overline{\Omega}^\tau} \sum_{t'\in\overline{\Omega}^\tau,\, t'\neq t} \frac{\|\psi(\cdot,t)-\psi(\cdot,t')\|_h^2}{|t-t'|^2}$$

$$+\tau\sum_{t\in\Omega^\tau}\left(\frac{1}{t}+\frac{1}{T-t}\right)\|\psi(\cdot,t)\|_h^2 + \sum_{i=1}^{2}\|\eta_{ij}\|^2_{i,h\tau} + \|\eta\|^2_{h\tau}\Bigg]. \tag{3.103}$$

The task of deriving an error bound for (3.101) is thus reduced to estimating the terms on the right-hand side of (3.103).

Theorem 3.4 *Suppose that the solution u of the initial-boundary-value problem* (3.100) *belongs to $W_2^{s,s/2}(Q)$, $1 < s \leq 3$, $f \in W_2^{s-2,s/2-1}(Q)$ and let the coefficients a_{ij} and a satisfy the assumptions from Sect.* 3.4.1. *Suppose also that $c_1 h^2 \leq \tau \leq c_2 h^2$. Then, the global error of the finite difference scheme* (3.101) *is bounded in the mesh-dependent $W_2^{1,1/2}$ norm as follows*:

$$\|u-U\|_{W_2^{1,1/2}(Q_h^\tau)} \leq Ch^{s-1}\Bigg(\max_{i,j}\|a_{ij}\|_{W_2^{s-1}(\Omega)}$$

$$+\|a\|_{W_2^{s-2}(\Omega)} + \sqrt{\log\frac{1}{h}}\Bigg)\|u\|_{W_2^{s,s/2}(Q)}, \quad \text{when } 2 < s \leq 3, \tag{3.104}$$

and

$$\|T_1^2T_2^2u - U\|_{W_2^{1,1/2}(Q_h^\tau)} \leq Ch^{s-1}\Big(\max_{i,j}\|a_{ij}\|_{W_p^{s-1+\delta}(\Omega)} + \|a_0\|_{L_{2+\varepsilon}(\Omega)}$$

$$+\max_i\|a_i\|_{W_p^{s-1+\delta}(\Omega)}\Big)\|u\|_{W_2^{s,s/2}(Q)}, \quad \text{when } 1 < s \leq 2, \tag{3.105}$$

where $C = C(s)$ is a positive constant, independent of h and τ.

Proof First of all, we decompose η_{ij} as follows:

$$\eta_{ij} = \eta_{ij1} + \eta_{ij2} + \eta_{ij3} + \eta_{ij4} + \eta_{ij5},$$

where, for $i, j = 1, 2$,

$$\eta_{ij1} := T_i^+ T_{3-i}^2 (a_{ij} T_t^- \partial_j u) - (T_i^+ T_{3-i}^2 a_{ij})(T_i^+ T_{3-i}^2 T_t^- \partial_j u),$$

$$\eta_{ij2} := \left[T_i^+ T_{3-i}^2 a_{ij} - \frac{1}{2}(a_{ij} + a_{ij}^{+i})\right](T_i^+ T_{3-i}^2 T_t^- \partial_j u),$$

$$\eta_{ij3} := \frac{1}{2}(a_{ij} + a_{ij}^{+i})\left\{T_i^+ T_{3-i}^2 T_t^- \partial_j u - \frac{1}{2}\left[D_{x_j}^+(Pu) + (D_{x_j}^-(Pu))^{+i}\right]\right\},$$

$$\eta_{ij4} := -\frac{1}{4}(a_{ij} - a_{ij}^{+i})\left[D_{x_j}^+(T_t^- u) - (D_{x_j}^-(T_t^- u))^{+i}\right],$$

$$\eta_{ij5} := -\frac{1}{4}(a_{ij} - a_{ij}^{+i})\left[D_{x_j}^+(Pu - T_t^- u) - (D_{x_j}^-(Pu - T_t^- u))^{+i}\right].$$

For $1 < s \le 2$ we let $\eta = \eta_0 + \eta_1 + \eta_2 + \eta_3 + \eta_4 + \eta_5$, where

$$\eta_0 := (T_1^2 T_2^2 a_0)(T_1^2 T_2^2 T_t^- u) - T_1^2 T_2^2 (a_0 T_t^- u),$$

$$\eta_1 := (T_1^2 T_2^2 a_0) T_1^2 T_2^2 (u - T_t^- u),$$

$$\eta_{2i} := (T_1^2 T_2^2 \partial_i a_i)(T_1^2 T_2^2 T_t^- u) - T_1^2 T_2^2 [(T_t^- u)\partial_i a_i], \quad i = 1, 2,$$

$$\eta_{2i+1} := (T_1^2 T_2^2 \partial_i a_i) T_1^2 T_2^2 (u - T_t^- u), \quad i = 1, 2.$$

For $2 < s \le 3$ we define $\eta = \eta_6 + \eta_7 + \eta_8 + \eta_9$, where

$$\eta_6 := (T_1^2 T_2^2 a)(T_t^- u - T_1^2 T_2^2 T_t^- u),$$

$$\eta_7 := (T_1^2 T_2^2 a)(T_1^2 T_2^2 u - T_1^2 T_2^2 T_t^- u),$$

$$\eta_8 := (T_1^2 T_2^2 a)(u - T_1^2 T_2^2 u - T_t^- u + T_1^2 T_2^2 T_t^- u),$$

$$\eta_9 := (T_1^2 T_2^2 a)(T_1^2 T_2^2 T_t^- u) - T_1^2 T_2^2 (a T_t^- u).$$

Let us introduce the elementary rectangles

$$K^0 = K^0(x) := \{y : |y_j - x_j| < h,\ j = 1, 2\},$$

$$K^i = K^i(x) := \{y : x_i < y_i < x_i + h,\ |y_{3-i} - x_{3-i}| < h\}, \quad i = 1, 2,$$

and the parallelepipeds

$$G^0 = G^0(x, t) := K^0 \times (t - \tau, t),$$

$$G^i = G^i(x, t) := K^i \times (t - \tau, t), \quad i = 1, 2.$$

For $2 < s \le 3$, η_{ij1}, $i, j = 1, 2$, satisfy the conditions under which an estimate of the form (2.174) holds:

$$\|\eta_{ij1}(\cdot, t)\|_{i,h} \le C h^{s-1} \|a_{ij}\|_{W_2^{s-1}(\Omega)} \|T_t^- u(\cdot, t)\|_{W_2^s(\Omega)}, \quad 2 < s \le 3.$$

Thus, summing over the nodes of the mesh Ω_+^τ we get

$$\|\eta_{ij1}\|_{i,h\tau} \le Ch^{s-1}\|a_{ij}\|_{W_2^{s-1}(\Omega)}\|u\|_{W_2^{s,s/2}(Q)}, \quad 2 < s \le 3. \tag{3.106}$$

Analogously, for $1 < s \le 2$, (2.173) implies that

$$\|\eta_{ij1}\|_{i,h\tau} \le Ch^{s-1}\|a_{ij}\|_{W_p^{s-1+\delta}(\Omega)}\|u\|_{W_2^{s,s/2}(Q)}, \quad 1 < s \le 2. \tag{3.107}$$

Similarly, using the estimates (2.174) and (2.173) for η_{ij2} and η_{ij4} we obtain bounds of the form (3.106), (3.107).

When $s > 1$, $\eta_{ij3}(x,t)$ is a bounded bilinear functional of $(a_{ij}, u) \in C(\overline{K}^i) \times W_2^{s,s/2}(G^i)$, which vanishes whenever u is a polynomial of degree two in x_1 and x_2 and of arbitrary degree in t with constant coefficients. Invoking the Bramble–Hilbert lemma we deduce that

$$\left|\eta_{ij3}(x,t)\right| \le Ch^{s-3}\|a_{ij}\|_{C(\overline{K}^i)}|u|_{\widehat{W}_2^{s,s/2}(G^i)}, \quad 1 < s \le 3.$$

Summing through the nodes of the mesh $Q_h^{\tau+}$ gives

$$\|\eta_{ij3}\|_{i,h\tau} \le Ch^{s-1}\|a_{ij}\|_{C(\overline{\Omega})}\|u\|_{W_2^{s,s/2}(Q)}, \quad 1 < s \le 3.$$

By noting the embeddings

$$W_2^{s-1}(\Omega) \hookrightarrow C(\overline{\Omega}) \quad \text{for } 2 < s \le 3$$

and

$$W_p^{s-1+\delta}(\Omega) \hookrightarrow C(\overline{\Omega}) \quad \text{for } 1 < s \le 2,$$

we obtain bounds of the form (3.106) and (3.107). The same argument applies to η_{ij5}.

The term η_0 satisfies the conditions under which a bound of the form (2.176) holds:

$$\left\|\eta_0(\cdot,t)\right\|_h \le Ch^{s-1}\|a_0\|_{L_{2+\varepsilon}(\Omega)}\left\|T_t^- u(\cdot,t)\right\|_{W_2^s(\Omega)}, \quad 1 < s \le 2.$$

Summing over the nodes of Ω_+^τ we thus get

$$\|\eta_0\|_{h\tau} \le Ch^{s-1}\|a_0\|_{L_{2+\varepsilon}(\Omega)}\|u\|_{W_2^{s,s/2}(Q)}, \quad 1 < s \le 2. \tag{3.108}$$

Analogously, using (2.173), for η_2 and η_4 we obtain bounds of the form (3.107), while using (2.175) yields the following bounds on η_6 and η_9:

$$\|\eta_6\|_{h\tau}, \|\eta_9\|_{h\tau} \le Ch^{s-1}\|a\|_{W_2^{s-2}(\Omega)}\|u\|_{W_2^{s,s/2}(Q)}, \quad 2 < s \le 3. \tag{3.109}$$

For $s > 1$ and $q \geq 1$, $\eta_1(x,t)$ is a bounded bilinear functional of $(a,u) \in L_q(K^0) \times W_2^{s,s/2}(G^0)$, which vanishes if u is a polynomial of degree one in x_1 and x_2 (with constant coefficients). The Bramble–Hilbert lemma gives

$$\left|\eta_1(x,t)\right| \leq C h^{s-2-2/q} \|a_0\|_{L_q(K^0)} |u|_{W_2^{s,s/2}(G^0)}, \quad 1 < s \leq 2.$$

By bounding the right-hand side from above further we deduce that

$$\left|\eta_1(x,t)\right| \leq C h^{s-2-2/q} \|a_0\|_{L_q(\Omega)} |u|_{W_2^{s,s/2}(G^0)}, \quad 1 < s \leq 2.$$

Summing through the nodes of the mesh thus yields

$$\|\eta_1\|_{h\tau} \leq C h^{s-2/q} \|a_0\|_{L_q(\Omega)} |u|_{W_2^{s,s/2}(Q)}, \quad 1 < s \leq 2,$$

and setting $q = 2 + \varepsilon$ then gives the desired bound:

$$\|\eta_1\|_{h\tau} \leq C h^{s-1} \|a_0\|_{L_{2+\varepsilon}(\Omega)} \|u\|_{W_2^{s,s/2}(Q)}, \quad 1 < s \leq 2. \tag{3.110}$$

For $s > 1$, $\eta_{2i+1}(x,t)$, $i = 1,2$, are bounded bilinear functionals of $(a_i, u) \in L_\infty(K^0) \times W_2^{s,s/2}(G^0)$, which vanish when u is a polynomial of degree one in x_1 and x_2 with constant coefficients. Similarly as before, we arrive at

$$|\eta_{2i+1}| \leq C h^{s-3} \|a_i\|_{L_\infty(\Omega)} |u|_{W_2^{s,s/2}(G^0)}, \quad 1 < s \leq 2,$$

and

$$\|\eta_{2i+1}\|_{h\tau} \leq C h^{s-1} \|a_i\|_{L_\infty(\Omega)} \|u\|_{W_2^{s,s/2}(Q)}, \quad 1 < s \leq 2.$$

Using the embedding $W_p^{s-1+\delta}(\Omega) \hookrightarrow L_\infty(\Omega)$ then yields

$$\|\eta_{2i+1}\|_{h\tau} \leq C h^{s-1} \|a_i\|_{W_p^{s-1+\delta}(\Omega)} \|u\|_{W_2^{s,s/2}(Q)}, \quad 1 < s \leq 2. \tag{3.111}$$

For $\lambda > 1/2$, $\eta_7(x,t)$ is a bounded bilinear functional of $(a, T_1^2 T_2^2 u) \in L_q(K^0) \times W_2^\lambda(t-\tau, t)$, which vanishes if $T_1^2 T_2^2 u$ is a constant function. By applying the Bramble–Hilbert lemma we obtain

$$\left|\eta_7(x,t)\right| \leq C h^{2\lambda-1-2/q} \|a\|_{L_q(K^0)} \left|T_1^2 T_2^2 u\right|_{W_2^\lambda(t-\tau,t)}, \quad 1/2 < \lambda \leq 1.$$

For $1/2 < \lambda < 1$,

$$\begin{aligned}
\left|T_1^2 T_2^2 u\right|_{W_2^\lambda(t-\tau,t)} &= \left(\int_{t-\tau}^{t} \int_{t-\tau}^{t} \frac{|T_1^2 T_2^2 u(\cdot,t') - T_1^2 T_2^2 u(\cdot,t'')|^2}{|t'-t''|^{1+2\lambda}}\, \mathrm{d}t'\, \mathrm{d}t'' \right)^{1/2} \\
&\leq C h^{-2/r} \left(\int_{t-\tau}^{t} \int_{t-\tau}^{t} \frac{\|u(\cdot,t') - u(\cdot,t'')\|_{L_r(K^0)}^2}{|t'-t''|^{1+2\lambda}}\, \mathrm{d}t'\, \mathrm{d}t'' \right)^{1/2}.
\end{aligned}$$

Setting $r = 2q/(q-2)$ with $q > 2$ yields

$$|\eta_7(x,t)| \le Ch^{2\lambda-2}\|a\|_{L_q(K^0)}\left(\int_{t-\tau}^{t}\int_{t-\tau}^{t}\frac{\|u(\cdot,t')-u(\cdot,t'')\|^2_{L_{\frac{2q}{q-2}}(K^0)}}{|t'-t''|^{1+2\lambda}}\,\mathrm{d}t'\,\mathrm{d}t''\right)^{1/2}.$$

Summing through the nodes of the mesh and using Hölder's inequality gives

$$\|\eta_7\|_{h\tau} \le Ch^{2\lambda}\|a\|_{L_q(\Omega)}\left(\int_0^T\int_0^T\frac{\|u(\cdot,t')-u(\cdot,t'')\|^2_{L_{\frac{2q}{q-2}}(\Omega)}}{|t'-t''|^{1+2\lambda}}\,\mathrm{d}t'\,\mathrm{d}t''\right)^{1/2}.$$

Let us choose q such that:

$$W_2^{s-2}(\Omega) \hookrightarrow L_q(\Omega) \quad \text{and} \quad W_2^1(\Omega) \hookrightarrow L_{2q/(q-2)}(\Omega).$$

For $2 < s \le 3$ this can be achieved by selecting q such that $2 < q < 2/(3-s)$. We then obtain the following bound on $\|\eta_7\|_{h\tau}$:

$$\begin{aligned}\|\eta_7\|_{h\tau} &\le Ch^{2\lambda}\|a\|_{W_2^{s-2}(\Omega)}\left(\int_0^T\int_0^T\frac{\|u(\cdot,t')-u(\cdot,t'')\|^2_{W_2^1(\Omega)}}{|t'-t''|^{1+2\lambda}}\,\mathrm{d}t'\,\mathrm{d}t''\right)^{1/2}\\ &\le Ch^{2\lambda}\|a\|_{W_2^{s-2}(\Omega)}\|u\|_{\widehat{W}_2^{2\lambda+1,\lambda+1/2}(Q)}, \quad 2 < s \le 3,\ 1/2 < \lambda < 1.\end{aligned}$$

The same result holds for $\lambda = 1$; then,

$$\int_{t-\tau}^{t}\int_{t-\tau}^{t}\frac{\mathrm{d}t'\,\mathrm{d}t''}{|t'-t''|^{1+2\lambda}} \quad \text{is replaced in the argument above by} \quad \int_{t-\tau}^{t}\mathrm{d}t',$$

and

$$u(\cdot,t') - u(\cdot,t'') \quad \text{is replaced by} \quad \frac{\partial u(\cdot,t')}{\partial t'}.$$

Setting $s = 2\lambda + 1$ we finally obtain the following bound on η_7:

$$\|\eta_7\|_{h\tau} \le Ch^{s-1}\|a\|_{W_2^{s-2}(\Omega)}\|u\|_{W_2^{s,s/2}(Q)}, \quad 2 < s \le 3. \tag{3.112}$$

When $a \in L_2(\Omega)$ and $s > 2$, we have that $\eta_8(x,t)$ is a bounded linear functional of $u \in W_2^{s,s/2}(G^0)$, which vanishes on all polynomials of degree two in x_1 and x_2 and on all polynomials of degree one in t (with constant coefficients). By the Bramble–Hilbert lemma,

$$\begin{aligned}|\eta_8| &\le Ch^{s-3}\|a_i\|_{L_2(K^0)}|u|_{\widehat{W}_2^{s,s/2}(G^0)}\\ &\le Ch^{s-3}\|a_i\|_{L_2(\Omega)}|u|_{\widehat{W}_2^{s,s/2}(G^0)}, \quad 2 < s \le 3.\end{aligned}$$

Summing through the nodes of $Q_h^{\tau+}$ and noting the embedding $W_2^{s-2}(\Omega) \hookrightarrow L_2(\Omega)$, $s > 2$, yields

$$\|\eta_8\|_{h\tau} \le C h^{s-1} \|a\|_{W_2^{s-2}(\Omega)} \|u\|_{W_2^{s,s/2}(Q)}, \quad 2 < s \le 3. \tag{3.113}$$

Now let us consider the terms in (3.103) containing $\psi := u - T_1^2 T_2^2 u$. Obviously,

$$\psi = 0 \quad \text{for } 1 < s \le 2. \tag{3.114}$$

When $2 < s \le 3$, analogously as in Sect. 3.3, we have that

$$\begin{aligned}
&\tau^2 \sum_{t\in\overline{\Omega}^\tau} \sum_{t'\in\overline{\Omega}^\tau,\, t'\ne t} \frac{\|\psi(\cdot,t) - \psi(\cdot,t')\|_h^2}{|t-t'|^2} \\
&= 2\tau^2 \sum_{m=1}^{M} \sum_{k=0}^{m-1} \frac{\|\psi^m - \psi^k\|_h^2}{|t_m - t_k|^2} \\
&\le 6\tau^2 \sum_{m=1}^{M} \sum_{k=0}^{m-1} \frac{\|\psi^m - T_t^- \psi^m\|_h^2}{|t_m - t_k|^2} + 6\tau^2 \sum_{m=1}^{M} \sum_{k=0}^{m-1} \frac{\|T_t^- \psi^m - T_t^+ \psi^k\|_h^2}{|t_m - t_k|^2} \\
&\quad + 6\tau^2 \sum_{m=1}^{M} \sum_{k=0}^{m-1} \frac{\|T_t^+ \psi^k - \psi^k\|_h^2}{|t_m - t_k|^2} =: J_1 + J_2 + J_3.
\end{aligned}$$

We shall bound each of the terms J_1, J_2 and J_3. Similarly as in Sect. 3.3 we have

$$J_1 = 6 \sum_{m=1}^{M} \sum_{k=0}^{m-1} \frac{\|\psi^m - T_t^- \psi^m\|_h^2}{(m-k)^2} \le \frac{\pi^2}{\tau} \tau \sum_{m=1}^{M} \|\psi^m - T_t^- \psi^m\|_h^2,$$

and hence, by estimating $\psi^m - T_t^- \psi^m$ using the Bramble–Hilbert lemma and summing over the nodes of the mesh, we obtain

$$J_1 \le C h^{2s-2} \|u\|_{W_2^{s,s/2}(Q)}^2, \quad 2 < s \le 4. \tag{3.115}$$

An identical bound holds for J_3.

Further, analogously as in Sect. 3.3 we have that

$$J_2 \le 6 h^2 \tau^{2\lambda-1} \sum_{x\in\Omega^h} \int_0^T \int_0^T \frac{[\psi(x,t) - \psi(x,t')]^2}{|t-t'|^{1+2\lambda}}\, dt'\, dt,$$

where $0 < \lambda \le 1/2$. Using the integral representation

$$\begin{aligned}
\psi(x,t) &= u(x,t) - T_1^2 T_2^2 u(x,t) \\
&= h^{-2} \int_{x_1-h}^{x_1+h} \int_{x_2-h}^{x_2+h} \int_{x_1'}^{x_1} \int_{x_2'}^{x_2} \left(1 - \frac{|x_1' - x_1|}{h}\right)\left(1 - \frac{|x_2' - x_2|}{h}\right)
\end{aligned}$$

$$\times \frac{\partial^2 u}{\partial x_1 \partial x_2}(x_1'', x_2'', t)\, dx_2''\, dx_1''\, dx_2'\, dx_1'$$

$$- h^{-2} \int_0^h \int_0^{x_1'} \int_{x_1 - x_1''}^{x_1 + x_1''} \int_{x_2 - h}^{x_2 + h} \left(1 - \frac{x_1'}{h}\right)\left(1 - \frac{|x_2' - x_2|}{h}\right)$$

$$\times \frac{\partial^2 u}{\partial x_1^2}(x_1''', x_2', t)\, dx_2'\, dx_1'''\, dx_1''\, dx_1'$$

$$- h^{-2} \int_{x_1 - h}^{x_1 + h} \int_0^h \int_0^{x_2'} \int_{x_2 - x_2''}^{x_2 + x_2''} \left(1 - \frac{|x_1' - x_1|}{h}\right)\left(1 - \frac{x_2'}{h}\right)$$

$$\times \frac{\partial^2 u}{\partial x_2^2}(x_1', x_2''', t)\, dx_2'''\, dx_2''\, dx_2'\, dx_1'$$

and the Cauchy–Schwarz inequality then yields

$$J_2 \le C h^{2+4\lambda} |u|^2_{\widehat{W}_2^{2+2\lambda, 1+\lambda}(Q)}, \quad 0 < \lambda \le 1/2.$$

By writing $s = 2 + 2\lambda$ we thus obtain the following bound:

$$J_2 \le C h^{2s-2} \|u\|^2_{W_2^{s,s/2}(Q)}, \quad 2 < s \le 3. \tag{3.116}$$

Using the Bramble–Hilbert lemma and the trace theorem (Theorem 1.44), we have that

$$\tau \sum_{t \in \Omega^\tau} \left(\frac{1}{t} + \frac{1}{T - t}\right) \|\psi(\cdot, t)\|_h^2$$

$$\le C h^{2s-2} \tau \sum_{t \in \Omega^\tau} \left(\frac{1}{t} + \frac{1}{T - t}\right) |u(\cdot, t)|^2_{W_2^{s-1}(\Omega)}$$

$$\le C h^{2s-2} \log \frac{1}{h} \|u\|^2_{W_2^{s,s/2}(Q)}, \quad \text{when } 2 < s \le 3. \tag{3.117}$$

Finally, the desired error bound follows by combining (3.103) with (3.106)–(3.117). □

Remark 3.1 The error bounds (3.104) and (3.105) have been proved under the assumption $\tau \asymp h^2$. As we have noted in Sect. 3.3.3, this condition is of technical nature and may be avoided by a more careful analysis of the truncation error.

Remark 3.2 Similar results hold when the coefficients a_{ij} and a depend on t (see Samarskiĭ [159] for error bounds in the case of classical solutions).

Remark 3.3 Error bounds in the discrete $W_2^{2,1}$ and L_2 norms can be established analogously as in the elliptic case, provided that the discrete 'second fundamental inequality' (2.179) holds. Error bounds in discrete C^k and $C^{k,\alpha}$ norms follow directly from our error bounds in discrete $W_2^{m,m/2}$ norms by invoking discrete counterparts of embedding and trace theorems.

3.4.4 Factorized Scheme

The finite difference scheme (3.101) involves the solution of a system of linear algebraic equations at each time level. The linear system has a sparse, banded matrix, a typical row of which contains at most seven nonzero entries. In the special case when $a_{12} = a_{21} = 0$, a typical row of the matrix contains five nonzero entries. Either way, the computational cost of solving such a linear system is higher than in the one-dimensional case. Indeed, the matrix of the linear system that arises from the implicit finite difference approximation of the one-dimensional heat equation considered in Sect. 3.2.1 is much simpler: it is tridiagonal. Our objective here is to replace (3.101), without loss of accuracy, by a more economical scheme, which at each time level involves the solution of systems of linear algebraic equations with tridiagonal matrices only. To this end, we consider the following factorized finite difference scheme:

$$(I_h + \sigma\tau\Lambda_1)(I_h + \sigma\tau\Lambda_2)D_t^- U + \mathcal{L}_h\check{U} = T_1^2 T_2^2 T_t^- f, \tag{3.118}$$

with the same initial and boundary conditions as in (3.101). Here σ is a positive real parameter, $\Lambda_i U = -D_{x_i}^- D_{x_i}^+ U$, $i = 1, 2$, and I_h is the identity operator. According to (3.35) the finite difference scheme (3.118) is stable if the operator

$$(I_h + \sigma\tau\Lambda_1)(I_h + \sigma\tau\Lambda_2) - \frac{1}{2}\tau\mathcal{L}_h$$

is positive definite, uniformly with respect to the discretization parameters. This condition is satisfied if, for example,

$$\sigma \geq \max_{i,j} \|a_{ij}\|_{C(\overline{\Omega})},$$

and the mesh-size h is sufficiently small, i.e.

$$h < 3(c_2\|a\|_{L_2(\Omega)})^{-1}, \quad \text{when } 2 < s \leq 3,$$

or

$$h < \frac{1}{2}\left[c_2\left(\|a_0\|_{L_{2+\varepsilon}(\Omega)} + \|a_1\|_{L_p(\Omega)} + \|a_2\|_{L_p(\Omega)}\right)\right]^{-\frac{p}{p-2}}, \quad \text{when } 1 < s \leq 2,$$

with the same assumptions on ε and p as in Sect. 3.4.1.

In contrast with (3.101), the factorized scheme (3.118) is *economical* in the sense that the linear operators $(I_h + \sigma\tau\Lambda_i)$, $i = 1, 2$, that need to be inverted at each time level are represented by *tridiagonal* matrices.

The global error $e := Pu - U$ satisfies the following finite difference scheme:

$$(I_h + \sigma\tau\Lambda_1)(I_h + \sigma\tau\Lambda_2)D_t^- e + \mathcal{L}_h \check{e} = \sum_{i,j=1}^{2} D_{x_i}^- \eta'_{ij} + \eta' + D_t^- \psi \quad \text{in } Q_h^{\tau+},$$

$$e = 0 \quad \text{on } \Gamma^h \times \Omega_+^\tau \text{ and } \overline{\Omega}^h \times \{0\},$$

where

$$\eta'_{ij} := \eta_{ij} + \frac{1}{2}\tau D_t^- \left[a_{ij} D_{x_j}^+(Pu) + a_{ij}^{+i}\left(D_{x_j}^-(Pu)\right)^{+i}\right]$$

$$- \frac{1}{2}\sigma\tau\delta_{ij} D_t^- \left[D_{x_j}^+(Pu) + \left(D_{x_j}^-(Pu)\right)^{+i}\right]$$

$$+ \frac{1}{2}\sigma^2\tau^2(1-\delta_{ij}) D_{x_j}^+ D_{x_j}^- D_{x_i}^+ D_t^-(Pu),$$

$$\eta' := \eta - \tau\left(T_1^2 T_2^2 a\right) D_t^-(Pu),$$

with P as defined in Sect. 3.4.2, and δ_{ij} is the Kronecker delta. The a priori estimate (3.103) still holds if η_{ij} and η are replaced by η'_{ij} and η'. Using the techniques developed above it is easy to show that the factorized scheme (3.118) satisfies the error bounds (3.104) and (3.105); in other words, no accuracy has been lost compared to the implicit finite difference scheme considered in the previous section.

3.5 A Parabolic Interface Problem

Let $\Omega = (0,1)^2$, $\Gamma = \partial\Omega$, and let Σ be the intersection of the line segment $x_2 = \xi$, $0 < \xi < 1$, and $\overline{\Omega}$. We consider the following parabolic interface problem:

$$\begin{aligned}(1 + k\delta_\Sigma)\frac{\partial u}{\partial t} + \mathcal{L}u = f(x,t) \quad &\text{in } Q := \Omega \times (0,T],\\ u = 0 \quad &\text{on } \Gamma \times (0,T], \\ u(x,0) = u_0(x) \quad &\text{on } \Omega,\end{aligned} \tag{3.119}$$

where $\delta_\Sigma(x) := \delta(x_2 - \xi)$ is the Dirac distribution concentrated on Σ, $k(x) = k(x_1)$ and $\mathcal{L}$ is the same symmetric elliptic operator as in (2.166) and (3.100):

$$\mathcal{L}u := -\sum_{i,j=1}^{2} \partial_i(a_{ij}\partial_j) + au.$$

Clearly, (3.119) is a parabolic initial-boundary-value problem with a 'concentrated capacity' involved in the coefficient of the time derivative. In one space-dimension, similar problems were considered by Jovanović and Vulkov [94, 95]. Following [94], the solution of the initial-boundary-value problem (3.119) will be sought in the function space $\widetilde{W}_2^{s,s/2}(Q) := L_2((0,T); \widetilde{W}_2^s(\Omega)) \cap W_2^{s/2}((0,T); \widetilde{L}_2(\Omega))$, where $\widetilde{W}_2^s(\Omega)$ and $\widetilde{L}_2(\Omega)$ are the same as in Sect. 2.8.

When the source term $f = f(x,t)$ is sufficiently regular in the sense that it does not contain a 'concentrated load' such as $\delta_\Sigma(x)$, it is easily verified that (3.119) is equivalent to the following initial-boundary-value problem with transmission (conjugation) conditions on the interface Σ:

$$\begin{aligned} &\frac{\partial u}{\partial t} + \mathcal{L}u = f(x,t) \quad \text{in } Q^- \cup Q^+, \\ &u = 0 \quad \text{on } \Gamma \times (0,T], \\ &u(x,0) = u_0(x) \quad \text{on } \Omega, \\ &[u]_\Sigma = 0, \quad \left[\sum_{j=1}^{2} a_{2j}\partial_j u\right]_\Sigma = k\frac{\partial u}{\partial t}\bigg|_\Sigma, \end{aligned} \tag{3.120}$$

where $Q^\pm := \Omega^\pm \times (0,T]$, $\Omega^- := (0,1) \times (0,\xi)$, $\Omega^+ := (0,1) \times (\xi,1)$, and $[u]_\Sigma := u(x_1, \xi+0, t) - u(x_1, \xi-0, t)$.

Lemma 3.5 *Let the coefficients a_{ij}, a and k satisfy the assumptions of Lemma* 2.71 *and suppose that*

$$f \in L_2\big((0,T); W_2^{-1}(\Omega)\big) \quad \textit{and} \quad u_0 \in \widetilde{L}_2(\Omega).$$

Then, there exists a unique solution $u \in \widetilde{W}_2^{1,1/2}(Q)$ to the initial-boundary-value problem (3.120), *and there is a positive constant C such that the following a priori estimate holds*:

$$\|u\|^2_{\widetilde{W}_2^{1,1/2}(Q)} \le C\left(\|u_0\|^2_{\widetilde{L}_2(\Omega)} + \|u_0\|^2_{\widetilde{L}_2(\Sigma)} + \int_0^T \|f(\cdot,t)\|^2_{W_2^{-1}(\Omega)}\,\mathrm{d}t \right).$$

The proof is analogous to that of Theorem 26.1 in Wloka [199].

3.5.1 Finite Difference Approximation

Using the same notational conventions as in Sects. 2.8.1 and 3.4 we approximate the initial-boundary-value problem (3.119) on the mesh $\overline{Q}_h^\tau$ by the following implicit finite difference scheme:

$$\begin{aligned} (1+k\delta_{\Sigma^h})D_t^- U + \mathcal{L}_h U &= T_1^2 T_2^2 T_t^- f \quad \text{in } Q_h^{\tau+}, \\ U &= 0 \quad \text{on } \Gamma^h \times \overline{\Omega}^\tau, \\ U(x,0) &= u_0(x) \quad \text{on } \Omega^h, \end{aligned} \tag{3.121}$$

where the operator $\mathcal{L}_h$ is defined in the same way as in Sects. 2.8.1 and 3.4; i.e.

$$\mathcal{L}_h U := -\frac{1}{2} \sum_{i,j=1}^{2} \left[D_{x_i}^+ (a_{ij} D_{x_j}^- U) + D_{x_i}^- (a_{ij} D_{x_j}^+ U) \right] + (T_1^2 T_2^2 a) U,$$

and

$$\delta_{\Sigma^h}(x) = \delta_h(x_2 - \xi) := \begin{cases} 0 & \text{for } x \in \Omega^h \setminus \Sigma^h, \\ 1/h & \text{for } x \in \Sigma^h, \end{cases}$$

is the discrete Dirac delta-function concentrated on Σ^h. For the sake of simplicity, we shall assume that ξ is a rational number and ξ/h is an integer.

Besides the norms defined in Sects. 2.8.1 and 3.4, we shall also consider the following mesh-dependent norms and seminorms:

$$\|U\|^2_{L_2(\Sigma^h \times \Omega^\tau)} := \tau \sum_{t \in \Omega_+^\tau} \|U(\cdot,t)\|^2_{L_2(\Sigma^h)},$$

$$|U|^2_{L_2(\Omega^\tau; W_2^{1/2}(\Sigma^h))} := \tau \sum_{t \in \Omega_+^\tau} |U(\cdot,t)|^2_{W_2^{1/2}(\Sigma^h)},$$

$$|U|^2_{W_2^{1/2}(\Omega^\tau; L_2(\Omega^h))} := \tau^2 \sum_{t \in \overline{\Omega}^\tau} \sum_{t' \in \overline{\Omega}^\tau, t' \neq t} \frac{\|U(\cdot,t) - U(\cdot,t')\|^2_{L_2(\Omega^h)}}{|t-t'|^2},$$

$$|U|^2_{W_2^{1/2}(\Omega^\tau; L_2(\Sigma^h))} := \tau^2 \sum_{t \in \overline{\Omega}^\tau} \sum_{t' \in \overline{\Omega}^\tau, t' \neq t} \frac{\|U(\cdot,t) - U(\cdot,t')\|^2_{L_2(\Sigma^h)}}{|t-t'|^2},$$

$$\|U\|^2_{\ddot{W}_2^{1/2}(\Omega^\tau; L_2(\Omega^h))} := |U|^2_{W_2^{1/2}(\Omega^\tau; L_2(\Omega^h))} + \tau \sum_{t \in \overline{\Omega}^\tau} \left(\frac{1}{t} + \frac{1}{T-t} \right) \|U(\cdot,t)\|^2_{L_2(\Omega^h)},$$

$$\|U\|^2_{\ddot{W}_2^{1/2}(\Omega^\tau; L_2(\Sigma^h))} := |U|^2_{W_2^{1/2}(\Omega^\tau; L_2(\Sigma^h))} + \tau \sum_{t \in \overline{\Omega}^\tau} \left(\frac{1}{t} + \frac{1}{T-t} \right) \|U(\cdot,t)\|^2_{L_2(\Sigma^h)},$$

$$\begin{aligned} \|U\|^2_{\widetilde{W}_2^{1,1/2}(Q_h^\tau)} := \tau \sum_{t \in \Omega^{\tau+}} \|U(\cdot,t)\|^2_{W_2^1(\Omega^h)} + |U|^2_{W_2^{1/2}(\Omega^\tau; L_2(\Omega^h))} \\ + |U|^2_{W_2^{1/2}(\Omega^\tau; L_2(\Sigma^h))}. \end{aligned}$$

Let u be the solution of the initial-boundary-value problem (3.120) and let U denote the solution of the finite difference scheme (3.121). Then, the global error $e := u - U$ satisfies the finite difference scheme

$$
\begin{aligned}
(1 + k\delta_{\Sigma^h})D_t^- e + \mathcal{L}_h e = \varphi \quad & \text{in } \Omega_h^{\tau+},\\
e = 0 \quad & \text{on } \Gamma^h \times \overline{\Omega}_\tau,\\
e(x,0) = 0 \quad & \text{on } \Omega^h,
\end{aligned}
\tag{3.122}
$$

where

$$
\begin{aligned}
\varphi &:= \sum_{i,j=1}^{2} D_{x_i}^- \eta_{ij} + \eta + D_t^- \psi + \delta_{\Sigma^h} D_t^- \mu,\\
\eta_{ij} &:= T_i^+ T_{3-i}^2 T_t^- (a_{ij}\partial_j u) - \frac{1}{2}\big(a_{ij} D_{x_j}^+ u + a_{ij}^{+i} D_{x_j}^- u^{+i}\big),\\
\eta &:= (T_1^2 T_2^2 a)u - T_1^2 T_2^2 T_t^- (au),\\
\psi &:= u - T_1^2 T_2^2 u,\\
\mu &:= ku - T_1^2(ku).
\end{aligned}
$$

We consider the decompositions

$$
\eta_{1j} := \tilde{\eta}_{1j} + \delta_{\Sigma^h}\hat{\eta}_{1j}, \qquad \eta := \tilde{\eta} + \delta_{\Sigma^h}\hat{\eta}, \qquad \psi := \tilde{\psi} + \delta_{\Sigma^h}\hat{\psi},
$$

where

$$
\tilde{\eta}_{1j} := \eta_{1j} - \delta_{\Sigma^h}\hat{\eta}_{1j}, \qquad \tilde{\eta} := \eta - \delta_{\Sigma^h}\hat{\eta}, \qquad \tilde{\psi} := \psi - \delta_{\Sigma^h}\hat{\psi},
$$

and

$$
\begin{aligned}
\hat{\eta}_{11} &:= \frac{1}{6}h^2 T_1^+ T_t^- \big([a_{11}\partial_1\partial_2 u + \partial_2 a_{11}\partial_1 u]_\Sigma\big),\\
\hat{\eta}_{12} &:= \frac{1}{6}h^2 T_1^+ T_t^- \big([a_{12}\partial_2^2 u + \partial_2 a_{12}\partial_2 u]_\Sigma\big) - \frac{1}{4}h^2 T_1^+ T_t^- \big([\partial_1(a_{12}\partial_2 u)]_\Sigma\big),\\
\hat{\eta} &:= -\frac{1}{6}h^2\big[(T_1^2 a)(T_1^2 T_t^- \partial_2 u)\big]_\Sigma,\\
\hat{\psi} &:= \frac{1}{6}h^2\big[T_1^2\partial_2 u\big]_\Sigma.
\end{aligned}
$$

By applying (3.18) and (3.19) we obtain the a priori estimate

$$\|e\|_{\widetilde{W}_2^{1,1/2}(Q_h^\tau)} \le C\Bigg[\sum_{j=1}^{2}\Big(\|\eta_{2j}\|_{2,h\tau} + \|\tilde{\eta}_{1j}\|_{1,h\tau} + |\hat{\eta}_{1j}|_{L_2(\Omega^\tau;W_2^{1/2}(\Sigma^h))}\Big)$$

$$+ \|\tilde{\eta}\|_{L_2(Q_h^\tau)} + \|\hat{\eta}\|_{L_2(\Sigma^h\times\Omega^\tau)} + \|\tilde{\psi}\|_{\ddot{W}_2^{1/2}(\Omega^\tau;L_2(\Omega^h))}$$

$$+ \|\hat{\psi}\|_{\ddot{W}_2^{1/2}(\Omega^\tau;L_2(\Sigma^h))} + \|\mu\|_{\ddot{W}_2^{1/2}(\Omega^\tau;L_2(\Sigma^h))}\Bigg]. \qquad (3.123)$$

Thus, in order to estimate the rate of convergence of the finite difference scheme (3.121) it suffices to bound the terms appearing on the right-hand side of (3.123). We shall suppose for the sake of simplicity that $\tau \asymp h^2$.

Theorem 3.6 *Let the solution u of the initial-boundary-value problem* (3.120) *belong to $\widetilde{W}_2^{s,s/2}(Q)$, $a_{ij} \in W_2^{s-1}(\Omega^\pm)$, and suppose that $a \in W_2^{s-1}(\Omega^\pm)$ and $k \in W_2^{s-1}(\Sigma)$, $5/2 < s \le 3$. Then, assuming that $\tau \asymp h^2$, the global error of the finite difference scheme* (3.121) *satisfies the following error bound:*

$$\|u - U\|_{\widetilde{W}_2^{1,1/2}(Q_h^\tau)} \le Ch^{s-1}\Big(\max_{i,j}\|a_{ij}\|_{W_2^{s-1}(\Omega^+)} + \max_{i,j}\|a_{ij}\|_{W_2^{s-1}(\Omega^-)}$$

$$+ \|a\|_{W_2^{s-2}(\Omega^+)} + \|a\|_{W_2^{s-2}(\Omega^-)} + \|k\|_{W_2^{s-1}(\Sigma)}$$

$$+ \sqrt{\log\tfrac{1}{h}}\Big)\|u\|_{\widetilde{W}_2^{s,s/2}(Q)},$$

for $5/2 < s \le 3$, where $C = C(s)$ is a positive constant, independent of h and τ.

Proof The terms η_{2j}, $j = 1, 2$, were bounded in Sect. 3.4.3. After summation over the mesh we obtain

$$\|\eta_{2j}\|_{2,h\tau} \le Ch^{s-1}\Big(\|a_{2j}\|_{W_2^{s-1}(\Omega^-)}\|u\|_{W_2^{s,s/2}(Q^-)}$$

$$+ \|a_{2j}\|_{W_2^{s-1}(\Omega^+)}\|u\|_{W_2^{s,s/2}(Q^+)}\Big), \quad 2 < s \le 3. \qquad (3.124)$$

The terms $\tilde{\eta}_{1j}$, $j = 1, 2$, and $\tilde{\eta}$ for $x \notin \Sigma^h$ were bounded in Sect. 3.4.3; for $x \in \Sigma^h$ they can be handled analogously. Hence,

$$\|\tilde{\eta}_{1j}\|_{1,h\tau} \le Ch^{s-1}\Big(\|a_{1j}\|_{W_2^{s-1}(\Omega^-)}\|u\|_{W_2^{s,s/2}(Q^-)}$$

$$+ \|a_{1j}\|_{W_2^{s-1}(\Omega^+)}\|u\|_{W_2^{s,s/2}(Q^+)}\Big), \quad 5/2 < s \le 3, \qquad (3.125)$$

and

$$\|\tilde{\eta}\|_{L_2(Q_{h\tau})} \le Ch^{s-1}\Big(\|a\|_{W_2^{s-2}(\Omega^-)}\|u\|_{W_2^{s,s/2}(Q^-)}$$

$$+ \|a\|_{W_2^{s-2}(\Omega^+)}\|u\|_{W_2^{s,s/2}(Q^+)}\Big), \quad 2 < s \le 3. \qquad (3.126)$$

It follows from (2.251) and (2.252) that

$$\big|\hat{\eta}_{1j}(\cdot,t)\big|_{W_2^{1/2}(\Sigma^h)} \le Ch^{s-1}\big(\|a_{1j}\|_{W_2^{s-1}(\Omega^-)}\|T_t^- u(\cdot,t)\|_{W_2^s(\Omega^-)} + \|a_{1j}\|_{W_2^{s-1}(\Omega^+)}\big\|T_t^- u(\cdot,t)\big\|_{W_2^s(\Omega^+)}\big), \quad 5/2 < s \le 3,$$

and therefore

$$|\hat{\eta}_{1j}|_{L_2(\Omega^\tau;W_2^{1/2}(\Sigma^h))} \le Ch^{s-1}\big(\|a_{1j}\|_{W_2^{s-1}(\Omega^-)}\|u\|_{W_2^{s,s/2}(Q^-)} + \|a_{1j}\|_{W_2^{s-1}(\Omega^+)}\|u\|_{W_2^{s,s/2}(Q^+)}\big), \quad 5/2 < s \le 3. \quad (3.127)$$

The term $\hat{\eta}$ can be bounded directly as follows:

$$\|\hat{\eta}\|_{L_2(\Sigma^h\times\Omega^\tau)} \le Ch^2\big(\|a\|_{W_2^{s-2}(\Omega^-)}\|u\|_{W_2^{s,s/2}(Q^-)} + \|a\|_{W_2^{s-2}(\Omega^+)}\|u\|_{W_2^{s,s/2}(Q^+)}\big), \quad s > 5/2. \quad (3.128)$$

The term $\tilde{\psi}$, for $x \notin \Sigma^h$, was bounded in Sect. 3.4.3. For $x \in \Sigma^h$ we consider the decomposition $\tilde{\psi} := \psi^+ + \psi^-$, where

$$\begin{aligned}
\psi^+ :=& \frac{1}{2}\Big(u - T_1^2T_2^{2+}u + \frac{h}{3}T_1^2\partial_2 u\Big)\Big|_{(x_1,\xi+0)} \\
=& -\frac{1}{h^2}\int_{x_1-h}^{x_1+h}\int_{\xi}^{\xi+h}\Big(1-\frac{|x_1-x_1'|}{h}\Big)\Big(1-\frac{x_2'-\xi}{h}\Big) \\
&\times\int_{x_1}^{x_1'}\int_{x_1}^{x_1''}\partial_1^2 u(x_1''',x_2')\,dx_1'''\,dx_1''\,dx_2'\,dx_1' \\
&+\frac{1}{h^2}\int_{x_1-h}^{x_1+h}\int_{\xi}^{\xi+h}\Big(1-\frac{|x_1-x_1'|}{h}\Big)\Big(1-\frac{x_2'-\xi}{h}\Big) \\
&\times\int_{\xi}^{x_2'}\int_{x_1}^{x_1'}\partial_1\partial_2 u(x_1'',x_2'')\,dx_1''\,dx_2''\,dx_2'\,dx_1' \\
&-\frac{1}{h^2}\int_{x_1-h}^{x_1+h}\int_{\xi}^{\xi+h}\Big(1-\frac{|x_1-x_1'|}{h}\Big)\Big(1-\frac{x_2'-\xi}{h}\Big) \\
&\times\int_{\xi}^{x_2'}\int_{\xi}^{x_2''}\partial_2^2 u(x_1',x_2''')\,dx_2'''\,dx_2''\,dx_2'\,dx_1',
\end{aligned}$$

and ψ^- is defined analogously. By bounding $\psi^\pm$ in the same way as $\tilde{\psi}$ for $x \notin \Sigma^h$ and combining the bounds, we get:

$$\|\tilde{\psi}\|_{\ddot{W}_2^{1/2}(\Omega^\tau;L_2(\Omega^h))} \le Ch^{s-1}\sqrt{\log\tfrac{1}{h}}\big(\|u\|_{W_2^{s,s/2}(Q^-)} + \|u\|_{W_2^{s,s/2}(Q^+)}\big), \quad 5/2 < s \le 3. \tag{3.129}$$

Analogously, one obtains

$$\|\mu\|_{\ddot{W}_2^{1/2}(\Omega^\tau;L_2(\Sigma^h))} \le Ch^{s-1}\sqrt{\log\frac{1}{h}}\,\|k\|_{W_2^{s-1}(\Sigma)}\|u\|_{W_2^{s,s/2}(\Sigma\times(0,T))}, \tag{3.130}$$

for $2 < s \le 3$, while $\hat{\psi}$ can be bounded directly, yielding

$$\|\hat{\psi}\|_{\ddot{W}_2^{1/2}(\Omega^\tau;L_2(\Sigma^h))} \le Ch^2\sqrt{\log\frac{1}{h}}\,\|u\|_{W_2^{2,1}(\Sigma\times(0,T))}. \tag{3.131}$$

The assertion then follows from (3.123)–(3.131). □

3.5.2 Factorized Scheme

Analogously as in Sect. 3.4.4, we shall construct here a factorized, unconditionally stable version of the finite difference approximation (3.121) of the initial-boundary-value problem (3.119). For the sake of simplicity we shall suppose that $k = \text{Const.} > 0$ and consider the following finite difference scheme (with the same definitions of the meshes, mesh-functions and finite difference operators as in Sect. 3.5.1):

$$\begin{aligned} (I_h + \theta\tau\Lambda_1)(B_h + \theta\tau\Lambda_2)D_t^+ U + \mathcal{L}_h U &= T_1^2 T_2^2 T_t^+ f \quad \text{in } \Omega_h^{\tau-}, \\ U &= 0 \quad \text{on } \Gamma^h \times \overline{\Omega}^\tau, \\ U(x,0) &= u_0(x) \quad \text{on } \Omega^h, \end{aligned} \tag{3.132}$$

where

$$\Lambda_i U := -D_{x_i}^+ D_{x_i}^- U,$$
$$B_h U := (1 + k\delta_{\Sigma^h})U,$$

I_h is the identity operator and θ is a real parameter. Obviously, when the values of U for some fixed $t = t' \in \Omega^\tau$ are known, the values of U on the next time level $t = t' + \tau$ can be computed by inverting the operators $I_h + \theta\tau\Lambda_1$ and $B_h + \theta\tau\Lambda_2$. Since these operators can be represented by tridiagonal matrices, the solution of (3.132) can be computed very efficiently by successively solving systems of linear algebraic equations, each having a tridiagonal matrix. We note that the operator $(I_h + \theta\tau\Lambda_1)(B_h + \theta\tau\Lambda_2)$ is symmetric and when $s > 2$ and $\tau \asymp h^2$ the operator-inequality

$$(I_h + \theta\tau\Lambda_1)(B_h + \theta\tau\Lambda_2) - \tau\mathcal{L}_h \ge cB_h, \quad \text{with } 0 < c < 1,$$

holds for sufficiently large positive θ and sufficiently small h (independent of the size of the time step τ).

Let u be the solution of the initial-boundary-value problem (3.120) and let U denote the solution of the finite difference scheme (3.132). The global error $e := u - U$ satisfies the finite difference scheme

$$\begin{aligned}(I_h + \theta\tau\Lambda_1)(B_h + \theta\tau\Lambda_2)D_t^+ e + \mathcal{L}_h e &= \overline{\varphi} \quad \text{in } \Omega_h^{\tau-},\\ e &= 0 \quad \text{on } \Gamma^h \times \overline{\Omega}^\tau,\\ e(x,0) &= 0 \quad \text{on } \Omega^h,\end{aligned} \tag{3.133}$$

where

$$\begin{aligned}\overline{\varphi} &:= \sum_{i,j=1}^{2} \overline{\eta}_{ij,\bar{x}_i} + \overline{\eta} + D_t^+\overline{\psi} + \delta_{\Sigma^h} D_t^+\overline{\mu},\\ \overline{\eta}_{ij} &:= T_i^+ T_{3-i}^2 T_t^+(a_{ij}\partial_j u) - \frac{1}{2}\left(a_{ij} D_{x_j}^+ u + a_{ij}^{+i} D_{x_j}^- u^{+i}\right),\\ \overline{\eta} &:= \left(T_1^2 T_2^2 a\right)u - T_1^2 T_2^2 T_t^+(au),\\ \overline{\psi} &:= u - T_1^2 T_2^2 u - \theta\tau\left(D_{x_1}^+ D_{x_1}^- u + D_{x_2}^+ D_{x_2}^- u\right) + \theta^2\tau^2 D_{x_1}^+ D_{x_1}^- D_{x_2}^+ D_{x_2}^- u,\\ \overline{\mu} &:= k\left(u - T_1^2 u - \theta\tau D_{x_1}^+ D_{x_1}^- u\right).\end{aligned}$$

Let us consider the decompositions

$$\overline{\eta}_{1j} = \tilde{\overline{\eta}}_{1j} + \delta_{\Sigma^h}\hat{\overline{\eta}}_{1j}, \qquad \overline{\eta} = \tilde{\overline{\eta}} + \delta_{\Sigma^h}\hat{\overline{\eta}}, \qquad \overline{\psi} = \tilde{\overline{\psi}} + \delta_{\Sigma^h}\hat{\overline{\psi}},$$

where

$$\tilde{\overline{\eta}}_{1j} := \overline{\eta}_{1j} - \delta_{\Sigma^h}\hat{\overline{\eta}}_{1j}, \qquad \tilde{\overline{\eta}} := \overline{\eta} - \delta_{\Sigma^h}\hat{\overline{\eta}}, \qquad \tilde{\overline{\psi}} := \overline{\psi} - \delta_{\Sigma^h}\hat{\overline{\psi}},$$

and

$$\begin{aligned}\hat{\overline{\eta}}_{11} &:= \frac{1}{6}h^2 T_1^+ T_t^+\left([a_{11}\partial_1\partial_2 u + \partial_2 a_{11}\partial_1 u]_\Sigma\right),\\ \hat{\overline{\eta}}_{12} &:= \frac{1}{6}h^2 T_1^+ T_t^+\left(\left[a_{12}\partial_2^2 u + \partial_2 a_{12}\partial_2 u\right]_\Sigma\right) - \frac{1}{4}h^2 T_1^+ T_t^+\left(\left[\partial_1(a_{12}\partial_2 u)\right]_\Sigma\right),\\ \hat{\overline{\eta}} &:= -\frac{1}{3}h^2\left[\left(T_1^2 a\right)\left(T_1^2 T_t^- \partial_2 u\right)\right]_\Sigma,\\ \hat{\overline{\psi}} &:= \left(\frac{1}{6}h^2 - \theta\tau\right)\left[T_1^2(\partial_2 u)\right]_\Sigma.\end{aligned}$$

By applying (3.18) and (3.19) to (3.133), we obtain the following a priori estimate, which represents the starting point for the error analysis of the finite difference scheme (3.132):

$$\|e\|_{\widetilde{W}_2^{1,1/2}(Q_h^\tau)} \le C\Bigg[\sum_{j=1}^{2}\Big(\|\overline{\eta}_{2j}\|_{2,h\tau} + \|\tilde{\overline{\eta}}_{1j}\|_{1,h\tau} + |\hat{\overline{\eta}}_{1j}|_{L_2(\Omega^\tau;W_2^{1/2}(\Sigma^h))}\Big)$$
$$+ \|\tilde{\overline{\eta}}\|_{L_2(Q_h^\tau)} + \|\hat{\overline{\eta}}\|_{L_2(\Sigma^h\times\Omega^\tau)} + \|\tilde{\overline{\psi}}\|_{\ddot{W}_2^{1/2}(\Omega^\tau;L_2(\Omega^h))}$$
$$+ \|\hat{\overline{\psi}}\|_{\ddot{W}_2^{1/2}(\Omega^\tau;L_2(\Sigma^h))} + \|\overline{\mu}\|_{\ddot{W}_2^{1/2}(\Omega^\tau;L_2(\Sigma^h))}\Bigg]. \qquad (3.134)$$

For $\overline{\eta}_{ij}$ and $\overline{\eta}$ the same bounds hold as for η_{ij} and η, while $\overline{\psi}$ and $\overline{\mu}$ can be bounded analogously to ψ and μ. Thus we obtain the following result.

Theorem 3.7 *Let* $k = \text{Const.} > 0$; *then, under the same assumptions as in the statement of Theorem* 3.6, *the global error of the finite difference scheme* (3.132) *satisfies the following error bound*:

$$\|u - U\|_{\widetilde{W}_2^{1,1/2}(Q_h^\tau)}$$
$$\le Ch^{s-1}\Big(\max_{i,j}\|a_{ij}\|_{W_2^{s-1}(\Omega^+)} + \max_{i,j}\|a_{ij}\|_{W_2^{s-1}(\Omega^-)}$$
$$+ \|a\|_{W_2^{s-2}(\Omega^+)} + \|a\|_{W_2^{s-2}(\Omega^-)} + \sqrt{\log\tfrac{1}{h}}\Big)\|u\|_{\widetilde{W}_2^{s,s/2}(Q)}, \quad 5/2 < s \le 3,$$

where $C = C(s)$ *is a positive constant, independent of* h *and* τ.

3.6 A Parabolic Transmission Problem

In this section we focus our attention on transmission problems whose solutions are defined in two (or more) disconnected domains. Such a situation may occur when the solution in the intermediate region is known, or can be determined from a simpler equation. The effect of the intermediate region can be modelled by means of nonlocal jump conditions across the intermediate region (see, Tikhonov [180], Kačur et al. [113], Datta [27], Givoli [55, 56], Qatanani et al. [152], Druet [35], Jovanović and Vulkov [102, 104]).

As a model example, we consider the following initial-boundary-value problem: find two functions, $u_1(x, y, t)$ and $u_2(x, y, t)$, that satisfy the system of parabolic equations

$$\frac{\partial u_1}{\partial t} - \Delta u_1 = f_1(x, y, t), \quad (x, y) \in \Omega_1 := (a_1, b_1) \times (c_1, d_1),\ t \in (0, T], \quad (3.135)$$

$$\frac{\partial u_2}{\partial t} - \Delta u_2 = f_2(x, y, t), \quad (x, y) \in \Omega_2 := (a_2, b_2) \times (c_2, d_2),\ t \in (0, T], \quad (3.136)$$

where $-\infty < a_1 < b_1 < a_2 < b_2 < +\infty$, and, e. g., $c_2 < c_1 < d_1 < d_2$, the internal transmission conditions of nonlocal Robin–Dirichlet type

$$\frac{\partial u_1}{\partial x}(b_1, y, t) + \alpha_1(y)u_1(b_1, y, t) = \int_{c_2}^{d_2} \beta_1(y, y')u_2(a_2, y', t)\mathrm{d}y', \quad y \in (c_1, d_1),\ t \in (0, T], \tag{3.137}$$

$$\frac{\partial u_2}{\partial x}(a_2, y, t) + \alpha_2(y)u_2(a_2, y, t) = \int_{c_1}^{d_1} \beta_2(y, y')u_1(b_1, y', t)\mathrm{d}y', \quad y \in (c_2, d_2),\ t \in (0, T], \tag{3.138}$$

the simplest external Dirichlet boundary conditions for $t \in (0, T]$:

$$\begin{aligned} u_1(x, c_1, t) = u_1(x, d_1, t) = 0, &\quad x \in (a_1, b_1), \\ u_2(x, c_2, t) = u_2(x, d_2, t) = 0, &\quad x \in (a_2, b_2), \\ u_1(a_1, y, t) = 0, \quad y \in (c_1, d_1); \qquad u_2(b_2, y, t) = 0, &\quad y \in (c_2, d_2), \end{aligned} \tag{3.139}$$

and the initial conditions

$$\begin{aligned} u_1(x, y, 0) = u_{10}(x, y), &\quad (x, y) \in \Omega_1, \\ u_2(x, y, 0) = u_{20}(x, y), &\quad (x, y) \in \Omega_2. \end{aligned} \tag{3.140}$$

Note that for a special choice of α_i and β_i such an initial-boundary-value problem models linearized radiative heat transfer in a system of absolutely black bodies (see Amosov [3]).

In the sequel we shall assume that the data satisfy the regularity conditions

$$\alpha_i \in L_\infty(c_i, d_i), \qquad \beta_i \in L_\infty\big((c_i, d_i) \times (c_{3-i}, d_{3-i})\big), \quad i = 1, 2. \tag{3.141}$$

In physical problems (see Amosov [3]) we also often have that

$$\alpha_i > 0, \qquad \beta_i > 0, \quad i = 1, 2.$$

3.6.1 Weak Solutions and Function Spaces

We introduce the product space

$$L := L_2(\Omega_1) \times L_2(\Omega_2) = \big\{v = (v_1, v_2) : v_i \in L_2(\Omega_i)\big\},$$

equipped with the inner product and the associated norm

$$(u, v)_L := (u_1, v_1)_{L_2(\Omega_1)} + (u_2, v_2)_{L_2(\Omega_2)}, \quad \|v\|_L = (v, v)_L^{1/2},$$

where $u = (u_1, u_2)$ and $v = (v_1, v_2)$. We also define the spaces

$$W_2^k := \{v = (v_1, v_2) : v_i \in W_2^k(\Omega_i)\}, \quad k = 1, 2, \dots$$

equipped with the inner products and norms

$$(u, v)_{W_2^k} := (u_1, v_1)_{W_2^k(\Omega_1)} + (u_2, v_2)_{W_2^k(\Omega_2)}, \quad \|v\|_{W_2^k} = (v, v)_{W_2^k}^{1/2}.$$

In particular, we let

$$\mathring{W}_2^1 := \{v = (v_1, v_2) \in W_2^1 : v_i = 0 \text{ on } \Gamma_i, i = 1, 2\},$$

where

$$\Gamma_1 := \partial\Omega_1 \setminus \{(b_1, y) : y \in (c_1, d_1)\},$$
$$\Gamma_2 := \partial\Omega_2 \setminus \{(a_2, y) : y \in (c_2, d_2)\}.$$

Finally, with $u = (u_1, u_2)$ and $v = (v_1, v_2)$ we define the bilinear functional:

$$\begin{aligned}
a(u, v) := {} & \int_{\Omega_1} \left(\frac{\partial u_1}{\partial x}\frac{\partial v_1}{\partial x} + \frac{\partial u_1}{\partial y}\frac{\partial v_1}{\partial y}\right) \mathrm{d}x\,\mathrm{d}y \\
& + \int_{\Omega_2} \left(\frac{\partial u_2}{\partial x}\frac{\partial v_2}{\partial x} + \frac{\partial u_2}{\partial y}\frac{\partial v_2}{\partial y}\right) \mathrm{d}x\,\mathrm{d}y \\
& + \int_{c_1}^{d_1} \alpha_1(y) u_1(b_1, y) v_1(b_1, y)\,\mathrm{d}y \\
& + \int_{c_2}^{d_2} \alpha_2(y) u_2(a_2, y) v_2(a_2, y)\,\mathrm{d}y \\
& - \int_{c_2}^{d_2} \int_{c_1}^{d_1} \beta_1(y, y') u_2(a_2, y') v_1(b_1, y)\,\mathrm{d}y\,\mathrm{d}y' \\
& - \int_{c_2}^{d_2} \int_{c_1}^{d_1} \beta_2(y', y) u_1(b_1, y) v_2(a_2, y')\,\mathrm{d}y\,\mathrm{d}y'. \qquad (3.142)
\end{aligned}$$

The following coercivity result holds (cf. Jovanović and Vulkov [102]).

Lemma 3.8 *Under the conditions* (3.141) *the bilinear functional a, defined by* (3.142), *is bounded on $W_2^1 \times W_2^1$ and satisfies the following Gårding inequality on $\mathring{W}_2^1$: there exist positive constants m and κ such that*

$$a(u, u) + \kappa\|u\|_L^2 \geq m\|u\|_{W_2^1}^2 \quad \forall u \in \mathring{W}_2^1.$$

Proof The boundedness of the bilinear functional a follows from (3.141) and the trace theorem, according to which

$$\|u_i\|_{L_2(\partial\Omega_i)} \leq C\|u_i\|_{W_2^1(\Omega_i)}, \quad i = 1, 2.$$

From the Friedrichs inequality (1.23) we immediately obtain that

$$\sum_{i=1}^{2}\int_{\Omega_i}\left[\left(\frac{\partial u_i}{\partial x}\right)^2+\left(\frac{\partial u_i}{\partial y}\right)^2\right]\mathrm{d}x\,\mathrm{d}y \geq c_0\|u\|^2_{W_2^1},$$

where c_0 is a positive constant, which depends on Ω_1 and Ω_2. The remaining terms in $a(u,u)$ can be bounded by $\varepsilon\|u\|^2_{W_2^1}+\frac{C}{\varepsilon}\|u\|^2_L$; indeed, using the trace inequality (cf. Theorem A.2 on p. 122 in [139]) it follows that:

$$\|u_i\|_{L_2(\partial\Omega_i)} \leq \varepsilon\|\nabla u_i\|_{L_2(\Omega_i)}+\frac{C}{\varepsilon}\|u_i\|_{L_2(\Omega_i)}, \quad \varepsilon>0,\ i=1,2.$$

The stated result then follows for a sufficiently small $\varepsilon>0$. □

Let $W_2^{-1}=(\mathring{W}_2^1)^*$ be the dual space of $\mathring{W}_2^1$, and let $\langle\cdot,\cdot\rangle$ denote the associated duality pairing. The spaces $\mathring{W}_2^1$, L and W_2^{-1} form a Gelfand triple: i.e. $\mathring{W}_2^1 \hookrightarrow L \hookrightarrow W_2^{-1}$, with continuous and dense embeddings. We also introduce the space

$$\mathcal{W}(0,T):=\left\{u : u\in L_2\big((0,T),\mathring{W}_2^1\big),\ \frac{\mathrm{d}u}{\mathrm{d}t}\in L_2\big((0,T),W_2^{-1}\big)\right\}$$

with inner product

$$(u,v)_{\mathcal{W}(0,T)}:=\int_0^T\left[(u,v)_{W_2^1}+\left(\frac{\mathrm{d}u}{\mathrm{d}t},\frac{\mathrm{d}v}{\mathrm{d}t}\right)_{W_2^{-1}}\right]\mathrm{d}t.$$

The weak formulation of problem (3.135)–(3.139) is then:

$$\left\langle\frac{\mathrm{d}u}{\mathrm{d}t},v\right\rangle+a(u,v)=\langle f,v\rangle \quad \forall v\in W_2^1. \tag{3.143}$$

The problem (3.143) fits into the general theory of parabolic differential operators in Hilbert spaces (see Wloka [199]). By applying Theorem 3.2 (cf. also Theorem 26.1 from Wloka [199]) to (3.143) we obtain the following assertion.

Theorem 3.9 *Let the assumptions* (3.141) *hold and suppose that* $u_0=(u_{10},u_{20})\in L$, $f=(f_1,f_2)\in L_2((0,T),W_2^{-1})$. *Then, for* $0<T<+\infty$, *the initial-boundary-value problem* (3.135)–(3.140) *has a unique weak solution* $u\in\mathcal{W}(0,T)$; *moreover* u *depends continuously on* f *and* u_0.

Because the norm $\|\cdot\|_{W_2^{-1}}$ is not computable, following Lions and Magenes [127] we shall, instead, consider the initial-boundary-value problem (3.135)–(3.140) with right-hand sides f_i, $i=1,2$, of the form:

$$f_i(x,y,t) = f_{i0}(x,y,t) + \frac{\partial(\varrho_i(x) f_{i1}(x,y,t))}{\partial x} + \frac{\partial f_{i2}(x,y,t)}{\partial y}$$
$$+ \int_0^T \frac{f_{i3}(x,y,t,t') - f_{i3}(x,y,t',t)}{|t-t'|} \mathrm{d}t', \quad i = 1,2, \tag{3.144}$$

where $f_{i0}, f_{i1}, f_{i2} \in L_2((0,T), L_2(\Omega_i)) = L_2(Q_i)$, $Q_i = \Omega_i \times (0,T)$, $f_{i3} \in L_2((0,T)^2, L_2(\Omega_i)) = L_2(R_i)$, $R_i = \Omega_i \times (0,T)^2$, $\varrho_i \in C([a_i,b_i])$ and

$$\gamma_1(b_1 - x) \le \varrho_1(x) \le C_1(b_1 - x), \quad x \in (a_1,b_1), \ C_1 \ge \gamma_1 > 0,$$
$$\gamma_2(x - a_2) \le \varrho_2(x) \le C_2(x - a_2), \quad x \in (a_2,b_2), \ C_2 \ge \gamma_2 > 0.$$

We shall also consider the case

$$f_i(x,t) = \frac{\partial g_i(x,t)}{\partial t}, \quad i = 1,2, \tag{3.145}$$

where $g_i \in \ddot{W}_2^{1/2}((0,T), L_2(\Omega_i))$, $i = 1,2$. The norm in $\ddot{W}_2^{1/2}(0,T)$ is defined by

$$\|\varphi\|^2_{\ddot{W}_2^{1/2}(0,T)} := |\varphi|^2_{W_2^{1/2}(0,T)} + \int_0^T \left(\frac{1}{t} + \frac{1}{T-t}\right) \varphi^2(t)\,\mathrm{d}t.$$

We also define the space $W_2^{1,1/2} := L_2((0,T), W_2^1) \cap W_2^{1/2}((0,T), L)$.

The next two theorems follow from the results of Sect. 3.1.2, and in particular from (3.18) and (3.19).

Theorem 3.10 *Suppose that the hypotheses* (3.141) *hold and let* $u_{i0} \in L_2(\Omega_i)$, $f_{i0}, f_{i1} f_{i2} \in L_2(Q_i)$, $f_{i3} \in L_2(R_i)$, $i = 1,2$. *Then, the initial-boundary-value problem* (3.135)–(3.140), (3.144) *has a unique weak solution* $u = (u_1,u_2) \in W_2^{1,1/2}$ *and the following a priori estimate holds*:

$$\|u\|^2_{W_2^{1,1/2}} \le C \sum_{i=1}^2 \big(\|u_{i0}\|^2_{L_2(\Omega_i)} + \|f_{i0}\|^2_{L_2(Q_i)}$$
$$+ \|f_{i1}\|^2_{L_2(Q_i)} + \|f_{i2}\|^2_{L_2(Q_i)} + \|f_{i3}\|^2_{L_2(R_i)}\big). \tag{3.146}$$

Theorem 3.11 *Let the hypotheses* (3.141) *hold and let* $u_{i0} \in L_2(\Omega_i)$, $g_i \in \ddot{W}_2^{1/2}((0,T), L_2(\Omega_i))$, $i = 1,2$. *Then, the initial-boundary-value problem* (3.135)–(3.140), (3.145) *has a unique weak solution* $u = (u_1,u_2) \in W_2^{1,1/2}$ *and the following a priori estimate holds*:

$$\|u\|^2_{W_2^{1,1/2}} \le C\big(\|u_0\|^2_L + \|g\|^2_{\ddot{W}_2^{1/2}((0,T),L)}\big). \tag{3.147}$$

In both cases C is a computable constant depending on T.

3.6.2 Finite Difference Approximation

Let $\overline{\Omega}_i^{h_i}$ be a uniform mesh on $[a_i, b_i]$ with mesh-size $h_i := (b_i - a_i)/N_i$, $i = 1, 2$. We consider

$$\Omega_i^{h_i} := \overline{\Omega}_i^{h_i} \cap (a_i, b_i), \qquad \Omega_{i-}^{h_i} := \Omega_i^{h_i} \cup \{a_i\}, \qquad \Omega_{i+}^{h_i} := \Omega_i^{h_i} \cup \{b_i\}, \quad i = 1, 2.$$

Analogously, we consider the uniform mesh $\overline{\Omega}_i^{k_i}$ on $[c_i, d_i]$ with mesh-size $k_i := (d_i - c_i)/M_i$ and its submeshes

$$\Omega_i^{k_i} := \overline{\Omega}_i^{k_i} \cap (c_i, d_i), \qquad \Omega_{i-}^{k_i} := \Omega_i^{k_i} \cup \{c_i\}, \qquad \Omega_{i+}^{k_i} := \Omega_i^{k_i} \cup \{d_i\}, \quad i = 1, 2.$$

We shall assume that $h_1 \asymp h_2 \asymp k_1 \asymp k_2$ and define $h := \max\{h_1, h_2, k_1, k_2\}$. Finally, we introduce a uniform mesh $\overline{\Omega}^\tau$ on $[0, T]$ with the step size $\tau := T/M$, $M \geq 1$, and its submeshes Ω^τ and $\Omega^\tau_\pm$ (see Sect. 3.1.4). We shall consider vector-functions of the form $V = (V_1, V_2)$ where V_i is a mesh-function defined on $\overline{\Omega}_i^{h_i} \times \overline{\Omega}_i^{k_i} \times \overline{\Omega}^\tau$, $i = 1, 2$. We define the difference quotients:

$$D_{x,i}^+ V_i := \frac{V_i(x + h_i, y, t) - V_i(x, y, t)}{h_i} =: D_{x,i}^- V_i(x + h_i, y, t), \quad i = 1, 2,$$

with $D_{y,i}^\pm V_i$ and $D_t^\pm V_i$ defined analogously. We shall use the notations

$$D_x^\pm V = \big(D_{x,1}^\pm V_1, D_{x,2}^\pm V_2\big), \qquad D_y^\pm V = \big(D_{y,1}^\pm V_1, D_{y,2}^\pm V_2\big),$$
$$D_t^\pm V = \big(D_t^\pm V_1, D_t^\pm V_2\big).$$

Further, we define the Steklov mollifiers in the usual way:

$$T_{x,i} f_i(x, y, t) = T_{x,i}^\mp f_i\left(x \pm \frac{1}{2} h_i, y, t\right) := \frac{1}{h_i} \int_{x-h_i/2}^{x+h_i/2} f_i\big(x', y, t\big)\, \mathrm{d}x', \quad i = 1, 2,$$

with $T_{y,i}$, $T_{y,i}^\pm$, T_t and $T_t^\pm$ defined analogously. For $x = b_1$ and $x = a_2$ we also require the following asymmetric mollifiers:

$$T_{x,1}^{2-} f_1(b_1, y, t) := \frac{2}{h_1} \int_{b_1 - h_1}^{b_1} \left(1 - \frac{b_1 - x'}{h_1}\right) f_1\big(x', y, t\big)\, \mathrm{d}x',$$
$$T_{x,2}^{2+} f_2(a_2, y, t) := \frac{2}{h_2} \int_{a_2}^{a_2 + h_2} \left(1 - \frac{x' - a_2}{h_2}\right) f_2\big(x', y, t\big)\, \mathrm{d}x'.$$

With the notational conventions $\hbar_i := h_i$, $x \in \Omega_i^{h_i}$, $i = 1, 2$, $\hbar_1(b_1) := h_1/2$, $\hbar_2(a_2) := h_2/2$, we introduce the discrete inner products

$$(V, W)_{L_h} := k_1 \sum_{x \in \Omega_{1+}^{h_1}} \sum_{y \in \Omega_1^{k_1}} V_1 W_1 \hbar_1 + k_2 \sum_{x \in \Omega_{2-}^{h_2}} \sum_{y \in \Omega_2^{k_2}} V_2 W_2 \hbar_2,$$

$$(V, W)_{L_{h'}} := h_1 k_1 \sum_{x\in\Omega_{1+}^{h_1}} \sum_{y\in\Omega_1^{k_1}} V_1 W_1 + h_2 k_2 \sum_{x\in\Omega_{2+}^{h_2}} \sum_{y\in\Omega_2^{k_2}} V_2 W_2,$$

$$(V, W)_{L_{h''}} := k_1 \sum_{x\in\Omega_{1+}^{h_1}} \sum_{y\in\Omega_{1+}^{k_1}} V_1 W_1 \hbar_1 + k_2 \sum_{x\in\Omega_{2-}^{h_2}} \sum_{y\in\Omega_{2+}^{k_2}} V_2 W_2 \hbar_2$$

and the associated mesh-dependent norms

$$\|V\|_{L_h}^2 := (V, V)_{L_h}, \qquad \|V\|_{L_{h'}}^2 := (V, V)_{L_{h'}}, \qquad \|V\|_{L_{h''}}^2 := (V, V)_{L_{h''}}.$$

We also define the following mesh-dependent norms:

$$\|V_i\|_{L_2(\Omega_\pm^\tau)}^2 := \tau \sum_{t\in\Omega_\pm^\tau} V_i^2, \qquad \|V\|_{L_2(\Omega^k)}^2 := k_1 \sum_{y\in\Omega_1^{k_1}} V_1^2 + k_2 \sum_{y\in\Omega_2^{k_2}} V_2^2,$$

$$\|V\|_{L_2(\Omega_\pm^\tau, H)}^2 := \tau \sum_{t\in\Omega_\pm^\tau} \|V(\cdot, t)\|_H^2, \quad \text{where } H := L_h, L_{h'}, L_{h''}, L_2(\Omega^k),$$

as well as

$$\|V\|_{\tilde{L}_2(\Omega^\tau, L_h)}^2 := \tau \sum_{t\in\Omega^\tau} \left(\frac{1}{t} + \frac{1}{T-t}\right) \|V(\cdot, t)\|_{L_h}^2,$$

$$|V|_{W_2^{1/2}(\overline{\Omega}^\tau, L_h)}^2 := \tau^2 \sum_{t\in\overline{\Omega}^\tau} \sum_{t'\in\overline{\Omega}^\tau, t'\neq t} \frac{\|V(\cdot, t) - V(\cdot, t')\|_{L_h}^2}{|t - t'|^2},$$

and

$$\|V\|_{\ddot{W}_2^{1/2}(\overline{\Omega}^\tau, L_h)}^2 := |V|_{W_2^{1/2}(\overline{\Omega}^\tau, L_h)}^2 + \|V\|_{\tilde{L}_2(\Omega^\tau, L_h)}^2,$$

$$\|V\|_{W_{2,h\tau}^{1,1/2}}^2 := \|D_x^- V\|_{L_2(\Omega_+^\tau, L_{h'})}^2 + \|D_y^- V\|_{L_2(\Omega_+^\tau, L_{h''})}^2 + \|V\|_{L_2(\Omega_+^\tau, L_h)}^2 + |V|_{W_2^{1/2}(\overline{\Omega}^\tau, L_h)}^2.$$

We shall assume in what follows that u_i belongs to $W_2^{s,s/2}(Q_i)$, $i = 1, 2$, with $s \leq 3$, while $\alpha_i \in W_2^{s-1}(c_i, d_i)$ and $\beta_i \in W_2^{s-1}((c_i, d_i) \times (c_{3-i}, d_{3-i}))$, $i = 1, 2$, $s \leq 3$. Consequently, $f_i \in W_2^{s-2,(s-2)/2}(Q_i)$, $i = 1, 2$, $s \leq 3$, need not be continuous functions. We therefore approximate the initial-boundary-value problem (3.135)–(3.140) with the following implicit finite difference scheme with mollified data:

$$D_t^- U_1 - D_{x,1}^- D_{x,1}^+ U_1 - D_{y,1}^- D_{y,1}^+ U_1 = \bar{f}_1,$$

$$x \in \Omega_1^{h_1}, \ y \in \Omega_1^{k_1}, \ t \in \Omega_+^\tau, \tag{3.148}$$

$$D_t^- U_1(b_1,y,t) + \frac{2}{h_1}\Big[D_{x,1}^- U_1(b_1,y,t) + \alpha_1(y)U_1(b_1,y,t)$$
$$- k_2 \sum_{y'\in\Omega_2^{k_2}} \beta_1(y,y')U_2(a_2,y',t)\Big] - D_{y,1}^- D_{y,1}^+ U_1(b_1,y,t)$$
$$= \bar{f}_1(b_1,y,t), \quad y\in\Omega_1^{k_1}, t\in\Omega_+^\tau, \tag{3.149}$$

$$D_t^- U_2 - D_{x,2}^- D_{x,2}^+ U_2 - D_{y,2}^- D_{y,2}^+ U_2 = \bar{f}_2,$$
$$x\in\Omega_2^{h_2},\ y\in\Omega_2^{k_2},\ t\in\Omega_+^\tau, \tag{3.150}$$
$$D_t^- U_2(a_2,y,t) - \frac{2}{h_2}\Big[D_{x,2}^+ U_2(a_2,y,t) - \alpha_2(y)U_2(a_2,y,t)$$
$$+ k_1 \sum_{y'\in\Omega_1^{k_1}} \beta_2(y,y')U_1(b_1,y',t)\Big] - D_{y,2}^- D_{y,2}^+ U_2(a_2,y,t)$$
$$= \bar{f}_2(a_2,y,t), \quad y\in\Omega_2^{k_2},\ t\in\Omega_+^\tau, \tag{3.151}$$

subject to the boundary conditions

$$\begin{aligned}
&U_1(x,c_1,t) = U_1(x,d_1,t) = 0, \quad x\in\overline{\Omega}_1^{h_1},\ t\in\overline{\Omega}^\tau,\\
&U_2(x,c_2,t) = U_2(x,d_2,t) = 0, \quad x\in\overline{\Omega}_2^{h_2},\ t\in\overline{\Omega}^\tau,\\
&U_1(a_1,y,t) = 0, \quad y\in\Omega_1^{k_1}; \qquad U_2(b_2,y,t) = 0, \quad y\in\Omega_2^{k_2},
\end{aligned} \tag{3.152}$$

and the initial conditions

$$U_i(x,y,0) = u_{i0}(x,y), \quad x\in\Omega_{i\pm}^{h_i},\ y\in\Omega_i^{k_i},\ i=1,2, \tag{3.153}$$

where

$$\bar{f}_i(x,y,t) := T_{x,i}^2 T_{y,i}^2 T_t^- f_i(x,y,t), \quad x\in\Omega_i^{h_i},\ y\in\Omega_i^{k_i},\ t\in\Omega_+^\tau,\ i=1,2,$$

$$\bar{f}_1(b_1,y,t) := T_{x,1}^{2-} T_{y,1}^2 T_t^- f_1(b_1,y,t), \qquad \bar{f}_2(a_2,y,t) := T_{x,2}^{2+} T_{y,2}^2 T_t^- f_2(a_2,y,t).$$

The finite difference scheme (3.148)–(3.153) fits into the general framework (3.38), where $\mathcal{H}^h$ is the space of mesh-functions $U=(U_1,U_2)$, with U_i defined on the mesh $\overline{\Omega}_i^{h_i}\times\overline{\Omega}_i^{k_i}$, $i=1,2$, where

$$U_1 = 0 \quad \text{for } x=b_1 \quad \text{and} \quad U_2 = 0 \quad \text{for } x=a_2,$$

and

$$\bar{B}_h = I_h, \qquad A_h = A_{0h} + A_{1h}, \qquad A_{0h}U = (A_{01h}U_1, A_{02h}U_2),$$

with

$$A_{01h}U_1 := \begin{cases} -D_{x,1}^- D_{x,1}^+ U_1 - D_{y,1}^- D_{y,1}^+ U_1, & x \in \Omega_1^{h_1},\ y \in \Omega_1^{k_1}, \\ \frac{2}{\hbar_1} D_{x,1}^- U_1 - D_{y,1}^- D_{y,1}^+ U_1, & x = b_1,\ y \in \Omega_1^{k_1}, \end{cases}$$

and $A_{02h}U_2$ is defined analogously. Hence,

$$(A_{0h}U, U)_{L_h} = \left\| D_x^- U \right\|_{L_{h'}}^2 + \left\| D_y^- U \right\|_{L_{h''}}^2 \geq C_3 \|U\|_{L_h}^2, \quad C_3 > 0,$$

thanks to the discrete Friedrichs inequality; and, by the Cauchy–Schwarz inequality,

$$\begin{aligned}(A_{1h}U, V)_{L_h} &= k_1 \sum_{y \in \Omega_1^{k_1}} \alpha_1(y) U_1(b_1, y) V_1(b_1, y) \\ &\quad + k_2 \sum_{y \in \Omega_2^{k_2}} \alpha_2(y) U_2(a_2, y) V_2(a_2, y) \\ &\quad - k_1 k_2 \sum_{y \in \Omega_1^{k_1}} \sum_{y' \in \Omega_2^{k_2}} \beta_1(y, y') U_2(a_2, y') V_1(b_1, y) \\ &\quad - k_1 k_2 \sum_{y \in \Omega_1^{k_1}} \sum_{y' \in \Omega_2^{k_2}} \beta_2(y', y) U_1(b_1, y) V_2(a_2, y') \\ &\leq C_4 \Bigg(k_1 \sum_{y \in \Omega_1^{k_1}} U_1^2(b_1, y) + k_2 \sum_{y \in \Omega_2^{k_2}} U_2^2(a_2, y) \Bigg)^{1/2} \\ &\quad \times \Bigg(k_1 \sum_{y \in \Omega_1^{k_1}} V_1^2(b_1, y) + k_2 \sum_{y \in \Omega_2^{k_2}} V_2^2(a_2, y) \Bigg)^{1/2}.\end{aligned}$$

Further, we have that

$$\begin{aligned}k_1 \sum_{y \in \Omega_1^{k_1}} U_1^2(b_1, y) &= k_1 h_1 \sum_{y \in \Omega_1^{k_1}} \sum_{x \in \Omega_{1+}^{h_1}} D_{x,1}^- \big(U_1^2(x, y)\big) \\ &= k_1 h_1 \sum_{y \in \Omega_1^{k_1}} \sum_{x \in \Omega_{1+}^{h_1}} D_{x,1}^- \big(U_1(x, y)\big) \big[U_1(x, y) + U_1(x - h_1, y)\big] \\ &\leq 2 \Bigg(k_1 h_1 \sum_{y \in \Omega_1^{k_1}} \sum_{x \in \Omega_{1+}^{h_1}} \big(D_{x,1}^- U_1\big)^2 \Bigg)^{1/2} \Bigg(k_1 \sum_{y \in \Omega_1^{k_1}} \sum_{x \in \Omega_{1+}^{h_1}} U_1^2 \hbar_1 \Bigg)^{1/2},\end{aligned}$$

with analogous bounds for the other summands. Thus,

$$\begin{aligned}(A_{1h}U, V)_{L_h}^2 &\leq 2C_4^2 \|U\|_{L_h} \left\| D_x^- U \right\|_{L_{h'}} \|V\|_{L_h} \left\| D_x^- V \right\|_{L_{h'}} \\ &\leq 2C_4^2 \|U\|_{L_h} \|U\|_{A_{0h}} \|V\|_{L_h} \|V\|_{A_{0h}}.\end{aligned}$$

Hence, the finite difference scheme (3.148)–(3.153) is unconditionally stable and satisfies an a priori estimate of the form (3.45).

Let $u = (u_1, u_2)$ be the solution of the initial-boundary-value problem (3.135)–(3.140) and let $U = (U_1, U_2)$ be the solution of the finite difference scheme (3.148)–(3.153). We define $e_i := u_i - U_i$ for $i = 1, 2$, and $e := (e_1, e_2)$. Then, the global error $e = u - U$ satisfies the following finite difference scheme:

$$
\begin{aligned}
&D_t^- e_1 - D_{x,1}^- D_{x,1}^+ e_1 - D_{y,1}^- D_{y,1}^+ e_1 = D_t^- \psi_1 + D_{x,1}^+ \eta_1 + D_{y,1}^+ \zeta_1,\\
&\quad x \in \Omega_1^{h_1},\ y \in \Omega_1^{k_1},\ t \in \Omega_+^\tau, \qquad (3.154)\\
&D_t^- e_1(b_1, y, t) + \frac{2}{h_1}\Big[D_{x,1}^- e_1(b_1, y, t) + \alpha_1(y) e_1(b_1, y, t)\\
&\quad - k_2 \sum_{y' \in \Omega_2^{k_2}} \beta_1(y, y') e_2(a_2, y', t) \Big] - D_{y,1}^- D_{y,1}^+ e_1(b_1, y, t)\\
&\quad = D_t^- \psi_1(b_1, y, t) + D_{y,1}^+ \zeta_1(b_1, y, t) - \frac{2}{h_1} \eta_1(b_1, y, t)\\
&\quad + \frac{2}{h_1} \mu_1(y, t), \quad y \in \Omega_1^{k_1}, t \in \Omega_+^\tau, \qquad (3.155)\\
&D_t^- e_2 - D_{x,2}^- D_{x,2}^+ e_2 - D_{y,2}^- D_{y,2}^+ e_2 = D_t^- \psi_2 + D_{x,2}^+ \eta_2 + D_{y,2}^+ \zeta_2,\\
&\quad x \in \Omega_2^{h_2},\ y \in \Omega_2^{k_2},\ t \in \Omega_+^\tau, \qquad (3.156)\\
&D_t^- e_2(a_2, y, t) - \frac{2}{h_2}\Big[D_{x,2}^+ e_2(a_2, y, t) - \alpha_2(y) e_2(a_2, y, t)\\
&\quad + k_1 \sum_{y' \in \Omega_1^{k_1}} \beta_2(y, y') e_1(b_1, y', t) \Big] - D_{y,2}^- D_{y,2}^+ e_2(a_2, y, t)\\
&\quad = D_t^- \psi_2(a_2, y, t) + D_{y,2}^+ \zeta_2(a_2, y, t) + \frac{2}{h_2} \eta_2(a_2 + h_2, y, t)\\
&\quad + \frac{2}{h_2} \mu_2(y, t), \quad y \in \Omega_2^{k_2}, t \in \Omega_+^\tau, \qquad (3.157)
\end{aligned}
$$

with

$$
\begin{aligned}
&e_1(x, c_1, t) = e_1(x, d_1, t) = 0, \quad x \in \overline{\Omega}_1^{h_1},\ t \in \overline{\Omega}^\tau,\\
&e_2(x, c_2, t) = e_2(x, d_2, t) = 0, \quad x \in \overline{\Omega}_2^{h_2},\ t \in \overline{\Omega}^\tau, \qquad (3.158)\\
&e_1(a_1, y, t) = 0, \quad y \in \Omega_1^{k_1}; \qquad e_2(b_2, y, t) = 0, \quad y \in \Omega_2^{k_2},
\end{aligned}
$$

and the initial conditions

$$
e_i(x, y, 0) = 0, \quad x \in \Omega_{i\pm}^{h_i},\ y \in \Omega_i^{k_i},\ i = 1, 2, \qquad (3.159)
$$

where

$$\psi_i := u_i - T^2_{x,i} T^2_{y,i} u_i, \quad x \in \Omega_i^{h_i}, \ y \in \Omega_i^{k_i}, \ t \in \overline{\Omega}^\tau, \ i = 1, 2,$$

$$\psi_1 := u_1 - T^{2-}_{x,1} T^2_{y,1} u_1 - \frac{h_1}{3} T^2_{y,1} \frac{\partial u_1}{\partial x}, \quad x = b_1, \ y \in \Omega_1^{k_1}, \ t \in \overline{\Omega}^\tau,$$

$$\psi_2 := u_2 - T^{2+}_{x,2} T^2_{y,2} u_2 + \frac{h_2}{3} T^2_{y,2} \frac{\partial u_2}{\partial x}, \quad x = a_2, \ y \in \Omega_2^{k_2}, \ t \in \overline{\Omega}^\tau,$$

and

$$\eta_i := T^-_{x,i} T^2_{y,i} T^-_t \left(\frac{\partial u_i}{\partial x} \right) - D^-_{x,i} u_i, \quad x \in \Omega_{i+}^{h_i}, \ y \in \Omega_i^{k_i}, \ t \in \Omega_+^\tau, \ i = 1, 2,$$

$$\zeta_i := T^2_{x,i} T^-_{y,i} T^-_t \left(\frac{\partial u_i}{\partial y} \right) - D^-_{y,i} u_i, \quad x \in \Omega_i^{h_i}, \ y \in \Omega_{i+}^{k_i}, \ t \in \Omega_+^\tau, \ i = 1, 2,$$

$$\zeta_1 := T^{2-}_{x,1} T^-_{y,1} T^-_t \left(\frac{\partial u_1}{\partial y} \right) - D^-_{y,1} u_1 - \frac{h_1}{3} T^-_{y,1} T^-_t \left(\frac{\partial^2 u_1}{\partial x \partial y} \right), \quad x = b_1,$$

$$\zeta_2 := T^{2+}_{x,2} T^-_{y,2} T^-_t \left(\frac{\partial u_2}{\partial y} \right) - D^-_{y,2} u_2 + \frac{h_2}{3} T^-_{y,2} T^-_t \left(\frac{\partial^2 u_2}{\partial x \partial y} \right), \quad x = a_2,$$

together with

$$\begin{aligned}
\mu_1(y,t) := {} & \big[\alpha_1(y) u_1(b_1, y, t) - T^2_{y,1} T^-_t \big(\alpha_1(y) u_1(b_1, y, t)\big)\big] \\
& - \Bigg[k_2 \sum_{y' \in \Omega_2^{k_2}} \beta_1(y, y') u_2(a_2, y', t) \\
& - \int_c^d T^2_{y,1} T^-_t \big(\beta_1(y, y') u_2(a_2, y', t)\big) \, \mathrm{d}y' \Bigg] \\
& + \frac{h_1^2}{6} \left[T^2_{y,1} T^-_t \frac{\partial^2 u_1}{\partial x \partial t} + D^+_{y,1} \left(T^-_{y,1} T^-_t \frac{\partial^2 u_1}{\partial x \partial y} \right) \right]_{(b_1, y, t)}, \\
& y \in \Omega_1^{k_1}, \ t \in \Omega_+^\tau,
\end{aligned}$$

$$\begin{aligned}
\mu_2(y,t) := {} & \big[\alpha_2(y) u_2(a_2, y, t) - T^2_{y,2} T^-_t \big(\alpha_2(y) u_2(a_2, y, t)\big)\big] \\
& - \Bigg[k_1 \sum_{y' \in \Omega_1^{k_1}} \beta_2(y, y') u_1(b_1, y', t) \\
& - \int_c^d T^2_{y,2} T^-_t \big(\beta_2(y, y') u_1(b_1, y', t)\big) \, \mathrm{d}y' \Bigg] \\
& - \frac{h_2^2}{6} \left[T^2_{y,2} T^-_t \frac{\partial^2 u_2}{\partial x \partial t} - D^+_{y,2} \left(T^-_{y,2} T^-_t \frac{\partial^2 u_2}{\partial x \partial y} \right) \right]_{(a_2, y, t)}, \\
& y \in \Omega_2^{k_2}, \ t \in \Omega_+^\tau.
\end{aligned}$$

The relevant a priori estimate for the solution of the finite difference scheme (3.154)–(3.159) is given by the following lemma, whose proof follows directly from (3.45) and (3.46).

Lemma 3.12 *Suppose that the coefficients of the finite difference scheme* (3.154)–(3.159) *are well-defined at the mesh-points. Then, the global error e, which is the solution of the finite difference scheme* (3.154)–(3.159), *satisfies the a priori estimate*

$$\|e\|^2_{W^{1,1/2}_{2,h\tau}} \le C\Big(\|\psi\|^2_{\ddot{W}^{1/2}_2(\overline{\Omega}^\tau, L_h)} + \|\eta\|^2_{L_2(\Omega^\tau_+, L_{h'})} + \|\zeta\|^2_{L_2(\Omega^\tau_+, L_{h''})} + \|\mu\|^2_{L_2(\Omega^\tau_+, L_2(\Omega^k))}\Big), \tag{3.160}$$

where C is a positive constant depending on T, but independent of h and τ.

Thus, in order to derive an error bound for the finite difference scheme (3.148)–(3.153), it suffices to bound the terms appearing on the right-hand side of (3.160). For simplicity, we shall assume in what follows that $\tau \asymp h^2$.

Theorem 3.13 *Suppose that the solution of the initial-boundary-value problem* (3.135)–(3.140) *belongs to* $W^{s,s/2}_2$, $5/2 < s \le 3$, $\alpha_i \in W^{s-1}_2(c_i, d_i)$, $\beta_i \in W^{s-1}_2((c_i, d_i) \times (c_{3-i}, d_{3-i}))$, $i = 1, 2$, *and let* $\tau \asymp h^2$. *Then, the solution U of the finite difference scheme* (3.148)–(3.153) *converges to the solution u of the initial-boundary-value problem* (3.135)–(3.140), *and the following error bound holds*:

$$\|u - U\|_{W^{1,1/2}_{2,h\tau}} \le Ch^{s-1}\Big(\sqrt{\log \tfrac{1}{h}} + \max_i \|\alpha_i\|_{W^{s-1}_2(c_i,d_i)} + \max_i \|\beta_i\|_{W^{s-1}_2((c_i,d_i)\times(c_{3-i},d_{3-i}))}\Big)\|u\|_{W^{s,s/2}_2}, \quad 5/2 < s \le 3, \tag{3.161}$$

where $C = C(s)$ *is a positive constant, independent of h.*

Proof The term ψ_1, for $x \in \Omega^{h_1}_1$, can be bounded in the same way as the analogous term ψ in Sect. 3.4.3. The same is true of $\psi_1(b_1, y, t)$ if $u_1 \in W^{s,s/2}_2(\Omega_1)$, $s > 5/2$; we note that for smaller values of s this term is not necessarily well-defined. Hence, from (3.115)–(3.117), we immediately obtain the bounds

$$\sum_{x\in\Omega^{h_1}_{1+}} \hbar_1 k_1 \sum_{y\in\Omega^{k_1}_1} \tau^2 \sum_{t\in\overline{\Omega}^\tau} \sum_{t'\in\overline{\Omega}^\tau, t'\ne t} \frac{|\psi_1(x, y, t) - \psi_1(x, y, t')|^2}{|t - t'|^2} \le Ch^{2(s-1)}\|u_1\|^2_{W^{s,s/2}_2(Q_1)}, \quad 5/2 < s \le 3,$$

and

$$\sum_{x\in\Omega_{1+}^{h_1}} \hbar_1 k_1 \sum_{y\in\Omega_1^{k_1}} \tau \sum_{t\in\Omega^\tau} \left(\frac{1}{t}+\frac{1}{T-t}\right)\psi_1^2(x,y,t)$$

$$\le Ch^{2(s-1)}\log\frac{1}{h}\|u_1\|^2_{W_2^{s,s/2}(Q_1)}, \quad 5/2<s\le 3.$$

Analogous results hold for ψ_2, whereby

$$\|\psi\|_{\ddot{W}_2^{1/2}(\overline{\Omega}^\tau,L_h)} \le Ch^{s-1}\sqrt{\log\frac{1}{h}}\|u_1\|_{W_2^{s,s/2}}, \quad 5/2<s\le 3, \tag{3.162}$$

where we have used the notation

$$\|u\|^2_{W_2^{s,s/2}} := \|u_1\|^2_{W_2^{s,s/2}(Q_1)} + \|u_2\|^2_{W_2^{s,s/2}(Q_2)}.$$

When $s>2$, $\eta_1(x,y,t)$ is a bounded linear functional of $u_1\in W_2^{s,s/2}(G)$, where G is the elementary cell $(x-h_1,x)\times(y-k_1,y+k_1)\times(t-\tau,t)$, which vanishes on the monomials $1,x,y,t,x^2,xy$ and y^2. Invoking the Bramble–Hilbert lemma and summing over the nodes of the mesh we obtain

$$h_1k_1\tau \sum_{x\in\Omega_{1+}^{h_1}}\sum_{y\in\Omega_1^{k_1}}\sum_{t\in\Omega_+^\tau} \eta_1^2(x,y,t)\le Ch^{2(s-1)}\|u_1\|^2_{W_2^{s,s/2}(Q_1)}, \quad 2<s\le 3.$$

An analogous inequality holds for η_2, and hence

$$\|\eta\|_{L_2(\Omega_+^\tau,L_{h'})} \le Ch^{s-1}\|u\|_{W_2^{s,s/2}}, \quad 2<s\le 3. \tag{3.163}$$

Furthermore, an analogous result holds for ζ, assuming that $s>5/2$; again we note that for smaller values of s the expressions $\zeta_1(b_1,y,t)$ and $\zeta_2(a_2,y,t)$ are not necessarily well-defined. Hence,

$$\|\zeta\|_{L_2(\Omega_+^\tau,L_{h'})} \le Ch^{s-1}\|u\|_{W_2^{s,s/2}}, \quad 5/2<s\le 3. \tag{3.164}$$

The term μ_1 can be decomposed as

$$\mu_1=\mu_{11}+\mu_{12}+\mu_{13}+\mu_{14}+\mu_{15}+\mu_{16},$$

where

$$\mu_{11} := \alpha_1(y)u_1(b_1,y,t) - T^2_{y,1}T_t^-\big(\alpha_1(y)u_1(b_1,y,t)\big),$$

$$\mu_{12} := k_2\sum_{y'\in\Omega_2^{k_2}} \nu_{12}(y,y',t),$$

$$\mu_{13} := k_2 \sum_{y' \in \Omega_2^{k_2}} \nu_{13}(y, y', t),$$

$$\mu_{14} := k_2 \sum_{y' \in \Omega_2^{k_2}} \nu_{14}(y, y', t),$$

$$\mu_{15} := \frac{1}{6} h_1^2 T_{y,1}^2 T_t^- \frac{\partial^2 u_1}{\partial x \partial t}(b_1, y, t),$$

$$\mu_{16} := \frac{1}{6} h_1^2 D_{y,1}^+ \left(T_{y,1}^- T_t^- \frac{\partial^2 u_1}{\partial x \partial y} \right)(b_1, y, t),$$

with

$$\nu_{12} := \big[T_{y,1}^2 \beta_1(y, y') - \beta_1(y, y')\big] u_2(a_2, y', t),$$

$$\nu_{13} := T_{y,1}^2 \beta_1(y, y') \big[T_t^- u_2(a_2, y', t) - u_2(a_2, y', t)\big],$$

$$\nu_{14} := T_{y',2}^- \big[T_{y,1}^2 \beta_1(y, y') T_t^- u_2(a_2, y', t)\big] - \frac{1}{2} \big[T_{y,1}^2 \beta_1(y, y') T_t^- u_2(a_2, y', t)\big]$$
$$- \frac{1}{2} \big[T_{y,1}^2 \beta_1(y, y' - k_2) T_t^- u_2(a_2, y' - k_2, t)\big].$$

When $s > 5/2$, $\mu_{11}(y, t)$ is a bounded linear functional of $v_1 := \alpha_1 u_1 \in W_2^{s-1,(s-1)/2}(M^1)$, where M^1 is the elementary cell $(y - k_1, y + k_1) \times (t - \tau, t)$, which vanishes if $v_1 = 1$ and $v_1 = y$. By applying the Bramble–Hilbert lemma, Lemmas 1.47 and 1.54 and the trace theorem we obtain

$$k_1 \tau \sum_{y \in \Omega_1^{k_1}} \sum_{t \in \Omega_+^{\tau}} \mu_{11}^2(y, t) \le C h^{2(s-1)} \|v_1\|^2_{W_2^{s-1,(s-1)/2}((c_1,d_1)\times(0,T))}$$
$$\le C h^{2(s-1)} \|\alpha_1\|^2_{W_2^{s-1}(c_1,d_1)} \|u_1\|^2_{W_2^{s-1,(s-1)/2}(\Gamma_1 \times (0,T))}$$
$$\le C h^{2(s-1)} \|\alpha_1\|^2_{W_2^{s-1}(c_1,d_1)} \|u_1\|^2_{W_2^{s,s/2}(Q_1)}, \qquad 5/2 < s \le 3.$$

When $u_2 \in C(\overline{Q}_2)$, ν_{12} is a bounded linear functional of $\beta_1 \in W_2^{s-1}(K)$, $s > 2$, where K is the elementary cell $(y - k_1, y + k_1) \times (y' - k_2, y' + k_2)$, which vanishes if $\beta_1(y, y')$ is a polynomial of degree 1. Invoking, again, the Bramble–Hilbert lemma and the Sobolev embedding theorem, we obtain

$$k_1 \tau \sum_{y \in \Omega_1^{k_1}} \sum_{t \in \Omega_+^{\tau}} \mu_{12}^2(y, t) \le C h^{2(s-1)} \|\beta_1\|^2_{W_2^{s-1}((c_1,d_1)\times(c_2,d_2))} \|u_2\|^2_{C(\overline{Q}_2)}$$
$$\le C h^{2(s-1)} \|\beta_1\|^2_{W_2^{s-1}((c_1,d_1)\times(c_2,d_2))} \|u_2\|^2_{W_2^{s,s/2}(Q_2)}, \qquad 2 < s \le 3.$$

If $\beta_1 \in C([c_1, d_1] \times [c_2, d_2])$, ν_{13} is a bounded linear functional of $u_2(a_2, \cdot, \cdot) \in W_2^{s-1,(s-1)/2}(M^2)$, $s > 5/2$, where M^2 is the elementary cell $(y' - k_2, y' + k_2) \times$

$(t-\tau, t)$, which vanishes if $u_2 = 1$ and $u_2 = y'$. By applying the Bramble–Hilbert lemma, the embedding theorem and the trace theorem, we obtain

$$k_1\tau \sum_{y\in\Omega_1^{k_1}} \sum_{t\in\Omega_+^\tau} \mu_{13}^2(y,t)$$
$$\leq Ch^{2(s-1)} \|\beta_1\|^2_{C([c_1,d_1]\times[c_2,d_2])} \|u_2\|^2_{W_2^{s-1,\frac{s-1}{2}}(\Gamma_2\times(0,T))}$$
$$\leq Ch^{2(s-1)} \|\beta_1\|^2_{W_2^{s-1}((c_1,d_1)\times(c_2,d_2))} \|u_2\|^2_{W_2^{s,s/2}(Q_2)}, \quad 5/2 < s \leq 3.$$

Next, ν_{14} is a bounded linear functional of $v(y') = T^2_{y,1}\beta_1(y,y')T_t^- u_2(a_2,y',t) \in W_2^{s-1}(E^2)$, $s > 3/2$, where E^2 is the elementary interval $(y'-k_2, y'+k_2)$, which vanishes if $v = 1$ and $v = y'$. By applying the Bramble–Hilbert lemma, Lemma 1.52 and the trace theorem, we obtain

$$k_1\tau \sum_{y\in\Omega_1^{k_1}} \sum_{t\in\Omega_+^\tau} \mu_{14}^2(y,t)$$
$$\leq Ch^{2(s-1)} k_1\tau \sum_{y\in\Omega_1^{k_1}} \sum_{t\in\Omega_+^\tau} \|(T^2_{y,1}\beta_1)(T_t^- u_2)\|^2_{W_2^{s-1}(c_2,d_2)}$$
$$\leq Ch^{2(s-1)} k_1 \sum_{y\in\Omega_1^{k_1}} \|T^2_{y1}\beta_1\|^2_{W_2^{s-1}(c_2,d_2)} \tau \sum_{t\in\Omega_+^\tau} \|T_t^- u_2\|^2_{W_2^{s-1}(c_2,d_2)}$$
$$\leq Ch^{2(s-1)} \|\beta_1\|^2_{W_2^{s-1}((c_1,d_1)\times(c_2,d_2))} \|u_2\|^2_{W_2^{s-1,(s-1)/2}(\Gamma_2\times(0,T))}$$
$$\leq Ch^{2(s-1)} \|\beta_1\|^2_{W_2^{s-1}((c_1,d_1)\times(c_2,d_2))} \|u_2\|^2_{W_2^{s,s/2}(Q_2)}, \quad 3/2 < s \leq 3.$$

The term μ_{15} may be estimated directly, yielding

$$k_1\tau \sum_{y\in\Omega_1^{k_1}} \sum_{t\in\Omega_+^\tau} \mu_{15}^2(y,t) \leq Ch^4 \left\| \frac{\partial^2 u_1}{\partial x \partial t} \right\|^2_{L_2(\Gamma_1\times(0,T))},$$

and thus, by noting the transmission condition (3.137), the Sobolev embedding theorem and the trace theorem, it follows that

$$k_1\tau \sum_{y\in\Omega_1^{k_1}} \sum_{t\in\Omega_+^\tau} \mu_{15}^2(y,t)$$
$$\leq Ch^4 \Big(\|\alpha_1\|^2_{W_2^{s-1}(c_1,d_1)} \|u_1\|^2_{W_2^{s,s/2}(Q_1)}$$
$$+ \|\beta_1\|^2_{W_2^{s-1}((c_1,d_1)\times(c_2,d_2))} \|u_2\|^2_{W_2^{s,s/2}(Q_2)} \Big), \quad 5/2 < s \leq 3.$$

The term μ_{16} is a bounded linear functional of $v(y) = T_t^- \frac{\partial^2 u_1}{\partial x \partial y} \in W_2^{s-2}(E^1)$, $s \geq 2$, where E^1 is the elementary interval $(y - k_1, y + k_1)$, which vanishes if $v = 1$ and $v = y$. By applying the Bramble–Hilbert lemma we obtain

$$k_1 \tau \sum_{y \in \Omega_1^{k_1}} \sum_{t \in \Omega_+^\tau} \mu_{16}^2(y,t) \leq C h^{2(s-1)} \tau \sum_{t \in \Omega_+^\tau} \left\| T_t^- \frac{\partial^2 u_1}{\partial x \partial y} \right\|^2_{W_2^{s-2}(c_1,d_1)}, \qquad 2 \leq s \leq 3,$$

and hence, by noting the transmission condition (3.137), Lemma 1.52, the Sobolev embedding theorem and the trace theorem, it follows that

$$\begin{aligned} k_1 \tau &\sum_{y \in \Omega_1^{k_1}} \sum_{t \in \Omega_+^\tau} \mu_{16}^2(y,t) \\ &\leq C h^{2(s-1)} \left(\|\alpha_1\|^2_{W_2^{s-1}(c_1,d_1)} \|u_1\|^2_{W_2^{s,s/2}(Q_1)} \right. \\ &\quad \left. + \|\beta_1\|^2_{W_2^{s-1}((c_1,d_1)\times(c_2,d_2))} \|u_2\|^2_{W_2^{s,s/2}(Q_2)} \right), \qquad 5/2 < s \leq 3. \end{aligned}$$

By collecting the bounds above and noting that analogous bounds hold for μ_2, we have that

$$\begin{aligned} \|\mu\|^2_{L_2(\Omega_+^\tau, L_2(\Omega^k))} \leq C h^{2(s-1)} &\left(\max_i \|\alpha_1\|^2_{W_2^{s-1}(c_1,d_1)} \right. \\ &\left. + \max_i \|\beta_1\|^2_{W_2^{s-1}((c_1,d_1)\times(c_2,d_2))} \right) \|u\|^2_{W_2^{s,s/2}}, \qquad 5/2 < s \leq 3. \end{aligned} \tag{3.165}$$

Finally, from (3.160)–(3.165) we obtain (3.161), with $C = C(s)$ signifying a positive constant, independent of h. ☐

In the next section we briefly consider a factorized version of the implicit finite difference scheme (3.148)–(3.153).

3.6.3 Factorized Scheme

The implicit finite difference scheme (3.148)–(3.153) is not economical, because on each time level it requires the solution of a two-dimensional elliptic difference problem. To overcome this practical shortfall, we consider here the following factorized counterpart of the finite difference scheme (3.148)–(3.153):

$$\begin{aligned} (I_h + \sigma\tau\Lambda_{11})(I_h + \sigma\tau\Lambda_{12}) D_t^+ U_1 - D_{x,1}^- D_{x,1}^+ U_1 - D_{y,1}^- D_{y,1}^+ U_1 \\ = \hat{f}_1, \quad x \in \Omega_1^{h_1},\ y \in \Omega_1^{k_1},\ t \in \Omega_-^\tau, \end{aligned} \tag{3.166}$$

$$(I_h+\sigma\tau\Lambda_{11})(I_h+\sigma\tau\Lambda_{12})D_t^+U_1(b_1,y,t)+\frac{2}{h_1}\Big[D_{x,1}^-U_1(b_1,y,t)$$

$$+\alpha_1(y)U_1(b_1,y,t)-k_2\sum_{y'\in\Omega_2^{k_2}}\beta_1\big(y,y'\big)U_2\big(a_2,y',t\big)\Big]$$

$$-D_{y,1}^-D_{y,1}^+U_1(b_1,y,t)=\hat{f}_1(b_1,y,t),\quad y\in\Omega_1^{k_1},\ t\in\Omega_-^\tau, \tag{3.167}$$

$$(I_h+\sigma\tau\Lambda_{21})(I_h+\sigma\tau\Lambda_{22})D_t^+U_2-D_{x,2}^-D_{x,2}^+U_2-D_{y,2}^-D_{y,2}^+U_2$$

$$=\hat{f}_2,\quad x\in\Omega_2^{h_2},\ y\in\Omega_2^{k_2},\ t\in\Omega_-^\tau, \tag{3.168}$$

$$(I_h+\sigma\tau\Lambda_{21})(I_h+\sigma\tau\Lambda_{22})D_t^+U_2(a_2,y,t)-\frac{2}{h_2}\Big[D_{x,2}^+U_2(a_2,y,t)$$

$$-\alpha_2(y)U_2(a_2,y,t)+k_1\sum_{y'\in\Omega_1^{k_1}}\beta_2\big(y,y'\big)U_1\big(b_1,y',t\big)\Big]$$

$$-D_{y,2}^-D_{y,2}^+U_2(a_2,y,t)=\hat{f}_2(a_2,y,t),\quad y\in\Omega_2^{k_2},\ t\in\Omega_-^\tau, \tag{3.169}$$

subject to the initial and boundary conditions (3.152)–(3.153), where we have used the notation $\hat{f}_i(\cdot,\cdot,t):=\bar{f}_i(\cdot,\cdot,t+\tau)$, and

$$\Lambda_{i2}U_i:=-D_{y,i}^-D_{y,i}^+U_i,\qquad \Lambda_{i1}U_i:=-D_{x,i}^-D_{x,i}^+U_i,\quad x\in\Omega_i^{h_i},\ y\in\Omega_i^{k_i},$$

$$\Lambda_{11}U_1(b_1,\cdot,\cdot)=\frac{2}{h_1}D_{x,1}^-U_1(b_1,\cdot,\cdot),\qquad \Lambda_{21}U_2(a_2,\cdot,\cdot)=-\frac{2}{h_2}D_{x,2}^+U_2(a_2,\cdot,\cdot).$$

Let us define

$$B_hU:=(B_{1h}U_1,B_{2h}U_2)\quad\text{and}\quad A_{0h}U:=(A_{01h}U_1,A_{02h}U_2),$$

where

$$B_{ih}:=(I_h+\sigma\tau\Lambda_{i1})(I_h+\sigma\tau\Lambda_{i2})\quad\text{and}\quad A_{0ih}:=\Lambda_{i1}+\Lambda_{i2}.$$

Then,

$$\begin{aligned}(\bar{B}_hU,U)_{L_h}&=\big((B_h-\tau A_{0h})U,U\big)_{L_h}\\&=\|U\|_{L_h}^2+(\sigma-1)\tau\|U\|_{A_{0h}}^2\\&\quad+\sigma^2\tau^2\sum_{i=1}^{2}h_ik_i\sum_{x\in\Omega_{i+}^{h_i}}\sum_{y\in\Omega_{i+}^{k_i}}\big(D_{x,i}^-D_{y,i}^-U_i\big)^2.\end{aligned}$$

Hence, $\bar{B}_h$ is positive definite for $\sigma\ge 1$, uniformly with respect to the discretization parameters. Assuming that $\tau\asymp h^2$, we also have that

$$\|U\|_{\bar{B}_h}\asymp\|U\|_{L_h}.$$

In contrast with (3.148)–(3.153), the finite difference scheme (3.166)–(3.169), (3.152), (3.153) is computationally efficient, since on each time level the set of unknowns may be computed by solving systems of linear algebraic equations that have tridiagonal matrices.

Let $u = (u_1, u_2)$ be the solution of the initial-boundary-value problem (3.135)–(3.140) and let $U = (U_1, U_2)$ be the solution of the difference scheme (3.166)–(3.169), (3.152), (3.153). We define $e_i := u_i - U_i$ for $i = 1, 2$, and $e := (e_1, e_2)$. Then, the global error $e = u - U$ satisfies the following finite difference scheme:

$$(I_h + \sigma\tau\Lambda_{11})(I_h + \sigma\tau\Lambda_{12})D_t^+ e_1 - D_{x,1}^- D_{x,1}^+ e_1 - D_{y,1}^- D_{y,1}^+ e_1$$
$$= D_t^+ \tilde{\psi}_1 + D_{x,1}^+ \tilde{\eta}_1 + D_{y,1}^+ \tilde{\zeta}_1, \quad x \in \Omega_1^{h_1},\ y \in \Omega_1^{k_1},\ t \in \Omega_-^\tau, \tag{3.170}$$

$$(I_h + \sigma\tau\Lambda_{11})(I_h + \sigma\tau\Lambda_{12})D_t^+ e_1(b_1, y, t) + \frac{2}{h_1}\Big[D_{x,1}^- e_1(b_1, y, t)$$
$$+ \alpha_1(y)e_1(b_1, y, t) - k_2 \sum_{y' \in \Omega_2^{k_2}} \beta_1(y, y')e_2(a_2, y', t)\Big]$$
$$- D_{y,1}^- D_{y,1}^+ e_1(b_1, y, t)$$
$$= D_t^+ \tilde{\psi}_1(b_1, y, t) + D_{y,1}^+ \tilde{\zeta}_1(b_1, y, t)$$
$$- \frac{2}{h_1}\tilde{\eta}_1(b_1, y, t) + \frac{2}{h_1}\tilde{\mu}_1(y, t), \quad y \in \Omega_1^{k_1},\ t \in \Omega_-^\tau, \tag{3.171}$$

$$(I_h + \sigma\tau\Lambda_{21})(I_h + \sigma\tau\Lambda_{22})D_t^+ e_2 - D_{x,2}^- D_{x,2}^+ e_2 - D_{y,2}^- D_{y,2}^+ e_2$$
$$= D_t^+ \tilde{\psi}_2 + D_{x,2}^+ \tilde{\eta}_2 + D_{y,2}^+ \tilde{\zeta}_2, \quad x \in \Omega_2^{h_2},\ y \in \Omega_2^{k_2},\ t \in \Omega_-^\tau, \tag{3.172}$$

$$(I_h + \sigma\tau\Lambda_{21})(I_h + \sigma\tau\Lambda_{22})D_t^+ e_2(a_2, y, t) - \frac{2}{h_2}\Big[D_{x,2}^+ e_2(a_2, y, t)$$
$$- \alpha_2(y)e_2(a_2, y, t) + k_1 \sum_{y' \in \Omega_1^{k_1}} \beta_2(y, y')e_1(b_1, y', t)\Big]$$
$$- D_{y,2}^- D_{y,2}^+ e_2(a_2, y, t)$$
$$= D_t^+ \tilde{\psi}_2(a_2, y, t) + D_{y,2}^+ \tilde{\zeta}_2(a_2, y, t)$$
$$+ \frac{2}{h_2}\tilde{\eta}_2(a_2 + h_2, y, t) + \frac{2}{h_2}\tilde{\mu}_2(y, t), \quad y \in \Omega_2^{k_2},\ t \in \Omega_-^\tau, \tag{3.173}$$

with the boundary and initial conditions (3.158)–(3.159), where, for $i = 1, 2$,

$$\tilde{\psi}_i(\cdot, \cdot, t) := \psi_i(\cdot, \cdot, t),$$
$$\tilde{\eta}_i(\cdot, \cdot, t) := \big[\eta_i + (1-\sigma)\tau D_{x,i}^- D_t^- u_i + \sigma^2\tau^2 D_{x,i}^- D_{y,i}^+ D_{y,i}^- D_t^- u_i\big]\big|_{(\cdot,\cdot,t+\tau)},$$

$$\tilde{\zeta}_i(\cdot,\cdot,t) := \left[\zeta_i + (1-\sigma)\tau D_{y,i}^- D_t^- u_i\right]\Big|_{(\cdot,\cdot,t+\tau)},$$

$$\tilde{\mu}_i(\cdot,t) := \left(\mu_i + \tau k_{3-i} \sum_{y' \in \Omega_{3-i}^{k_{3-i}}} \beta_i D_t^- u_{3-i}\right)\Bigg|_{(\cdot,t+\tau)}.$$

From (3.45) and (3.46) we deduce an a priori estimate analogous to (3.160), with ψ, η, ζ and μ replaced by $\tilde{\psi}$, $\tilde{\eta}$, $\tilde{\zeta}$ and $\tilde{\mu}$, respectively.

When $\tau \asymp h^2$ it is easily seen that $\tilde{\psi}$, $\tilde{\eta}$, $\tilde{\zeta}$ and $\tilde{\mu}$ satisfy the same bounds as the corresponding terms ψ, η, ζ and μ. Hence, the factorized finite difference scheme (3.166)–(3.169), (3.152), (3.153) also satisfies the error bound (3.161).

3.7 Bibliographical Notes

In this chapter we have derived error bounds for finite difference approximations of some model initial-boundary-value problems for second-order linear parabolic partial differential equations. The procedure was based on the Bramble–Hilbert lemma and its generalizations, and can be seen as a further development of the methodology presented in Chap. 2.

As we have already mentioned, in the case of second-order linear parabolic partial differential equations a complete theory of existence and uniqueness of weak solutions to initial-boundary-value problems has been developed in the anisotropic Sobolev spaces $W_2^{s,s/2}(Q)$. We therefore chose to use analogous mesh-dependent norms in our analysis of finite difference approximations of the various initial-boundary-value problems considered.

Similarly to the elliptic case discussed in Chap. 2, for a finite difference approximation of a second-order parabolic partial differential equation an error bound of the form

$$\|u - U\|_{W_2^{r,r/2}(Q_h^\tau)} \le C(h + \sqrt{\tau})^{s-r} \|u\|_{W_2^{s,s/2}(Q)}, \quad r < s, \tag{3.174}$$

is said to be *optimal*, or *compatible with the smoothness of the solution* of the initial-boundary-value problem. Here usually $0 \le r \le 1 < s$. If the mesh-sizes h and τ satisfy the relation $\tau \asymp h^2$, i.e. $c_1 h^2 \le \tau \le c_2 h^2$, then the error bound (3.174) reduces to

$$\|u - U\|_{W_2^{r,r/2}(Q_h^\tau)} \le C h^{s-r} \|u\|_{W_2^{s,s/2}(Q)}. \tag{3.175}$$

In the case of linear parabolic equations with variable coefficients, the constant C depends on norms of the coefficients. For example, if the coefficients are independent of t, then one obtains an error bound of the form

$$\|u - U\|_{W_2^{r,r/2}(Q_h^\tau)} \le C h^{s-r} \left(\max_{i,j} \|a_{ij}\|_{W_p^{s-1}(\Omega)} + \|a\|_{W_p^{s-2}(\Omega)}\right) \|u\|_{W_2^{s,s/2}(Q)}. \tag{3.176}$$

For parabolic equations with constant coefficients, error bounds of the form (3.174) were obtained by Lazarov [122] for $r = 0$ and $s = 2$. A similar error bound in a discrete L_p norm (for $s = 2$) was derived in the work of Godev and Lazarov [58].

The case of fractional values of s was studied by Ivanović, Jovanović and Süli [74, 105]. Estimates of the form (3.175) were obtained for $2 \le s \le 4$, $r = 0, 2$. For $r = 1$ the estimate was derived in the discrete $W_2^{1,0}$ norm rather than in the, more natural, discrete $W_2^{1,1/2}$ norm.

Dražić [32] obtained error bounds of the form (3.174) and (3.175); he also stated certain conditions under which the step sizes h and τ appearing in the error bounds may be chosen independently of each other.

In the papers by Scott and Seward [164] and Seward, Kasibhatla and Fairweather [165] the influence of mollifying the initial datum on the convergence rate of the difference scheme was investigated.

In each of those publications the Bramble–Hilbert lemma was used in the derivation of the error bounds. We note that some error bounds for finite difference approximations of parabolic problems with weak solutions were obtained much earlier using different, more classical, techniques based on Fourier series (see e.g. Juncosa and Young [112]). We also highlight here the more recent work of Carter and Giles [25], where sharp estimates of the error arising from explicit and implicit approximations of the constant-coefficient one-dimensional convection-diffusion equation with Dirac initial datum were derived. The study of this particular model problem was motivated by applications in computational finance and the desire to prove convergence of approximations to adjoint partial differential equations. The error analysis in [25] was based on Fourier analysis and asymptotic approximation of the integrals resulting from the application of an inverse Fourier transform.

For early developments concerning the use of Besov space theory and techniques from harmonic analysis in the stability and convergence analysis of finite difference approximations of pure initial-value problems for parabolic equations with nonsmooth initial data, we refer to the monograph of Brenner, Thomée and Wahlbin [24], and the references cited therein.

Finite difference approximations of parabolic equations with variable coefficients were considered by Weinelt, Lazarov and Streit [197], and Kuzik and Makarov [117]—for integer values of s, and by Jovanović [82, 85]—for fractional values of s.

Finite difference schemes for various nonstandard parabolic problems with interfaces and/or dynamic boundary conditions were studied by Jovanović and Vulkov [94, 95, 97–99] and Bojović and Jovanović [17]. Parabolic transmission problems in disjoint domains were investigated by Jovanović and Vulkov [102, 104]. Second-order convergence in the mesh-dependent W_2^1 norm for non-Fickian diffusion models was proved by Barbeiro, Ferreira and Pinto [7].

Finite-difference schemes on time-adaptive grids for parabolic equations with generalized solutions were studied by Samarskiĭ et al. [161].

The application of function space interpolation theory was considered by Bojović and Jovanović [16] for the derivation of error bounds for finite difference approximations of parabolic problems.

Variational-difference schemes also satisfy error bounds of the form (3.174)–(3.176) (see Jovanović [78]). However, for those schemes, error bounds involving the 'continuous' rather than the discrete $W_2^{r,r/2}$ norm on the left-hand side are more common. See, for example, Zlotnik [201, 202], Hackbusch [65], Amosov and Zlotnik [4]). A similar comment applies to finite element methods for parabolic problems (see, for example, the monograph of Thomée [177]).

Besides the error bounds described above, for parabolic problems one can also derive error bounds in the norms of the function spaces

$$L_\infty\big((0,T), L_2(\Omega)\big) \quad \text{and} \quad L_\infty\big((0,T), W_2^1(\Omega)\big);$$

see, for example, Douglas and Dupont [30], Douglas, Dupont and Wheeler [31], Rannacher [153], Thomée and Wahlbin [178], Wheeler [198], Zlamal [200]; for error estimates in negative norms, we refer to Thomée [176, 177].

Chapter 4
Finite Difference Approximation of Hyperbolic Problems

This chapter is devoted to finite difference methods for time-dependent problems governed by linear second-order hyperbolic equations. In the next section we discuss the question of well-posedness of initial-boundary-value problems for linear second-order hyperbolic partial differential equations. In Sect. 4.2 we review some classical results concerning standard finite difference approximations of the wave equation. Section 4.3 is devoted to finite difference schemes for the wave equation with nonsmooth initial data and source term. In Sect. 4.4 we extend the analysis to a linear second-order hyperbolic equation with variable coefficients: error bounds are derived in the discrete $W^1_\infty((0,T); L_2(\Omega)) \cap L_\infty((0,T); W^1_2(\Omega))$ norm denoted by $\|\cdot\|^{(1)}_{2,\infty}$. In Sects. 4.5 and 4.6 we shall be concerned with the finite difference approximation of interface problems and transmission problems for second-order linear hyperbolic equations. The chapter closes with bibliographical notes.

4.1 Hyperbolic Equations

Similarly as in the case of second-order parabolic equations, we shall begin our considerations with a brief discussion concerning the existence and uniqueness of solutions to initial-boundary-value problems for linear second-order hyperbolic equations. More specifically, we shall consider the equation

$$\frac{\mathrm{d}^2 u}{\mathrm{d}t^2} + A(t)u = f(t), \quad t \in (0, T], \tag{4.1}$$

subject to the initial conditions

$$u(0) = u_0, \qquad \frac{\mathrm{d}u}{\mathrm{d}t}(0) = u_1, \tag{4.2}$$

where u is a function of the independent variables $x \in \Omega$ and $t \in [0, T]$, with $T > 0$, and Ω and the operator $A(t)$ satisfy the same conditions as in the previous chapter

B.S. Jovanović, E. Süli, *Analysis of Finite Difference Schemes*,
Springer Series in Computational Mathematics 46,
DOI 10.1007/978-1-4471-5460-0_4, © Springer-Verlag London 2014

(cf. Sect. 3.1). Problem (4.1), (4.2) can be viewed as a second-order ordinary differential equation in a Hilbert space. The technical details of this viewpoint will be discussed in the next section.

4.1.1 Abstract Hyperbolic Initial-Value Problems

Let us consider the Gelfand triple

$$\mathcal{V} \hookrightarrow \mathcal{H} \equiv \mathcal{H}' \hookrightarrow \mathcal{V}'$$

with continuous and dense embeddings, where $\mathcal{V}$ and $\mathcal{H}$ are separable real Hilbert spaces, $\mathcal{H}$ is identified with its dual space $\mathcal{H}'$ via the Riesz representation theorem, and $\mathcal{V}'$ denotes the dual space of $\mathcal{V}$. We shall denote by $(\cdot,\cdot)$ and $\|\cdot\| = \|\cdot\|_{\mathcal{H}}$ the inner product and norm of the Hilbert space $\mathcal{H}$; the norms in $\mathcal{V}$ and $\mathcal{V}'$ will be denoted by $\|\cdot\|_{\mathcal{V}}$ and $\|\cdot\|_{\mathcal{V}'}$, respectively, and $\langle\cdot,\cdot\rangle$ will signify the duality pairing between $\mathcal{V}'$ and $\mathcal{V}$. For $t \in [0,T]$ we consider the bilinear functional $(v,w) \in \mathcal{V} \times \mathcal{V} \mapsto a(t; v, w) \in \mathbb{R}$, which satisfies the conditions (a), (b) and (c) from Sect. 3.1.1. In addition, $a(t;\cdot,\cdot)$ will be assumed to be symmetric for each $t \in [0,T]$, i.e.

$$a(t; v, w) = a(t; w, v) \quad \forall v, w \in \mathcal{V},\ \forall t \in [0,T], \tag{4.3}$$

continuously differentiable with respect to $t \in [0,T]$, and such that

$$\left|\frac{\mathrm{d}}{\mathrm{d}t} a(t; v, w)\right| \le c\|v\|_{\mathcal{V}}\|w\|_{\mathcal{V}} \quad \forall v, w \in \mathcal{V},\ \forall t \in [0,T], \tag{4.4}$$

where the constant $c > 0$ is independent of $t \in [0,T]$. In the present setting, by viewing $A(t) : \mathcal{V} \to \mathcal{V}'$ as a bounded linear operator on $\mathcal{V}$, we have that $a(t; v, w) = \langle A(t)v, w\rangle$ for all $v, w \in \mathcal{V}$ and $t \in [0,T]$.

We consider the following problem (**H**): *given that* $f \in L_2((0,T);\mathcal{H})$, $u_0 \in \mathcal{V}$ *and* $u_1 \in \mathcal{H}$, *find*

$$u \in L_2\big((0,T);\mathcal{V}\big) \quad \textit{with} \quad \frac{\mathrm{d}u}{\mathrm{d}t} \in L_2\big((0,T);\mathcal{H}\big),\ \frac{\mathrm{d}^2u}{\mathrm{d}t^2} \in L_2\big((0,T);\mathcal{V}'\big),$$

which satisfies equation (4.1) *in* $\mathcal{V}'$, *that is*

$$\left\langle \frac{\mathrm{d}^2u}{\mathrm{d}t^2}, v\right\rangle + \big\langle A(t)u, v\big\rangle = \big\langle f(t), v\big\rangle \quad \forall v \in \mathcal{V},$$

in the sense of distributions on $(0,T)$, *and the initial conditions* (4.2).

We state the following existence and uniqueness result for problem (H) (see Wloka [199], Theorem 29.1 on p. 397).

Theorem 4.1 *Suppose that* (4.3), (4.4) *and hypotheses* (a), (b) *and* (c) *from Sect.* 3.1.1 *hold. Then, problem* (H) *has a unique solution, and the map*

$$\{f, u_0, u_1\} \mapsto \left\{u, \frac{\mathrm{d}u}{\mathrm{d}t}, \frac{\mathrm{d}^2u}{\mathrm{d}t^2}\right\}$$

is continuous and linear from $L_2((0,T);\mathcal{H}) \times \mathcal{V} \times \mathcal{H}$ *into* $L_2((0,T);\mathcal{V}) \times L_2((0,T);\mathcal{H}) \times L_2((0,T);\mathcal{V}')$.

4.1.2 Some a Priori Estimates

We shall now embark on the derivation of energy estimates for problem (H). Our arguments in this section, performed in infinite-dimensional Hilbert spaces, will be largely formal, their main purpose being to motivate their counterparts in finite-dimensional Hilbert spaces, which we shall rigorously prove in the next section. The discrete energy inequalities established there will then play a crucial role in the error analysis of finite difference schemes for problem (H).

We shall focus our attention on the special case when the operator $A = A(t)$ is independent of t. In contrast with the previous section where A was viewed as a *bounded* linear operator from the real separable Hilbert space $\mathcal{V}$ into its dual space $\mathcal{V}'$, here we shall suppose that A is an *unbounded* selfadjoint positive definite linear operator in the real separable Hilbert space $\mathcal{H}$, whose domain of definition $D(A)$ is dense in $\mathcal{H}$. Analogously as in Sect. 3.1.2, we introduce the *energy spaces* $\mathcal{V} = \mathcal{H}_A$ and $\mathcal{V}' = \mathcal{H}_{A^{-1}}$. Then, the bilinear functional $a(t;\cdot,\cdot) = a(\cdot,\cdot) = (\cdot,\cdot)_A$, defined on $\mathcal{V} \times \mathcal{V}$ by extending $(A\cdot,\cdot)$ from $D(A) \times D(A)$, satisfies the conditions (a), (b) and (c) from Sect. 3.1.2 with $c_0 = c_1 = 1$ and $\lambda = 0$, as well as conditions (4.3) and (4.4).

Assuming that $u_0 \in \mathcal{V}$, $u_1 \in \mathcal{H}$, $f \in L_2((0,T);\mathcal{H})$, Theorem 4.1 implies the existence of a unique solution u to problem (H), with $u \in L_2((0,T);\mathcal{V})$, $\frac{\mathrm{d}u}{\mathrm{d}t} \in L_2((0,T);\mathcal{H})$ and $\frac{\mathrm{d}^2u}{\mathrm{d}t^2} \in L_2((0,T);\mathcal{V}')$. Let us assume for the sake of simplicity that $\frac{\mathrm{d}^2u}{\mathrm{d}t^2}(t)$ and $Au(t)$ both belong to $\mathcal{H}$ for a.e. $t \in [0,T]$. By taking the inner product of (4.1) with $2\,\frac{\mathrm{d}u}{\mathrm{d}t}$, we then obtain

$$2\left(\frac{\mathrm{d}^2u}{\mathrm{d}t^2}, \frac{\mathrm{d}u}{\mathrm{d}t}\right) + 2\left(Au, \frac{\mathrm{d}u}{\mathrm{d}t}\right) = 2\left(f(t), \frac{\mathrm{d}u}{\mathrm{d}t}\right).$$

As the operator A has been assumed to be independent of t, by noting that

$$2\left(\frac{\mathrm{d}^2u}{\mathrm{d}t^2}, \frac{\mathrm{d}u}{\mathrm{d}t}\right) = \frac{\mathrm{d}}{\mathrm{d}t}\left(\left\|\frac{\mathrm{d}u}{\mathrm{d}t}\right\|^2\right), \qquad 2\left(Au, \frac{\mathrm{d}u}{\mathrm{d}t}\right) = \frac{\mathrm{d}}{\mathrm{d}t}\left(\|u\|_A^2\right),$$

and applying the Cauchy–Schwarz inequality, we deduce that

$$\frac{\mathrm{d}}{\mathrm{d}t}\left(\left\|\frac{\mathrm{d}u}{\mathrm{d}t}\right\|^2 + \|u\|_A^2\right) \le 2\|f(t)\|\left\|\frac{\mathrm{d}u}{\mathrm{d}t}\right\| \le 2\|f(t)\|\left(\left\|\frac{\mathrm{d}u}{\mathrm{d}t}\right\|^2 + \|u\|_A^2\right)^{1/2}.$$

By writing $\|\mathrm{d}u/\mathrm{d}t\|^2 + \|u\|_A^2 = [(\|\mathrm{d}u/\mathrm{d}t\|^2 + \|u\|_A^2)^{1/2}]^2$ on the left-hand side, differentiating with respect to t, simplifying and then integrating the resulting inequality between 0 and t and using the initial conditions (4.2) we obtain

$$\left[\left\|\frac{\mathrm{d}u}{\mathrm{d}t}(t)\right\|^2 + \|u(t)\|_A^2\right]^{1/2} \le \left(\|u_1\|^2 + \|u_0\|_A^2\right)^{1/2} + \int_0^t \|f(s)\|\,\mathrm{d}s, \quad (4.5)$$

for a.e. $t \in [0, T]$. Hence,

$$\mathrm{ess.sup}_{t\in[0,T]}\left[\left\|\frac{\mathrm{d}u}{\mathrm{d}t}(t)\right\|^2 + \|u(t)\|_A^2\right]^{1/2} \le \left(\|u_1\|^2 + \|u_0\|_A^2\right)^{1/2} + \|f\|_{L_1((0,T);\mathcal{H})}.$$

Under the additional assumptions $Au_0 \in \mathcal{H}$, $u_1 \in \mathcal{V}$ and $f \in L_1((0,T);\mathcal{V})$, and by applying to (4.1) the operator $A^{1/2}$ and noting (4.5), we (formally) obtain, for a.e. $t \in [0, T]$,

$$\left[\left\|\frac{\mathrm{d}u}{\mathrm{d}t}(t)\right\|_A^2 + \|Au(t)\|^2\right]^{1/2} \le \left(\|u_1\|_A^2 + \|Au_0\|^2\right)^{1/2} + \int_0^t \|f(s)\|_A\,\mathrm{d}s.$$

Hence, by noting the obvious inequality

$$\left\|\frac{\mathrm{d}^2u}{\mathrm{d}t^2}\right\| \le \|f\| + \|Au\|,$$

we deduce that, for a.e. $t \in [0, T]$,

$$\left[\left\|\frac{\mathrm{d}^2u}{\mathrm{d}t^2}(t)\right\|^2 + \left\|\frac{\mathrm{d}u}{\mathrm{d}t}(t)\right\|_A^2 + \|Au(t)\|^2\right]^{1/2}$$
$$\le 2\left(\|u_1\|_A^2 + \|Au_0\|^2\right)^{1/2} + 2\int_0^t \|f(s)\|_A\,\mathrm{d}s + \|f(t)\|. \quad (4.6)$$

Similarly, by applying to (4.1) the operator $A^{-1/2}$, noting (4.5) and omitting the term $\|\mathrm{d}u/\mathrm{d}t\|_{A^{-1}}$ we (formally) obtain, for a.e. $t \in [0, T]$,

$$\|u(t)\| \le \left(\|u_1\|_{A^{-1}}^2 + \|u_0\|^2\right)^{1/2} + \int_0^t \|f(s)\|_{A^{-1}}\,\mathrm{d}s. \quad (4.7)$$

Analogously as in Sect. 3.1.2, we shall also consider the more general equation

$$B\frac{\mathrm{d}^2u}{\mathrm{d}t^2} + Au = f(t), \quad t \in (0, T], \quad (4.8)$$

subject to the initial conditions (4.2), where B and A are unbounded selfadjoint densely defined linear operators on $\mathcal{H}$. Let us suppose that $A = A_0 + A_1$, where A_1 and A_2 are densely defined selfadjoint linear operators on $\mathcal{H}$, and there exist positive constants $m_i > 0$, $i = 1, 2, 3$, such that

$$\begin{aligned}
&(Bu, u) \geq m_1 \|u\|^2, \quad u \in D(B);\\
&(A_0 u, u) \geq m_2 (Bu, u), \quad u \in D(A_0) \cap D(B);\\
&(A_1 u, v)^2 \leq m_3 \|u\| \, \|u\|_{A_0} \|v\| \, \|v\|_{A_0}, \quad u \in D(A_0) \cap D(A_1),\ v \in D(A_0).
\end{aligned} \tag{4.9}$$

By applying $B^{-1/2}$ to (4.8), we obtain

$$\frac{\mathrm{d}^2 \tilde{u}}{\mathrm{d}t^2} + \tilde{A}\tilde{u} = \tilde{f}(t), \quad t \in (0, T],$$

where we have used the notations

$$\tilde{u} := B^{1/2} u, \qquad \tilde{A} := B^{-1/2} A B^{-1/2}, \qquad \tilde{f} := B^{-1/2} f.$$

Let us further define $\tilde{A}_i := B^{-1/2} A_i B^{-1/2}$, $i = 0, 1$. We observe that the linear operator $\tilde{A}_0$ is positive definite on $\mathcal{H}$; indeed,

$$\begin{aligned}
(\tilde{A}_0 v, v) &= \big(B^{-1/2} A_0 B^{-1/2} v, v\big) = \big(A_0 B^{-1/2} v, B^{-1/2} v\big)\\
&\geq m_2 \big(B B^{-1/2} v, B^{-1/2} v\big) = m_2 \|v\|^2, \quad v \in D(\tilde{A}_0).
\end{aligned}$$

We shall therefore take $\mathcal{V} = \mathcal{H}_{\tilde{A}_0}$ and $a(t; v, w) = (\tilde{A} v, w)$, $v \in D(\tilde{A})$, $w \in \mathcal{H}$. In addition, for any $v \in D(\tilde{A}_0) \cap D(\tilde{A}_1)$, $w \in D(\tilde{A}_0)$,

$$\begin{aligned}
(\tilde{A}_1 v, w)^2 &= \big(A_1 B^{-1/2} v, B^{-1/2} w\big)^2\\
&\leq m_3 \big\|B^{-1/2} v\big\| \, \big\|B^{-1/2} v\big\|_{A_0} \big\|B^{-1/2} w\big\| \, \big\|B^{-1/2} w\big\|_{A_0}\\
&\leq \frac{m_3}{m_1 m_2} \big\|B^{-1/2} v\big\|_{A_0}^2 \big\|B^{-1/2} w\big\|_{A_0}^2 = \frac{m_3}{m_1 m_2} \|v\|_{\tilde{A}_0}^2 \|w\|_{\tilde{A}_0}^2
\end{aligned}$$

and

$$\begin{aligned}
\big|(\tilde{A}_1 v, v)\big| &\leq \sqrt{m_3} \big\|B^{-1/2} v\big\| \, \big\|B^{-1/2} v\big\|_{A_0} \leq \sqrt{\frac{m_3}{m_1}} \|v\| \, \big\|B^{-1/2} v\big\|_{A_0}\\
&= \sqrt{\frac{m_3}{m_1}} \|v\| \|v\|_{\tilde{A}_0} \leq \frac{1}{2} \|v\|_{\tilde{A}_0}^2 + \frac{m_3}{2m_1} \|v\|^2,
\end{aligned}$$

which, after continuously extending $a(t; \cdot, \cdot)$ to $\mathcal{V} \times \mathcal{V}$, imply that conditions (b) and (c) from Sect. 3.1.2 hold with $c_1 = 1 + \sqrt{\frac{m_3}{m_1 m_2}}$, $c_0 = 1/2$ and $\lambda = \frac{m_3}{2m_1}$, and that condition (a) is trivially satisfied. Returning to our original notation, we deduce that in the case of $u_0 \in \mathcal{H}_{A_0}$, $u_1 \in \mathcal{H}_B$ and $f \in L_2((0, T); \mathcal{H}_{B^{-1}})$ problem (4.8), (4.2) has a unique solution $u \in L_2((0, T); \mathcal{H}_{A_0})$, with $\frac{\mathrm{d}u}{\mathrm{d}t} \in L_2((0, T); \mathcal{H}_B)$ and $B \frac{\mathrm{d}^2 u}{\mathrm{d}t^2} \in L_2((0, T); \mathcal{H}_{A_0^{-1}})$.

Let us rewrite (4.8) as

$$B\frac{\mathrm{d}^2u}{\mathrm{d}t^2} + \bar{A}u = \frac{m_3}{2}u + f(t), \quad t \in (0, T], \tag{4.10}$$

where we have used the notation $\bar{A} := A + \frac{m_3}{2}I$, with I signifying the identity operator on $\mathcal{H}$. It can be directly verified that the linear operator $\bar{A}$ is selfadjoint, densely defined and positive definite on $\mathcal{H}$, and that

$$\frac{1}{2}(A_0u, u) \le (\bar{A}u, u) \le \left(1 + \sqrt{\frac{m_3}{m_1m_2}} + \frac{m_3}{2m_1m_2}\right)(A_0u, u). \tag{4.11}$$

By taking (once again, formally) the inner product of (4.10) with $2\frac{\mathrm{d}u}{\mathrm{d}t}$, similarly as previously, we deduce that

$$\begin{aligned}\frac{\mathrm{d}}{\mathrm{d}t}\left(\left\|\frac{\mathrm{d}u}{\mathrm{d}t}\right\|_B^2 + \|u\|_{\bar{A}}^2\right) &\le m_3\|u\|_{B^{-1}}\left\|\frac{\mathrm{d}u}{\mathrm{d}t}\right\|_B + 2\|f\|_{B^{-1}}\left\|\frac{\mathrm{d}u}{\mathrm{d}t}\right\|_B \\ &\le \frac{m_3}{m_1}\sqrt{\frac{2}{m_2}}\|u\|_{\bar{A}}\left\|\frac{\mathrm{d}u}{\mathrm{d}t}\right\|_B + 2\|f\|_{B^{-1}}\left\|\frac{\mathrm{d}u}{\mathrm{d}t}\right\|_B \\ &\le C_1\left(\left\|\frac{\mathrm{d}u}{\mathrm{d}t}\right\|_B^2 + \|u\|_{\bar{A}}^2\right) + \|f\|_{B^{-1}}^2,\end{aligned}$$

where $C_1 := \frac{m_3}{m_1\sqrt{2m_2}} + 1$. Multiplying the resulting inequality by e^{-C_1t} yields

$$\frac{\mathrm{d}}{\mathrm{d}t}\left[\mathrm{e}^{-C_1t}\left(\left\|\frac{\mathrm{d}u}{\mathrm{d}t}(t)\right\|_B^2 + \|u(t)\|_{\bar{A}}^2\right)\right] \le \mathrm{e}^{-C_1t}\|f(t)\|_{B^{-1}}^2,$$

which after integration and an obvious majorization gives

$$\left\|\frac{\mathrm{d}u}{\mathrm{d}t}(t)\right\|_B^2 + \|u(t)\|_{\bar{A}}^2 \le \mathrm{e}^{C_1t}\left(\|u_1\|_B^2 + \|u_0\|_{\bar{A}}^2 + \int_0^t \|f(s)\|_{B^{-1}}^2\,\mathrm{d}s\right).$$

Finally, using (4.11) we deduce that, for a.e. $t \in [0, T]$,

$$\left\|\frac{\mathrm{d}u}{\mathrm{d}t}(t)\right\|_B^2 + \|u(t)\|_{A_0}^2 \le C\left(\|u_1\|_B^2 + \|u_0\|_{A_0}^2 + \int_0^t \|f(s)\|_{B^{-1}}^2\,\mathrm{d}s\right), \tag{4.12}$$

where $C = C_2\mathrm{e}^{C_1t}$ is a computable constant, which depends on t.

By taking the inner product of (4.10) with $2\bar{A}^{-1}B\frac{\mathrm{d}u}{\mathrm{d}t}$, a similar argument implies the existence of a constant $C = C(T) > 0$, such that, for a.e. $t \in [0, T]$,

$$\|u(t)\|_B^2 \le C\left(\|Bu_1\|_{A_0^{-1}}^2 + \|u_0\|_B^2 + \int_0^t \|f(s)\|_{A_0^{-1}}^2\,\mathrm{d}s\right). \tag{4.13}$$

4.1.3 Abstract Three-Level Operator-Difference Schemes

Suppose that $\mathcal{H}^h$ is a finite-dimensional real Hilbert space with inner product $(\cdot,\cdot)_h$ and norm $\|\cdot\|_h := \|\cdot\|_{\mathcal{H}^h}$ and $\overline{\Omega}^\tau$ is a uniform mesh with step size $\tau := T/M$, $M \geq 2$, on the interval $[0,T]$. With the notations from Sect. 3.1.4, we consider the family of three-level operator difference schemes

$$\begin{aligned} B_h\big(D_t^+D_t^-U\big) + D_h\big(D_t^0U\big) + A_hU = F, \quad & t \in \Omega^\tau, \\ U(0) = U^0, \qquad U(\tau) = U^1. & \end{aligned} \tag{4.14}$$

Here, the function $F : \Omega^\tau \to \mathcal{H}^h$ is given, as are $U^0 \in \mathcal{H}^h$ and $U^1 \in \mathcal{H}^h$; $U : \overline{\Omega}^\tau \to \mathcal{H}^h$ is the unknown function,

$$D_t^0U := \frac{1}{2}\big(D_t^+U + D_t^-U\big) = (\hat{U} - \check{U})/2\tau$$

is the symmetric first difference quotient, A_h and B_h are linear selfadjoint positive definite operators on $\mathcal{H}^h$, uniformly with respect to h, and D_h is a linear selfadjoint nonnegative operator on $\mathcal{H}^h$.

We take the inner product of (4.14) with $2\tau D_t^0U = \hat{U} - \check{U}$. Noting that

$$U = \frac{1}{4}(\hat{U} + 2U + \check{U}) - \frac{1}{4}\tau^2 D_t^+D_t^-U,$$
$$\hat{U} - \check{U} = \tau\big(D_t^+U + D_t^-U\big) = (\hat{U} + U) - (U + \check{U}),$$

we obtain

$$\begin{aligned} &\left(\left(B_h - \frac{\tau^2}{4}A_h\right)D_t^+U, D_t^+U\right)_h - \left(\left(B_h - \frac{\tau^2}{4}A_h\right)D_t^-U, D_t^-U\right)_h \\ &+ 2\tau\big\|D_t^0U\big\|_{D_h}^2 + \left\|\frac{\hat{U}+U}{2}\right\|_{A_h}^2 - \left\|\frac{U+\check{U}}{2}\right\|_{A_h}^2 = 2\tau\big(F, D_t^0U\big)_h. \end{aligned} \tag{4.15}$$

Thus, if $F = 0$ and

$$D_h \geq 0, \qquad \bar{B}_h := B_h - \frac{1}{4}\tau^2 A_h \geq 0, \tag{4.16}$$

we deduce from (4.15) that

$$\big\|D_t^+U\big\|_{\bar{B}_h}^2 + \left\|\frac{\hat{U}+U}{2}\right\|_{A_h}^2 \leq \big\|D_t^-U\big\|_{\bar{B}_h}^2 + \left\|\frac{U+\check{U}}{2}\right\|_{A_h}^2.$$

Summing over $t \in \Omega^\tau$ we have that

$$\big\|D_t^+U^m\big\|_{\bar{B}_h}^2 + \left\|\frac{U^{m+1}+U^m}{2}\right\|_{A_h}^2 \leq \big\|D_t^+U^0\big\|_{\bar{B}_h}^2 + \left\|\frac{U^1+U^0}{2}\right\|_{A_h}^2.$$

This inequality expresses stability of the homogeneous operator-difference scheme (4.14) with respect to perturbations of the initial data, under the (sufficient) conditions (4.16); (cf. Samarskiĭ [159], Sect. 6.3).

When $F \neq 0$, we shall suppose instead of (4.16) that the following, slightly stronger, condition holds:

$$D_h \geq 0, \qquad \bar{B}_h := B_h - \frac{1}{4}\tau^2 A_h > 0. \tag{4.17}$$

Then,

$$\begin{aligned} 2\tau\big(F, D_t^0 U\big)_h &= \tau\big(F, D_t^+ U + D_t^- U\big)_h \\ &\leq \tau \|F\|_{\bar{B}_h^{-1}}\big(\|D_t^+ U\|_{\bar{B}_h} + \|D_t^- U\|_{\bar{B}_h}\big) \leq \tau \|F\|_{\bar{B}_h^{-1}}(J + \check{J}), \end{aligned}$$

where we have used the notation

$$J = J(t) := \left(\|D_t^+ U\|_{\bar{B}_h}^2 + \left\|\frac{\hat{U} + U}{2}\right\|_{A_h}^2\right)^{1/2}, \qquad \check{J} := J(t - \tau).$$

Thus we deduce from (4.15) that

$$J^2 - \check{J}^2 \leq \tau \|F\|_{\bar{B}_h^{-1}}(J + \check{J}).$$

Hence, after dividing both sides by $J + \check{J}$ and summing over the points of the mesh Ω^τ, we obtain

$$J^m \leq J^0 + \tau \sum_{k=1}^{m} \|F^k\|_{\bar{B}_h^{-1}};$$

that is,

$$\begin{aligned} &\left(\|D_t^+ U^m\|_{\bar{B}_h}^2 + \left\|\frac{U^{m+1} + U^m}{2}\right\|_{A_h}^2\right)^{1/2} \\ &\leq \left(\|D_t^+ U^0\|_{\bar{B}_h}^2 + \left\|\frac{U^1 + U^0}{2}\right\|_{A_h}^2\right)^{1/2} + \tau \sum_{k=1}^{m} \|F^k\|_{\bar{B}_h^{-1}}. \end{aligned} \tag{4.18}$$

The inequality (4.18) can be seen as the discrete analogue of the a priori estimate (4.5).

When $D_h = 0$, similarly as above, by taking the inner product of (4.14) with $2\tau \bar{B}_h^{-1} A_h (D_t^0 U)$ we obtain

$$\left(\|D_t^+ U^m\|_{A_h}^2 + \left\|A_h \frac{U^{m+1} + U^m}{2}\right\|_{\bar{B}_h^{-1}}^2\right)^{1/2}$$

$$\leq \left(\| D_t^+ U^0 \|^2_{A_h} + \left\| A_h \frac{U^1 + U^0}{2} \right\|^2_{\bar{B}_h^{-1}} \right)^{1/2} + \tau \sum_{k=1}^{m} \| \bar{B}_h^{-1} F^k \|_{A_h}.$$

Hence, using (4.14) we deduce that

$$\left[\| D_t^+ D_t^- U^m \|^2_{\bar{B}_h} + \| D_t^+ U^m \|^2_{A_h} + \left\| A_h \frac{U^{m+1} + U^m}{2} \right\|^2_{\bar{B}_h^{-1}} \right]^{1/2}$$
$$\leq 2 \left(\| D_t^+ U^0 \|^2_{A_h} + \left\| A_h \frac{U^1 + U^0}{2} \right\|^2_{\bar{B}_h^{-1}} \right)^{1/2}$$
$$+ 2\tau \sum_{k=1}^{m} \| \bar{B}_h^{-1} F^k \|_{A_h} + \| F^m \|_{\bar{B}_h^{-1}}. \tag{4.19}$$

Under the same hypothesis ($D_h = 0$) as above, by taking the inner product of (4.14) with $2\tau A_h^{-1} \bar{B}_h (D_t^0 U)$, an analogous argument yields that

$$\left\| \frac{U^{m+1} + U^m}{2} \right\|_{\bar{B}_h} \leq \left(\| \bar{B}_h (D_t^+ U^0) \|^2_{A_h^{-1}} + \left\| \frac{U^1 + U^0}{2} \right\|^2_{\bar{B}_h} \right)^{1/2} + \tau \sum_{k=1}^{m} \| F^k \|_{A_h^{-1}}. \tag{4.20}$$

The inequalities (4.19) and (4.20) are discrete analogues of the a priori estimates (4.6) and (4.7).

Let us finally consider the case when the operator A_h is *not* positive definite. Suppose that A_h, B_h and D_h are still selfadjoint linear operators on $\mathcal{H}^h$ and let us assume that the following conditions hold:

$$D_h \geq 0, \qquad A_h = A_{0h} + A_{1h}, \qquad A_{ih}^* = A_{ih}, \quad i = 0, 1,$$
$$(B_h U, U)_h \geq m_1 \| U \|_h^2, \qquad (A_{0h} U, U)_h \geq m_2 (B_h U, U)_h,$$
$$(A_{1h} U, V)_h^2 \leq m_3 \| U \|_h \| U \|_{A_{0h}} \| V \|_h \| V \|_{A_{0h}}.$$

Equation (4.14) can be rewritten in the form

$$\bar{B}_h (D_t^+ D_t^- U) + D_h (D_t^0 U) + \frac{1}{4} \bar{A}_h (\hat{U} + 2U + \check{U})$$
$$= \frac{1}{8} m_3 (\hat{U} + 2U + \check{U}) + F, \tag{4.21}$$

where we have used the notations

$$\bar{B}_h := B_h - \frac{\tau^2}{4} A_h, \qquad \bar{A}_h := A_h + \frac{m_3}{2} I_h, \quad I_h \text{ is the identity operator on } \mathcal{H}^h.$$

It is then shown, analogously as in the 'continuous' case, that

$$\frac{1}{2} (A_{0h} U, U)_h \leq (\bar{A}_h U, U)_h \leq \left(1 + \sqrt{\frac{m_3}{m_1 m_2}} + \frac{m_3}{2 m_1 m_2} \right) (A_{0h} U, U)_h,$$

which implies the positive definiteness of $\bar{A}_h$, uniformly in h. Let us suppose that the operator $\bar{B}_h$ is positive definite, uniformly with respect to h; i.e. that

$$(\bar{B}_h U, U)_h \geq m_4 \|U\|_h^2, \quad m_4 > 0,$$

where the constant m_4 is independent of h. Since

$$\begin{aligned}
(\bar{B}_h U, U)_h &= (B_h U, U)_h - \frac{\tau^2}{4}(A_h U, U)_h \\
&\geq \left[m_1 - \frac{\tau^2}{4}\left(1 + \sqrt{\frac{m_3}{m_1 m_2}}\right)\|A_{0h}\|\right]\|U\|_h^2,
\end{aligned}$$

the positive definiteness of $\bar{B}_h$, uniformly in h, will be ensured with $m_4 < m_1$ once we have taken τ sufficiently small so that

$$\tau^2 \leq \frac{4(m_1 - m_4)}{1 + \sqrt{\frac{m_3}{m_1 m_2}}}.$$

By taking the inner product of (4.21) with $2\tau D_t^0 U = \hat{U} - \check{U}$, similarly as before, we get that

$$\begin{aligned}
&\|D_t^+ U\|_{\bar{B}_h}^2 - \|D_t^- U\|_{\bar{B}_h}^2 + 2\tau\|D_t^0 U\|_{D_h}^2 + \left\|\frac{\hat{U}+U}{2}\right\|_{\bar{A}_h}^2 - \left\|\frac{U+\check{U}}{2}\right\|_{\bar{A}_h}^2 \\
&= \frac{m_3}{4}\tau\left(\frac{\hat{U}+U}{2} + \frac{U+\check{U}}{2}, D_t^+ U + D_t^- U\right)_h + \tau\left(F, D_t^+ U + D_t^- U\right)_h \\
&\leq \frac{m_3}{4}\tau\left(\left\|\frac{\hat{U}+U}{2}\right\|_{\bar{B}_h^{-1}} + \left\|\frac{U+\check{U}}{2}\right\|_{\bar{B}_h^{-1}}\right)\left(\|D_t^+ U\|_{\bar{B}_h} + \|D_t^- U\|_{\bar{B}_h}\right) \\
&\quad + \tau\|F\|_{\bar{B}_h^{-1}}\left(\|D_t^+ U\|_{\bar{B}_h} + \|D_t^- U\|_{\bar{B}_h}\right).
\end{aligned}$$

Using the inequality

$$\|U\|_{\bar{B}_h^{-1}} \leq \sqrt{\frac{2}{m_1 m_2 m_4}}\|U\|_{\bar{A}_h}$$

in conjunction with the Cauchy–Schwarz inequality, after some obvious majorizations we obtain

$$\bar{J} - \check{\bar{J}} \leq \frac{C_1 \tau}{2}(\bar{J} + \check{\bar{J}}) + \tau\|F\|_{\bar{B}_h^{-1}}^2,$$

where we have used the notations

$$\bar{J} := \|D_t^+ U\|_{\bar{B}_h}^2 + \left\|\frac{\hat{U}+U}{2}\right\|_{\bar{A}_h}^2, \qquad C_1 := \frac{m_3}{\sqrt{2 m_1 m_2 m_4}}.$$

For $\tau < 2/C_1$,

$$\bar{J} \le \frac{1+\frac{C_1\tau}{2}}{1-\frac{C_1\tau}{2}}\check{J} + \frac{\tau}{1-\frac{C_1\tau}{2}}\|F\|^2_{\bar{B}_h^{-1}},$$

and therefore, by induction,

$$\begin{aligned}\bar{J}^m &\le \left(\frac{1+\frac{C_1\tau}{2}}{1-\frac{C_1\tau}{2}}\right)^m \bar{J}^0 + \frac{\tau}{1-\frac{C_1\tau}{2}}\sum_{k=1}^{m}\left(\frac{1+\frac{C_1\tau}{2}}{1-\frac{C_1\tau}{2}}\right)^{m-k}\|F^k\|^2_{\bar{B}_h^{-1}}\\ &\le \left(\frac{1+\frac{C_1\tau}{2}}{1-\frac{C_1\tau}{2}}\right)^m\left(\bar{J}^0 + \tau\sum_{k=1}^{m}\|F^k\|^2_{\bar{B}_h^{-1}}\right), \quad m = 1,\dots,M-1.\end{aligned}$$

By bounding from above further, we deduce that, for all $m = 1,\dots,M-1$,

$$\begin{aligned}&\|D_t^+U^m\|^2_{\bar{B}_h} + \left\|\frac{U^{m+1}+U^m}{2}\right\|^2_{\bar{A}_h}\\ &\quad\le C_2 e^{C_1T}\left(\|D_t^+U^0\|^2_{\bar{B}_h} + \left\|\frac{U^1+U^0}{2}\right\|^2_{\bar{A}_h} + \tau\sum_{k=1}^{M}\|F^k\|^2_{\bar{B}_h^{-1}}\right).\end{aligned} \tag{4.22}$$

If $D_h = 0$, by taking the inner product of (4.21) with $2\tau\bar{A}_h^{-1}\bar{B}_h(D_t^0U)$, a similar argument yields the a priori bound

$$\begin{aligned}\left\|\frac{U^{m+1}+U^m}{2}\right\|^2_{\bar{B}_h} &\le C_2e^{C_1T}\Bigg(\|\bar{B}_h(D_t^+U^0)\|^2_{\bar{A}_h^{-1}}\\ &\quad + \left\|\frac{U^1+U^0}{2}\right\|^2_{\bar{B}_h} + \tau\sum_{k=1}^{M}\|F^k\|^2_{\bar{A}_h^{-1}}\Bigg).\end{aligned} \tag{4.23}$$

The inequalities (4.22) and (4.23) are discrete analogues of (4.12) and (4.13).

4.2 Classical Difference Schemes for the Wave Equation

This section is devoted to a summary of some well-known results concerning standard finite difference approximations of the wave equation; we shall assume here that the solution possesses a sufficient number of continuous partial derivatives. Later on, in Sects. 4.3 and 4.4, we shall relax the regularity requirements on the solution and we shall extend these results to second-order hyperbolic equations with variable coefficients.

4.2.1 Explicit and Weighted Schemes

Our first model problem is the wave equation in the domain $Q = \Omega \times (0, T]$, where $\Omega = (0, 1)$, $T > 0$:

$$\begin{aligned}
&\text{find } u(x,t) \quad \text{such that}\\
&\frac{\partial^2 u}{\partial t^2} = \frac{\partial^2 u}{\partial x^2} + f(x,t), \quad x \in (0,1),\ t \in (0,T],\\
&u(0,t) = 0, \qquad u(1,t) = 0, \quad t \in (0,T],\\
&u(x,0) = u_0(x), \qquad \frac{\partial u}{\partial t}(x,0) = u_1(x), \quad x \in [0,1].
\end{aligned} \tag{4.24}$$

Physically, $u(x,t)$ represents the displacement at a point x and time t of an elastic string of unit length, subject to the initial displacement $u_0(x)$, the initial velocity $u_1(x)$, and body forces whose density, in space and time, is described by the function f. We shall assume for the moment that f is a smooth function of $(x,t) \in \overline{Q}$ and that u_0 and u_1 are smooth functions of $x \in \overline{\Omega}$, compatible with the boundary conditions at $x = 0$ and $x = 1$; i.e. $u_0(0) = 0$, $u_0(1) = 0$, $u_1(0) = 0$, $u_1(1) = 0$.

4.2.1.1 The Explicit Scheme

Similarly as in Sect. 3.2, we consider the uniform mesh

$$\overline{Q}_h^\tau = \overline{\Omega}^h \times \overline{\Omega}^\tau = \{(x_j, t_m) : 0 \le j \le N; 0 \le m \le M\},$$

with mesh-sizes $h := 1/N$ and $\tau := T/M$, $N, M \ge 2$, in the region $\overline{Q} = [0,1] \times [0,T]$. On $\overline{Q}_h^\tau$ we approximate (4.24) by the following finite difference scheme:

$$\begin{aligned}
&\text{find } U_j^m,\ j = 0, \dots, N,\ m = 0, \dots, M, \quad \text{such that}\\
&D_t^+ D_t^- U_j^m = D_x^+ D_x^- U_j^m + f(x_j, t_m),\\
&\quad j = 1, \dots, N-1,\ m = 1, \dots, M-1,\\
&U_0^m = 0, \qquad U_N^m = 0, \quad m = 2, \dots, M,\\
&U_j^0 = u_0(x_j), \qquad U_j^1 = u_0(x_j) + \tau u_1(x_j) + \frac{1}{2}\tau^2\big[u_0''(x_j) + f(x_j, 0)\big],\\
&\quad j = 0, \dots, N,
\end{aligned} \tag{4.25}$$

where U_j^m represents the approximation of $u(x_j, t_m)$, the value of the analytical solution u at the mesh-point (x_j, t_m), $D_t^+ D_t^- U_j^m$ is the second divided (central) difference in the t-direction and $D_x^+ D_x^- U_j^m$ is the second divided (central) difference

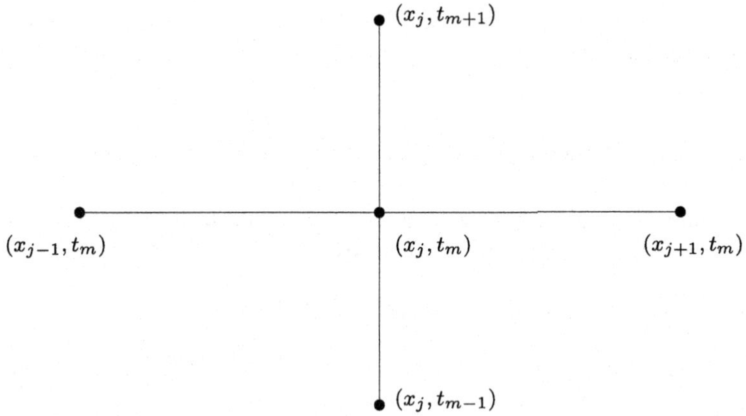

Fig. 4.1 Five-point stencil for the explicit scheme

in the x-direction. Clearly, (4.25) is a five-point difference scheme involving the values of U at the mesh-points

$$(x_j, t_{m-1}), \quad (x_{j-1}, t_m), \quad (x_j, t_m), \quad (x_{j+1}, t_m), \quad (x_j, t_{m+1}),$$

shown in Fig. 4.1. The scheme (4.25) is applied as follows. First we set $m = 1$. Since $U_j^0, U_{j-1}^1, U_j^1, U_{j+1}^1$ are given by the initial conditions, the values U_j^2, $j = 0, \dots, N$, can be computed from (4.25):

$$\begin{aligned} U_j^2 &= 2U_j^1 - U_j^0 + \frac{\tau^2}{h^2}(U_{j+1}^1 - 2U_j^1 + U_{j-1}^1) + \tau^2 f(x_j, t_2), \\ j &= 1, \dots, N-1, \\ U_0^2 &= 0, \qquad U_N^2 = 0. \end{aligned}$$

Suppose that we have already calculated U_j^k, $j = 0, \dots, N$, the values of U on time level $t_k = k\tau$ for all $k \le m$. The values of U on the next time level $t_{m+1} = (m+1)\tau$ can then be obtained from (4.25) by rewriting it as

$$\begin{aligned} U_j^{m+1} &= 2U_j^m - U_j^{m-1} + \frac{\tau^2}{h^2}(U_{j+1}^m - 2U_j^m + U_{j-1}^m) + \tau^2 f(x_j, t_m), \\ j &= 1, \dots, N-1, \\ U_0^{m+1} &= 0, \qquad U_N^{m+1} = 0, \end{aligned}$$

for any m, $1 \le m \le M-1$.

The values of U at $t = t_{m+1}$ can be calculated explicitly from those of U^m, U^{m-1} and the data; hence the name *explicit scheme*. In fact, (4.25) is just a special case of a two-parameter family of three-level finite difference schemes for the wave equation. The two parameters are frequently referred to as the weights in the scheme, and hence the name *weighted scheme*. The analysis of the weighted scheme is the subject of the next section.

4.2.1.2 Weighted Scheme

The explicit scheme for the wave equation can be embedded in the following two-parameter family of finite difference schemes:

$$
\begin{aligned}
&\text{find } U_j^m,\ j=0,\dots,N,\ m=0,\dots,M, \quad \text{such that}\\
&D_t^+D_t^-U_j^m = D_x^+D_x^-\big[\theta_1 U_j^{m+1} + (1-\theta_1-\theta_2)U_j^m + \theta_2 U_j^{m-1}\big] + f(x_j,t_m),\\
&\quad j=1,\dots,N-1,\ m=1,\dots,M-1,\\
&U_0^m = 0, \qquad U_N^m = 0, \quad m=2,\dots,M,\\
&U_j^0 = u_0(x_j), \qquad U_j^1 = u_0(x_j) + \tau u_1(x_j) + \frac{1}{2}\tau^2\big[u_0''(x_j) + f(x_j,0)\big],\\
&\quad j=0,\dots,N,
\end{aligned}
\tag{4.26}
$$

where θ_1 and θ_2 are nonnegative 'weights'. The explicit scheme considered in the previous subsection corresponds to the choice $\theta_1 = \theta_2 = 0$. In general, the weighted scheme is a nine-point finite difference scheme, involving the values of U at the mesh-points

$$
(x_{j-1},t_{m-1}),\quad (x_j,t_{m-1}),\quad (x_{j+1},t_{m-1}),\quad (x_{j-1},t_m),\quad (x_j,t_m),\quad (x_{j+1},t_m),
$$
$$
(x_{j-1},t_{m+1}),\quad (x_j,t_{m+1}),\quad (x_{j+1},t_{m+1}),
$$

shown in Fig. 4.2. Unlike the explicit scheme in which the data and the values of the approximate solution U at two previous time levels provide an explicit expression for the values of U on the next time level, the difference scheme (4.26) necessitates the solution of a system of linear equations on each time level to determine the values of U at the mesh-points on that time level. More precisely, (4.26) can be rewritten as follows:

$$
\begin{aligned}
&-\frac{\theta_1\tau^2}{h^2}U_{j-1}^{m+1} + \left(1+\frac{2\theta_1\tau^2}{h^2}\right)U_j^{m+1} - \frac{\theta_1\tau^2}{h^2}U_{j+1}^{m+1}\\
&= 2U_j^m - U_j^{m-1} + (1-\theta_1-\theta_2)\frac{\tau^2}{h^2}\left(U_{j-1}^m - 2U_j^m + U_{j+1}^m\right)\\
&\quad + \frac{\theta_2\tau^2}{h^2}\left(U_{j-1}^{m-1} - 2U_j^{m-1} + U_{j+1}^{m-1}\right) + \tau^2 f(x_j,t_m),\\
&j=1,\dots,N-1,\\
&U_0^{m+1} = 0, \qquad U_N^{m+1} = 0.
\end{aligned}
$$

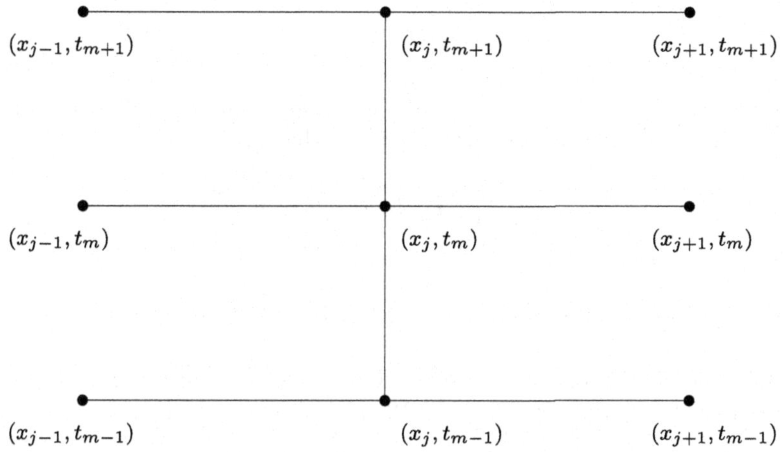

Fig. 4.2 Nine-point stencil for the weighted scheme

Thus, when $\theta_1 \neq 0$, starting from the values U_j^0 and U_j^1, $j = 0, \ldots, N$, on the first two time levels, which are specified by the initial conditions, the values U_j^{m+1}, $j = 0, \ldots, N$, on each subsequent time level $t = t_{m+1}$, $m = 1, \ldots, M - 1$, can be computed by solving a system of linear algebraic equations with a tridiagonal matrix of size $(N - 1) \times (N - 1)$.

4.2.2 Stability of the Weighted Difference Scheme

Let, as before, $\mathcal{S}_0^h$ denote the linear space of real-valued mesh-functions defined on the mesh $\overline{\Omega}^h$, which vanish on $\overline{\Omega}^h \setminus \Omega^h$. We shall equip $\mathcal{S}_0^h$ with the inner product

$$(V, W)_h := \sum_{j=1}^{N-1} h V_j W_j$$

and the corresponding induced norm

$$\|V\|_h := (V, V)_h^{1/2}.$$

Let us also consider the norm

$$\|V]|_h := \left(\sum_{j=0}^{N-1} h |V_j|^2 \right)^{1/2}.$$

Using the identity

$$\theta_1 U^{m+1} + (1 - \theta_1 - \theta_2) U^m + \theta_2 U^{m-1}$$

$$= U^m + \frac{\tau^2}{2}(\theta_1 + \theta_2) D_t^+ D_t^- U^m + \tau(\theta_1 - \theta_2) D_t^0 U^m,$$

the weighted scheme (4.26) can be reformulated as an operator difference scheme of the form (4.14) in the linear space $\mathcal{H}^h = \mathcal{S}_0^h$, where

$$A_h U = \Lambda U := \begin{cases} -D_x^+ D_x^- U & \text{for } x \in \Omega^h, \\ 0 & \text{for } x \in \overline{\Omega}^h \setminus \Omega^h, \end{cases}$$

$$D_h := (\theta_1 - \theta_2)\tau\Lambda, \qquad B_h := I_h + \frac{\tau^2}{2}(\theta_1 + \theta_2)\Lambda,$$

and I_h is the identity operator on $\mathcal{S}_0^h$. The operator Λ is positive definite on $\mathcal{S}_0^h$, uniformly with respect to h, and (cf. (2.18), (2.22)) we have that

$$8\|V\|_h^2 \leq (\Lambda V, V)_h < \frac{4}{h^2}\|V\|_h^2, \quad \text{i.e. } 8I_h \leq \Lambda < \frac{4}{h^2} I_h. \tag{4.27}$$

We shall consider two distinct ranges of the parameters θ_1 and θ_2.

❶ When

$$\theta_1 \geq \theta_2 \quad \text{and} \quad \theta_1 + \theta_2 \geq \frac{1}{2}, \tag{4.28}$$

the conditions (4.17) are trivially satisfied. The inequality (4.18) reduces to

$$\begin{aligned} &\left(\|D_t^+ U^m\|^2_{I_h + \frac{1}{2}(\theta_1+\theta_2-\frac{1}{2})\tau^2\Lambda} + \left\| \frac{U^{m+1} + U^m}{2} \right\|^2_\Lambda \right)^{1/2} \\ &\leq \left(\|D_t^+ U^0\|^2_{I_h + \frac{1}{2}(\theta_1+\theta_2-\frac{1}{2})\tau^2\Lambda} + \left\| \frac{U^1 + U^0}{2} \right\|^2_\Lambda \right)^{1/2} \\ &\quad + \tau \sum_{k=1}^m \|F^k\|_{(I_h + \frac{1}{2}(\theta_1+\theta_2-\frac{1}{2})\tau^2\Lambda)^{-1}}. \end{aligned} \tag{4.29}$$

Further, by Lemma 2.10, we have that

$$\|U\|_\Lambda = \|[D_x^+ U\|_h. \tag{4.30}$$

Using the relations

$$\|U\|^2_{I_h + \frac{1}{2}(\theta_1+\theta_2-\frac{1}{2})\tau^2\Lambda} = \|U\|_h^2 + \frac{1}{2}\left(\theta_1 + \theta_2 - \frac{1}{2}\right)\tau^2 \|U\|_\Lambda^2 \geq \|U\|_h^2,$$

$$\|U\|_{(I_h + \frac{1}{2}(\theta_1+\theta_2-\frac{1}{2})\tau^2\Lambda)^{-1}} = \sup_{V \in \mathcal{S}_0^h} \frac{|(U, V)_h|}{\|V\|_{I_h + \frac{1}{2}(\theta_1+\theta_2-\frac{1}{2})\tau^2\Lambda}} \leq \|U\|_h,$$

we deduce from (4.29) that

$$\left(\|D_t^+U^m\|_h^2+\left\|\left[D_x^+\frac{U^{m+1}+U^m}{2}\right\|_h^2\right)^{1/2}$$
$$\leq\left[\|D_t^+U^0\|_h^2+\left\|\left[D_x^+\frac{U^1+U^0}{2}\right\|_h^2\right.$$
$$\left.+\frac{1}{2}\left(\theta_1+\theta_2-\frac{1}{2}\right)\tau^2\|[D_t^+D_x^+U^0\|_h^2\right]^{1/2}+\tau\sum_{k=1}^{m}\|F^k\|_h. \tag{4.31}$$

The inequality (4.31) expresses the *unconditional stability* of the finite difference scheme (4.26).

❷ When

$$\theta_1\geq\theta_2 \quad \text{and} \quad \theta_1+\theta_2<\frac{1}{2}, \tag{4.32}$$

we shall suppose that the mesh-sizes h and τ satisfy the additional condition

$$\tau\leq h\sqrt{\frac{1-\varepsilon}{1-2(\theta_1+\theta_2)}}, \qquad 0<\varepsilon<1. \tag{4.33}$$

Using (4.27), we then have that

$$I_h+\frac{1}{2}\left(\theta_1+\theta_2-\frac{1}{2}\right)\tau^2\Lambda\geq\left[1-\frac{1}{2}\left(\frac{1}{2}-\theta_1-\theta_2\right)\frac{4\tau^2}{h^2}\right]I_h\geq\varepsilon I_h;$$

i.e. the conditions (4.17) are again satisfied. Furthermore,

$$\varepsilon\|U\|_h^2\leq\|U\|^2_{I_h+\frac{1}{2}(\theta_1+\theta_2-\frac{1}{2})\tau^2\Lambda}\leq\|U\|_h^2,$$

and

$$\|U\|_h^2\leq\|U\|^2_{(I_h+\frac{1}{2}(\theta_1+\theta_2-\frac{1}{2})\tau^2\Lambda)^{-1}}\leq\frac{1}{\varepsilon}\|U\|_h^2,$$

and hence we deduce from (4.29) that

$$\left(\|D_t^+U^m\|_h^2+\left\|\left[D_x^+\frac{U^{m+1}+U^m}{2}\right\|_h^2\right)^{1/2}$$
$$\leq C_\varepsilon\left[\left(\|D_t^+U^0\|_h^2+\left\|\left[D_x^+\frac{U^1+U^0}{2}\right\|_h^2\right)^{1/2}+\tau\sum_{k=1}^{m}\|F^k\|_h\right], \tag{4.34}$$

where C_ε is a computable constant, which depends on ε. The inequality (4.34) expresses the *conditional stability* of the finite difference scheme (4.26) under the hypothesis that the mesh-sizes h and τ satisfy (4.33). Note that (4.33) is less restrictive than the corresponding condition (3.65) for the conditional stability of the explicit finite difference scheme for the heat equation. .

4.2.3 Error Analysis of Difference Schemes for the Wave Equation

In this section we investigate the accuracy of the finite difference scheme (4.26) for the numerical solution of the initial-boundary-value problem (4.24). We begin our considerations by defining the truncation error of the weighted scheme (4.26) as follows:

$$\begin{aligned}\varphi_j^m &:= D_t^+ D_t^- u(x_j, t_m) - D_x^+ D_x^- \big[\theta_1 u(x_j, t_{m+1}) \\ &\quad + (1-\theta_1-\theta_2) u(x_j, t_m) + \theta_2 u(x_j, t_{m-1})\big] - f(x_j, t_m),\end{aligned}$$

for $j = 1, \ldots, N-1$, $m = 1, \ldots, M-1$. The global error is defined by

$$e_j^m := u(x_j, t_m) - U_j^m,$$

for $j = 0, \ldots, N$, $m = 0, \ldots, M$. It is easily seen that e_j^m satisfies the following finite difference scheme:

$$\begin{aligned}&D_t^+ D_t^- e_j^m - D_x^+ D_x^- \big[\theta_1 e_j^{m+1} + (1-\theta_1-\theta_2) e_j^m + \theta_2 e_j^{m-1}\big] = \varphi_j^m, \\ &\quad 1 \le j \le N-1, 1 \le m \le M-1, \\ &e_0^m = 0, \qquad e_N^m = 0, \quad 0 \le m \le M, \\ &e_j^0 = 0, \quad 1 \le j \le N-1, \\ &e_j^1 = \eta_j, \quad 1 \le j \le N-1,\end{aligned}$$

where

$$\eta_j := u(x_j, \tau) - u_0(x_j) - \tau u_1(x_j) - \frac{1}{2}\tau^2\big[u_0''(x_j) + f(x_j, 0)\big].$$

Thanks to the stability result (4.31) established in the previous section,

$$\begin{aligned}&\max_{0 \le m \le M-1} \left(\| D_t^+ e^m \|_h^2 + \left\| \left[D_x^+ \frac{e^{m+1} + e^m}{2} \right\|_h^2 \right)^{1/2} \\ &\le \left[\frac{1}{\tau^2} \|\eta\|_h^2 + \frac{1}{2}(\theta_1 + \theta_2) \big\| \big[D_x^+ \eta \big\|_h^2 \right]^{1/2} + \tau \sum_{k=1}^{M-1} \|\varphi^k\|_h, \qquad (4.35)\end{aligned}$$

provided that (4.28) holds; if on the other hand (4.32) holds, then, by (4.34), we have that

$$\max_{0 \le m \le M-1} \left(\| D_t^+ e^m \|_h^2 + \left\| \left[D_x^+ \frac{e^{m+1} + e^m}{2} \right\|_h^2 \right)^{1/2}$$

$$\le C_\varepsilon\left[\left(\frac{1}{\tau^2}\|\eta\|_h^2+\frac{1}{4}[\![D_x^+\eta]\!]_h^2\right)^{1/2}+\tau\sum_{k=1}^{M-1}\|\varphi^k\|_h\right], \tag{4.36}$$

provided that (4.32) and (4.33) hold. In each case we have to bound $\|\eta\|_h$, $[\![D_x^+\eta]\!]_h$ and $\|\varphi^k\|_h$ to complete the error analysis. Using the differential equation, φ_j^k (with k replaced by m) can be rewritten as

$$\begin{aligned}\varphi_j^m &= D_t^+D_t^-u(x_j,t_m)-D_x^+D_x^-\big[\theta_1u(x_j,t_{m+1})\\ &\quad+(1-\theta_1-\theta_2)u(x_j,t_m)+\theta_2u(x_j,t_{m-1})\big]-\frac{\partial^2u}{\partial t^2}(x_j,t_m)+\frac{\partial^2u}{\partial x^2}(x_j,t_m)\\ &=\left(D_t^+D_t^-u(x_j,t_m)-\frac{\partial^2u}{\partial t^2}(x_j,t_m)\right)\\ &\quad+\left(\frac{\partial^2u}{\partial x^2}(x_j,t_m)-D_x^+D_x^-u(x_j,t_m)\right)\\ &\quad-\frac{\theta_1+\theta_2}{2}\tau^2D_t^+D_t^-D_x^+D_x^-u(x_j,t_m)-(\theta_1-\theta_2)\tau D_t^0D_x^+D_x^-u(x_j,t_m).\end{aligned}$$

In order to estimate the truncation error φ_j^m, we shall expand the various terms involved in it into Taylor series with remainder terms. By noting that

$$\begin{aligned}u(x_j,t_{m\pm1})&=u(x_j,t_m)\pm\tau\frac{\partial u}{\partial t}(x_j,t_m)+\frac{\tau^2}{2}\frac{\partial^2u}{\partial t^2}(x_j,t_m)\\ &\quad\pm\frac{\tau^3}{6}\frac{\partial^3u}{\partial t^3}(x_j,t_m)+\frac{\tau^4}{24}\frac{\partial^4u}{\partial t^4}(x_j,t'_\pm),\end{aligned}$$

where $t'_-\in(t_{m-1},t_m)$ and $t'_+\in(t_m,t_{m+1})$, we deduce that

$$D_t^+D_t^-u(x_j,t_m)-\frac{\partial^2u}{\partial t^2}(x_j,t_m)=\frac{\tau^2}{12}\frac{\partial^4u}{\partial t^4}(x_j,t'),$$

where $t'\in(t_{m-1},t_{m+1})$. Similarly, we obtain that

$$\frac{\partial^2u}{\partial x^2}(x_j,t_m)-D_x^+D_x^-u(x_j,t_m)=-\frac{h^2}{12}\frac{\partial^4u}{\partial x^4}(x',t_m),$$

where $x'\in(x_{j-1},x_{j+1})$. Further,

$$D_t^+D_t^-D_x^+D_x^-u(x_j,t_m)=\frac{\partial^4u}{\partial x^2\partial t^2}(x'',t''),$$

where $x''\in(x_{j-1},x_{j+1})$ and $t''\in(t_{m-1},t_{m+1})$, and

$$D_t^0D_x^+D_x^-u(x_j,t_m)=\frac{\partial^3u}{\partial x^2\partial t}(x''',t'''),$$

where $x''' \in (x_{j-1}, x_{j+1})$ and $t''' \in (t_{m-1}, t_{m+1})$. It follows from these expansions that

$$|\varphi_j^m| \le |\theta_1 - \theta_2| \tau M_{2x1t} + \frac{\tau^2}{12}\big[M_{4t} + 6(\theta_1 + \theta_2) M_{2x2t}\big] + \frac{h^2}{12} M_{4x}, \quad (4.37)$$

where we have used the notation

$$M_{kxlt} := \max_{(x,t)\in\overline{Q}} \left| \frac{\partial^{k+l}}{\partial x^k \partial t^l} u(x,t) \right|,$$

with $M_{lt} := M_{0xlt}$ and $M_{kx} := M_{kx0t}$. Now, (4.24) and (4.26) imply that

$$\eta_j = u(x_j, \tau) - u(x_j, 0) - \tau \frac{\partial u}{\partial t}(x_j, 0) - \frac{\tau^2}{2} \frac{\partial^2 u}{\partial t^2}(x_j, 0) = \frac{\tau^3}{6} \frac{\partial^3 u}{\partial t^3}(x_j, \tilde{t}),$$

where $\tilde{t} \in (0, \tau)$. Thus we deduce that

$$D_x^+ \eta_j = \frac{\tau^3}{6} \frac{\partial^4 u}{\partial x \partial t^3}(\tilde{x}, \tilde{t}), \quad \tilde{x} \in (x_j, x_{j+1}).$$

The last two equalities then imply that

$$|\eta_j| \le \frac{\tau^3}{6} M_{3t}, \qquad |D_x^+ \eta_j| \le \frac{\tau^3}{6} M_{1x3t}. \quad (4.38)$$

Finally, (4.37), (4.38) and the a priori estimates (4.35) and (4.36) yield the desired bound on the global error of the finite difference scheme (4.26):

$$\max_{0\le m\le M-1} \left(\|D_t^+ e^m\|_h^2 + \left\| \left[D_x^+ \frac{e^{m+1} + e^m}{2} \right\|_h^2 \right)^{1/2} \le C_1(h^2 + \tau^2), \qquad \theta_1 = \theta_2, \quad (4.39)$$

$$\max_{0\le m\le M-1} \left(\|D_t^+ e^m\|_h^2 + \left\| \left[D_x^+ \frac{e^{m+1} + e^m}{2} \right\|_h^2 \right)^{1/2} \le C_2(h^2 + \tau), \qquad \theta_1 > \theta_2. \quad (4.40)$$

The constants C_1 and C_2 featuring in these error bounds depend on M_{2x1t}, M_{4t}, M_{2x2t}, M_{4x}, M_{3t}, M_{1x3t} (which we have *assumed* to be finite) and T, but they are independent of h and τ. If θ_1 and θ_2 satisfy (4.32), then the bounds (4.39) and (4.40) hold under the condition (4.33), and the constants C_1 and C_2 then also depend on ε.

The error bounds (4.39) and (4.40) have been derived under very restrictive assumptions on the smoothness of the solution. In the next section we shall be concerned with the error analysis of difference schemes for the wave equation under less demanding hypotheses on the regularity of the data and of the corresponding solution of the initial-boundary-value problem.

4.3 The Wave Equation with Nonsmooth Data

4.3.1 The Initial-Boundary-Value Problem and Its Discretization

We consider the initial-boundary-value problem (4.24) in the space-time domain $Q := (0, 1) \times (0, T]$. The mesh $\overline{Q}_h^\tau$ is defined in the same way as in Sect. 4.2. We shall also retain the various other pieces of notation that were introduced in Sect. 4.2.

The initial-boundary-value problem (4.24) will be approximated by the symmetric weighted finite difference scheme, with weights $\theta_1 = \theta_2 = \frac{1}{4}$, and with a mollified source term:

$$
\begin{aligned}
&\text{find } U_j^m,\ j=0,\dots,N,\ m=0,\dots,M, \quad \text{such that}\\
&D_t^+ D_t^- U_j^m = \frac{1}{4} D_x^+ D_x^- \left(U_j^{m+1} + 2U_j^m + U_j^{m-1}\right) + T_x T_t f(x_j, t_m),\\
&\quad j = 1,\dots,N-1,\ m = 1,\dots,M-1,\\
&U_0^m = 0, \qquad U_N^m = 0, \quad m = 0,\dots,M, \\
&U_j^0 = u_0(x_j),\\
&U_j^1 = u_0(x_j) + \tau T_x u_1(x_j) + \frac{1}{2}\tau^2\left[D_x^+ D_x^- u_0(x_j) + T_x \tilde{T}_t^+ f(x_j, 0)\right],\\
&\quad j = 1,\dots,N-1,
\end{aligned}
\tag{4.41}
$$

where the mollifiers T_x, T_t and $\tilde{T}_t^\pm$ are defined by

$$
T_x v(x,t) := \frac{1}{h}\int_{x-h/2}^{x+h/2} v(x',t)\,\mathrm{d}x', \qquad T_t v(x,t) := \frac{1}{\tau}\int_{t-\tau/2}^{t+\tau/2} v(x,t')\,\mathrm{d}t',
$$

and

$$
\tilde{T}_t^+ v(x,t) := \frac{2}{\tau}\int_{t}^{t+\tau/2} v(x,t')\,\mathrm{d}t', \qquad \tilde{T}_t^- v(x,t) := \frac{2}{\tau}\int_{t-\tau/2}^{t} v(x,t')\,\mathrm{d}t'.
$$

If v is a distribution on Q, then T_x, T_t and $\tilde{T}_t^\pm$ should be interpreted as convolutions with suitable piecewise constant functions. According to the stability results that were proved in the previous section the finite difference scheme (4.41) is unconditionally stable.

4.3.2 Error Analysis

It is easily seen that the global error

$$
e_j^m := u(x_j, t_m) - U_j^m
$$

of the finite difference scheme (4.41) satisfies the following equalities

$$\begin{aligned}
&D_t^+ D_t^- e_j^m - \frac{1}{4} D_x^+ D_x^- \left(e_j^{m+1} + 2e_j^m + e_j^{m-1}\right) = \varphi_j^m, \\
&\quad 1 \le j \le N-1,\ 1 \le m \le M-1, \\
&e_0^m = 0, \qquad e_N^m = 0, \quad 0 \le m \le M, \\
&e_j^0 = 0, \quad 1 \le j \le N-1, \\
&e_j^1 = \zeta_j, \quad 1 \le j \le N-1,
\end{aligned} \tag{4.42}$$

where

$$\varphi := D_t^+ D_t^- u - T_x T_t \frac{\partial^2 u}{\partial t^2} - D_x^+ D_x^- u + T_x T_t \frac{\partial^2 u}{\partial x^2} - \frac{\tau^2}{4} D_x^+ D_x^- D_t^+ D_t^- u,$$

$$\zeta := u(x,\tau) - u(x,0) - \frac{\tau}{h} \int_{x-h/2}^{x+h/2} \frac{\partial u}{\partial t}\left(x',0\right) \mathrm{d}x' - \frac{\tau^2}{2} D_x^+ D_x^- u(x,0)$$

$$- \frac{\tau}{h} \int_{x-h/2}^{x+h/2} \int_0^{\tau/2} \left[\frac{\partial^2 u}{\partial t^2}(x',t') - \frac{\partial^2 u}{\partial x^2}(x',t')\right] \mathrm{d}t' \, \mathrm{d}x'.$$

Let us define the following seminorm and norm, respectively:

$$\|V\|_{2,\infty,h\tau}^{(0)} := \max_{t \in \Omega_-^\tau} \left\| \frac{V(\cdot,t+\tau) + V(\cdot,t)}{2} \right\|_h,$$

$$\|V\|_{2,\infty,h\tau}^{(1)} := \max_{t \in \Omega_-^\tau} \left\{ \|D_t^+ V(\cdot,t)\|_h^2 + \left\| \left[D_x^+ \left(\frac{V(\cdot,t+\tau) + V(\cdot,t)}{2} \right) \right\|_h^2 \right\}^{1/2}.$$

The finite difference scheme (4.42) can be restated as a three-level operator-difference scheme (4.14), where $\mathcal{H}^h = \mathcal{S}_0^h$, $A_h = \Lambda$, $D_h = 0$ and $B_h = I_h + \frac{1}{4}\tau^2 \Lambda$. It follows from (4.18), (4.30) and the initial conditions (4.42) that the following a priori estimate holds:

$$\|e\|_{2,\infty,h\tau}^{(1)} \le \tau^{-1} \|\zeta\|_h + \frac{1}{2} \| [D_x^+ \zeta \|_h + \tau \sum_{k=1}^{M-1} \|\varphi^k\|_h. \tag{4.43}$$

In order to complete the error analysis of the scheme in the seminorm $\|\cdot\|_{2,\infty,h\tau}^{(0)}$ we require two auxiliary lemmas, which we now state and prove.

Lemma 4.2 *The solution of the finite difference scheme*

$$D_t^+ D_t^- V_j^m - \frac{1}{4} D_x^+ D_x^- \left(V_j^{m+1} + 2V_j^m + V_j^{m-1}\right) = D_x^- \xi_j^m,$$

$$1 \le j \le N-1,\ 1 \le m \le M-1,$$

subject to homogeneous Dirichlet boundary conditions at $j = 0$ and $j = N$, and with V^0 and V^1 specified, satisfies the a priori estimate

$$\|V\|^{(0)}_{2,\infty,h\tau} \leq \|D_t^+ V^0\|_{\Lambda^{-1}} + \left\| \frac{V^1 + V^0}{2} \right\|_h + \tau \sum_{k=1}^{M-1} [\![\xi^k\|_h.$$

Proof The inequality follows directly from (4.20), (4.30) and the relation

$$\|D_x^- \xi\|_{\Lambda^{-1}} = \sup_{Z \in \mathcal{S}_0^h} \frac{|(D_x^- \xi, Z)_h|}{\|Z\|_\Lambda} \leq \sup_{Z \in \mathcal{S}_0^h} \frac{[\![\xi\|_h [\![D_x^+ Z\|_h}{[\![D_x^+ Z\|_h} = [\![\xi\|_h. \quad (4.44)$$

□

Lemma 4.3 *The solution of the finite difference scheme*

$$D_t^+ D_t^- W_j^m - \frac{1}{4} D_x^+ D_x^- \left(W_j^{m+1} + 2W_j^m + W_j^{m-1}\right) = D_t^- \eta_j^m,$$
$$1 \leq j \leq N-1, \ 1 \leq m \leq M-1,$$

subject to homogeneous Dirichlet boundary conditions at $j = 0$ and $j = N$, and with W^0 and W^1 given, satisfies the a priori estimate

$$\|W\|^{(0)}_{2,\infty,h\tau} \leq \|D_t^+ W^0 - \eta^0\|_{\Lambda^{-1}} + \left\| \frac{W^1 + W^0}{2} \right\|_h + \tau \sum_{k=1}^{M-1} \left\| \frac{\eta^k + \eta^{k-1}}{2} \right\|_h.$$

Proof By applying the operator Λ^{-1} to the difference scheme and taking the inner product of the resulting difference equation with $W^{m+1} - W^{m-1}$ yields

$$\|D_t^+ W^m\|^2_{\Lambda^{-1}} + \left\| \frac{W^{m+1} + W^m}{2} \right\|_h^2 - \|D_t^+ W^{m-1}\|^2_{\Lambda^{-1}}$$
$$- \left\| \frac{W^m + W^{m-1}}{2} \right\|_h^2 = \left(D_t^- \eta^m, W^{m+1} - W^{m-1}\right)_{\Lambda^{-1}}.$$

Further, we have that

$$\begin{aligned}
\left(D_t^- \eta^m, W^{m+1} - W^{m-1}\right)_{\Lambda^{-1}} &= \left(\eta^m - \eta^{m-1}, D_t^+ W^m + D_t^+ W^{m-1}\right)_{\Lambda^{-1}} \\
&= 2\left(\eta^m, D_t^+ W^m\right)_{\Lambda^{-1}} - 2\left(\eta^{m-1}, D_t^+ W^{m-1}\right)_{\Lambda^{-1}} \\
&\quad - \tau\left(\eta^m + \eta^{m-1}, D_t^+ D_t^- W^m\right)_{\Lambda^{-1}}.
\end{aligned}$$

Using again the difference equation from the statement of the lemma satisfied by the mesh-function W, we obtain

$$
\begin{aligned}
&-\tau\big(\eta^m+\eta^{m-1}, D_t^+D_t^-W^m\big)_{\Lambda^{-1}}\\
&\quad=-\tau\big(\eta^m+\eta^{m-1}, D_t^-\eta^m\big)_{\Lambda^{-1}}\\
&\qquad+\tau\left(\eta^m+\eta^{m-1}, \frac{1}{4}\big(W^{m+1}+2W^m+W^{m-1}\big)\right)_h\\
&\quad=-\|\eta^m\|^2_{\Lambda^{-1}}+\|\eta^{m-1}\|^2_{\Lambda^{-1}}\\
&\qquad+\tau\left(\frac{\eta^m+\eta^{m-1}}{2}, \frac{W^{m+1}+W^m}{2}+\frac{W^m+W^{m-1}}{2}\right)_h.
\end{aligned}
$$

These relations imply that

$$
\big(J^m\big)^2-\big(J^{m-1}\big)^2=\tau\left(\frac{\eta^m+\eta^{m-1}}{2}, \frac{W^{m+1}+W^m}{2}+\frac{W^m+W^{m-1}}{2}\right)_h,
$$

where we have used the notation

$$
\begin{aligned}
\big(J^m\big)^2&:=\big\|D_t^+W^m\big\|^2_{\Lambda^{-1}}-2\big(\eta^m, D_t^+W^m\big)^2_{\Lambda^{-1}}+\big\|\eta^m\big\|^2_{\Lambda^{-1}}+\left\|\frac{W^{m+1}+W^m}{2}\right\|^2_h\\
&=\big\|D_t^+W^m-\eta^m\big\|^2_{\Lambda^{-1}}+\left\|\frac{W^{m+1}+W^m}{2}\right\|^2_h.
\end{aligned}
$$

Hence,

$$
\begin{aligned}
\big(J^m\big)^2-\big(J^{m-1}\big)^2&\le\tau\left\|\frac{\eta^m+\eta^{m-1}}{2}\right\|_h\\
&\quad\times\left(\left\|\frac{W^{m+1}+W^m}{2}\right\|_h+\left\|\frac{W^m+W^{m-1}}{2}\right\|_h\right)\\
&\le\tau\left\|\frac{\eta^m+\eta^{m-1}}{2}\right\|_h\big(J^m+J^{m-1}\big),
\end{aligned}
$$

which, after dividing both sides by J^m+J^{m-1} and summing over m, yields the desired inequality. □

The right-hand side φ of the finite difference scheme (4.42) can be represented as

$$
\varphi=D_x^-\xi+D_t^-\eta,
$$

where

$$
\begin{aligned}
\xi_j^m&:=T_t\frac{\partial u}{\partial x}(x_j+h/2,t_m)-D_x^+u(x_j,t_m)-\frac{\tau^2}{4}D_x^+D_t^+D_t^-u(x_j,t_m),\\
\eta_j^m&:=D_t^+u(x_j,t_m)-T_x\frac{\partial u}{\partial t}(x_j,t_m+\tau/2).
\end{aligned}
$$

By applying Lemmas 4.2 and 4.3 and the initial conditions we obtain

$$\|e\|^{(0)}_{2,\infty,h\tau} \leq \big\|\zeta/\tau - \eta^0\big\|_{\Lambda^{-1}} + \|\zeta/2\|_h + \tau \sum_{k=1}^{M-1} \big[\!\big[\xi^k\big\|_h + \tau \sum_{k=0}^{M-1} \big\|\eta^k\big\|_h .$$

Further, $\zeta/\tau - \eta^0 = D_x^- \chi$, where

$$\chi_j := \int_0^{\tau/2} \frac{\partial u}{\partial x}(x_j + h/2, t)\,\mathrm{d}t - \frac{\tau}{2} D_x^+ u(x_j, 0),$$

and therefore, by using the relation (4.44), we finally obtain the following a priori estimate:

$$\|e\|^{(0)}_{2,\infty,h\tau} \leq [\![\chi\|_h + \frac{1}{2}\|\zeta\|_h + \tau \sum_{k=1}^{M-1} \big[\!\big[\xi^k\big\|_h + \tau \sum_{k=0}^{M-1} \big\|\eta^k\big\|_h . \tag{4.45}$$

Thus, in order to complete the convergence analysis of the finite difference scheme (4.41) in the mesh-dependent norm $\|\cdot\|^{(1)}_{2,\infty,h\tau}$ and the seminorm $\|\cdot\|^{(0)}_{2,\infty,h\tau}$, it suffices to bound the right-hand sides of the inequalities (4.43) and (4.45). We shall suppose to this end that the mesh-sizes h and τ are linked by the condition $\tau \asymp h$, i.e.

$$c_1 h \leq \tau \leq c_2 h, \quad c_1, c_2 = \text{Const.} > 0. \tag{4.46}$$

It is easily seen that φ_j^m is a bounded linear functional of $u \in W_2^s(G_j^m)$, $s > 2$, $G_j^m = (x_{j-1}, x_{j+1}) \times (t_{m-1}, t_{m+1})$, which vanishes on all cubic polynomials. By applying the Bramble–Hilbert lemma we obtain

$$\big|\varphi_j^m\big| \leq C h^{s-3} |u|_{W_2^s(G_j^m)}, \quad 2 < s \leq 4,$$

which, after summation over the mesh-points, yields that

$$\tau \sum_{m=1}^{M-1} \big\|\varphi^m\big\|_h \leq \sqrt{T}\left(\tau \sum_{m=1}^{M-1} \big\|\varphi^m\big\|_h^2\right)^{1/2} = \sqrt{T}\left(h\tau \sum_{j=1}^{N-1}\sum_{m=1}^{M-1} |\varphi_j^m|^2\right)^{1/2}$$

$$\leq C h^{s-2} \|u\|_{W_2^s(Q)}, \quad 2 < s \leq 4. \tag{4.47}$$

Analogously, ζ_j is a bounded linear functional of $u \in W_2^s(G_j^{0+})$, $s > 3/2$, $G_j^{0+} = (x_{j-1}, x_{j+1}) \times (0, \tau)$, which vanishes on all quadratic polynomials, and therefore by the Bramble–Hilbert lemma we have that

$$|\zeta_j| \leq C h^{s-1} |u|_{W_2^s(G_j^{0+})}, \quad 3/2 < s \leq 3.$$

Thus, summing over j we deduce that

$$\|\zeta\|_h \leq C h^{s-1/2} |u|_{W_2^s(Q_\tau)}, \quad 3/2 < s \leq 3, \tag{4.48}$$

where $Q_\tau := (0,1) \times (0,\tau)$. From (4.48), with $s = 3$ and using the inequality (2.199), we further have that

$$\|\zeta\|_h \le Ch^{5/2+\min(s'-3,1/2)} |\log h|^{1-|\operatorname{sgn}(s'-7/2)|} \|u\|_{W_2^{s'}(Q)}, \tag{4.49}$$

where $3 < s' \le 4$. Hence, by further majorization, (4.48) and (4.49) yield the bound

$$\frac{1}{\tau}\|\zeta\|_h \le Ch^{s-2}\|u\|_{W_2^s(Q)}, \quad 2 < s \le 4. \tag{4.50}$$

Also,

$$\|\!\![D_x^+ \zeta \|_h \le \frac{2}{h}\|\zeta\|_h;$$

thus, from (4.46) and (4.50) it immediately follows that

$$\|\!\![D_x^+ \zeta \|_h \le Ch^{s-2}\|u\|_{W_2^s(Q)}, \quad 2 < s \le 4. \tag{4.51}$$

Finally, from (4.43), (4.47), (4.50) and (4.51) we obtain the following error bound for the finite difference scheme (4.41) in the norm $\|\cdot\|^{(1)}_{2,\infty,h\tau}$:

$$\|u - U\|^{(1)}_{2,\infty,h\tau} \le Ch^{s-2}\|u\|_{W_2^s(Q)}, \quad 2 < s \le 4, \tag{4.52}$$

where $C = C(s)$ is a positive constant, independent of h.

Let us now turn our attention to the case when $3/2 < s \le 3$. We begin by noting that ξ_j^m is a bounded linear functional of $u \in W_2^s(G_{j+}^m)$, $s > 3/2$, $G_{j+}^m := (x_j, x_{j+1}) \times (t_{m-1}, t_{m+1})$, which vanishes on all quadratic polynomials. The Bramble–Hilbert lemma therefore implies that

$$|\xi_j^m| \le Ch^{s-2}|u|_{W_2^s(G_{j+}^m)}, \quad 3/2 < s \le 3,$$

which, after summation over the mesh-points, yields

$$\tau\sum_{m=1}^{M-1} \|\!\![\xi^m \|_h \le Ch^{s-1}\|u\|_{W_2^s(Q)}, \quad 3/2 < s < 3. \tag{4.53}$$

The quantity η_j^m is a bounded linear functional of $u \in W_2^s(G_{j*}^{m+})$, $s > 2$, $G_{j*}^{m+} := (x_j - h/2, x_j + h/2) \times (t_m, t_{m+1})$, which vanishes on all quadratic polynomials. Similarly as in the previous case we obtain that

$$\tau\sum_{m=0}^{M-1} \|\eta^m\|_h \le Ch^{s-1}\|u\|_{W_2^s(Q)}, \quad 3/2 < s < 3. \tag{4.54}$$

Further, χ_j is a bounded linear functional of $u \in W_2^s(G_{j+}^{0*})$, $s > 3/2$, with $G_{j+}^{0*} := (x_j, x_{j+1}) \times (0, \tau/2)$, which vanishes on all linear polynomials. Thus, by the Bramble–Hilbert lemma and summation over the mesh-points,

$$[\![\chi\|_h \le Ch^{s-1/2}|u|_{W_2^s(Q_\tau)}, \quad 3/2 < s < 2.$$

By taking $s = 2$ and using the inequality (2.199), we further deduce that

$$[\![\chi\|_h \le Ch^{3/2+\min(s'-2,1/2)}|\log h|^{1-|\mathrm{sgn}(s'-5/2)|}\|u\|_{W_2^{s'}(Q)}, \quad 2 < s' \le 3.$$

After further majorization the last two inequalities imply that

$$[\![\chi\|_h \le Ch^{s-1}\|u\|_{W_2^s(Q)}, \quad 3/2 < s < 3. \tag{4.55}$$

Finally, from (4.45), (4.53), (4.54), (4.55) and (4.48) we obtain the following bound on the global error of the finite difference scheme (4.41) in the seminorm $\|\cdot\|^{(0)}_{2,\infty,h\tau}$: there exists a positive constant $C = C(s)$, independent of h, such that

$$\|u - U\|^{(0)}_{2,\infty,h\tau} \le Ch^{s-1}\|u\|_{W_2^s(Q)}, \quad 3/2 < s \le 3. \tag{4.56}$$

Two remarks are in order. First, we note that the error bound (4.56) has been shown to hold for all s in the range $3/2 < s \le 3$, but not for $1 < s \le 3/2$. The reason for this is that the right-hand side $T_x T_t f$ of the finite difference scheme (4.41) need not be a continuous function when $s \le 3/2$, and therefore the scheme is not meaningful as stated for $s \le 3/2$. For this latter range of s a stronger mollification of f is necessary (e.g. $T_x^2 T_t^2 f$).

Our second remark is concerned with the requirement that the mesh-sizes h and τ be linked by the condition (4.46). Since the difference scheme under consideration is unconditionally stable (cf. the last sentence of Sect. 4.3.1), linking τ to h in the convergence analysis of the scheme, as was done above, is unnatural. Although, admittedly, (4.46) is less demanding than the corresponding condition (3.88) in the parabolic case, we shall show that by careful estimation of the various functionals that are responsible for the emergence of the condition (4.46) in the convergence analysis, (4.46) can be completely avoided, at least in some cases. Suppose, for example, that $s = 4$. It is easily seen that φ_j^m can be represented as follows:

$$\begin{aligned}\varphi_j^m = &\frac{1}{h\tau}\int_{x_j-h/2}^{x_j+h/2}\int_{x'}^{x_j}\int_{x_j}^{x''}\int_{t_{m-1}}^{t_{m+1}}\left(1 - \frac{|t'-t_m|}{\tau}\right)\\ &\times \frac{\partial^4 u}{\partial x^2 \partial t^2}(x''', t')\,\mathrm{d}t'\,\mathrm{d}x'''\,\mathrm{d}x''\,\mathrm{d}x'\\ &+ \frac{1}{h\tau^2}\int_{x_j-h/2}^{x_j+h/2}\int_{t_{m-1}}^{t_{m+1}}\int_{t_m-\tau/2}^{t_m+\tau/2}\int_{t''}^{t'}\int_{t_m}^{t'''}\left(1 - \frac{|t'-t_m|}{\tau}\right)\\ &\times \frac{\partial^4 u}{\partial t^4}(x', t'''')\,\mathrm{d}t''''\,\mathrm{d}t'''\,\mathrm{d}t''\,\mathrm{d}t'\,\mathrm{d}x'\end{aligned}$$

$$
\begin{aligned}
&-\frac{1}{h\tau}\int_{t_m-\tau/2}^{t_m+\tau/2}\int_{t'}^{t_m}\int_{t_m}^{t''}\int_{x_{j-1}}^{x_{j+1}}\left(1-\frac{|x'-x_j|}{h}\right)\\
&\qquad\times\frac{\partial^4 u}{\partial x^2\partial t^2}(x',t''')\,\mathrm{d}x'\,\mathrm{d}t'''\,\mathrm{d}t''\,\mathrm{d}t'\\
&-\frac{1}{h^2\tau}\int_{t_m-\tau/2}^{t_m+\tau/2}\int_{x_{j-1}}^{x_{j+1}}\int_{x_j-h/2}^{x_j+h/2}\int_{x''}^{x'}\int_{x_j}^{x'''}\left(1-\frac{|x'-x_j|}{h}\right)\\
&\qquad\times\frac{\partial^4 u}{\partial x^4}(x'''',t')\,\mathrm{d}x''''\,\mathrm{d}x'''\,\mathrm{d}x''\,\mathrm{d}x'\,\mathrm{d}t'\\
&-\frac{\tau}{4h}\int_{x_{j-1}}^{x_{j+1}}\int_{t_{m-1}}^{t_{m+1}}\left(1-\frac{|x'-x_j|}{h}\right)\left(1-\frac{|t'-t_m|}{\tau}\right)\\
&\qquad\times\frac{\partial^4 u}{\partial x^2\partial t^2}(x',t')\,\mathrm{d}t'\,\mathrm{d}x'.
\end{aligned}
$$

Thus we directly have that

$$
|\varphi_j^m|\le\frac{C(h^2+\tau^2)}{\sqrt{h\tau}}|u|_{W_2^4(G_j^m)}
$$

and

$$
\tau\sum_{m=1}^{M-1}\|\varphi^m\|_h\le C(h^2+\tau^2)\|u\|_{W_2^4(Q)}.
$$

The integral representation

$$
\begin{aligned}
\zeta_j=&\frac{1}{h}\int_{x_j-h/2}^{x_j+h/2}\int_{x'}^{x_j}\int_{x_j}^{x''}\int_0^{\tau}\frac{\partial^3 u}{\partial x^2\partial t}(x''',t')\,\mathrm{d}t'\,\mathrm{d}x'''\,\mathrm{d}x''\,\mathrm{d}x'\\
&+\frac{\tau}{h}\int_{x_{j-1}}^{x_{j+1}}\int_0^{\tau}\int_0^{t'}\left(1-\frac{|x'-x_j|}{h}\right)\frac{\partial^3 u}{\partial x^2\partial t}(x',t'')\,\mathrm{d}t''\,\mathrm{d}t'\,\mathrm{d}x'\\
&+\frac{1}{h}\int_{x_j-h/2}^{x_j+h/2}\int_0^{\tau}\int_0^{t'}(\tau-t')\frac{\partial^3 u}{\partial t^3}(x',t'')\,\mathrm{d}t''\,\mathrm{d}t'\,\mathrm{d}x'\\
&-\frac{\tau}{h}\int_{x_j-h/2}^{x_j+h/2}\int_0^{\tau/2}\int_0^{t'}\frac{\partial^3 u}{\partial t^3}(x',t'')\,\mathrm{d}t''\,\mathrm{d}t'\,\mathrm{d}x'\\
&+\frac{\tau}{h}\int_{x_j-h/2}^{x_j+h/2}\int_{x_j}^{x'}\int_0^{\tau/2}\frac{\partial^3 u}{\partial x^2\partial t}(x'',t')\,\mathrm{d}t'\,\mathrm{d}x''\,\mathrm{d}x'\\
&-\frac{\tau}{h}\int_{x_{j-1}}^{x_{j+1}}\int_{x_j}^{x'}\int_0^{\tau/2}\left(1-\frac{|x'-x_j|}{h}\right)\frac{\partial^3 u}{\partial x^2\partial t}(x'',t')\,\mathrm{d}t'\,\mathrm{d}x''\,\mathrm{d}x'
\end{aligned}
$$

yields that

$$|\zeta_j| \le C\big(h^2+\tau^2\big)\sqrt{\tau/h}\,|u|_{W_2^3(G_j^{0+})}.$$

Hence, using the inequality (2.199) we obtain

$$\frac{1}{\tau}\|\zeta\|_h \le \frac{C(h^2+\tau^2)}{\sqrt{\tau}}|u|_{W_2^3(Q_\tau)} \le C\big(h^2+\tau^2\big)\|u\|_{W_2^4(Q)}.$$

Similarly, by noting the integral representation

$$\begin{aligned}
D_x^+\zeta_j &= \frac{1}{h^2}\int_{x_j-h/2}^{x_j+h/2}\int_{x'}^{x_j}\int_{x_j}^{x''}\int_{x'''}^{x'''+h}\int_0^{\tau}\frac{\partial^4 u}{\partial x^3\partial t}\big(x'''',t'\big)\,\mathrm{d}t'\,\mathrm{d}x''''\,\mathrm{d}x'''\,\mathrm{d}x''\,\mathrm{d}x' \\
&+\frac{\tau}{h^2}\int_{x_{j-1}}^{x_{j+1}}\int_{x'}^{x'+h}\int_0^{\tau}\int_0^{t'}\bigg(1-\frac{|x'-x_j|}{h}\bigg)\frac{\partial^4 u}{\partial x^3\partial t}\big(x'',t''\big)\,\mathrm{d}t''\,\mathrm{d}t'\,\mathrm{d}x''\,\mathrm{d}x' \\
&+\frac{1}{h^2}\int_{x_j-h/2}^{x_j+h/2}\int_{x'}^{x'+h}\int_0^{\tau}\int_0^{t'}\big(\tau-t'\big)\frac{\partial^4 u}{\partial x\partial t^3}\big(x'',t''\big)\,\mathrm{d}t''\,\mathrm{d}t'\,\mathrm{d}x''\,\mathrm{d}x' \\
&-\frac{\tau}{h^2}\int_{x_j-h/2}^{x_j+h/2}\int_{x'}^{x'+h}\int_0^{\tau/2}\int_0^{t'}\frac{\partial^4 u}{\partial x\partial t^3}\big(x'',t''\big)\,\mathrm{d}t''\,\mathrm{d}t'\,\mathrm{d}x''\,\mathrm{d}x' \\
&+\frac{\tau}{h^2}\int_{x_j-h/2}^{x_j+h/2}\int_{x_j}^{x'}\int_{x''}^{x''+h}\int_0^{\tau/2}\frac{\partial^4 u}{\partial x^3\partial t}\big(x''',t'\big)\,\mathrm{d}t'\,\mathrm{d}x'''\,\mathrm{d}x''\,\mathrm{d}x' \\
&-\frac{\tau}{h^2}\int_{x_{j-1}}^{x_{j+1}}\int_{x_j}^{x'}\int_{x''}^{x''+h}\int_0^{\tau/2}\bigg(1-\frac{|x'-x_j|}{h}\bigg) \\
&\times\frac{\partial^4 u}{\partial x^3\partial t}\big(x''',t'\big)\,\mathrm{d}t'\,\mathrm{d}x'''\,\mathrm{d}x''\,\mathrm{d}x',
\end{aligned}$$

we have that

$$\big|D_x^+\zeta_j\big| \le C\big(h^2+\tau^2\big)\sqrt{\tau/h}\,|u|_{W_2^4(G_j^{0+}\cup G_{j+1}^{0+})}$$

and

$$\big\|D_x^+\zeta\big\|_h \le C\big(h^2+\tau^2\big)\sqrt{\tau}\,|u|_{W_2^4(Q_\tau)} \le C\big(h^2+\tau^2\big)\|u\|_{W_2^4(Q)}.$$

From these bounds and the a priori estimate (4.43) we obtain the following error bound in which the mesh-sizes h and τ are not linked:

$$\|u-U\|_{2,\infty,h\tau}^{(1)} \le C\big(h^2+\tau^2\big)\|u\|_{W_2^4(Q)}.$$

We close this section with an interesting error bound, which is derived using function space interpolation (see Zlotnik [204], Jovanović [87, 88]). For the sake of simplicity we consider the initial-boundary-value problem (4.24) with $f=0$ and $u_1=0$. Suppose that $u_0 \in \mathfrak{W}_2^s(0,1)$, $1 \le s \le 4$, where

$$\mathfrak{W}_2^s(0,1) := \left\{ v \in W_2^s(0,1) : u_0^{(2i)}(0) = u_0^{(2i)}(1) = 0, \text{ for } 0 \le i < \frac{2s-1}{4} \right\}. \tag{4.57}$$

For $s=1$ and $s=4$ in particular this assumption enables us to extend u_0 as an odd function outside the interval $(0,1)$ by preserving its Sobolev class, W_2^s. From (4.5) we directly have the following a priori estimate:

$$\max_{t\in[0,T]} \left(\left\| \frac{\partial u}{\partial t}(\cdot,t) \right\|^2_{L_2(0,1)} + \left\| \frac{\partial u}{\partial x}(\cdot,t) \right\|^2_{L_2(0,1)} \right)^{1/2}$$
$$\le \left(\left\| \frac{\partial u}{\partial t}(\cdot,0) \right\|^2_{L_2(0,1)} + \left\| \frac{\partial u}{\partial x}(\cdot,0) \right\|^2_{L_2(0,1)} \right)^{1/2} = \|u_0'\|_{L_2(0,1)}. \tag{4.58}$$

By differentiating the equation

$$\frac{\partial^2 u}{\partial t^2} = \frac{\partial^2 u}{\partial x^2}$$

with respect to x and t and by applying (4.58) we obtain that

$$\max_{t\in[0,T]} \left\| \frac{\partial^k u}{\partial x^i \partial t^{k-i}}(\cdot,t) \right\|_{L_2(0,1)} \le \|u_0^{(k)}\|_{L_2(0,1)}, \tag{4.59}$$

where $1 \le k \le [s]$, and $0 \le i \le k$. In the case of $s=4$ we have from (4.59) on noting (4.57) that

$$\|u\|_{W_2^4(Q)} \le C\|u_0\|_{W_2^4(0,1)}, \qquad u_0 \in \mathfrak{W}_2^4(0,1). \tag{4.60}$$

Let $\mathcal{B}_h^\tau := C(\Omega^\tau, W_2^1(\Omega^h))$ denote the linear space of all real-valued mesh-functions defined on $\overline{Q}_h^\tau$ that vanish at $x=0$ and $x=1$, with the seminorm

$$\|V\|_{C(\Omega^\tau,W_2^1(\Omega^h))} := \max_{0\le m\le M-1} \left\| \frac{V^{m+1}+V^m}{2} \right\|_{W_2^1(\Omega^h)},$$

where we have used the notation

$$\|W\|^2_{W_2^1(\Omega^h)} := \|W\|_h^2 + \|D_x^+ W]|_h^2.$$

Let us suppose that the mesh-sizes h and τ satisfy the condition (4.46). We have from (4.52), (4.60) and the discrete Friedrichs inequality (2.26) that

$$\|u-U\|_{C(\Omega^\tau,W_2^1(\Omega^h))} \le Ch^2\|u_0\|_{W_2^4(0,1)}, \qquad u_0 \in \mathfrak{W}_2^4(0,1). \tag{4.61}$$

Next we shall derive a similar bound under the assumption $u_0 \in \mathfrak{W}_2^1(0,1)$ and will then interpolate between that bound and (4.61). Obviously,

$$\left\| D_x^+ \left(\frac{e^{m+1} + e^m}{2} \right) \right\|_h \leq \left\| D_x^+ \left(\frac{u^{m+1} + u^m}{2} \right) \right\|_h + \left\| D_x^+ \left(\frac{U^{m+1} + U^m}{2} \right) \right\|_h .$$

Further, we have that

$$\left\| D_x^+ \left(\frac{u^{m+1} + u^m}{2} \right) \right\|_h \leq \frac{1}{2} (\| D_x^+ u^{m+1} \|_h + \| D_x^+ u^m \|_h),$$

and

$$\| D_x^+ u^m \|_h^2 = h \sum_{j=0}^{N-1} \left| \frac{u(x_{j+1}, t_m) - u(x_j, t_m)}{h} \right|^2$$

$$= h \sum_{j=0}^{N-1} \left| \frac{1}{h} \int_{x_j}^{x_{j+1}} \frac{\partial u}{\partial x}(x, t_m) \, \mathrm{d}x \right|^2 \leq \int_0^1 \left| \frac{\partial u}{\partial x}(x, t_m) \right|^2 \mathrm{d}x.$$

By noting (4.58), we then have from the last inequality that

$$\| D_x^+ u^m \|_h \leq \| u_0' \|_{L_2(0,1)} \leq \| u_0 \|_{W_2^1(0,1)}.$$

An analogous inequality holds for $D_x^+ u^{m+1}$, and hence we obtain

$$\left\| D_x^+ \left(\frac{u^{m+1} + u^m}{2} \right) \right\|_h \leq \| u_0 \|_{W_2^1(0,1)}.$$

The solution of the finite difference scheme (4.41) (in the case of $f = 0$ and $u_1 = 0$) satisfies the a priori estimate (4.18), which implies that

$$\left\| D_x^+ \left(\frac{U^{m+1} + U^m}{2} \right) \right\|_h \leq \| D_t^+ U^0 \|_h + \left\| D_x^+ \left(\frac{U^1 + U^0}{2} \right) \right\|_h$$

$$= \frac{1}{2} \tau \| D_x^+ D_x^- u_0 \|_h + \left\| D_x^+ \left(u_0 + \frac{1}{4} \tau^2 D_x^+ D_x^- u_0 \right) \right\|_h$$

$$\leq C \| D_x^+ u_0 \|_h \leq C \| u_0' \|_{L_2(0,1)} \leq C \| u_0 \|_{W_2^1(0,1)}.$$

We thus deduce by applying a triangle inequality that

$$\| u - U \|_{C(\Omega^\tau, W_2^1(\Omega^h))} \leq C \| u_0 \|_{W_2^1(0,1)}, \quad u_0 \in \mathfrak{W}_2^1(0,1). \tag{4.62}$$

We are now ready to interpolate between (4.61) and (4.62). The linear space $\mathfrak{W}_2^s(0,1)$, equipped with the norm of $W_2^s(0,1)$, is a Banach space. Let us consider the mapping $A_h^\tau : u_0 \mapsto u - U$. Clearly, A_h^τ is linear, and (4.61) implies that the operator $A_h^\tau : \mathfrak{W}_2^4(0,1) \to \mathcal{B}_h^\tau$ is bounded, with

$$\|A_h^\tau\|_{\mathfrak{W}_2^4(0,1)\to\mathcal{B}_h^\tau} \le Ch^2.$$

Similarly, by (4.62), the linear operator $A_h^\tau : \mathfrak{W}_2^1(0,1) \to \mathcal{B}_h^\tau$ is bounded, with

$$\|A_h^\tau\|_{\mathfrak{W}_2^1(0,1)\to\mathcal{B}_h^\tau} \le C.$$

By function space interpolation (cf. Sect. 1.1.5) we deduce that the linear operator $A_h^\tau : (\mathfrak{W}_2^1(0,1), \mathfrak{W}_2^4(0,1))_{\theta,2} \to \mathcal{B}_h^\tau$ is also bounded, with

$$\|A_h^\tau\|_{(\mathfrak{W}_2^1(0,1),\mathfrak{W}_2^4(0,1))_{\theta,2}\to\mathcal{B}_h^\tau} \le Ch^{2\theta}, \quad 0 < \theta < 1,$$

which, by taking $s = 3\theta + 1$, yields the error bound

$$\|u - U\|_{C(\Omega^\tau, W_2^1(\Omega^h))} \le Ch^{\frac{2}{3}(s-1)} \|u_0\|_{(\mathfrak{W}_2^1(0,1),\mathfrak{W}_2^4(0,1))_{(s-1)/3,2}} \tag{4.63}$$

for $1 < s < 4$, where $C = C(s)$ is a positive constant, independent of h.

The inequality (4.63) guarantees convergence of the scheme under a less restrictive assumption on u_0 than (4.52) (i.e. even for $1 < s \le 2$, whereas in (4.52) $s > 2$ was needed; we refer to the notes at the end of this chapter for further comments on the error bound (4.63)).

4.4 Hyperbolic Problems with Variable Coefficients

4.4.1 Formulation of the Problem

Let us consider, in $Q := \Omega \times (0,T] = (0,1)^2 \times (0,T]$, the initial-boundary-value problem for a symmetric second-order hyperbolic equation with variable coefficients:

$$\begin{aligned} \frac{\partial^2 u}{\partial t^2} + \mathcal{L}u = f, &\qquad (x,t) = (x_1,x_2,t) \in Q, \\ u = 0, &\qquad (x,t) \in \Gamma \times (0,T] = \partial\Omega \times (0,T], \\ u(x,0) = u_0(x), &\qquad \frac{\partial u}{\partial t}(x,0) = u_1(x), \quad x \in \Omega, \end{aligned} \tag{4.64}$$

where

$$\mathcal{L}u := -\sum_{i,j=1}^{2} \partial_i(a_{ij}\partial_j u) + au, \quad \partial_i = \frac{\partial}{\partial x_i}.$$

We shall suppose that the solution of (4.64) belongs to the Sobolev space $W_2^s(Q)$, $2 < s \le 4$, and the coefficients $a_{ij} = a_{ij}(x)$ and $a = a(x)$ satisfy the conditions

$$a_{ij} \in W_2^{s-1}(\Omega), \quad a_{ij} = a_{ji},$$
$$\exists c_0 > 0 \ \forall x \in \Omega \ \forall \xi \in \mathbb{R}^2 : \quad \sum_{i,j=1}^{2} a_{ij}(x)\xi_i\xi_j \ge c_0 \sum_{i=1}^{2} \xi_i^2,$$
$$a \in W_2^{s-2}(\Omega), \quad a(x) \ge 0 \quad \text{a.e. in } \Omega.$$

These conditions ensure that the coefficients of the scheme belong to appropriate spaces of multipliers; that is,

$$a_{ij} \in M\big(W_2^{s-1}(Q)\big), \quad a \in M\big(W_2^s(Q) \to W_2^{s-2}(Q)\big).$$

4.4.2 The Finite Difference Scheme

Let $N, M \in \mathbb{N}$, $N, M \ge 2$, $h := 1/N$ and $\tau := T/M$. We introduce the uniform mesh Ω^h with mesh-size h in Ω and the uniform mesh Ω^τ with mesh-size τ on $(0, T)$. Using the notations from Sects. 2.6, 3.1.4 and 3.4, we define $Q_h^\tau := \Omega^h \times \Omega^\tau$ and $\overline{Q}_h^\tau := \overline{\Omega}^h \times \overline{\Omega}^\tau$. It will be assumed that the mesh-sizes h and τ are linked by the condition (4.46):

$$c_1 h \le \tau \le c_2 h, \quad c_1, c_2 = \text{Const.} > 0.$$

For a function V defined on $\overline{Q}_h^\tau$ we consider the divided differences $D_{x_i}^{\pm} V$ (see Sect. 2.6) and $D_t^{\pm} V$ (see Sect. 3.1.4), the Steklov mollifier T_i in the x_i direction (see Sect. 2.6), and the mollifiers T_t, $\tilde{T}_t^{\pm}$ in the t direction (see Sect. 4.3).

The initial-boundary-value problem (4.64) will be approximated on $\overline{Q}_h^\tau$ by the finite difference scheme

$$\begin{aligned} D_t^+ D_t^- U + \frac{1}{4}\mathcal{L}_h(\hat{U} + 2U + \check{U}) &= T_1 T_2 T_t f \quad \text{in } Q_h^\tau, \\ U &= 0 \quad \text{on } \Gamma^h \times \overline{\Omega}^\tau, \\ U &= u_0 \quad \text{on } \Omega^h \times \{0\}, \end{aligned} \tag{4.65}$$

with

$$\hat{U} := u_0 + \tau T_1 T_2 u_1 + \frac{\tau^2}{2}\big(-\mathcal{L}_h u_0 + T_1 T_2 \tilde{T}_t^+ f\big) \quad \text{on } \Omega^h \times \{0\},$$

where

$$\mathcal{L}_h U := -\frac{1}{2} \sum_{i,j=1}^{2} \big[D_{x_i}^+\big(a_{ij} D_{x_j}^- U\big) + D_{x_i}^-\big(a_{ij} D_{x_j}^+ U\big)\big] + (T_1 T_2 a) U.$$

The scheme (4.65) is a standard symmetric finite difference scheme with weights $\theta_1 = \theta_2 = 1/4$ (see Samarskiĭ [159]) and mollified right-hand side and lowest coefficient. According to the results obtained in Sect. 4.1.3, the finite difference scheme (4.65) is unconditionally stable.

4.4.3 Convergence of the Finite Difference Scheme

Let u be the solution of the initial-boundary-value problem (4.64) and let U denote the solution of the difference scheme (4.65). The global error $e := u - U$ satisfies the following finite difference scheme:

$$\begin{aligned} &D_t^+ D_t^- e + \frac{1}{4}\mathcal{L}_h(\hat{e} + 2e + \check{e}) = \varphi \quad \text{in } Q_h^\tau, \\ &e = 0 \quad \text{on } \Gamma^h \times \overline{\Omega}^\tau, \\ &e = 0, \qquad \hat{e} = \tau\upsilon + \frac{1}{2}\tau^2\tilde{\varphi} \quad \text{in } \Omega^h \times \{0\}, \end{aligned} \tag{4.66}$$

where

$$\begin{aligned} \varphi &:= \sum_{i,j=1}^{2} \eta_{ij} + \eta + \zeta + \chi, \\ \eta_{ij} &:= T_1T_2T_t\partial_i(a_{ij}\partial_j u) - \frac{1}{2}\big[D_{x_i}^-(a_{ij}D_{x_j}^+ u) + D_{x_i}^+(a_{ij}D_{x_j}^- u)\big], \\ \eta &:= (T_1T_2a)u - T_1T_2T_t(au), \\ \zeta &:= D_t^+D_t^- u - T_1T_2T_t\frac{\partial^2 u}{\partial t^2}, \\ \chi &:= \frac{\tau^2}{4}\mathcal{L}_h D_t^+D_t^- u, \\ \upsilon &:= \left(D_t^+ u - T_1T_2\frac{\partial u}{\partial t} - \frac{\tau}{2}T_1T_2\tilde{T}_t^+\frac{\partial^2 u}{\partial t^2}\right)\bigg|_{t=0}, \\ \tilde{\varphi} &:= \sum_{i,j=1}^{2} \tilde{\eta}_{ij} + \tilde{\eta}, \\ \tilde{\eta}_{ij} &:= \left\{T_1T_2\tilde{T}_t^+\partial_i(a_{ij}\partial_j u) - \frac{1}{2}\big[D_{x_i}^-(a_{ij}D_{x_j}^+ u) + D_{x_i}^+(a_{ij}D_{x_j}^- u)\big]\right\}\bigg|_{t=0}, \\ \tilde{\eta} &:= \big[(T_1T_2a)u - T_1T_2\tilde{T}_t^+(au)\big]\big|_{t=0}. \end{aligned}$$

Let us define the norm

$$\|V\|_{2,\infty,h\tau}^{(1)} := \max_{t\in\Omega_-^\tau}\left\{\left\|D_t^+ V(\cdot,t)\right\|_h^2 + \sum_{i=1}^{2}\left\|D_{x_i}^+\left(\frac{V(\cdot,t+\tau)+V(\cdot,t)}{2}\right)\right\|_{i,h}^2\right\}^{1/2}.$$

We deduce from (4.18) and (4.66), using the relations

$$\|V\|_{\mathcal{L}_h}^2 = (\mathcal{L}_h V, V)_h \ge c_0\big(\|D_{x_1}^+ V\|_{1,h}^2 + \|D_{x_2}^+ V\|_{2,h}^2\big),$$

$$\|V\|_{\mathcal{L}_h}^2 \le C_0\big(\|D_{x_1}^+ V\|_{1,h}^2 + \|D_{x_2}^+ V\|_{2,h}^2 + \|V\|_h^2\big) \le \frac{C_1}{h^2}\|V\|_h^2,$$

that

$$\|e\|_{2,\infty,h\tau}^{(1)} \le C\left[\left(\frac{1}{\tau}+\frac{1}{h}\right)\|e(\cdot,\tau)\|_h + \tau\sum_{t\in\Omega^\tau}\|\varphi(\cdot,t)\|_h\right].$$

Hence, by recalling the condition (4.46), we obtain the a priori bound

$$\|e\|_{2,\infty,h\tau}^{(1)} \le C\left(\|\upsilon\|_h + h\|\tilde{\varphi}\|_h + \tau\sum_{t\in\Omega^\tau}\|\varphi(\cdot,t)\|_h\right). \tag{4.67}$$

Thus the problem of error estimation is reduced to bounding the right-hand side of (4.67); we shall accomplish this in the proof of the next theorem.

Theorem 4.4 *Let the solution u of* (4.64) *belong to the Sobolev space $W_2^s(Q)$, $2 < s \le 4$, and suppose that $a_{ij} \in W_2^{s-1}(\Omega)$ and $a \in W_2^{s-2}(\Omega)$. Let also $c_1 h \le \tau \le c_2 h$, where $c_2 \ge c_1 > 0$. Then, the finite difference scheme* (4.65) *converges in the mesh-dependent norm $\|\cdot\|_{2,\infty,h\tau}^{(1)}$, and the global error of the scheme is bounded as follows: there exists a positive constant $C = C(s)$, independent of h, such that*

$$\|u-U\|_{2,\infty,h\tau}^{(1)} \le Ch^{s-2}\Big(\max_{i,j}\|a_{ij}\|_{W_2^{s-1}(\Omega)} + \|a\|_{W_2^{s-2}(\Omega)}\Big)\|u\|_{W_2^s(Q)},$$
$$\text{for } 2 < s \le 4. \tag{4.68}$$

Proof First of all, we decompose η_{ij} as follows:

$$\eta_{ij} = \eta_{ij1}+\eta_{ij2}+\eta_{ij3}+\eta_{ij4}+\eta_{ij5}+\eta_{ij6}+\eta_{ij7},$$

$$\eta_{ij1} := T_1T_2T_t(a_{ij}\partial_i\partial_j u) - (T_1T_2a_{ij})(T_1T_2T_t\partial_i\partial_j u),$$

$$\eta_{ij2} := (T_1T_2a_{ij})\left[T_1T_2T_t\partial_i\partial_j u - \frac{1}{2}\big(D_{x_i}^- D_{x_j}^+ u + D_{x_i}^+ D_{x_j}^- u\big)\right],$$

$$\eta_{ij3} := \frac{1}{2}(T_1T_2a_{ij} - a_{ij})\big(D_{x_i}^- D_{x_j}^+ u + D_{x_i}^+ D_{x_j}^- u\big),$$

$$\eta_{ij4} := T_1T_2T_t(\partial_i a_{ij}\partial_j u) - (T_1T_2\partial_i a_{ij})(T_1T_2T_t\partial_j u),$$

$$\eta_{ij5} := \left[T_1T_2\partial_i a_{ij} - \frac{1}{2}\big(D_{x_i}^+ a_{ij} + D_{x_i}^- a_{ij}\big)\right](T_1T_2T_t\partial_j u),$$

$$\eta_{ij6} := \frac{1}{2}\big(D_{x_i}^+ a_{ij} + D_{x_i}^- a_{ij}\big)\left[T_1T_2T_t\partial_j u - \frac{1}{2}\big(D_{x_j}^+ u^{-i} + D_{x_j}^- u^{+i}\big)\right],$$

$$\eta_{ij7} := \frac{1}{4}\big(D_{x_i}^+ a_{ij} - D_{x_i}^- a_{ij}\big)\big(D_{x_j}^+ u^{-i} - D_{x_j}^- u^{+i}\big).$$

Let us also define

$$\chi := \sum_{i,j=1}^{2}(\chi_{ij1} + \chi_{ij2}) + \chi_0 \quad \text{and} \quad \eta := \eta_1 + \eta_2,$$

where

$$\chi_{ij1} := -\frac{1}{8}\tau^2\big(a_{ij}^{-i} D_{x_i}^- D_{x_j}^+ D_t^+ D_t^- u + a_{ij}^{+i} D_{x_i}^+ D_{x_j}^- D_t^+ D_t^- u\big),$$

$$\chi_{ij2} := -\frac{1}{8}\tau^2\big[\big(D_{x_i}^- a_{ij}\big)D_{x_j}^+ D_t^+ D_t^- u + \big(D_{x_i}^+ a_{ij}\big)D_{x_j}^- D_t^+ D_t^- u\big],$$

$$\chi_0 := \frac{1}{4}\tau^2(T_1T_2a)D_t^+ D_t^- u,$$

$$\eta_1 := (T_1T_2a)(u - T_1T_2T_t u),$$

$$\eta_2 := (T_1T_2a)(T_1T_2T_t u) - T_1T_2T_t(au).$$

Analogously, for $t = 0$ we set

$$\tilde{\eta}_{ij} := \tilde{\eta}_{ij1} + \tilde{\eta}_{ij2} + \tilde{\eta}_{ij3} + \tilde{\eta}_{ij4} + \tilde{\eta}_{ij5} + \tilde{\eta}_{ij6} + \tilde{\eta}_{ij7} \quad \text{and} \quad \tilde{\eta} := \tilde{\eta}_1 + \tilde{\eta}_2,$$

where

$$\tilde{\eta}_{ij1} := T_1T_2\tilde{T}_t^+(a_{ij}\partial_i\partial_j u) - (T_1T_2a_{ij})\big(T_1T_2\tilde{T}_t^+\partial_i\partial_j u\big),$$

$$\tilde{\eta}_{ij2} := (T_1T_2a_{ij})\left[T_1T_2\tilde{T}_t^+\partial_i\partial_j u - \frac{1}{2}\big(D_{x_i}^- D_{x_j}^+ u + D_{x_i}^+ D_{x_j}^- u\big)\right],$$

$$\tilde{\eta}_{ij4} := T_1T_2\tilde{T}_t^+(\partial_i a_{ij}\partial_j u) - (T_1T_2\partial_i a_{ij})\big(T_1T_2\tilde{T}_t^+\partial_j u\big),$$

$$\tilde{\eta}_{ij5} := \left[T_1T_2\partial_i a_{ij} - \frac{1}{2}\big(D_{x_i}^+ a_{ij} + D_{x_i}^- a_{ij}\big)\right]\big(T_1T_2\tilde{T}_t^+\partial_j u\big),$$

$$\tilde{\eta}_{ij6} := \frac{1}{2}\big(D_{x_i}^+ a_{ij} + D_{x_i}^- a_{ij}\big)\left[T_1T_2\tilde{T}_t^+\partial_j u - \frac{1}{2}\big(D_{x_j}^+ u^{-i} + D_{x_j}^- u^{+i}\big)\right],$$

$$\tilde{\eta}_{ijl} := \eta_{ijl}, \quad l = 3 \text{ and } 7,$$

$$\tilde{\eta}_1 := (T_1T_2a)\big(u - T_1T_2\tilde{T}_t^+ u\big),$$

$$\tilde{\eta}_2 := (T_1T_2a)\big(T_1T_2\tilde{T}_t^+ u\big) - T_1T_2\tilde{T}_t^+(au).$$

The values of η_{ij1}, η_{ij3}, η_{ij6} and η_{ij7} at the mesh-point $(x,t) \in Q_h^\tau$ are bounded bilinear functionals of $(a_{ij}, u) \in W_q^\lambda(K^0) \times W_{2q/(q-2)}^\mu(G)$, where $K^0 = K^0(x) := (x_1 - h, x_1 + h) \times (x_2 - h, x_2 + h)$ and $G = G(x,t) := K^0 \times (t - \tau, t + \tau)$. For η_{ij1} we have $\lambda \geq 0$, $\mu \geq 2$ and $q \geq 2$, while for η_{ij3}, η_{ij6} and η_{ij7} we have that $\lambda > 2/q$, $\mu > 3/2 - 3/q$ and $q \geq 2$. Furthermore η_{ij1} and η_{ij6} vanish whenever a_{ij} is a constant or if u is a quadratic polynomial; η_{ij3} and η_{ij7} vanish if a_{ij} and u are linear polynomials. By applying the bilinear version of the Bramble–Hilbert lemma we obtain the bound

$$|\eta_{ij1}(x,t)| \leq C(h)|a_{ij}|_{W_q^\lambda(K^0)}|u|_{W_{2q/(q-2)}^\mu(G)},$$

where $C(h) = Ch^{\lambda+\mu+1/q-7/2}$, $0 \leq \lambda \leq 1$ and $2 \leq \mu \leq 3$. Summing through the points of the mesh Q_h^τ yields

$$\tau \sum_{t \in \Omega^\tau} \|\eta_{ij1}(\cdot,t)\|_h \leq Ch^{\lambda+\mu-2}\|a_{ij}\|_{W_q^\lambda(\Omega)}\|u\|_{W_{2q/(q-2)}^\mu(Q)},$$
$$0 \leq \lambda \leq 1, \quad 2 \leq \mu \leq 3. \tag{4.69}$$

The following embeddings hold

$$W_2^{\lambda+\mu-1}(\Omega) \hookrightarrow W_q^\lambda(\Omega) \quad \text{for } \mu \geq 2 - 2/q, \tag{4.70}$$

and

$$W_2^{\lambda+\mu}(Q) \hookrightarrow W_{2q/(q-2)}^\mu(Q) \quad \text{for } \lambda \geq 3/q. \tag{4.71}$$

Setting $\lambda + \mu = s$, (4.69)–(4.71) imply, for $q > 3$, that

$$\tau \sum_{t \in \Omega^\tau} \|\eta_{ij1}(\cdot,t)\|_h \leq Ch^{s-2}\|a_{ij}\|_{W_2^{s-1}(\Omega)}\|u\|_{W_2^s(Q)},$$
$$\text{for } 2 + 3/q \leq s \leq 4. \tag{4.72}$$

The bound (4.72) holds for any $q > 3$; thus, letting $q \to \infty$ we deduce that it holds for $2 < s \leq 4$.

In the same way, η_{ij6} satisfies a bound of the form (4.69) for $2/q < \lambda \leq 1$ and $3/2 - 3/q < \mu \leq 3$. By setting $\lambda + \mu = s$ and noting the embeddings (4.70) and (4.71) we obtain

$$\tau \sum_{t \in \Omega^\tau} \|\eta_{ij6}(\cdot,t)\|_h \leq Ch^{s-2}\|a_{ij}\|_{W_2^{s-1}(\Omega)}\|u\|_{W_2^s(Q)},$$
$$\text{for } 2 + 1/q \leq s \leq 4. \tag{4.73}$$

In fact, because q is arbitrary, (4.73) holds for $2 < s \leq 4$. The term χ_{ij2} is bounded in the same way.

The terms η_{ij3} and η_{ij7} satisfy bounds of the form (4.69) for $2/q < \lambda \le 2$ and $3/2 - 3/q < \mu \le 2$. Thus, in the same way as in the previous cases,

$$\tau \sum_{t\in\Omega^\tau} \left(\left\| \eta_{ij3}(\cdot,t) \right\|_h + \left\| \eta_{ij7}(\cdot,t) \right\|_h \right) \le Ch^{s-2} \|a_{ij}\|_{W_2^{s-1}(\Omega)} \|u\|_{W_2^s(Q)}, \quad (4.74)$$

for $2 + 1/q \le s \le 4$, and thus for $2 < s \le 4$ as well.

When $a_{ij} \in L_\infty(K^0)$, $\eta_{ij2}(x,t)$ is a bounded linear functional of $u \in W_2^s(G)$, $s \ge 2$, which vanishes whenever u is a cubic polynomial. Using the Bramble–Hilbert lemma we obtain

$$\left| \eta_{ij2}(x,t) \right| \le Ch^{s-7/2} \|a_{ij}\|_{L_\infty(K^0)} |u|_{W_2^s(G)}, \quad 2 \le s \le 4.$$

Summing over the points of the mesh Q_h^τ yields

$$\tau \sum_{t\in\Omega^\tau} \left\| \eta_{ij2}(\cdot,t) \right\|_h \le Ch^{s-2} \|a_{ij}\|_{L_\infty(\Omega)} \|u\|_{W_2^s(Q)}, \quad 2 \le s \le 4.$$

Finally, by noting the embedding

$$W_2^{s-1}(\Omega) \hookrightarrow L_\infty(\Omega) \quad \text{for } s > 2,$$

we obtain the desired bound on η_{ij2}:

$$\tau \sum_{t\in\Omega^\tau} \left\| \eta_{ij2}(\cdot,t) \right\|_h \le Ch^{s-2} \|a_{ij}\|_{W_2^{s-1}(\Omega)} \|u\|_{W_2^s(Q)}, \quad 2 < s \le 4. \quad (4.75)$$

The term χ_{ij1} is estimated in the same way.

Further, $\eta_{ij4}(x,t)$ and $\eta_{ij5}(x,t)$ are bounded bilinear functionals of $(a_{ij}, T_t u) \in W_q^\lambda(K^0) \times W_{2q/(q-2)}^\mu(K^0)$, for $\lambda \ge 1$, $\mu \ge 1$, $q \ge 2$. Moreover, η_{ij4} vanishes when a_{ij} or $T_t u$ is a linear polynomial; η_{ij5} vanishes if a_{ij} is a quadratic polynomial, or if $T_t u$ is a constant. By applying the bilinear version of the Bramble–Hilbert lemma we get

$$\left| \eta_{ij4}(x,t) \right| \le Ch^{\lambda+\mu-3} |a_{ij}|_{W_q^\lambda(K^0)} \left| T_t u(\cdot,t) \right|_{W_{2q/(q-2)}^\mu(K^0)}, \quad 1 \le \lambda,\ \mu \le 2,$$

and summing through the mesh-points then yields that

$$\left\| \eta_{ij4}(\cdot,t) \right\|_h \le Ch^{\lambda+\mu-2} \|a_{ij}\|_{W_q^\lambda(\Omega)} \left\| T_t u(\cdot,t) \right\|_{W_{2q/(q-2)}^\mu(\Omega)}, \quad 1 \le \lambda,\ \mu \le 2. \quad (4.76)$$

By noting the embedding

$$W_2^{\lambda+\mu}(\Omega) \hookrightarrow W_{2q/(q-2)}^\mu(\Omega) \quad \text{for } \lambda \ge 2/q \quad (4.77)$$

and setting $\lambda + \mu = s$, from (4.76), (4.70) and (4.77) we obtain

$$\left\| \eta_{ij4}(\cdot,t) \right\|_h \le Ch^{s-2} \|a_{ij}\|_{W_2^{s-1}(\Omega)} \left\| T_t u(\cdot,t) \right\|_{W_2^s(\Omega)}, \quad 3 - 2/q \le s \le 4. \quad (4.78)$$

Letting $q \to 2+0$ we deduce that (4.78) holds for $2 < s \le 4$. Since

$$\tau \sum_{t\in\Omega^\tau} \left\| T_t u(\cdot,t)\right\|_{W_2^s(\Omega)} \le T^{1/2}\left(\tau \sum_{t\in\Omega^\tau} \left\| T_t u(\cdot,t)\right\|^2_{W_2^s(\Omega)}\right)^{1/2} \le C\|u\|_{W_2^s(Q)},$$

summation of the inequality (4.78) through $t \in \Omega^\tau$ yields that

$$\tau \sum_{t\in\Omega^\tau} \left\|\eta_{ij4}(\cdot,t)\right\|_h \le Ch^{s-2}\|a_{ij}\|_{W_2^{s-1}(\Omega)}\|u\|_{W_2^s(Q)}, \quad 2 < s \le 4. \tag{4.79}$$

Similarly, by the Bramble–Hilbert lemma, we obtain a bound on η_{ij5} of the form (4.76) for $\mu = 1$ and $1 \le \lambda \le 3$. Setting $q = 2$, $\lambda = s-1$ and using the embedding

$$W_2^s(\Omega) \hookrightarrow W_\infty^1(\Omega) \quad \text{for } s > 2,$$

we obtain the bound

$$\left\|\eta_{ij5}(\cdot,t)\right\|_h \le Ch^{s-2}\|a_{ij}\|_{W_2^{s-1}(\Omega)}\left\|T_t u(\cdot,t)\right\|_{W_2^s(\Omega)}, \quad 2 < s \le 4,$$

and further

$$\tau \sum_{t\in\Omega^\tau} \left\|\eta_{ij5}(\cdot,t)\right\|_h \le Ch^{s-2}\|a_{ij}\|_{W_2^{s-1}(\Omega)}\|u\|_{W_2^s(Q)}, \quad 2 < s \le 4. \tag{4.80}$$

The terms $\chi_0(x,t)$ and $\eta_1(x,t)$ are bounded bilinear functionals of $(a,u) \in L_q(K^0) \times W^\mu_{2q/(q-2)}(G)$, for $\mu > 3/2 - 3/q$, $q \ge 2$, which vanish when u is a linear polynomial. By applying the Bramble–Hilbert lemma,

$$|\chi_0|, |\eta_1| \le Ch^{\mu+1/q-3/2}\|a\|_{L_q(K^0)}|u|_{W^\mu_{2q/(q-2)}(G)}, \quad 3/2 - 3/q < \mu \le 2.$$

Thus, by summing through the points of the mesh Q_h^τ, we get

$$\tau \sum_{t\in\Omega^\tau} \left(\left\|\chi_0(\cdot,t)\right\|_h + \left\|\eta_1(\cdot,t)\right\|_h\right) \le Ch^\mu\|a\|_{L_q(\Omega)}\|u\|_{W^\mu_{2q/(q-2)}(Q)},$$
$$3/2 - 3/q < \mu \le 2.$$

By choosing $\mu = s-2$ and noting the embeddings

$$W_2^s(Q) \hookrightarrow W^{s-2}_{2q/(q-2)}(Q) \quad \text{for } q > 2 \quad \text{and} \quad W_2^{s-2}(\Omega) \hookrightarrow L_q(\Omega),$$

where $2 \le q < 2/(3-s)$ for $2 < s < 3$, $2 \le q < \infty$ for $s = 3$, and q is arbitrary for $s > 3$, we obtain the bound

$$\tau \sum_{t\in\Omega^\tau} \left(\left\|\chi_0(\cdot,t)\right\|_h + \left\|\eta_1(\cdot,t)\right\|_h\right) \le Ch^{s-2}\|a\|_{W_2^{s-2}(\Omega)}\|u\|_{W_2^s(Q)},$$
$$7/2 - 3/q < s \le 4. \tag{4.81}$$

Since $q > 2$ is arbitrary, letting $q \to 2+0$, we deduce that (4.81) holds for $2 < s \le 4$.

Next, $\eta_2(x,t)$ is a bounded bilinear functional of $(a,u) \in W_q^\lambda(K^0) \times W^\mu_{2q/(q-2)}(G)$ for $\lambda \ge 0$, $\mu \ge 0$, $q \ge 2$, which vanishes if a or u is a constant function. Similarly as in the previous case we have that

$$\big|\eta_2(x,t)\big| \le Ch^{\lambda+\mu+1/q-3/2}\|a\|_{L_q(K^0)}|u|_{W^\mu_{2q/(q-2)}(G)}, \quad 0 \le \lambda,\ \mu \le 1,$$

and

$$\tau \sum_{t\in\Omega^\tau} \big\|\eta_2(\cdot,t)\big\|_h \le Ch^{\lambda+\mu}\|a\|_{W_q^\lambda(\Omega)}\|u\|_{W^\mu_{2q/(q-2)}(Q)}, \quad 0 \le \lambda,\ \mu \le 1.$$

Letting $\lambda + \mu = s - 2$ and using the embeddings

$$W_2^{\lambda+\mu}(\Omega) \hookrightarrow W_q^\lambda(\Omega) \quad \text{for } \mu \ge 1 - 2/q \tag{4.82}$$

and

$$W_2^{\lambda+\mu+2}(Q) \hookrightarrow W^\mu_{2q/(q-2)}(Q), \tag{4.83}$$

we get the bound

$$\tau \sum_{t\in\Omega^\tau} \big\|\eta_2(\cdot,t)\big\|_h \le Ch^{s-2}\|a\|_{W_2^{s-2}(\Omega)}\|u\|_{W_2^s(Q)},$$
$$3 - 2/q \le s \le 4. \tag{4.84}$$

Letting $q \to 2+0$, we deduce that (4.84) holds for $2 < s \le 4$.

The term $\zeta(x,t)$ is a bounded linear functional of $u \in W_2^s(G)$ for $s > 2$. Moreover, ζ vanishes on cubic polynomials. By applying the Bramble–Hilbert lemma and summing through the mesh-points, we get the bound

$$\tau \sum_{t\in\Omega^\tau} \big\|\zeta(\cdot,t)\big\|_h \le Ch^{s-2}\|u\|_{W_2^s(Q)}, \quad 2 < s \le 4. \tag{4.85}$$

Analogously, $\upsilon(x)$ is a bounded linear functional of $u \in W_2^s(G^{0+})$, where $G^{0+} = G^{0+}(x) := K^0 \times (0,\tau)$, for $s > 2$. Moreover, $\upsilon(x)$ vanishes on quadratic polynomials. By applying the Bramble–Hilbert lemma we get

$$\|\upsilon\|_h \le Ch^{s-3/2}|u|_{W_2^s(Q_\tau)}, \quad 2 < s \le 3,$$

where $Q_\tau := \Omega \times (0,\tau)$. Setting $s = 3$ and using the inequality (2.199) we deduce that

$$\begin{aligned}\|\upsilon\|_h &\le Ch^{3/2}|u|_{W_2^3(Q_\tau)}\\ &\le Ch^{3/2+\min(s'-3,1/2)}|\log h|^{1-|\operatorname{sgn}(s'-7/2)|}\|u\|_{W_2^{s'}(Q)}, \quad 3 < s' \le 4.\end{aligned}$$

By combining the bounds above on $\|v\|_h$ we then have that

$$\|v\|_h \leq Ch^{s-2}|u|_{W_2^s(Q_\tau)}, \quad 2 < s \leq 4. \tag{4.86}$$

The terms $\tilde{\eta}_{ijl}$ $(l = 1,3,4,5,6,7)$ have analogous properties to η_{ijl} in the domain G^{0+}, and are bounded in the same way. Thus, for example, $\tilde{\eta}_{ij1}(x)$ is a bounded bilinear functional of $(a_{ij}, u) \in W_q^\lambda(K^0) \times W_{2q/(q-2)}^\mu(G^{0+})$, for $\lambda \geq 0$, $\mu \geq 2$, $q \geq 2$, which vanishes whenever a_{ij} is a constant or if u is a quadratic polynomial. By applying the bilinear version of the Bramble–Hilbert lemma we obtain the bound

$$\left|\tilde{\eta}_{ij1}(x)\right| \leq Ch^{\lambda+\mu+1/q-7/2}|a_{ij}|_{W_q^\lambda(K^0)}|u|_{W_{2q/(q-2)}^\mu(G^{0+})}.$$

Summation over the mesh-points then yields

$$\begin{aligned} h\|\tilde{\eta}_{ij1}\|_h &\leq Ch^{\lambda+\mu+1/q-3/2}\|a_{ij}\|_{W_q^\lambda(\Omega)}\|u\|_{W_{2q/(q-2)}^\mu(Q_\tau)} \\ &\leq Ch^{\lambda+\mu-2}\|a_{ij}\|_{W_q^\lambda(\Omega)}\|u\|_{W_{2q/(q-2)}^\mu(Q)}. \end{aligned}$$

Thus, by setting $\lambda + \mu = s$ and using the embeddings (4.70) and (4.71), we obtain

$$h\|\tilde{\eta}_{ij1}\|_h \leq Ch^{s-2}\|a_{ij}\|_{W_2^{s-1}(\Omega)}\|u\|_{W_2^s(Q)}, \quad 2 \leq s \leq 4. \tag{4.87}$$

The terms $\tilde{\eta}_{ijl}$, with $l = 3, \ldots, 7$, are bounded in the same way.

When $a_{ij} \in L_\infty(K^0)$, $\tilde{\eta}_{ij2}(x)$ is a bounded linear functional of $u \in W_2^s(G^{0+})$, $s \geq 2$, which vanishes whenever u is a quadratic polynomial. Using the Bramble–Hilbert lemma we obtain

$$\left|\eta_{ij2}(x)\right| \leq Ch^{s-7/2}\|a_{ij}\|_{L_\infty(K^0)}|u|_{W_2^s(G^{0+})}, \quad 2 \leq s \leq 3.$$

Summing over the points of the mesh Ω^h yields that

$$h\left\|\eta_{ij2}(\cdot,t)\right\|_h \leq Ch^{s-3/2}\|a_{ij}\|_{L_\infty(\Omega)}|u|_{W_2^s(Q_\tau)}, \quad 2 \leq s \leq 3.$$

Hence, by noting the embedding

$$W_2^{s-1}(\Omega) \hookrightarrow L_\infty(\Omega) \quad \text{for } s > 2$$

and the inequality (2.199), in the same way as in the estimation of $\|v\|_h$, we obtain that

$$h\left\|\eta_{ij2}(\cdot,t)\right\|_h \leq Ch^{s-2}\|a_{ij}\|_{L_\infty(\Omega)}\|u\|_{W_2^s(Q)}, \quad 2 \leq s \leq 4. \tag{4.88}$$

The term $\tilde{\eta}_1(x)$ is bounded directly. If $a \in L_2(\Omega)$ and $u \in C(\overline{Q})$, then

$$\left|\tilde{\eta}_1(x)\right| \leq Ch^{-1}\|a\|_{L_2(K^0)}\|u\|_{C(\overline{Q})},$$

which implies that

$$h\|\tilde{\eta}_1\|_h \le Ch\|a\|_{L_2(\Omega)}\|u\|_{C(\overline{Q})}.$$

Hence, by using the embeddings

$$W_2^{s-2}(\Omega) \hookrightarrow L_2(\Omega), \qquad W_2^s(Q) \hookrightarrow C(\overline{Q}), \quad s > 2,$$

we get, after further majorization, that

$$h\|\tilde{\eta}_1\|_h \le Ch^{s-2}\|a\|_{W_2^{s-2}(\Omega)}\|u\|_{W_2^s(Q)}, \quad 2 < s \le 3. \tag{4.89}$$

Similarly, if $a \in L_2(\Omega)$ and $u \in C^1(\overline{Q})$, then

$$\left|\tilde{\eta}_1(x)\right| \le C\|a\|_{L_2(K^0)}\|u\|_{C^1(\overline{Q})},$$

which further yields that

$$h\|\tilde{\eta}_1\|_h \le Ch^2\|a\|_{L_2(\Omega)}\|u\|_{C^1(\overline{Q})}.$$

Hence, by the previous embeddings, together with

$$W_2^s(Q) \hookrightarrow C^1(\overline{Q}), \quad s > 3,$$

we obtain, after further majorization, that

$$h\|\tilde{\eta}_1\|_h \le Ch^{s-2}\|a\|_{W_2^{s-2}(\Omega)}\|u\|_{W_2^s(Q)}, \quad 3 < s \le 4. \tag{4.90}$$

Finally, $\tilde{\eta}_2(x)$ is a bounded bilinear functional of $(a, u) \in W_q^\lambda(K^0) \times W_{2q/(q-2)}^\mu(G^{0+})$ for $\lambda \ge 0$, $\mu \ge 0$, $q \ge 2$, which vanishes if a or u is a constant function. By applying the Bramble–Hilbert lemma we obtain

$$\left|\tilde{\eta}_2(x)\right| \le Ch^{\lambda+\mu+1/q-3/2}\|a\|_{L_q(K^0)}|u|_{W_{2q/(q-2)}^\mu(G^{0+})}, \quad 0 \le \lambda,\ \mu \le 1,$$

and

$$h\|\tilde{\eta}_2\|_h \le Ch^{\lambda+\mu+1/q+1/2}\|a\|_{W_q^\lambda(\Omega)}\|u\|_{W_{2q/(q-2)}^\mu(Q_\tau)}, \quad 0 \le \lambda,\ \mu \le 1.$$

Setting $\lambda + \mu = s - 2$, using the embeddings (4.82) and (4.83) and letting $q \to 2+0$, we get the bound

$$h\|\tilde{\eta}_2\|_h \le Ch^{s-2}\|a\|_{W_2^{s-2}(\Omega)}\|u\|_{W_2^s(Q)}, \quad 2 < s \le 4. \tag{4.91}$$

By combining (4.67), (4.72)–(4.75), (4.79)–(4.81) and (4.84)–(4.91) we then arrive at the error bound (4.68). □

Remark 4.1 Similar error bounds can be derived when the coefficients a_{ij} and a depend on t. However, the proof of an a priori bound of the form (4.67) is then more complicated (see Samarskiĭ [159]).

4.4.4 Factorized Scheme

As in the parabolic case (see Sect. 3.4.4) the finite difference scheme (4.65) can be replaced by a, more economical, factorized scheme, which is in this case

$$(I_h + \sigma\tau^2 \Lambda_1)(I_h + \sigma\tau^2 \Lambda_2) D_t^+ D_t^- U + \mathcal{L}_h U = T_1 T_2 T_t f \quad \text{in } Q_h^\tau, \qquad (4.92)$$

with the same initial and boundary conditions as in (4.65). Here σ is a positive real parameter, $\Lambda_i U = -D_{x_i}^- D_{x_i}^+ U$, $i = 1, 2$, and I_h is the identity operator. According to (4.17), the finite difference scheme (4.92) is stable if the operator

$$(I_h + \sigma\tau^2 \Lambda_1)(I_h + \sigma\tau^2 \Lambda_2) - \frac{1}{4}\tau^2 \mathcal{L}_h$$

is positive definite, uniformly with respect to the discretization parameters. This condition holds, for example, if

$$\sigma \geq \frac{1}{2} \max_{i,j} \|a_{ij}\|_{C(\overline{\Omega})}$$

and

$$h < 4c_2^{-2} \|a\|_{L_2(\Omega)}^{-1}.$$

In contrast with (4.65), the factorized scheme (4.92) is *economical* in the sense that only systems of linear algebraic equations with *tridiagonal* matrices have to be solved on each time level, corresponding to the operators $(I_h + \sigma\tau^2 \Lambda_i)$, $i = 1, 2$.

The global error $e := u - U$ satisfies the following equalities:

$$\begin{aligned}
&(I_h + \sigma\tau^2 \Lambda_1)(I_h + \sigma\tau^2 \Lambda_2) D_t^+ D_t^- e + \mathcal{L}_h e = \varphi' \quad \text{on } Q_h^\tau, \\
&e = 0 \quad \text{in } \Gamma^h \times \overline{\Omega}^\tau, \\
&e = 0, \qquad \hat{e} = \tau\upsilon + \frac{1}{2}\tau^2 \tilde{\varphi} \quad \text{on } \Omega^h \times \{0\},
\end{aligned}$$

where

$$\varphi' := \sum_{i,j=1}^{2} \eta_{ij} + \eta + \zeta + \chi',$$

and

$$\begin{aligned}
\chi' := &-\sigma\tau^2 (D_{x_1}^+ D_{x_1}^- D_t^+ D_t^- u + D_{x_2}^+ D_{x_2}^- D_t^+ D_t^- u) \\
&+ \sigma^2 \tau^4 D_{x_1}^+ D_{x_1}^- D_{x_2}^+ D_{x_2}^- D_t^+ D_t^- u.
\end{aligned}$$

The a priori estimate (4.67) holds if φ is replaced by φ'. It is easy to show that

$$\tau \sum_{t\in\Omega^\tau} \|\chi'\|_h \le Ch^{s-2}\|u\|_{W_2^s(Q)}, \quad 2 < s \le 4,$$

which implies that the factorized scheme (4.92) satisfies the error bound (4.68).

4.5 Hyperbolic Interface Problem

Using the notations from Sect. 4.4, we shall consider the following hyperbolic interface problem:

$$\begin{aligned} (1+k\delta_\Sigma)\frac{\partial^2 u}{\partial t^2} + \mathcal{L}u = f, \quad & (x,t) = (x_1,x_2,t) \in Q, \\ u = 0, \quad & (x,t) \in \Gamma \times (0,T] = \partial\Omega \times [0,T], \\ u(x,0) = u_0(x), \quad \frac{\partial u}{\partial t}(x,0) = u_1(x), \quad & x \in \Omega, \end{aligned} \tag{4.93}$$

where, as in the parabolic case (cf. Sect. 3.5), Σ is the intersection of the line segment $x_2 = \xi$, $0 < \xi < 1$, and $\overline{\Omega}$, $\delta_\Sigma(x) := \delta(x_2 - \xi)$ is the Dirac distribution concentrated on Σ, $k(x) = k(x_1)$, and $\mathcal{L}$ is the symmetric elliptic operator

$$\mathcal{L}u := -\sum_{i,j=1}^{2} \partial_i(a_{ij}\partial_j u) + au,$$

satisfying the usual ellipticity and regularity properties (see Sect. 4.4.1).

Problem (4.93) is a hyperbolic initial-boundary-value problem with "concentrated mass" in the coefficient of the time derivative. In the one-dimensional case, analogous problems were considered by Jovanović and Vulkov [93, 96].

When $f = f(x,t)$ does not include in its definition a Dirac function concentrated on Σ, (4.93) reduces to

$$\frac{\partial^2 u}{\partial t^2} + \mathcal{L}u = f(x,t) \quad \text{in } Q^- \cup Q^+,$$

where $Q^\pm = \Omega^\pm \times (0,T]$, $\Omega^- = (0,1)\times(0,\xi)$, $\Omega^+ = (0,1)\times(\xi,1)$, with the following transmission (conjugation) conditions on the interface Σ:

$$[u]_\Sigma := u(x_1,\xi+0,t) - u(x_1,\xi-0,t) = 0, \quad \left[\sum_{j=1}^{2} a_{2j}\partial_j u\right]_\Sigma = k\frac{\partial^2 u}{\partial t^2}\bigg|_\Sigma. \tag{4.94}$$

We shall now construct a finite difference approximation of this problem.

4.5.1 Finite Difference Approximation

The initial-boundary-value problem (4.93) will be approximated on $\overline{Q}_h^\tau$ by the finite difference scheme

$$
\begin{aligned}
&(1+k\delta_{\Sigma^h})D_t^+D_t^-U+\frac{1}{4}\mathcal{L}_h(\hat{U}+2U+\check{U})=T_1T_2T_tf \quad \text{in } Q_h^\tau,\\
&U=0 \quad \text{on } \Gamma^h\times\overline{\Omega}^\tau,\\
&U=u_0 \quad \text{on } \Omega^h\times\{0\}, \qquad\qquad (4.95)\\
&(1+k\delta_{\Sigma^h})D_t^+U=T_1T_2\big[(1+k\delta_\Sigma)u_1\big]\\
&\qquad\qquad +\frac{1}{2}\tau\big(-\mathcal{L}_hu_0+T_1T_2\tilde{T}_t^+f\big) \quad \text{on } \Omega^h\times\{0\},
\end{aligned}
$$

where the operator $\mathcal{L}_h$ is defined as in Sect. 4.4.2; i.e.

$$
\mathcal{L}_hU:=-\frac{1}{2}\sum_{i,j=1}^{2}\big[D_{x_i}^+\big(a_{ij}D_{x_j}^-U\big)+D_{x_i}^-\big(a_{ij}D_{x_j}^+U\big)\big]+(T_1T_2a)U;
$$

and

$$
\delta_{\Sigma^h}(x)=\delta_h(x_2-\xi):=\begin{cases}0 & \text{when } x\in\Omega^h\setminus\Sigma^h,\\ 1/h & \text{when } x\in\Sigma^h,\end{cases}
$$

is the discrete Dirac delta-function concentrated on Σ^h. For the sake of simplicity, we assume that ξ is a rational number and ξ/h is an integer.

The finite difference scheme (4.95) fits into the canonical form (4.14), with $A_h=\mathcal{L}_h$, $D_h=0$ and

$$
B_hU=(1+k\delta_{\Sigma^h})U+\frac{1}{4}\tau^2\mathcal{L}_hU.
$$

Hence, the conditions (4.17) are satisfied and the difference scheme (4.95) is therefore unconditionally stable.

Let u be the solution of the initial-boundary-value problem (4.93) and let U denote the solution of the difference scheme (4.95). The global error $e:=u-U$ is then a solution of the following finite difference scheme:

$$
\begin{aligned}
&(1+k\delta_{\Sigma^h})D_t^+D_t^-e+\frac{1}{4}\mathcal{L}_h(\hat{e}+2e+\check{e})=\varphi+\mu\delta_{\Sigma^h} \quad \text{in } Q_h^\tau,\\
&e=0 \quad \text{on } \Gamma^h\times\overline{\Omega}^\tau, \qquad\qquad (4.96)\\
&e=0, \qquad (1+k\delta_{\Sigma^h})D_t^+e=\upsilon+\tilde{\upsilon}\delta_{\Sigma^h}+\frac{1}{2}\tau\tilde{\varphi} \quad \text{in } \Omega^h\times\{0\},
\end{aligned}
$$

where

$$\varphi := \sum_{i,j=1}^{2} \eta_{ij} + \eta + \zeta + \chi,$$

$$\eta_{ij} := T_1T_2T_t\partial_i(a_{ij}\partial_j u) - \frac{1}{2}\big[D_{x_i}^-(a_{ij}D_{x_j}^+u) + D_{x_i}^+(a_{ij}D_{x_j}^-u)\big],$$

$$\eta := (T_1T_2a)u - T_1T_2T_t(au),$$

$$\zeta := D_t^+D_t^-u - T_1T_2T_t\frac{\partial^2 u}{\partial t^2},$$

$$\chi := \frac{\tau^2}{4}\mathcal{L}_h D_t^+D_t^-u = \sum_{i,j=1}^{2}\chi_{ij} + \chi_0,$$

$$\chi_{ij} := -\frac{\tau^2}{8}\big[D_{x_i}^-(a_{ij}D_{x_j}^+D_t^+D_t^-u) + D_{x_i}^+(a_{ij}D_{x_j}^-D_t^+D_t^-u)\big],$$

$$\chi_0 := \frac{\tau^2}{4}(T_1T_2a)u,$$

$$\mu := kD_t^+D_t^-u - T_1T_t\left(k\frac{\partial^2 u}{\partial t^2}\right)$$

$$\upsilon := \left(D_t^+u - T_1T_2\frac{\partial u}{\partial t} - \frac{\tau}{2}T_1T_2\tilde{T}_t^+\frac{\partial^2 u}{\partial t^2}\right)\bigg|_{t=0},$$

$$\tilde{\upsilon} := \left[kD_t^+u - T_1\left(k\frac{\partial u}{\partial t}\right) - \frac{\tau}{2}T_1\tilde{T}_t^+\left(k\frac{\partial^2 u}{\partial t^2}\right)\right]\bigg|_{t=0},$$

$$\tilde{\varphi} := \sum_{i,j=1}^{2}\tilde{\eta}_{ij} + \tilde{\eta},$$

$$\tilde{\eta}_{ij} := \left\{T_1T_2\tilde{T}_t^+\partial_i(a_{ij}\partial_j u) - \frac{1}{2}\big[D_{x_i}^-(a_{ij}D_{x_j}^+u) + D_{x_i}^+(a_{ij}D_{x_j}^-u)\big]\right\}\bigg|_{t=0},$$

$$\tilde{\eta} := \big[(T_1T_2a)u - T_1T_2\tilde{T}_t^+(au)\big]\big|_{t=0}.$$

Let us define the norm

$$\|V\|_{2,\infty,h\tau}^{(\tilde{1})} := \max_{t\in\Omega_-^\tau}\Bigg[\|D_t^+V(\cdot,t)\|_h^2 + \|D_t^+V(\cdot,t)\|_{L_2(\Sigma^h)}^2 + \sum_{i=1}^{2}\left\|D_{x_i}^+\left(\frac{V(\cdot,t+\tau)+V(\cdot,t)}{2}\right)\right\|_{i,h}^2\Bigg]^{1/2},$$

where the norm $\|V\|_{L_2(\Sigma^h)}$ has been introduced in Sect. 2.8.1. The a priori estimate (4.18) yields that

$$\|e\|^{(\tilde{1})}_{2,\infty,h\tau} \le C\bigg(\|\upsilon\|_{L_2(\Omega^h\setminus\Sigma^h)} + h\|\tilde{\varphi}\|_{L_2(\Omega^h\setminus\Sigma^h)} + \tau\sum_{t\in\Omega^\tau}\|\varphi\|_{L_2(\Omega^h\setminus\Sigma^h)} + h\|\upsilon\|_{L_2(\Sigma^h)} + \|\tilde{\upsilon}\|_{L_2(\Sigma^h)} + h^2\|\tilde{\varphi}\|_{L_2(\Sigma^h)} + h\tau\sum_{t\in\Omega^\tau}\|\varphi\|_{L_2(\Sigma^h)} + \tau\sum_{t\in\Omega^\tau}\|\mu\|_{L_2(\Sigma^h)}\bigg), \quad (4.97)$$

where we have used the notation

$$\|V\|^2_{L_2(\Omega^h\setminus\Sigma^h)} := h^2\sum_{x\in\Omega^h\setminus\Sigma^h} V^2(x).$$

Thus, in order to derive an error bound for the finite difference scheme (4.96) it suffices to bound the terms on the right-hand side of (4.97). As in the previous section, we shall assume for the sake of simplicity that $\tau \asymp h$.

Theorem 4.5 *Suppose that the solution u of* (4.93) *belongs to the Sobolev space $W_2^s(Q^\pm)$, $7/2 < s \le 4$, $a_{ij} \in W_2^{s-1}(\Omega^\pm)$, $a \in W_2^{s-2}(\Omega^\pm)$ and $k \in W_2^2(\Sigma)$. Let also $c_1 h \le \tau \le c_2 h$, where $c_2 \ge c_1 > 0$. Then, the finite difference scheme* (4.95) *converges in the mesh-dependent norm $\|\cdot\|^{(\tilde{1})}_{2,\infty,h\tau}$, and the following error bound holds for $7/2 < s \le 4$: there exists a positive constant $C = C(s)$, independent of h, such that*

$$\|u - U\|^{(\tilde{1})}_{2,\infty,h\tau} \le Ch^{s-2}\Big[\Big(\max_{i,j}\|a_{ij}\|_{W_2^{s-1}(\Omega^+)} + \|a\|_{W_2^{s-2}(\Omega^+)} + \|k\|_{W_2^2(\Sigma)}\Big)\|u\|_{W_2^s(Q^+)} + \Big(\max_{i,j}\|a_{ij}\|_{W_2^{s-1}(\Omega^-)} + \|a\|_{W_2^{s-2}(\Omega^-)} + \|k\|_{W_2^2(\Sigma)}\Big)\|u\|_{W_2^s(Q^-)}\Big]. \quad (4.98)$$

Proof The terms υ, $\tilde{\varphi}$ and φ for $x \notin \Sigma^h$ have been bounded in Sect. 4.4. After summation over the mesh-points we obtain

$$\|\upsilon\|_{L_2(\Omega^h\setminus\Sigma^h)} + h\|\tilde{\varphi}\|_{L_2(\Omega^h\setminus\Sigma^h)} + \tau\sum_{t\in\Omega^\tau}\|\varphi\|_{L_2(\Omega^h\setminus\Sigma^h)} \le Ch^{s-2}\Big[\Big(\max_{i,j}\|a_{ij}\|_{W_2^{s-1}(\Omega^+)} + \|a\|_{W_2^{s-2}(\Omega^+)}\Big)\|u\|_{W_2^s(Q^+)} + \Big(\max_{i,j}\|a_{ij}\|_{W_2^{s-1}(\Omega^-)} + \|a\|_{W_2^{s-2}(\Omega^-)}\Big)\|u\|_{W_2^s(Q^-)}\Big], \quad 2 < s \le 4. \quad (4.99)$$

The term υ for $x \in \Sigma^h$ can be represented as

$$\upsilon = \upsilon^+ + \upsilon^- =: \frac{1}{2}\Big(T_t \frac{\partial u}{\partial t} - T_1 \tilde{T}_2^+ \frac{\partial u}{\partial t}\Big)\Big|_{t=\tau/2} + \frac{1}{2}\Big(T_t \frac{\partial u}{\partial t} - T_1 \tilde{T}_2^- \frac{\partial u}{\partial t}\Big)\Big|_{t=\tau/2},$$

where the mollifiers $\tilde{T}_2^{\pm}$ are defined analogously to $\tilde{T}_t^{\pm}$:

$$\tilde{T}_2^+ v(x,t) := \frac{2}{h}\int_{x_2}^{x_2+h/2} v(x_1, x_2', t)\, dx_2',$$

$$\tilde{T}_2^- v(x,t) := \frac{2}{h}\int_{x_2-h/2}^{x_2} v(x_1, x_2', t)\, dx_2'.$$

The above representation implies that

$$|\upsilon^{\pm}| \le Ch\|u\|_{C^2(\overline{Q}^{\pm})} \le Ch\|u\|_{W_2^s(Q^{\pm})}, \quad s > 7/2,$$

whereby

$$h\|\upsilon\|_{L_2(\Sigma^h)} \le Ch^2\big(\|u\|_{W_2^s(Q^+)} + \|u\|_{W_2^s(Q^-)}\big), \quad s > 7/2. \tag{4.100}$$

The term $\tilde{\upsilon}$ can be represented as

$$\tilde{\upsilon}(x_1,\xi,t) = \frac{1}{h\tau}\int_0^{\tau}\int_{x_1-h/2}^{x_1+h/2}\Big[\int_{\tau/2}^{t'}\int_{\tau/2}^{t''}\Big(k\frac{\partial^3 u}{\partial t^3}\Big)(x_1',\xi,t''')\, dt'''\, dt''$$

$$- \int_{x_1}^{x_1'}\int_{x_1}^{x_1''}\frac{\partial^2}{\partial x_1^2}\Big(k\frac{\partial u}{\partial t}\Big)(x_1''',\xi,t')\, dx_1'''\, dx_1''\Big]\, dx_1'\, dt'.$$

Summing over Σ^h and using the inequality (2.199) we obtain

$$\|\tilde{\upsilon}\|_{L_2(\Sigma^h)} \le Ch^2\Bigg[\Big\|k\frac{\partial^3 u}{\partial t^3}\Big\|_{W_2^1((0,T),L_2(\Sigma))} + \Big\|\frac{\partial^2}{\partial x_1^2}\Big(k\frac{\partial u}{\partial t}\Big)\Big\|_{W_2^1((0,T),L_2(\Sigma))}\Bigg].$$

By expressing $k\frac{\partial^2 u}{\partial t^2}\big|_{\Sigma}$ from (4.94) and using the trace theorem for Sobolev spaces we finally obtain

$$\|\tilde{\upsilon}\|_{L_2(\Sigma^h)} \le Ch^2\Big[\Big(\|k\|_{W_2^2(\Sigma)} + \max_{i,j}\|a_{ij}\|_{W_2^{s-1}(\Omega^+)}\Big)\|u\|_{W_2^s(Q^+)}$$

$$+ \Big(\|k\|_{W_2^2(\Sigma)} + \max_{i,j}\|a_{ij}\|_{W_2^{s-1}(\Omega^-)}\Big)\|u\|_{W_2^s(Q^-)}\Big], \quad s > 7/2. \tag{4.101}$$

The term $\tilde{\varphi}$ can be estimated directly:

$$|\tilde{\varphi}| \le \max_{i,j}\|a_{ij}\|_{C^1(\overline{\Omega}^+)}\|u\|_{C^2(\overline{Q}^+)} + \|a\|_{C(\overline{\Omega}^+)}\|u\|_{C(\overline{Q}^+)}$$

$$+ \max_{i,j} \|a_{ij}\|_{C^1(\overline{\Omega}^-)} \|u\|_{C^2(\overline{Q}^-)} + \|a\|_{C(\overline{\Omega}^-)} \|u\|_{C(\overline{Q}^-)}.$$

Hence, by using the Sobolev embedding theorem, we have that

$$\|\tilde{\varphi}\|_{L_2(\Sigma^h)} \le C\Big[\Big(\max_{i,j} \|a_{ij}\|_{W_2^{s-1}(\Omega^+)} + \|a\|_{W_2^{s-2}(\Omega^+)}\Big) \|u\|_{W_2^s(Q^+)}$$
$$+ \Big(\max_{i,j} \|a_{ij}\|_{W_2^{s-1}(\Omega^-)} + \|a\|_{W_2^{s-2}(\Omega^-)}\Big) \|u\|_{W_2^s(Q^-)}\Big], \quad s > 7/2. \tag{4.102}$$

For $x \in \Sigma^h$ we define

$$\varphi := \frac{1}{2}\Bigg[\sum_{i,j=1}^{2} \big(\eta_{ij}^+ + \eta_{ij}^- + \chi_{ij}^+ + \chi_{ij}^-\big) + \eta^+ + \eta^- + \zeta^+ + \zeta^- + \chi_0^+ + \chi_0^-\Bigg],$$

where

$$\eta_{11}^+ := T_1 \tilde{T}_2^+ T_t \partial_1(a_{11}\partial_1 u) - \frac{1}{2}\big[D_{x_1}^-\big(a_{11} D_{x_1}^+ u\big) + D_{x_1}^+\big(a_{11} D_{x_1}^- u\big)\big],$$
$$\eta_{12}^+ := T_1 \tilde{T}_2^+ T_t \partial_1(a_{12}\partial_2 u) - D_{x_1}^-\big(a_{12} D_{x_2}^+ u\big),$$
$$\eta_{21}^+ := T_1 \tilde{T}_2^+ T_t \partial_2(a_{21}\partial_1 u) - D_{x_2}^+\big(a_{21} D_{x_1}^- u\big),$$
$$\eta_{22}^+ := T_1 \tilde{T}_2^+ T_t \partial_2(a_{22}\partial_2 u) - \frac{2}{h}\Bigg[\frac{a_{22}(x_1,\xi) + a_{22}(x_1,\xi+h)}{2} D_{x_2}^+ u$$
$$- T_1 T_t (a_{22}\partial_2 u)\Big|_{x_2=\xi+0}\Bigg],$$
$$\eta^+ := \big(T_1 \tilde{T}_2^+ a\big) u - T_1 \tilde{T}_2^+ T_t(au),$$
$$\zeta^+ := D_t^+ D_t^- u - T_1 \tilde{T}_2^+ T_t \frac{\partial^2 u}{\partial t^2},$$
$$\chi_{11}^+ := -\frac{\tau^2}{4h}\big(a_{11} D_{x_1}^+ D_t^+ D_t^- u\big) + \frac{\tau^2}{4h}\big(a_{11} D_{x_1}^+ D_t^+ D_t^- u\big)\Big|_{(x_1-h,\xi,t)},$$
$$\chi_{12}^+ := -\frac{\tau^2}{4h}\big(a_{12} D_{x_2}^+ D_t^+ D_t^- u\big) + \frac{\tau^2}{4h}\big(a_{12} D_{x_2}^+ D_t^+ D_t^- u\big)\Big|_{(x_1-h,\xi,t)},$$
$$\chi_{21}^+ := -\frac{\tau^2}{4h}\big(a_{21} D_{x_1}^- D_t^+ D_t^- u\big)\Big|_{(x_1,\xi+h,t)} + \frac{\tau^2}{4h}\big(a_{21} D_{x_1}^- D_t^+ D_t^- u\big),$$
$$\chi_{22}^+ := -\frac{\tau^2}{4h}\big[a_{22}(x_1,\xi) + a_{22}(x_1,\xi+h)\big]\big(D_{x_2}^+ D_t^+ D_t^- u\big),$$
$$\chi_0^+ := \frac{\tau^2}{4}\big(T_1 \tilde{T}_2^+ a\big) u,$$

with η_{ij}^-, η^-, ζ^-, χ_{ij}^- and χ_0^- being defined analogously.

The terms $\eta_{ij}^{\pm}$ can be bounded analogously to η_{ij} for $x \in \Omega^h \setminus \Sigma^h$. To this end we decompose η_{11}^+ as follows:

$$\eta_{11}^+ = \eta_{111}^+ + \eta_{112}^+ + \eta_{113}^+ + \eta_{114}^+ + \eta_{115}^+ + \eta_{116}^+ + \eta_{117}^+,$$

where

$$\eta_{111}^+ := T_1 \tilde{T}_2^+ T_t \big(a_{11} \partial_1^2 u\big) - \big(T_1 \tilde{T}_2^+ a_{11}\big)\big(T_1 \tilde{T}_2^+ T_t \partial_1^2 u\big),$$
$$\eta_{112}^+ := \big(T_1 \tilde{T}_2^+ a_{11}\big)\big(T_1 \tilde{T}_2^+ T_t \partial_1^2 u - D_{x_1}^- D_{x_1}^+ u\big),$$
$$\eta_{113}^+ := \big(T_1 \tilde{T}_2^+ a_{ij} - a_{ij}\big) D_{x_1}^- D_{x_1}^+ u,$$
$$\eta_{114}^+ := T_1 \tilde{T}_2^+ T_t (\partial_1 a_{11} \partial_1 u) - \big(T_1 \tilde{T}_2^+ \partial_1 a_{11}\big)\big(T_1 \tilde{T}_2^+ T_t \partial_1 u\big),$$
$$\eta_{115}^+ := \Big[T_1 \tilde{T}_2^+ \partial_1 a_{11} - \frac{1}{2}\big(D_{x_1}^+ a_{11} + D_{x_i}^- a_{11}\big)\Big]\big(T_1 \tilde{T}_2^+ T_t \partial_1 u\big),$$
$$\eta_{116}^+ := \frac{1}{2}\big(D_{x_1}^+ a_{11} + D_{x_1}^- a_{11}\big)\Big[T_1 \tilde{T}_2^+ T_t \partial_1 u - \frac{1}{2}\big(D_{x_1}^+ u + D_{x_1}^- u\big)\Big],$$
$$\eta_{117}^+ := \frac{1}{4}\big(D_{x_1}^+ a_{11} - D_{x_1}^- a_{11}\big)\big(D_{x_1}^+ u - D_{x_1}^- u\big).$$

Then, for $k \in \{1, 3, 4, 6, 7\}$,

$$|\eta_{11k}^+| \le Ch \|a_{11}\|_{C^1(\overline{\Omega}^+)} \|u\|_{C^2(\overline{Q}^+)}.$$

Thus, by summing over the mesh $\Sigma^h \times \Omega^\tau$ and using the Sobolev embedding theorem we obtain

$$h\tau \sum_{t \in \Omega^\tau} \|\eta_{11k}^+\|_{L_2(\Sigma^h)} \le Ch^2 \|a_{11}\|_{W_2^{s-1}(\Omega^+)} \|u\|_{W_2^s(Q^+)}, \quad s > 7/2. \qquad (4.103)$$

Inequality (4.103) holds also for η_{112}^+ and η_{115}^+. Here we have made use of the Bramble–Hilbert lemma to bound

$$T_1 \tilde{T}_2^+ T_t \partial_1^2 u - D_{x_1}^- D_{x_1}^+ u \quad \text{and} \quad T_1 \tilde{T}_2^+ \partial_1 a_{11} - \frac{1}{2}\big(D_{x_1}^+ a_{11} + D_{x_i}^- a_{11}\big),$$

and after summation over the mesh-points in $\Sigma^h \times \Omega^\tau$ we applied the inequality (2.199).

The remaining terms $\eta_{ij}^{\pm}$ and $\chi_{ij}^{\pm}$ can be estimated in the same manner, resulting in the bounds

$$h\tau \sum_{t \in \Omega^\tau} \|\eta_{ij}^{\pm}\|_{L_2(\Sigma^h)} \le Ch^2 \|a_{ij}\|_{W_2^{s-1}(\Omega^{\pm})} \|u\|_{W_2^s(Q^{\pm})}, \quad s > 7/2, \qquad (4.104)$$

$$h\tau \sum_{t\in\Omega^\tau} \|\chi_{ij}^{\pm}\|_{L_2(\Sigma^h)} \le Ch^2 \|a_{ij}\|_{W_2^{s-1}(\Omega^\pm)} \|u\|_{W_2^s(Q^\pm)}, \quad s > 7/2. \tag{4.105}$$

The terms $\zeta^\pm$ can be bounded analogously to η_{112}^{+}, using the Bramble–Hilbert lemma and the inequality (2.199):

$$h\tau \sum_{t\in\Omega^\tau} \|\zeta^{\pm}\|_{L_2(\Sigma^h)} \le Ch^2 \|u\|_{W_2^s(Q^\pm)}, \quad s > 7/2. \tag{4.106}$$

The bounds on $\eta^\pm$ and $\chi_0^\pm$ directly follow from the Sobolev embedding theorem:

$$\begin{aligned} h\tau \sum_{t\in\Omega^\tau} \|\eta^{\pm}\|_{L_2(\Sigma^h)} &\le Ch^2 \|a\|_{C(\overline{\Omega}^\pm)} \|u\|_{C^1(\overline{Q}^\pm)} \\ &\le Ch^2 \|a\|_{W_2^{s-2}(\Omega^\pm)} \|u\|_{W_2^s(Q^\pm)}, \quad s > 3, \end{aligned} \tag{4.107}$$

$$\begin{aligned} h\tau \sum_{t\in\Omega^\tau} \|\chi_0^{\pm}\|_{L_2(\Sigma^h)} &\le Ch^3 \|a\|_{C(\overline{\Omega}^\pm)} \|u\|_{C(\overline{Q}^\pm)} \\ &\le Ch^3 \|a\|_{W_2^{s-2}(\Omega^\pm)} \|u\|_{W_2^s(Q^\pm)}, \quad s > 3. \end{aligned} \tag{4.108}$$

By applying the Bramble–Hilbert lemma we obtain

$$\tau \sum_{t\in\Omega^\tau} \|\mu\|_{L_2(\Sigma^h)} \le Ch^2 \left\| k \frac{\partial^2 u}{\partial t^2} \right\|_{W_2^2(\Sigma)}.$$

Thus, using the transmission condition (4.94) and the trace theorem for Sobolev spaces, we deduce that

$$\begin{aligned} h\tau \sum_{t\in\Omega^\tau} \|\mu\|_{L_2(\Sigma^h)} \le Ch^2 \Big(&\max_{i,j} \|a_{ij}\|_{W_2^{s-1}(\Omega^+)} \|u\|_{W_2^2(Q^+)} \\ &+ \max_{i,j} \|a_{ij}\|_{W_2^{s-1}(\Omega^-)} \|u\|_{W_2^2(Q^-)} \Big), \quad s > 7/2. \end{aligned} \tag{4.109}$$

Combining (4.97)–(4.109) then yields (4.98). □

4.5.2 Factorized Scheme

Analogously as in the parabolic case (see Sect. 3.5) we shall now consider a factorized scheme, which has the following form for the hyperbolic interface problem considered in the previous section:

$$(I_h + \theta\tau^2 \Lambda_1)(B_h + \theta\tau^2 \Lambda_2) D_t^+ D_t^- U + \mathcal{L}_h U = T_1 T_2 T_t f \quad \text{in } Q_h^\tau, \tag{4.110}$$

with the same initial and boundary conditions as in (4.95). Here

$$\Lambda_i U := -D_{x_i}^+ D_{x_i}^- U, \qquad B_h U := (1 + k\delta_{\Sigma^h})U,$$

I_h is the identity operator and θ is a real parameter. For the sake of simplicity we shall assume that $k = \text{Const.} > 0$. In this case, the operator $(I_h + \theta\tau\Lambda_1)(B_h + \theta\tau\Lambda_2)$ is selfadjoint and the operator inequality

$$\big(I_h + \theta\tau^2\Lambda_1\big)\big(B_h + \theta\tau^2\Lambda_2\big) - \frac{1}{4}\tau^2\mathcal{L}_h \geq B_h \geq I_h > 0$$

holds for $s > 3$ and sufficiently large θ.

Let u be the solution of the initial-boundary-value problem (4.93) and let U denote the solution of the finite difference scheme (4.110). The global error $e := u - U$ satisfies the finite difference scheme

$$\begin{aligned}
&\big(I_h + \theta\tau^2\Lambda_1\big)\big(B_h + \theta\tau^2\Lambda_2\big)D_t^+ D_t^- e + \mathcal{L}_h e = \varphi' + \mu'\delta_{\Sigma^h} \quad \text{in } Q_h^\tau,\\
&e = 0 \quad \text{on } \Gamma^h \times \overline{\Omega}^\tau, \qquad\qquad\qquad\qquad\qquad\qquad\qquad\qquad\qquad\qquad (4.111)\\
&e = 0, \qquad (1 + k\delta_{\Sigma^h})D_t^+ e = \upsilon + \tilde{\upsilon}\delta_{\Sigma^h} + \frac{1}{2}\tau\tilde{\varphi} \quad \text{in } \Omega^h \times \{0\},
\end{aligned}$$

where

$$\begin{aligned}
\varphi' &:= \sum_{i,j=1}^{2} \eta_{ij} + \eta + \zeta + \chi',\\
\chi' &:= -\theta\tau^2 D_{x_1}^+ D_{x_1}^- D_t^+ D_t^- u - \theta\tau^2 D_{x_2}^+ D_{x_2}^- D_t^+ D_t^- u\\
&\quad + \theta^2\tau^4 D_{x_1}^+ D_{x_1}^- D_{x_2}^+ D_{x_2}^- D_t^+ D_t^- u,\\
\mu' &:= k\left[D_t^+ D_t^- u - T_1 T_t\left(\frac{\partial^2 u}{\partial t^2}\right) - \theta\tau^2 D_{x_1}^+ D_{x_1}^- D_t^+ D_t^- u\right].
\end{aligned}$$

The a priori estimate (4.97) holds for (4.111) if φ and μ are replaced by φ' and μ'. Under the assumption that $\tau \asymp h$ the terms χ' and μ' satisfy analogous bounds to χ and μ, which implies that the factorized scheme (4.110) satisfies the error bound (4.98) for $7/2 < s \leq 4$.

4.6 Hyperbolic Transmission Problem

In this section we shall investigate the finite difference approximation of a hyperbolic transmission problem in two disconnected domains. The problem is the hyperbolic counterpart of the parabolic transmission problem considered in Sect. 3.6.

As a model problem, we consider the following initial-boundary-value problem: find two functions, $u_1(x, y, t)$ and $u_2(x, y, t)$, that satisfy the system of hyperbolic equations

$$\frac{\partial^2 u_1}{\partial t^2} - \Delta u_1 = f_1(x, y, t), \quad (x, y) \in \Omega_1 := (a_1, b_1) \times (c_1, d_1), \ t \in (0, T], \tag{4.112}$$

$$\frac{\partial^2 u_2}{\partial t^2} - \Delta u_2 = f_2(x, y, t), \quad (x, y) \in \Omega_2 := (a_2, b_2) \times (c_2, d_2), \ t \in (0, T], \tag{4.113}$$

where $-\infty < a_1 < b_1 < a_2 < b_2 < +\infty$ and $c_2 < c_1 < d_1 < d_2$, the internal transmission conditions of nonlocal Robin–Dirichlet type

$$\frac{\partial u_1}{\partial x}(b_1, y, t) + \alpha_1(y) u_1(b_1, y, t) = \int_{c_2}^{d_2} \beta_1(y, y') u_2(a_2, y', t) \mathrm{d}y', \quad y \in (c_1, d_1), \ t \in (0, T], \tag{4.114}$$

$$\frac{\partial u_2}{\partial x}(a_2, y, t) + \alpha_2(y) u_2(a_2, y, t) = \int_{c_1}^{d_1} \beta_2(y, y') u_1(b_1, y', t) \mathrm{d}y', \quad y \in (c_2, d_2), \ t \in (0, T], \tag{4.115}$$

the simplest external Dirichlet boundary conditions for $t \in (0, T]$:

$$\begin{gathered} u_1(x, c_1, t) = u_1(x, d_1, t) = 0, \quad x \in (a_1, b_1), \\ u_2(x, c_2, t) = u_2(x, d_2, t) = 0, \quad x \in (a_2, b_2), \\ u_1(a_1, y, t) = 0, \quad y \in (c_1, d_1); \qquad u_2(b_2, y, t) = 0, \quad y \in (c_2, d_2), \end{gathered} \tag{4.116}$$

and the initial conditions

$$\begin{aligned} u_1(x, y, 0) &= u_{10}(x, y), & \frac{\partial u_1}{\partial t}(x, y, 0) &= u_{11}(x, y), \quad (x, y) \in \Omega_1, \\ u_2(x, y, 0) &= u_{20}(x, y), & \frac{\partial u_2}{\partial t}(x, y, 0) &= u_{21}(x, y), \quad (x, y) \in \Omega_2. \end{aligned} \tag{4.117}$$

We shall assume in what follows that the data satisfy the following conditions:

$$\beta_1(y, y') = \beta_2(y', y) = \beta(y, y') \quad \forall (y, y') \in (c_1, d_1) \times (c_2, d_2), \tag{4.118}$$

$f_i \in L_2((0, T); L_2(\Omega_i))$ for $i = 1, 2$, and

$$\alpha_i \in L_\infty(c_i, d_i), \quad i = 1, 2; \qquad \beta \in L_\infty\big((c_1, d_1) \times (c_2, d_2)\big). \tag{4.119}$$

Using the notations from Sect. 3.6, we deduce that under the condition (4.118) the bilinear functional (3.142) is symmetric and defines a symmetric bounded linear operator $A : \mathring{W}_2^1 \to W_2^{-1}$ by the formula

$$\langle Au, v\rangle := a(u, v), \quad u, v \in \mathring{W}_2^1,$$

where $\langle\cdot,\cdot\rangle$ is the duality pairing between W_2^{-1} and W_2^1.

The weak form of (4.112)–(4.116) is

$$\left\langle \frac{\mathrm{d}^2 u}{\mathrm{d}t^2}, v\right\rangle + a\big(u(\cdot,t), v\big) = \big(f(\cdot,t), v\big)_L \quad \forall v \in \mathring{W}_2^1, \tag{4.120}$$

or, in operator form,

$$\frac{\mathrm{d}^2 u}{\mathrm{d}t^2} + Au = f \quad \text{in } W_2^{-1}. \tag{4.121}$$

The problem (4.120) fits into the general framework of hyperbolic differential operators in Hilbert spaces. By applying Theorem 4.1 to (4.120) we obtain the following result.

Theorem 4.6 *Suppose that* (4.118) *and* (4.119) *hold and assume that* $u^0 = (u_{10}, u_{20}) \in \mathring{W}_2^1$, $u^1 = (u_{11}, u_{21}) \in L$, $f = (f_1, f_2) \in L_2((0,T), L)$. *Then, for* $0 < T < +\infty$, *the initial-boundary-value problem* (4.112)–(4.117) *has a unique weak solution* $u \in L_2((0,T), \mathring{W}_2^1) \cap W_2^1((0,T), L)$, *which depends continuously on* f, u^0 *and* u^1. *The solution* u *satisfies the a priori estimate*

$$\|u\|^2_{L_\infty((0,T),W_2^1)} + \left\|\frac{\mathrm{d}u}{\mathrm{d}t}\right\|^2_{L_\infty((0,T),L)} \le C(T)\big(\|u^0\|^2_{W_2^1} + \|u^1\|^2_L + \|f\|^2_{L_2((0,T),L)}\big),$$

where $C(T) = C_2 \mathrm{e}^{C_1 T}$ *is a computable constant depending on* T.

In the sequel we shall adopt the following notational conventions:

$$Q_i := \Omega_i \times (0, T], \quad i = 1, 2,$$
$$W_2^s(Q) := W_2^s(Q_1) \times W_2^s(Q_2),$$
$$\|u\|^2_{W_2^s(Q)} := \|u_1\|^2_{W_2^s(Q_1)} + \|u_2\|^2_{W_2^s(Q_2)}.$$

4.6.1 Finite Difference Approximation

With the notations from Sect. 3.6.2, we approximate the initial-boundary-value problem (4.112)–(4.118) by the following explicit three-level finite difference scheme:

$$D_t^- D_t^+ U_1 - D_{x,1}^- D_{x,1}^+ U_1 - D_{y,1}^- D_{y,1}^+ U_1 = f_1,$$
$$x \in \Omega_1^{h_1}, \; y \in \Omega_1^{k_1}, \; t \in \Omega^\tau, \tag{4.122}$$

$$D_t^- D_t^+ U_1(b_1, y, t) + \frac{2}{h_1}\Big[D_{x,1}^- U_1(b_1, y, t) + \alpha_1(y) U_1(b_1, y, t)$$

$$- k_2 \sum_{y' \in \Omega_2^{k_2}} \beta(y, y') U_2(a_2, y', t)\Big] - D_{y,1}^- D_{y,1}^+ U_1(b_1, y, t)$$

$$= f_1(b_1, y, t), \quad y \in \Omega_1^{k_1},\ t \in \Omega^\tau, \tag{4.123}$$

$$D_t^- D_t^+ U_2 - D_{x,2}^- D_{x,2}^+ U_2 - D_{y,2}^- D_{y,2}^+ U_2 = f_2,$$

$$x \in \Omega_2^{h_2},\ y \in \Omega_2^{k_2},\ t \in \Omega^\tau, \tag{4.124}$$

$$D_t^- D_t^+ U_2(a_2, y, t) - \frac{2}{h_2}\Big[D_{x,2}^+ U_2(a_2, y, t) - \alpha_2(y) U_2(a_2, y, t)$$

$$+ k_1 \sum_{y' \in \Omega_1^{k_1}} \beta(y', y) U_1(b_1, y', t)\Big] - D_{y,2}^- D_{y,2}^+ U_2(a_2, y, t)$$

$$= f_2(a_2, y, t), \quad y \in \Omega_2^{k_2},\ t \in \Omega^\tau, \tag{4.125}$$

$$U_1(x, c_1, t) = U_1(x, d_1, t) = 0, \quad x \in \overline{\Omega}_1^{h_1},\ t \in \overline{\Omega}^\tau,$$

$$U_2(x, c_2, t) = U_2(x, d_2, t) = 0, \quad x \in \overline{\Omega}_2^{h_2},\ t \in \overline{\Omega}^\tau, \tag{4.126}$$

$$U_1(a_1, y, t) = 0, \quad y \in \Omega_1^{k_1}; \qquad U_2(b_2, y, t) = 0, \quad y \in \Omega_2^{k_2},$$

$$U_i(x, y, 0) = u_{i0}(x, y),$$

$$U_i(x, y, \tau) = u_{i0}(x, y) + \tau u_{i1}(x, y) + \frac{1}{2}\tau^2 [\Delta u_{i0} + f_i(x, y, 0)], \tag{4.127}$$

$$x \in \Omega_{i\pm}^{h_i},\ y \in \Omega_i^{k_i},\ i = 1, 2.$$

We shall assume in what follows that $h_1 \asymp h_2 \asymp k_1 \asymp k_2$, and define $h := \max\{h_1, h_2, k_1, k_2\}$.

The finite difference scheme (4.122)–(4.127) fits into the general framework (4.14), where $\mathcal{H}^h$ is the space of mesh-functions $U = (U_1, U_2)$, U_i is defined on the mesh $\overline{\Omega}_i^{h_i} \times \overline{\Omega}_i^{k_i}$, $i = 1, 2$, $U_1 = 0$ for $x = b_1$, $U_2 = 0$ for $x = a_2$,

$$B_h := I_h, \qquad D_h := 0, \quad \text{and} \quad A_h := A_{0h} + A_{1h},$$

and the operators A_{ih}, $i = 1, 2$, have been defined in Sect. 3.6.2. If the time step τ is sufficiently small,

$$\tau \le C_0 \min\{h_1, h_2, k_1, k_2\}, \tag{4.128}$$

where C_0 is a computable constant depending on

$$\|\alpha_1\|_{C([c_1,d_1])}, \quad \|\alpha_2\|_{C([c_2,d_2])} \quad \text{and} \quad \|\beta\|_{C([c_1,d_1]\times[c_2,d_2])},$$

the finite difference scheme (4.122)–(4.127) is conditionally stable and satisfies an a priori estimate of the form (4.22).

Let $u = (u_1, u_2)$ be the solution of the initial-boundary-value problem (4.112)–(4.117) and let $U = (U_1, U_2)$ denote the solution of the finite difference scheme (4.122)–(4.127). Then, the global error $e := u - U$ satisfies the following finite difference scheme:

$$D_t^- D_t^+ e_1 - D_{x,1}^- D_{x,1}^+ e_1 - D_{y,1}^- D_{y,1}^+ e_1 = \varphi_1,$$
$$x \in \Omega_1^{h_1},\ y \in \Omega_1^{k_1},\ t \in \Omega^\tau, \tag{4.129}$$

$$D_t^- D_t^+ e_1(b_1, y, t) + \frac{2}{h_1}\Big[D_{x,1}^- e_1(b_1, y, t) + \alpha_1(y) e_1(b_1, y, t)$$
$$- k_2 \sum_{y' \in \Omega_2^{k_2}} \beta_1(y, y') e_2(a_2, y', t)\Big] - D_{y,1}^- D_{y,1}^+ e_1(b_1, y, t)$$
$$= \varphi_1(b_1, y, t),\ y \in \Omega_1^{k_1}, \quad t \in \Omega^\tau, \tag{4.130}$$

$$D_t^- D_t^+ e_2 - D_{x,2}^- D_{x,2}^+ e_2 - D_{y,2}^- D_{y,2}^+ e_2 = \varphi_2,$$
$$x \in \Omega_2^{h_2},\ y \in \Omega_2^{k_2},\ t \in \Omega^\tau, \tag{4.131}$$

$$D_t^- D_t^+ e_2(a_2, y, t) - \frac{2}{h_2}\Big[D_{x,2}^+ e_2(a_2, y, t) - \alpha_2(y) e_2(a_2, y, t)$$
$$+ k_1 \sum_{y' \in \Omega_1^{k_1}} \beta_2(y, y') e_1(b_1, y', t)\Big] - D_{y,2}^- D_{y,2}^+ e_2(a_2, y, t)$$
$$= \varphi_2(a_2, y, t), \quad y \in \Omega_2^{k_2},\ t \in \Omega^\tau, \tag{4.132}$$

$$e_1(x, c_1, t) = e_1(x, d_1, t) = 0, \quad x \in \overline{\Omega}_1^{h_1},\ t \in \overline{\Omega}^\tau,$$
$$e_2(x, c_2, t) = e_2(x, d_2, t) = 0, \quad x \in \overline{\Omega}_2^{h_2},\ t \in \overline{\Omega}^\tau, \tag{4.133}$$
$$e_1(a_1, y, t) = 0, \quad y \in \Omega_1^{k_1}; \qquad e_2(b_2, y, t) = 0, \quad y \in \Omega_2^{k_2},$$

$$e_i(x, y, 0) = 0, \qquad e_i(x, y, \tau) = \chi_i, \quad x \in \Omega_{i\pm}^{h_i},\ y \in \Omega_i^{k_i},\ i = 1, 2, \tag{4.134}$$

where

$$\varphi_i := \xi_i + \eta_i + \zeta_i, \quad i = 1, 2,$$

$$\xi_i := D_t^- D_t^+ u_i - \frac{\partial^2 u_i}{\partial t^2}, \quad x \in \Omega_{i\pm}^{h_i},\ y \in \Omega_i^{k_i},\ t \in \Omega^\tau,\ i = 1, 2,$$

$$\eta_i := \frac{\partial^2 u_i}{\partial x^2} - D^-_{x,i} D^+_{x,i} u_i, \quad x \in \Omega_i^{h_i}, \ y \in \Omega_i^{k_i}, \ t \in \Omega^\tau, \ i = 1, 2,$$

$$\eta_1 := \left[\frac{\partial^2 u_1}{\partial x^2} - \frac{2}{h_1}\left(\frac{\partial u_1}{\partial x} - D^-_{x,1} u_1\right)\right] - \frac{2}{h_1}\left[\int_{c_2}^{d_2} \beta(y, y') u_2(a_2, y', t) \mathrm{d}y' \right.$$

$$\left. - k_2 \sum_{y' \in \Omega_2^{k_2}} \beta(y, y') u_2(a_2, y', t)\right], \quad x = b_1, \ y \in \Omega_1^{k_1}, \ t \in \Omega^\tau,$$

$$\eta_2 := \left[\frac{\partial^2 u_2}{\partial x^2} + \frac{2}{h_2}\left(\frac{\partial u_2}{\partial x} - D^+_{x,2} u_2\right)\right] + \frac{2}{h_2}\left[\int_{c_1}^{d_1} \beta(y, y') u_1(b_1, y', t) \mathrm{d}y' \right.$$

$$\left. - k_1 \sum_{y' \in \Omega_1^{k_1}} \beta(y, y') u_1(b_1, y', t)\right], \quad x = a_2, \ y \in \Omega_2^{k_2}, \ t \in \Omega^\tau,$$

$$\zeta_i := \frac{\partial^2 u_i}{\partial y^2} - D^-_{y,i} D^+_{y,i} u_i, \quad x \in \Omega_{i\pm}^{h_i}, \ y \in \Omega_i^{k_i}, \ t \in \Omega^\tau, \ i = 1, 2,$$

$$\chi_i := \int_0^\tau \int_0^{t'} \left[\frac{\partial^2 u_i}{\partial t^2}(x, y, t'') - \frac{\partial^2 u_i}{\partial t^2}(x, y, 0)\right] \mathrm{d}t'' \mathrm{d}t',$$

$$x \in \Omega_{i\pm}^{h_i}, \ y \in \Omega_i^{k_i}, \ i = 1, 2.$$

The following a priori estimate for the solution of the finite difference scheme (4.129)–(4.134) follows immediately from (4.22):

$$\|e\|^{(1)}_{2,\infty,h\tau} \leq C(T)\Bigg(\|\chi\|_{W^1_{2,h}} + \frac{1}{\tau}\|\chi\|_{L_h}$$

$$+ \|\xi\|_{L_2(\Omega^\tau, L_h)} + \|\eta\|_{L_2(\Omega^\tau, L_h)} + \|\zeta\|_{L_2(\Omega^\tau, L_h)}\Bigg). \quad (4.135)$$

Here we have used the notations

$$\|V\|^2_{W^1_{2,h}} := \left\|D^-_x V\right\|^2_{L_{h'}} + \left\|D^-_y V\right\|^2_{L_{h''}}$$

and

$$\|V\|^{(1)}_{2,\infty,h\tau} := \max_{t \in \Omega^\tau_-}\left[\left\|D^+_t V(\cdot, t)\right\|^2_{L_h} + \left\|\frac{V(\cdot, t+\tau) + V(\cdot, t)}{2}\right\|^2_{W^1_{2,h}}\right]^{1/2}.$$

In order to derive an error bound for the finite difference scheme (4.122)–(4.127) it therefore suffices to bound the terms appearing on the right-hand side of the inequality (4.135). For simplicity we shall assume in what follows that $\tau \asymp h$.

Theorem 4.7 *Let* $\alpha_i \in W_2^{s-1}(c_i, d_i)$, $i = 1, 2$, $\beta \in W_2^{s-1}((c_1, d_1) \times (c_2, d_2))$, $u \in W_2^s(Q)$, *and let the assumptions* (4.128) *and* $\tau \asymp h$ *hold. Then, the solution* U *of the* *finite difference scheme* (4.122)–(4.127) *converges to the solution* u *of the initial-boundary-value problem* (4.112)–(4.118) *and the following error bound holds for* $s > 7/2$: *there exists a positive constant* $C = C(s)$, *independent of* h, *such that*

$$\|u - U\|^{(1)}_{2,\infty,h\tau} \le Ch^{3/2}\big(1 + \|\beta\|_{W_2^{s-1}((c_1,d_1)\times(c_2,d_2))}\big)\|u\|_{W_2^s(Q)}. \qquad (4.136)$$

Proof The value of ξ_i at the mesh-point $(x, y, t) \in \Omega_i^{h_i} \times \Omega_i^{k_i} \times \Omega^\tau$ is a bounded linear functional of $u_i \in W_2^s(G^i)$, where

$$G^i := (x - h_i, x + h_i) \times (y - k_i, y + k_i) \times (t - \tau, t + \tau)$$

and $s > 7/2$, which vanishes on polynomials of degree ≤ 3. Using the Bramble–Hilbert lemma we deduce that

$$\big|\xi_i(x, y, t)\big| \le C(h)|u_i|_{W_2^s(G^i)}, \quad 7/2 < s \le 4,$$

where $C(h) = Ch^{s-7/2}$. Analogous results hold for $x = b_1$ and $x = a_2$, with suitable modifications of G^i. By summation over the mesh we obtain the bound

$$\|\xi\|_{L_2(\Omega^\tau, L_h)} \le Ch^{s-2}\|u\|_{W_2^s(Q)}, \quad 7/2 < s \le 4. \qquad (4.137)$$

Analogously, one deduces that

$$\|\zeta\|_{L_2(\Omega^\tau, L_h)} \le Ch^{s-2}\|u\|_{W_2^s(Q)}, \quad 7/2 < s \le 4, \qquad (4.138)$$

and

$$\left(\sum_{i=1}^{2} h_i \sum_{x\in\Omega_i^{h_i}} k_i \sum_{y\in\Omega_i^{k_i}} \tau \sum_{t\in\Omega^\tau} \eta_i^2\right)^{1/2} \le Ch^{s-2}\|u\|_{W_2^s(Q)}, \quad 7/2 < s \le 4. \qquad (4.139)$$

For $x = b_1$ we decompose the term η_1 as follows:

$$\eta_1 = \eta_{11} + \eta_{12} + \eta_{13},$$

where

$$\eta_{11} := \frac{\partial^2 u_1}{\partial x^2} - \frac{2}{h_1}\left(\frac{\partial u_1}{\partial x} - D_x^- u_1\right) - T_x T_y\left[\frac{\partial^2 u_1}{\partial x^2} - \frac{2}{h_1}\left(\frac{\partial u_1}{\partial x} - D_x^- u_1\right)\right],$$

$$\eta_{12} := T_x T_y\left[\frac{\partial^2 u_1}{\partial x^2} - \frac{2}{h_1}\left(\frac{\partial u_1}{\partial x} - D_x^- u_1\right)\right],$$

$$\eta_{13} := \frac{2}{h_1}\left(k_2 \sum_{y'\in\Omega_2^{k_2}} \beta(y, y')u_2(a_2, y', t) - \int_{c_2}^{d_2} \beta(y, y')u_2(a_2, y', t)\,\mathrm{d}y'\right).$$

The value of η_{11} at a mesh-point is a bounded linear functional of $u_i \in W_2^s$, $s > 7/2$, which vanishes on polynomials of degree ≤ 3. By using the Bramble–Hilbert lemma we deduce that

$$\Big(\hbar_1 k_1 \sum_{y\in\Omega_1^{k_1}} \tau \sum_{t\in\Omega^\tau} \eta_{11}^2(b_1,y,t)\Big)^{1/2} \le C h^{s-2} \|u_1\|_{W_2^s(Q_1)}, \; 7/2 < s \le 4. \quad (4.140)$$

The value of η_{12} at a mesh-point is a bounded linear functional of $u_i \in W_2^s$, $s > 5/2$, which vanishes on polynomials of degree ≤ 2. By invoking the Bramble–Hilbert lemma and the inequality (2.199), we deduce that

$$\Big(\hbar_1 k_1 \sum_{y\in\Omega_1^{k_1}} \tau \sum_{t\in\Omega^\tau} \eta_{12}^2(b_1,y,t)\Big)^{1/2} \le C h \|u_1\|_{W_2^3((b_1-h_1,b_1)\times(c_1,d_1)\times(0,T))}$$
$$\le C h^{3/2} \|u_1\|_{W_2^s(Q_1)}, \quad s > 7/2. \quad (4.141)$$

By applying the error bound for the trapezium rule, we obtain

$$\Big(\hbar_1 k_1 \sum_{y\in\Omega_1^{k_1}} \tau \sum_{t\in\Omega^\tau} \eta_{13}^2(b_1,y,t)\Big)^{1/2}$$
$$\le C h^{3/2} \|\beta\|_{W_2^{s-1}((c_1,d_1)\times(c_2,d_2))} \|u_2\|_{W_2^s(Q_2)}, \quad s > 7/2. \quad (4.142)$$

Thanks to (4.139), (4.140)–(4.142) and analogous bounds on $\eta_2(a_2,\cdot,\cdot)$ we arrive at the following inequality:

$$\|\eta\|_{L_2(\Omega^\tau,L_h)} \le C h^{3/2}\big(1 + \|\beta\|_{W_2^{s-1}((c_1,d_1)\times(c_2,d_2))}\big) \|u\|_{W_2^s}, \quad s > 7/2. \quad (4.143)$$

In order to bound the term χ we consider the decomposition

$$\chi_i = \chi_{i1} + \chi_{i2},$$

where

$$\chi_{i1} := T_x T_y \chi_i, \quad x \in \Omega_i^{h_i}, \; y \in \Omega_i^{k_i},$$
$$\chi_{11} := \tilde{T}_x^- T_y \chi_1, \quad x = b_1, \; y \in \Omega_1^{k_1},$$
$$\chi_{21} := \tilde{T}_x^+ T_y \chi_2, \quad x = a_2, \; y \in \Omega_2^{k_2}.$$

The value of χ_{11} at a mesh-point is a bounded linear functional of $u_i \in W_2^s$, where $s > 5/2$, which vanishes on polynomials of degree ≤ 2. Using the Bramble–Hilbert lemma and the inequality (2.199) we obtain

$$\Big(\sum_{x\in\Omega_1^{h_1}} \hbar_1 k_1 \sum_{y\in\Omega_1^{k_1}} \chi_{11}^2\Big)^{1/2} \le C h^{5/2} \|u_1\|_{W_2^3(\Omega_1\times(0,\tau))} \le C h^3 \|u_1\|_{W_2^s(Q_1)},$$

for $s > 7/2$. Similarly, the value of $\chi_{12} = \chi_1 - \chi_{11}$ at a mesh-point is a bounded linear functional of $u_i \in W_2^s$, where $s > 7/2$, which vanishes on polynomials of degree ≤ 3. Using the Bramble–Hilbert lemma we obtain

$$\Big(\sum_{x \in \Omega_1^{h_1}} \hbar_1 k_1 \sum_{y \in \Omega_1^{k_1}} \chi_{12}^2 \Big)^{1/2} \leq Ch^{s-1/2} \|u_1\|_{W_2^s(\Omega_1 \times (0,\tau))} \leq Ch^3 \|u_1\|_{W_2^s(Q_1)},$$

for $7/2 < s \leq 4$. Analogous inequalities hold for χ_{21} and χ_{22}, whereby

$$\|\chi\|_{L_h} \leq Ch^3 \|u\|_{W_2^s(Q)}, \quad 7/2 < s \leq 4. \tag{4.144}$$

We also note the obvious inequality

$$\|\chi\|_{W_{2,h}^1} \leq \frac{1}{h} \|\chi\|_{L_h}. \tag{4.145}$$

Finally, from (4.135), (4.137), (4.138), (4.143), (4.144) and (4.145) we deduce the error bound (4.136). □

Remark 4.2 The error bound (4.136) exhibits a loss of half an order from the expected second order of convergence. This loss can be avoided by using a more accurate approximation of the equations at the mesh-points of the inner boundary. For the analysis of the finite difference approximation of the corresponding one-dimensional problem, see Jovanović and Vulkov [100] and Jovanović [90]. Second-order convergence can also be shown to hold in the weaker norm $\|\cdot\|_{2,\infty,h\tau}^{(0)}$ (see Jovanović and Vulkov [103]).

4.7 Bibliographical Notes

In this chapter we derived error bounds for finite difference approximations of certain model initial-boundary-value problems for second-order linear hyperbolic partial differential equations. The procedure was based on the Bramble–Hilbert lemma and its generalizations, in conjunction with discrete energy estimates. In the derivation of the relevant energy estimates, discrete analogues of the norms

$$\|u\|_{L_\infty((0,T);W_2^r(\Omega))} + \left\| \frac{\partial u}{\partial t} \right\|_{L_\infty((0,T);W_2^{r-1}(\Omega))}, \tag{4.146}$$

$$\|u\|_{L_\infty((0,T);W_2^r(\Omega))} + \left\| \frac{\partial^r u}{\partial t^r} \right\|_{L_\infty((0,T);L_2(\Omega))} \tag{4.147}$$

were used.

For finite difference approximations of the first initial-boundary-value problem for second-order linear hyperbolic partial differential equations with constant coefficients, when $\tau \asymp h$, i.e. $c_1 h \leq \tau \leq c_2 h$, error bounds of the form

$$\|u - U\|^{(r)}_{2,\infty,h\tau} \leq Ch^{s-r-1} \|u\|_{W^s_2(Q)} \tag{4.148}$$

were derived by Jovanović and Ivanović [91] and Jovanović, Ivanović and Süli [107] for $r = 0, 1, 2$ and $r + 1 < s \leq r + 3$. Here $\|\cdot\|^{(r)}_{2,\infty,h\tau}$ is the discrete counterpart of the norm (4.146). We note that, in contrast with elliptic and parabolic problems, the error bound (4.148) is not compatible with the smoothness of the data. Indeed, in the transition from the function $u \in W^s_2(Q)$ to its trace on $t =$ Const. one looses half an order of Sobolev regularity, in the estimates (4.148); this gives rise to the observed loss of compatibility.

Error bounds in discrete norms of the form (4.147) for $r = -1$ and $r = 1$, and for $-1 < r < 1$ by function space interpolation, were derived by Dzhuraev and Moskal'kov [41].

Equations with variable coefficients $a_{ij} \in W^{s-1}_\infty(\Omega)$, $a \in W^{s-2}_\infty(\Omega)$ were studied by Jovanović, Ivanović and Süli [109], and an error bound of the form (4.148) was derived for $r = 1$ and $2 < s \leq 4$. The constant C in those error bounds depends on norms of the coefficients. Analogous results, under weaker assumptions on the coefficients, were subsequently obtained by Jovanović [86]. Dzhuraev, Kolesnik and Makarov [42] also considered hyperbolic equations with variable coefficients, however a method of lines was used as their numerical approximation. An estimate of the form (4.148) was obtained for $r = 0$ and a fixed integer value: $s = 2$.

Jovanović and Vulkov extended these results to hyperbolic interface problems and transmission problems (see [93, 96] and [90, 100, 103]).

In certain cases function space interpolation techniques give sharper error bounds; for results in this direction we refer to the extensive paper by Zlotnik [204] and the papers of Jovanović [87, 88]. In this respect, we make one final observation on our error bound (4.63). Had we used, instead of the K-method of interpolation with interpolation functor $(\cdot,\cdot)_{\theta,q}$, $0 < \theta < 1$, $1 \leq q \leq \infty$, the complex method of interpolation with interpolation functor $[\cdot,\cdot]_\theta$, $0 < \theta < 1$, (see, for example, Chap. 4 in Bergh and J. Löfström [9]; Chap. 1, Sect. 2.1 in Lions and Magenes [123]; or Chap. 1, Sect. 1.9 in Triebel [182]) we would have arrived at an error bound analogous to (4.63), for $1 < s < 4$, with the norm $\|u_0\|_{(\mathfrak{W}^1_2(0,1),\mathfrak{W}^4_2(0,1))_{(s-1)/3,2}}$ replaced by the norm $\|u_0\|_{[\mathfrak{W}^1_2(0,1),\mathfrak{W}^4_2(0,1)]_{(s-1)/3}}$ on the right-hand side.

References

1. Adams, R.A.: Sobolev Spaces. Academic Press, New York (1975)
2. Adams, R.A., Fournier, J.J.F.: Sobolev Spaces, 2nd edn. Pure and Applied Mathematics Series, vol. 140. Elsevier/Academic Press, Amsterdam (2003)
3. Amosov, A.A.: Global solvability of a nonlinear nonstationary problem with a nonlocal boundary condition of radiation heat transfer type. Differ. Equ. **41**(1), 96–109 (2005)
4. Amosov, A.A., Zlotnik, A.A.: Difference schemes of second-order of accuracy for the equations of the one-dimensional motion of a viscous gas. USSR Comput. Math. Math. Phys. **27**(4), 46–57 (1987)
5. Barbeiro, S.: Supraconvergent cell-centered scheme for two dimensional elliptic problems. Appl. Numer. Math. **59**, 56–72 (2009)
6. Barbeiro, S., Ferreira, J.A., Grigorieff, R.D.: Supraconvergence of a finite difference scheme for solutions in $H^s(0, L)$. IMA J. Numer. Anal. **25**, 797–811 (2005)
7. Barbeiro, S., Ferreira, J.A., Pinto, L.: H^1-second order convergent estimates for non Fickian models. Appl. Numer. Math. **61**, 201–215 (2011)
8. Bartle, R.G.: The Elements of Integration and Lebesgue Measure. Wiley, New York (1995)
9. Bergh, J., Löfström, J.: Interpolation Spaces, an Introduction. Grundlehren der mathematischen Wissenschaften, vol. 228. Springer, Berlin (1976)
10. Berikelashvili, G.: The convergence in W_2^2 of the difference solution of the Dirichlet problem. USSR Comput. Math. Math. Phys. **30**(2), 89–92 (1990)
11. Berikelashvili, G.: Construction and analysis of difference schemes for some elliptic problems, and consistent estimates of the rate of convergence. Mem. Differ. Equ. Math. Phys. **38**, 1–131 (2006)
12. Berikelashvili, G., Gupta, M.M., Mirianashvili, M.: Convergence of fourth order compact difference schemes for three-dimensional convection-diffusion equations. SIAM J. Numer. Anal. **45**(1), 443–455 (2007)
13. Besov, O.V., Il'in, V.P., Nikol'skiĭ, S.M.: Integral Representations of Functions and Imbedding Theorems. Nauka, Moscow (1975). (Russian)
14. Blanc, X., Le Bris, C., Lions, P.-L.: From molecular models to continuum mechanics. Arch. Ration. Mech. Anal. **164**(4), 341–381 (2002)
15. Blanc, X., Le Bris, C., Lions, P.-L.: Atomistic to continuum limits for computational materials science. M2AN Math. Model. Numer. Anal. **41**(2), 391–426 (2007)
16. Bojović, D., Jovanović, B.S.: Application of interpolation theory to determination of convergence rate for finite difference schemes of parabolic type. Mat. Vestn. **49**, 99–107 (1997)
17. Bojović, D., Jovanović, B.S.: Convergence of finite difference method for the parabolic problem with concentrated capacity and variable operator. J. Comput. Appl. Math. **189**(1–2), 286–303 (2006)

B.S. Jovanović, E. Süli, *Analysis of Finite Difference Schemes*,
Springer Series in Computational Mathematics 46,
DOI 10.1007/978-1-4471-5460-0, © Springer-Verlag London 2014

18. Bondesson, M.: An interior a priori estimate for parabolic difference operators and an application. Math. Comput. **25**, 43–58 (1971)
19. Bondesson, M.: Interior a priori estimates in discrete L_p norms for solutions of parabolic and elliptic difference equations. Ann. Mat. Pura Appl. **95**(4), 1–43 (1973)
20. Bramble, J.H., Hilbert, S.R.: Estimation of linear functionals on Sobolev spaces with application to Fourier transform and spline interpolation. SIAM J. Numer. Anal. **7**, 112–124 (1970)
21. Bramble, J.H., Hilbert, S.R.: Bounds for a class of linear functionals with application to Hermite interpolation. Numer. Math. **16**, 362–369 (1971)
22. Brandt, A.: Interior Schauder estimates for parabolic differential- (or difference-) equations via the maximum principle. Isr. J. Math. **7**, 254–262 (1969)
23. Brenner, S.C., Scott, L.R.: The Mathematical Theory of Finite Element Methods, 3rd edn. Texts in Applied Mathematics, vol. 15. Springer, New York (2008)
24. Brenner, P., Thomée, V., Wahlbin, L.B.: Besov Spaces and Applications to Difference Methods for Initial Value Problems. Lecture Notes in Mathematics, vol. 434. Springer, Berlin (1975)
25. Carter, R., Giles, M.B.: Sharp error estimates for discretizations of the 1D convection-diffusion equation with Dirac initial data. IMA J. Numer. Anal. **27**, 406–425 (2007)
26. Ciarlet, P.G.: The Finite Element Method for Elliptic Problems. North-Holland, Amsterdam (1978)
27. Datta, A.K.: Biological and Bioenvironmental Heat and Mass Transfer. Dekker, New York (2002)
28. Dauge, M.: Elliptic Boundary Value Problems on Corner Domains. Lecture Notes in Mathematics. Springer, Berlin (1988)
29. Dobson, M., Luskin, M.: Analysis of a force-based quasicontinuum approximation. M2AN Math. Model. Numer. Anal. **42**(1), 113–139 (2008)
30. Douglas, J., Dupont, T.: A finite element collocation method for quasilinear parabolic equations. Math. Comput. **27**, 17–28 (1973)
31. Douglas, J., Dupont, T., Wheeler, M.F.: A quasi-projection analysis of Galerkin methods for parabolic and hyperbolic equations. Math. Comput. **32**, 345–362 (1978)
32. Dražić, M.: Convergence rates of difference approximations to weak solutions of the heat transfer equation. Technical Report 86/22, Oxford University Computing Laboratory, Numerical Analysis Group, Oxford (1986)
33. Drenska, N.T.: Convergence of a difference scheme of the finite element method for the Poisson equation in the L_p metric. Vestn. Mosc. Univ. Ser. XV Vychisl. Mat. Kibernet. **1984**(3), 19–22 (1984). (Russian)
34. Drenska, N.T.: Convergence in the L_p metric of a difference scheme of the finite element method for an elliptic equation with constant coefficients. Vestn. Mosc. Univ. Ser. XV Vychisl. Mat. Kibernet. **1985**(4), 9–13 (1985). (Russian)
35. Druet, P.E.: Weak solutions to a time-dependent heat equation with nonlocal radiation condition and right hand side in L^p $(p \geq 1)$. WIAS Preprint 1253 (2008)
36. Dunford, N., Schwartz, J.T.: Linear Operators. Part I: General Theory. Interscience, New York (1957)
37. Dupont, T., Scott, L.R.: Polynomial approximation of functions in Sobolev spaces. Math. Comput. **34**, 441–463 (1980)
38. Durán, R.: On polynomial approximation in Sobolev spaces. SIAM J. Numer. Anal. **20**(5), 985–988 (1983)
39. D'yakonov, E.G.: Difference Methods for Solving Boundary Value Problems. Publ. House of Moscow State Univ., Moscow (1971) (vol. I), 1972 (vol. II) (Russian)
40. Dyda, B.: A fractional order Hardy inequality. Ill. J. Math. **48**(2), 575–588 (2004)
41. Dzhuraev, I.N., Moskal'kov, M.N.: Convergence of the solution of a weighted difference scheme to a generalized solution in $W_2^2(Q_T)$ of the vibrating-string equation. Differ. Equ. **21**, 1453–1459 (1985)

42. Dzhuraev, I.N., Kolesnik, T.N., Makarov, V.L.: Accuracy of the method of straight lines for second-order quasilinear hyperbolic equations with a small parameter at the highest time derivative. Differ. Equ. **21**, 784–789 (1985)
43. Edwards, R.E.: Fourier Series: A Modern Introduction vol. 2, 2nd edn. Springer, New York (1982)
44. Emmrich, E.: Supraconvergence and supercloseness of a discretisation for elliptic third-kind boundary-value problems on polygonal domains. Comput. Methods Appl. Math. **7**, 135–162 (2007)
45. Emmrich, E., Grigorieff, R.D.: Supraconvergence of a finite difference scheme for elliptic boundary value problems of the third kind in fractional order Sobolev spaces. Comput. Methods Appl. Math. **6**, 154–177 (2006)
46. Federer, H.: Geometric Measure Theory. Die Grundlehren der mathematischen Wissenschaften, vol. 153. Springer, Berlin (1969)
47. Ferreira, J.A., Grigorieff, R.D.: Supraconvergence and supercloseness of a scheme for elliptic equations on nonuniform grids. Numer. Funct. Anal. Optim. **27**(5–6), 539–564 (2006)
48. Gagliardo, E.: Caratterizzazioni delle tracce sulla frontiera relative ad alcune classi di funzioni in n variabili. Rend. Semin. Mat. Univ. Padova **27**, 284–305 (1957)
49. Gavrilyuk, I.P., Sazhenyuk, V.S.: Estimates of the rate of convergence of the penalty and net methods for a class of fourth-order elliptic variational inequalities. USSR Comput. Math. Math. Phys. **26**(6), 20–26 (1986)
50. Gavrilyuk, I.P., Lazarov, R.D., Makarov, V.L., Pirnazarov, S.I.: Estimates of the rate of convergence of difference schemes for fourth-order elliptic equations. USSR Comput. Math. Math. Phys. **23**(2), 64–70 (1983)
51. Gavrilyuk, I.P., Prikazchikov, V.G., Himich, A.N.: The accuracy of the solution of a difference boundary value problem for a fourth-order elliptic operator with mixed boundary conditions. USSR Comput. Math. Math. Phys. **26**(6), 139–147 (1986)
52. Gel'fand, I.M., Shilov, G.E.: Generalized Functions, Vol. I: Properties and Operations. Academic Press, New York (1964)
53. Gilbarg, D., Trudinger, N.S.: Elliptic Partial Differential Equations of Second Order. Springer, Berlin (2001)
54. Giusti, E.: Minimal Surfaces and Functions of Bounded Variation. Birkhäuser, Boston (1984)
55. Givoli, D.: Exact representation on artificial interfaces and applications in mechanics. Appl. Mech. Rev. **52**, 333–349 (1999)
56. Givoli, D.: Finite element modeling of thin layers. Comput. Model. Eng. Sci. **5**(6), 497–514 (2004)
57. Godev, K.N., Lazarov, R.D.: On the convergence of the difference scheme for the second boundary problem for the biharmonic equation with solution from W_p^k. In: Samarskij, A.A., Katai, I. (eds.) Mathematical Models in Physics and Chemistry and Numerical Methods of Their Realization, Proc. Seminar held in Visegrád, Hungary, 1982. Teubner Texte zur Mathematik, vol. 61, pp. 130–141
58. Godev, K.N., Lazarov, R.D.: Error estimates of finite-difference schemes in L_p-metrics for parabolic boundary value problems. C. R. Acad. Bulgare Sci. **37**, 565–568 (1984)
59. Grafakos, L.: Classical Fourier Analysis, 2nd edn. Graduate Texts in Mathematics, vol. 249. Springer, New York (2008)
60. Grisvard, P.: Équations différentielles abstraites. Ann. Sci. Éc. Norm. Super. **4**(2), 311–395 (1969). (French)
61. Grisvard, P.: Behaviour of the solutions of an elliptic boundary value problem in polygonal or polyhedral domain. In: Hubbard, B. (ed.) Numerical Solution of Partial Differential Equations, III, Proc. Third Sympos. (SYNSPADE), Univ. Maryland, College Park, Md., 1975. pp. 207–274. Academic Press, New York (1975)
62. Grisvard, P.: Elliptic Problems in Non-smooth Domains. Pitman, London (1985)
63. Grisvard, P.: Singularities in Boundary Value Problems. Research in Applied Mathematics, vol. 22. Masson, Paris (1992)

64. Gustafsson, B., Kreiss, H.-O., Oliger, J.: Time Dependent Problems and Difference Methods. Pure and Applied Mathematics. A Wiley-Interscience Publication. Wiley, New York (1995)
65. Hackbusch, W.: Optimal $H^{p,p/2}$ error estimates for a parabolic Galerkin method. SIAM J. Numer. Anal. **18**, 681–692 (1981)
66. Hackbusch, W.: On the regularity of difference schemes. Ark. Mat. **19**(1), 71–95 (1981)
67. Hackbusch, W.: On the regularity of difference schemes. II. Regularity estimates for linear and nonlinear problems. Ark. Mat. **21**(1), 3–28 (1983)
68. Hackbusch, W.: Elliptic Differential Equations. Theory and Numerical Treatment. Springer Series in Computational Mathematics, vol. 18. Springer, Berlin (2010). Translated from the 1986 corrected German edition by Regine Fadiman and Patrick D.F. Ion, Reprint of the 1992 English edition
69. Haroske, D.D., Triebel, H.: Distributions, Sobolev Spaces, Elliptic Equations. European Mathematical Society, Zürich (2008)
70. Hell, T.: Compatibility conditions for elliptic boundary value problems on non-smooth domains. Master's Thesis in Mathematics, Faculty of Mathematics, Computer Science and Physics of the University of Innsbruck, January 2011
71. Hörmander, L.: Lectures on linear partial differential operators. Mimeographed Notes, Stanford University (1960)
72. Hörmander, L.: Linear Partial Differential Operators 4th edn. Die Grundlehren der mathematischen Wissenschaften, vol. 116. Springer, Berlin (1969)
73. Iovanovic, B.S., Matus, P.P.: Estimation of the convergence rate of difference schemes for elliptic problems. Comput. Math. Math. Phys. **39**, 56–64 (1999)
74. Ivanović, L.D., Jovanović, B.S., Süli, E.: On the rate of convergence of difference schemes for the heat transfer equation on the solutions from $W_2^{s,s/2}$. Mat. Vesn. **36**, 206–212 (1984)
75. Ivanović, L.D., Jovanović, B.S., Süli, E.: On the convergence of difference schemes for the Poisson equation. In: Vrdoljak, B. (ed.) IV Conference on Applied Mathematics, Proc. Conf. Held in Split 1984, University of Split, Split, pp. 45–49 (1985)
76. Ivanovich, L.D., Jovanovich, B.S., Shili, E.: The convergence of difference schemes for a biharmonic equation. USSR Comput. Math. Math. Phys. **26**, 87–90 (1986)
77. Jacobsen, N.: Basic Algebra, vol. II. Freeman, San Francisco (1980)
78. Jovanović, B.S.: Convergence of projection-difference schemes for the heat equation. Mat. Vesn. **6(19)** (34), 279–292 (1982). (Russian)
79. Jovanović, B.S.: A generalization of the Bramble-Hilbert lemma. Zb. Rad. (Kragujevac) **8**, 81–87 (1987). (Serbian)
80. Jovanović, B.S.: Sur la méthode des domaines fictifs pour une équation elliptique quasilinéaire du quatrième ordre. Publ. Inst. Math. **42**(56), 167–173 (1987)
81. Jovanović, B.S.: Schéma aux differénces finies pour une équation elliptique quasilinéaire dans un domaine arbitraire. Mat. Vestn. **40**, 31–40 (1988)
82. Jovanović, B.S.: On the convergence of finite-difference schemes for parabolic equations with variable coefficients. Numer. Math. **54**, 395–404 (1989)
83. Jovanović, B.S.: Finite-difference approximations of elliptic equations with non-smooth coefficients. In: Sendov, B., Lazarov, R., Dimov, I. (eds.) Numerical Methods and Applications, Proc. Conf. Held in Sofia 1988, Bulgar. Acad. Sci., Sofia pp. 207–211 (1989)
84. Jovanović, B.S.: Optimal error estimates for finite-difference schemes with variable coefficients. Z. Angew. Math. Mech. **70**, 640–642 (1990)
85. Jovanović, B.S.: Convergence of finite-difference schemes for parabolic equations with variable coefficients. Z. Angew. Math. Mech. **71**, 647–650 (1991)
86. Jovanović, B.S.: Convergence of finite-difference schemes for hyperbolic equations with variable coefficients. Z. Angew. Math. Mech. **72**, 493–496 (1992)
87. Jovanović, B.S.: On the convergence rate of finite-difference schemes for hyperbolic equations. Z. Angew. Math. Mech. **73**, 656–660 (1993)
88. Jovanović, B.S.: On the estimates of the convergence rate of the finite difference schemes for the approximation of solutions of hyperbolic problems (Part II). Publ. Inst. Math. **55**(69), 149–155 (1994)

89. Jovanović, B.S.: Interpolation technique and convergence rate estimates for finite difference method. In: Numerical Analysis and Its Applications, Rousse, 1996. Lecture Notes in Comput. Sci., vol. 1196, pp. 200–211. Springer, Berlin (1997)
90. Jovanović, B.S.: Finite difference approximation of a hyperbolic transmission problem. In: Gautschi, W., Mastroianni, G., Rassias, T. (eds.) Approximation and Computation. Springer Optimization and Its Applications, vol. 42, pp. 319–329 (2010)
91. Jovanović, B.S., Ivanović, L.D.: On the convergence of the difference schemes for the equation of vibrating string. In: Milovanović, G.V. (ed.) Numerical Methods and Approximation Theory, Proc. Conf. Held in Niš 1984, University of Niš, Niš, pp. 155–159 (1984)
92. Jovanović, B.S., Popović, B.Z.: Convergence of a finite difference scheme for the third boundary value problem for elliptic equation with variable coefficients. Comput. Methods Appl. Math. **1**(4), 356–366 (2001)
93. Jovanović, B.S., Vulkov, L.G.: On the convergence of difference schemes for the string equation with concentrated mass. In: Ciegis, R., Samarskiĭand, A., Sapagovas, M. (eds.) Finite-Difference Schemes: Theory and Applications, Proc. of 3rd Int. Conf. Held in Palanga (Lithuania) 2000, IMI, Vilnius, pp. 107–116 (2000)
94. Jovanović, B.S., Vulkov, L.G.: Operator approach to the problems with concentrated factors. In: Numerical Analysis and Its Applications, Rousse, 2000. Lecture Notes in Comput. Sci., pp. 439–450. Springer, Berlin (2001)
95. Jovanović, B.S., Vulkov, L.G.: On the convergence of finite difference schemes for the heat equation with concentrated capacity. Numer. Math. **89**(4), 715–734 (2001)
96. Jovanović, B.S., Vulkov, L.G.: On the convergence of difference schemes for hyperbolic problems with concentrated data. SIAM J. Numer. Anal. **41**(2), 516–538 (2003)
97. Jovanović, B.S., Vulkov, L.G.: Stability of difference schemes for parabolic equations with dynamical boundary conditions and conditions on conjugation. Appl. Math. Comput. **163**, 849–868 (2005)
98. Jovanović, B.S., Vulkov, L.G.: Energy stability for a class of two-dimensional interface linear parabolic problems. J. Math. Anal. Appl. **311**, 120–138 (2005)
99. Jovanović, B.S., Vulkov, L.G.: On the convergence of difference schemes for parabolic problems with concentrated data. Int. J. Numer. Anal. Model. **5**(3), 386–406 (2008)
100. Jovanović, B.S., Vulkov, L.G.: Numerical solution of a hyperbolic transmission problem. Comput. Methods Appl. Math. **8**(4), 374–385 (2008)
101. Jovanović, B.S., Vulkov, L.G.: Finite difference approximations for some interface problems with variable coefficients. Appl. Numer. Math. **59**(2), 349–372 (2009)
102. Jovanović, B.S., Vulkov, L.G.: Numerical solution of a two-dimensional parabolic transmission problem. Int. J. Numer. Anal. Model. **7**(1), 156–172 (2010)
103. Jovanović, B.S., Vulkov, L.G.: Numerical solution of a two-dimensional hyperbolic transmission problem. J. Comput. Appl. Math. **235**, 519–534 (2010)
104. Jovanović, B.S., Vulkov, L.G.: Numerical solution of a parabolic transmission problem. IMA J. Numer. Anal. **31**, 233–253 (2011)
105. Jovanović, B.S., Ivanović, L.D., Süli, E.: On the convergence rate of difference schemes for the heat transfer equation. In: Vrdoljak, B. (ed.) IV Conference on Applied Mathematics, Proc. Conf. Held in Split 1984, University of Split, Split pp. 41–44 (1985)
106. Jovanović, B.S., Ivanović, L.D., Süli, E.: Convergence of difference schemes for the equation $-\Delta u + cu = f$ on generalized solutions from $W^s_{2,*}$ $(-\infty < s < +\infty)$. Publ. Inst. Math. **37**(51), 129–138 (1985). (Russian)
107. Jovanović, B.S., Ivanović, L.D., Süli, E.: Sur la convergence des schémas aux differences finies pour l'équation des ondes. Z. Angew. Math. Mech. **66**, 308–309 (1986)
108. Jovanović, B.S., Süli, E., Ivanović, L.D.: On finite difference schemes of high order accuracy for elliptic equations with mixed derivatives. Mat. Vestn. **38**, 131–136 (1986)
109. Jovanović, B.S., Ivanović, L.D., Süli, E.: Convergence of a finite-difference scheme for second-order hyperbolic equations with variable coefficients. IMA J. Numer. Anal. **7**, 39–45 (1987)

110. Jovanović, B.S., Ivanović, L.D., Süli, E.: Convergence of finite-difference schemes for elliptic equations with variable coefficients. IMA J. Numer. Anal. **7**, 301–305 (1987)
111. Jovanovich, B.S.: Convergence of discrete solutions to generalized solutions of boundary value problems. In: Bakhvalov, N.S., Kuznecov, Yu.A. (eds.) Variational-Difference Methods in Mathematical Physics, vol. 2, Proc. conf. held in Moscow 1983, Dept. Numer. Math. Acad. Sci. USSR, Moscow, pp. 120–129 (1984). (Russian)
112. Junkosa, M.L., Young, D.M.: On the order of convergence of solutions of a difference equation to a solution of the diffusion equation. SIAM J. **1**, 111–135 (1953)
113. Kačur, J., Van Keer, R., West, J.: On the numerical solution to a semi-linear transient heat transfer problem in composite media with nonlocal transmission conditions. In: Lewis, R.W. (ed.) Numerical Methods in Thermal Problems, VIII, pp. 1508–1519. Pineridge Press, Swansea (1993)
114. Kalinin, V.M., Makarov, V.L.: An estimate, in the L_2-norm, of the rate of convergence, for solutions in $W_2^1(\Omega)$, of a difference scheme for a third boundary-value problem in axially symmetric elasticity theory. Differ. Equ. **23**(7), 811–819 (1987)
115. Kreiss, H.-O., Thomée, V., Widlund, O.: Smoothing of initial data and rates of convergence for parabolic difference equations. Commun. Pure Appl. Math. **23**(2), 241–259 (1970)
116. Kufner, A., John, O., Fučik, S.: Function Spaces. Nordhoff International Publ., Leyden (1977)
117. Kuzik, A.M., Makarov, V.L.: The rate of convergence of a difference scheme using the sum approximation method for generalized solutions. USSR Comput. Math. Math. Phys. **26**(3), 192–196 (1986)
118. Ladyzhenskaya, O.A., Ural'tseva, N.N.: Linear and Quasilinear Elliptic Equations. Nauka, Moscow (1964). (Russian); English edn.: Academic Press, New York (1968)
119. Lazarov, R.D.: Convergence of finite-difference schemes for generalized solutions of Poisson's equation. Differ. Equ. **17**, 829–836 (1982)
120. Lazarov, R.D.: Convergence of finite-difference solutions to generalized solutions of a biharmonic equation in a rectangle. Differ. Equ. **17**, 836–843 (1982)
121. Lazarov, R.D.: On the convergence of the finite difference schemes for the Poisson's equation in discrete norms L_p. Wiss. Beitr. IH Wismar **7**(1/82), 86–90 (1982)
122. Lazarov, R.D.: Convergence of difference method for parabolic equations with generalized solutions. Pliska Stud. Math. Bulg. **5**, 51–59 (1982). (Russian)
123. Lazarov, R.D., Makarov, V.L.: Difference scheme of second order of accuracy for the axisymmetric Poisson equation in generalized solutions. USSR Comput. Math. Math. Phys. **21**(5), 95–107 (1981)
124. Lazarov, R.D., Mokin, Yu.I.: On the convergence of difference schemes for Poisson equations in L_p-metrics. Sov. Math. Dokl. **24**, 590–594 (1981)
125. Lazarov, R.D., Makarov, V.L., Samarskiĭ, A.A.: Application of exact difference schemes to the construction and study of difference scheme for generalized solutions. Math. USSR Sb. **45**, 461–471 (1983)
126. Lazarov, R.D., Makarov, V.L., Weinelt, W.: On the convergence of difference schemes for the approximation of solutions $u \in W_2^m$ $(m > 0.5)$ of elliptic equations with mixed derivatives. Numer. Math. **44**, 223–232 (1984)
127. Lions, J.L., Magenes, E.: Problèmes aux limites non homogènes et applications. Dunod, Paris (1968)
128. Lizorkin, P.I.: Generalized Liouville differentiation and the functional spaces $L_p^r(E_n)$. imbedding theorems. Mat. Sb. (N. S.) **60(102)**(3), 325–353 (1963). (Russian)
129. Makarov, V.L., Kalinin, V.M.: Consistent estimates for the rate of convergence of difference schemes in L_2-norm for the third boundary value problem of elasticity theory. Differ. Uravnen. **22**, 1265–1268 (1986). (Russian)
130. Makarov, V.L., Ryzhenko, A.I.: Matched estimates of the rate of convergence of the net method for Poisson's equation in polar coordinates. USSR Comput. Math. Math. Phys. **27**(3), 147–152 (1987)

131. Makarov, V.L., Ryzhenko, A.I.: Matched convergence-rate estimates of the mesh method for the axisymmetric Poisson equation in spherical coordinates. USSR Comput. Math. Math. Phys. **27**(4), 195–197 (1987)
132. Makridakis, C., Süli, E.: Finite element analysis of Cauchy–Born approximations to atomistic models. Arch. Ration. Mech. Anal. (2012). doi:10.1007/s00205-012-0582-8. Published electronically
133. Makridakis, C., Ortner, C., Süli, E.: A priori error analysis of two force-based atomistic/continuum models of a periodic chain. Numer. Math. **119**, 83–121 (2011)
134. Marletta, M.: Supraconvergence of discretisations methods on nonuniform meshes. M.Sc. Thesis, University of Oxford (1988)
135. Massey, W.S.: Singular Homology Theory. Graduate Texts in Mathematics, vol. 70. Springer, New York (1980)
136. Maz'ya, V.G.: Sobolev Spaces with Applications to Elliptic Partial Differential Equations. Grundlehren der mathematischen Wissenschaften, vol. 342. Springer, Heidelberg (2011)
137. Maz'ya, V.G., Shaposhnikova, T.O.: Theory of Multipliers in Spaces of Differentiable Functions. Monographs and Studies in Mathematics, vol. 23. Pitman, Boston (1985)
138. Maz'ya, V.G., Shaposhnikova, T.O.: Multiplikatory v prostranstvakh differentsiruemykh funktsii. Leningrad. Univ., Leningrad (1986)
139. Melenk, J.M.: On approximation in meshless methods. In: Craig, A.W., Blowey, J.F. (eds.) Frontiers of Numerical Analysis, Durham, 2004. Universitext. Springer, Berlin (2005)
140. Mokin, Yu.I.: Discrete analog of the multiplicator theorem. USSR Comput. Math. Math. Phys. **11**(3), 253–257 (1971)
141. Mokin, Yu.I.: A network analogue of the imbedding theorem for classes of type W. USSR Comput. Math. Math. Phys. **11**(6), 1361–1373 (1971)
142. Narasimhan, R.: Several Complex Variables. Chicago Lectures in Mathematics. University of Chicago Press, Chicago (1995). Reprint of the 1971 original
143. Nečas, J.: Les méthodes directes en théorie des équations elliptiques. Academia, Prague (1967)
144. Nikol'skiĭ, S.M.: Approximation of Functions of Several Variables and Imbedding Theorems. Nauka, Moscow (1977). (Russian)
145. Oberguggenberger, M.: Multiplication of Distributions and Applications to Partial Differential Equations. Pitman Research Notes on Mathematics Series, vol. 259, pp. 269–3674. Longman Scientific and Technical, Harlow (1992)
146. Oberguggenberger, M.: Generalized functions in nonlinear models—a survey. Nonlinear Anal. **47**(8), 5029–5040 (2001)
147. Oevermann, M., Klein, R.: A Cartesian grid finite volume method for elliptic equations with variable coefficients and embedded interfaces. J. Comp. Physiol. **19**, 749–769 (2006)
148. Oganesyan, L.A., Rukhovets, L.A.: Variational-Difference Methods for Solving Elliptic Equations. Publ. House of Armenian Acad. Sci, Erevan (1979). (Russian)
149. Ortner, C.: A priori and a posteriori analysis of the quasinonlocal quasicontinuum method in 1D. Math. Comput. **80**(275), 1265–1285 (2011)
150. Ortner, C., Süli, E.: Analysis of a quasicontinuum method in one dimension. M2AN Math. Model. Numer. Anal. **42**(1), 57–91 (2008)
151. Prikazchikov, V.G., Himich, A.N.: The eigenvalue difference problem for the fourth order elliptic operator with mixed boundary conditions. USSR Comput. Math. Math. Phys. **25**(5), 137–144 (1985)
152. Qatanani, N., Barham, A., Heeh, Q.: Existence and uniqueness of the solution of the coupled conduction-radiation energy transfer on diffusive-gray surfaces. Surv. Math. Appl. **2**, 43–58 (2007)
153. Rannacher, R.: Finite element solution of diffusion problems with irregular data. Numer. Math. **43**, 309–327 (1984)
154. Reed, M., Simon, B.: Methods of Modern Mathematical Physics. I: Functional Analysis. Academic Press, San Diego (1980)

155. Renardy, M., Rogers, R.C.: An Introduction to Partial Differential Equations, 2nd edn. Springer, New York (2004)
156. Richtmyer, R.D., Morton, K.W.: Difference Methods for Initial-Value Problems. Krieger, Melbourne (1994)
157. Rudin, W.: Real and Complex Analysis, 3rd edn. McGraw-Hill, New York (1986)
158. Rudin, W.: Functional Analysis, 2nd edn. International Series in Pure and Applied Mathematics. McGraw-Hill, New York (1991)
159. Samarskiĭ, A.A.: The Theory of Difference Schemes. Nauka, Moscow (1983). (Russian); English edn.: Monographs and Textbooks in Pure and Applied Mathematics, vol. 240. Dekker, New York (2001)
160. Samarskiĭ, A.A., Lazarov, R.D., Makarov, L.: Finite Difference Schemes for Differential Equations with Weak Solutions. Visshaya Shkola Publ., Moscow (1987). (Russian)
161. Samarskiĭ, A.A., Iovanovich, B.S., Matus, P.P., Shcheglik, V.S.: Finite-difference schemes on adaptive time grids for parabolic equations with generalized solutions. Differ. Equ. **33**, 981–990 (1997)
162. Schmeisser, H.J., Triebel, H.: Topics in Fourier Analysis and Function Spaces. Wiley, Chichester (1987)
163. Schwartz, L.: Théorie des distributions I, II. Herman, Paris (1950/1951)
164. Scott, J.A., Seward, W.L.: Finite difference methods for parabolic problems with nonsmooth initial data. Technical Report 86/22, Oxford University Computing Laboratory, Numerical Analysis Group, Oxford (1987)
165. Seward, W.L., Kasibhatla, P.S., Fairweather, G.: On the numerical solution of a model air pollution problem with non-smooth initial data. Commun. Appl. Numer. Methods **6**, 145–156 (1990)
166. Shreve, D.C.: Interior estimates in l^p for elliptic difference operators. SIAM J. Numer. Anal. **10**, 69–80 (1973)
167. Stein, E.M.: Singular Integrals and Differentiability of Functions. Princeton Univ. Press, Princeton (1970)
168. Stein, E.M., Weiss, G.L.: Introduction to Harmonic Analysis on Euclidean Spaces. Princeton Mathematical Series. Princeton Univ. Press, Princeton (1971)
169. Strang, G., Fix, G.: An Analysis of the Finite Element Method. Prentice Hall, New York (1973)
170. Strikwerda, J.: Finite Difference Schemes and Partial Differential Equations. SIAM, Philadelphia (2004)
171. Süli, E.: Convergence of finite volume schemes for Poisson's equation on non-uniform meshes. SIAM J. Numer. Anal. **28**, 1419–1430 (1991)
172. Süli, E., Mayers, D.F.: An Introduction to Numerical Analysis, 2nd edn. Cambridge University Press, Cambridge (2006)
173. Süli, E., Jovanović, B.S., Ivanović, L.D.: Finite difference approximations of generalized solutions. Math. Comput. **45**, 319–327 (1985)
174. Süli, E., Jovanović, B.S., Ivanović, L.D.: On the construction of finite difference schemes approximating generalized solutions. Publ. Inst. Math. **37**(51), 123–128 (1985)
175. Thomée, V.: Discrete interior Schauder estimates for elliptic difference operators. SIAM J. Numer. Anal. **5**, 626–645 (1968)
176. Thomée, V.: Negative norm estimates and superconvergence in Galerkin methods for parabolic problems. Math. Comput. **34**, 93–113 (1980)
177. Thomée, V.: Galerkin Finite Element Methods for Parabolic Problems, 2nd edn. Springer Series in Computational Mathematics, vol. 25. Springer, Berlin (2006)
178. Thomée, V., Wahlbin, L.B.: Convergence rates of parabolic difference schemes for non-smooth data. Math. Comput. **28**(125), 1–13 (1974)
179. Thomée, V., Westergren, B.: Elliptic difference equations and interior regularity. Numer. Math. **11**, 196–210 (1968)
180. Tikhonov, A.N.: On functional equations of the Volterra type and their applications to some problems of mathematical physics. Byull. Mosc. Gos. Univ., Ser. Mat. Mekh. **1**(8), 1–25

(1938). (Russian)
181. Triebel, H.: Fourier Analysis and Function Spaces. Teubner, Leipzig (1977)
182. Triebel, H.: Interpolation Theory, Function Spaces, Differential Operators. Deutscher Verlag der Wissenschaften, Berlin (1978)
183. Triebel, H.: Theory of Function Spaces. Monographs in Mathematics, vol. 78. Birkhäuser, Basel (1983)
184. Vladimirov, V.S.: Equations of Mathematical Physics, 2nd English edn. Monographs and Textbooks in Pure and Applied Mathematics, vol. 3. Dekker, New York (1971). 2nd English edn.: Mir, Moscow (1983)
185. Vladimirov, V.S.: Generalized Functions in Mathematical Physics. Mir, Moscow (1979). (English edn.)
186. Voĭtsekhovskiĭ, S.A., Gavrilyuk, I.P.: Convergence of finite-difference solutions to generalized solutions of the first boundary-value problem for a fourth-order quasilinear equation in regions of arbitrary shape. Differ. Equ. **21**, 1081–1088 (1985)
187. Voĭtsekhovskiĭ, S.A., Kalinin, V.M.: Limit of the rate of convergence of difference schemes for the first boundary-value problem of elasticity theory in the anisotropic case. USSR Comput. Math. Math. Phys. **29**(4), 87–91 (1989)
188. Voĭtsekhovskiĭ, S.A., Novichenko, V.N.: Justification of a difference scheme of an increased order of accuracy for the Dirichlet problem for the Poisson equation in classes of generalized solutions. Differ. Uravnen. **24**, 1631–1633 (1988). (Russian)
189. Voĭtsekhovskiĭ, S.A., Makarov, V.L., Shabliĭ, T.G.: The convergence of difference solutions to the generalized solutions of the Dirichlet problem for the Helmholtz equation in a convex polygon. USSR Comput. Math. Math. Phys. **25**(5), 36–43 (1985)
190. Voĭtsekhovskiĭ, S.A., Gavrilyuk, I.P., Sazhenyuk, V.S.: Estimates of the rate of convergence of difference schemes for variational elliptic second-order inequalities in an arbitrary domain. USSR Comput. Math. Math. Phys. **26**(3), 113–120 (1986)
191. Voĭtsekhovskiĭ, S.A., Gavrilyuk, I.P., Makarov, V.L.: Convergence of finite-difference solutions to generalized solutions of a first boundary-value problem for a fourth-order elliptic operator in a domain of arbitrary form. Differ. Equ. **23**(8), 962–966 (1987)
192. Voĭtsekhovskiĭ, S.A., Makarov, V.L., Rybak, Yu.I.: Estimates of the convergence rate of difference approximations of the Dirichlet problem for the equation $-\Delta u + \sum_{|\alpha|\le 1}(-1)^{|\alpha|}D^{\alpha}q_{\alpha}(x)u = f(x)$ when $q_{\alpha}(x) \in W_{\infty}^{\lambda|\alpha|}(\Omega)$, $\lambda \in (0, 1]$. Differ. Equ. **24**(11), 1338–1344 (1988)
193. Volkov, E.A.: On differential properties of solutions of boundary value problems for the Laplace and Poisson equations on a rectangle. Tr. Mat. Inst. Steklova **77**, 89–112 (1965). (Russian)
194. Weinan, E., Ming, P.: Cauchy–Born rule and the stability of crystalline solids: static problems. Arch. Ration. Mech. Anal. **183**(2), 241–297 (2007)
195. Weinelt, W.: Untersuchungen zur Konvergenzgeschwindigkeit bei Differenzenverfahren. Z. THK **20**, 763–769 (1978)
196. Weinelt, W., Lazarov, R.D., Makarov, V.L.: Convergence of finite-difference schemes for elliptic equations with mixed derivatives having generalized solutions. Differ. Equ. **19**, 838–843 (1983)
197. Weinelt, W., Lazarov, R.D., Streit, U.: Order of convergence of finite-difference schemes for weak solutions of the heat conduction equation in an anisotropic inhomogeneous medium. Differ. Equ. **20**, 828–834 (1984)
198. Wheeler, M.F.: L_∞ estimates of optimal orders for Galerkin methods for one dimensional second order parabolic and hyperbolic problems. SIAM J. Numer. Anal. **10**, 908–913 (1973)
199. Wloka, J.: Partial Differential Equations. Cambridge Univ. Press, Cambridge (1987)
200. Zlamal, M.: Finite element methods for parabolic equations. Math. Comput. **28**, 393–404 (1974)
201. Zlotnik, A.A.: Convergence rate estimate in L_2 of projection-difference schemes for parabolic equations. USSR Comput. Math. Math. Phys. **18**(6), 92–104 (1978)

202. Zlotnik, A.A.: Estimation of the rate of convergence in $V_2(Q_T)$ of projection-difference schemes for parabolic equations. Mosc. Univ. Comput. Math. Cybern. **1**, 28–38 (1980). 1980
203. Zlotnik, A.A.: Some finite-element and finite-difference methods for solving mathematical physics problems with non-smooth data in an n-dimensional cube, Part I. Sov. J. Numer. Anal. Math. Model. **6**(5), 421–451 (1991)
204. Zlotnik, A.A.: Convergence rate estimates of finite-element methods for second-order hyperbolic equations. In: Marchuk, G.I. (ed.) Numerical Methods and Applications, pp. 155–220. CRC Press, Boca Raton (1994)
205. Zlotnik, A.A.: On superconvergence of a gradient for finite element methods for an elliptic equation with the nonsmooth right-hand side. Comput. Methods Appl. Math. **2**, 295–321 (2002)
206. Zygmund, A.: Trigonometric Series, 2nd edn. Cambridge University Press, Cambridge (1988), vols. 1 and 2 combined

Index

B.S. Jovanović, E. Süli, *Analysis of Finite Difference Schemes*,
Springer Series in Computational Mathematics 46,
DOI 10.1007/978-1-4471-5460-0, © Springer-Verlag London 2014

List of Symbols

B.S. Jovanović, E. Süli, *Analysis of Finite Difference Schemes*,
Springer Series in Computational Mathematics 46,
DOI 10.1007/978-1-4471-5460-0, © Springer-Verlag London 2014

Function Spaces and Spaces of Distributions

Mollifiers

Differential Operators

Meshes

Difference Operators

Spaces of Mesh Functions

Mesh-Dependent Norms

Printed in the United States
By Bookmasters

Printed in the United States
By Bookmasters